World Geography

ASIA

Regions | Physical Geography | Biogeography and Natural Resources
Human Geography | Economic Geography | Gazetteer

Second Edition

F L A G S   O F   T H E   W O R L D

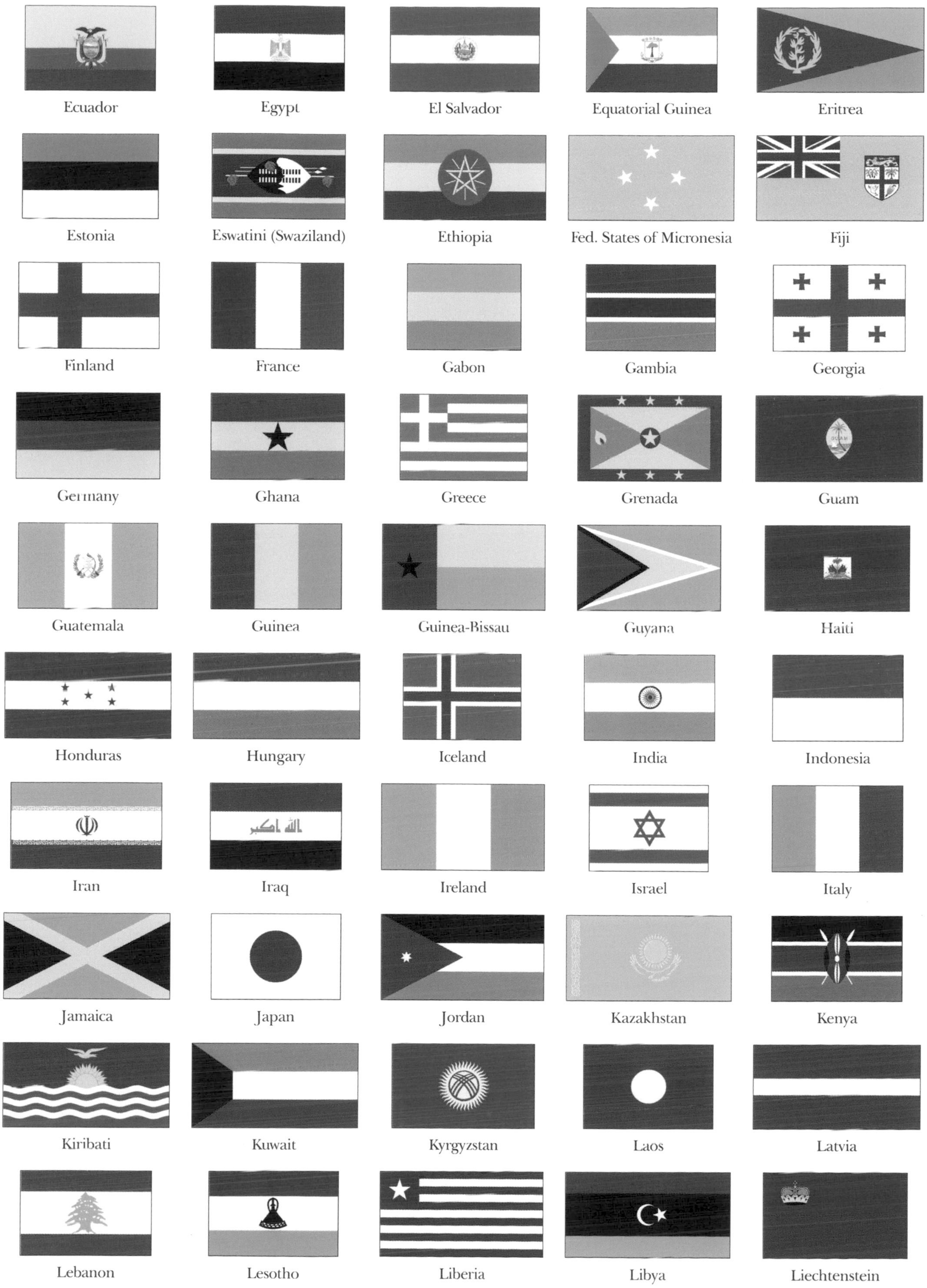

Ecuador

Egypt

El Salvador

Equatorial Guinea

Eritrea

Estonia

Eswatini (Swaziland)

Ethiopia

Fed. States of Micronesia

Fiji

Finland

France

Gabon

Gambia

Georgia

Germany

Ghana

Greece

Grenada

Guam

Guatemala

Guinea

Guinea-Bissau

Guyana

Haiti

Honduras

Hungary

Iceland

India

Indonesia

Iran

Iraq

Ireland

Israel

Italy

Jamaica

Japan

Jordan

Kazakhstan

Kenya

Kiribati

Kuwait

Kyrgyzstan

Laos

Latvia

Lebanon

Lesotho

Liberia

Libya

Liechtenstein

World Geography

ASIA

Regions | Physical Geography | Biogeography and Natural Resources
Human Geography | Economic Geography | Gazetteer

Second Edition

Volume 2

Editor
Joseph M. Castagno
Educational Reference Publishing, LLC

SALEM PRESS
A Division of EBSCO Information Services, Inc.
Ipswich, Massachusetts
GREY HOUSE PUBLISHING

Cover photo: Asia from outer space. Image by 1xpert.

Publisher's Cataloging-In-Publication Data
(Prepared by The Donohue Group, Inc.)

Names: Castagno, Joseph M., editor.
Title: World geography / editor, Joseph M. Castagno, Educational Reference Publishing, LLC.
Description: Second edition. | Ipswich, Massachusetts : Salem Press, a division of EBSCO Information Services, Inc. ; Amenia, NY : Grey House Publishing, [2020] | Interest grade level: High school. | Includes bibliographical references and index. | Summary: A six-volume geographic encyclopedia of the world, continents and countries of each continent. In addition to physical geography, the set also addresses human geography including population distribution, physiography and hydrology, biogeography and natural resources, economic geography, and political geography. | Contents: Volume 1. South & Central America — Volume 2. Asia — Volume 3. Europe — Volume 4. Africa — Volume 5. North America & the Caribbean — Volume 6. Australia, Oceania & the Antarctic.
Identifiers: ISBN 9781642654257 (set) | ISBN 9781642654288 (v. 1) | ISBN 9781642654318 (v. 2) | ISBN 9781642654301 (v. 3) | ISBN 9781642654295 (v. 4) | ISBN 9781642654271 (v. 5) | ISBN 9781642654325 (v. 6)
Subjects: LCSH: Geography—Encyclopedias, Juvenile. | CYAC: Geography—Encyclopedias. | LCGFT: Encyclopedias.
Classification: LCC G133 .W88 2020 | DDC 910/.3—dc23

First Printing
PRINTED IN CANADA

CONTENTS

The Challenge of COVID-19

As World Geography: Asia goes to press in April 2020, the entire globe is grappling with worst and most widespread pandemic in more than a century. The cause of the pandemic is a highly contagious viral condition known as Coronavirus Disease 2019, or COVID-19.

The first documented emergence COVID-19 occurred in December 2019 as an outbreak of pneumonia in Wuhan City, in Hubai Province, China. By January 2020, Chinese health officials had reported tens of thousands of infections and dozens of deaths. That same month, COVID-19 cases were appearing across Asia, Europe, and North America, and spreading rapidly. On March 11, the World Health Organization (WHO) declared COVID-19 viral disease a pandemic.

The rapid spread of COVID-19 viral disease has strong geographical components. The virus emerged in Wuhan, a huge, densely populated city. It spread rapidly among a population living largely indoors during the cold-weather months. But the most significant geographical factor in the spread of the virus may well be the globalization of transportation. Every day, thousands of people fly to destinations near and far. Each traveler carries the potential to unknowingly spread disease.

To curtail COVID-19's spread, countries have closed their borders, air travel has been drastically reduced, and sweeping mitigation policies have been inaugurated. Some densely populated Asian countries, including South Korea, Singapore, and Taiwan, enforced these measures very early, and have, as of the time of publication in April 2020, managed to dampen the effect of the virus and limit the number of confirmed cases and deaths due to COVID-19. Other places, slower to act, such as China and Iran, have been very hard hit.

Many COVID-19 questions remain: Where will the disease strike next? Will the onset of warmer weather reduce the communicability of the virus? Will a vaccine be available soon? Will there be a second wave of infection? When will life be back to normal?

Geographers will continue to play a unique role in answering these questions, applying the tools and techniques of their discipline to achieving the fullest possible epidemiological understanding of the pandemic.

PUBLISHER'S NOTE

North Americans have long thought of the field of geography as little more than the study of the names and locations of places. This notion is not without a basis in fact: Through much of the twentieth century, geography courses forced students to memorize names of states, capitals, rivers, seas, mountains, and countries. Both students and educators eventually rebelled against that approach, geography courses gradually fell out of favor, and the future of geography as a discipline looked doubtful. Happily, however, the field has undergone a remarkable transformation, starting in the 1990s. Geography now has a bright and pivotal significance at all levels of education.

While learning the locations of places remains an important part of geography studies, educators recognize that place-name recognition is merely the beginning of geographic understanding. Geography now places much greater emphasis on understanding the characteristics of, and interconnections among, places. Modern students address such questions as how the weather in Brazil can affect the price of coffee in the United States, why global warming threatens island nations, and how preserving endangered plant and animal species can conflict with the economic development of poor nation.

World Geography, Second Edition, addresses these and many other questions. Designed and written to meet the needs of high school students, while being accessible to both middle school and undergraduate college students, these six volumes take an integrated approach to the study of geography, emphasizing the connections among world regions and peoples. The set's six volumes concentrate on major world regions: South and Central America; Asia; Europe; Africa, North America; and Australia, Oceania, and the Antarctic. Each volume begins with common overview information related to the geography, maps and mapmaking. The core essays in the volumes begin with an overview section to provide global context and then goes on to examine important geographic aspects of the regions in that area of the world: its physical geography;

biogeography and natural resources; human geography (including its political geography); and economic geography. These essays range in length from three to ten pages. A gazetteer for the region indicates major political, geographic, and man-made features throughout the region.

A robust appendix found in each volume provides further information:

- The Earth in Space (The Solar System, Earth's Moon, The Sun and the Earth, The Seasons);
- Earth's Interior (Earths Internal Structure, Plate Tectonics, Volcanoes, Geologic Time Scale);
- Earth's Surface (Internal Geological Processes, External Processes, Fluvial and Karst Processes, Glaciation, Desert Landforms, Ocean Margins);
- Earth's Climates (The Atmosphere, Global Climates; Cloud Formation, Storms);
- Earth's Biological Systems (Biomes);
- Natural Resources (Soils, Water);
- Exploration and Transportation (Exploration and Historical Trade Routes, Road Transportation, Railways, Air Transportation);
- Energy and Engineering (Energy Sources, Alternative Energies, Engineering Projects);
- Industry and Trade (Manufacturing, Globalization of Manufacturing and Trade, Modern World Trade Patterns);
- Political Geography (Forms of Government, Political Geography, Geopolitics, National Park Systems);
- Boundaries and Time Zones (International Boundaries, Global Time and Time Zones);
- Global Education (Themes and Standards in Geography Education);
- Global Data (The World Gazetteer of Oceans and Continents, The World's Oceans and Seas, Major Land Areas of the World, Major Islands of the World, Countries of the World (including population and pollution density), Past and Projected World Population Growth, 1950-2050, The World's Largest Countries by Area, The World's Smallest Countries by Area, The World's Largest Countries by Population,

The World's Smallest Countries by Population, The World's Most Densely Populated Countries, The World's Least Densely Populated Countries, The World's Most Populous Cities, Major Lakes of the World, Major Rivers of the World, The Highest Peaks in Each Continent, Major Deserts of the World, Highest Waterfalls of the World).

- A Glossary, General Bibliography, and Index completes the backmatter.

The regional divisions in the set make it possible to study specific countries or parts of the world. Pairing the specific regional information, organized by regions, physical geography, biogeography and natural resources, human geography, economic geography, and a gazetteer, with global information makes it possible for students to see the connections not only between countries and places within the region, but also between the regions and the entire global system, all within a single volume.

To make this set as easy as possible to use, all of its volumes are organized in a similar fashion, with six major divisions—Regions (organized into subregions by volume), Physical Geography, Biogeography and Natural Resources, Human Geography, and Economic Geography. The number of subregions in each volume varies, depending upon the major world division being examined—Asia, for example, includes the following regions: China; Japan, Korea, and Taiwan; Southeast Asia; South Asia; Mongolia and Asian Russia; the Transcaucus; and the Middle East.

Physical geography considers a world region's physiography, hydrology, and climatology. Biogeography and natural resources explores renewable and nonrenewable resources, flora, and fauna. Human geography addresses the people, population distribution, culture regions, urbanization, and political geography of the area. Economic geography considers the regions agriculture, industries, engineering projects, transportation, trade, and communications.

Gazetteers include descriptive entries on hundreds of important places, especially those mentioned in the volume's essays. A typical entry gives the place name and location, indicating the category into which the place falls (mountain, river, city, country, lake, etc). The entries also include statistics relevant to the categories of place (height of mountains, length of rivers, population of cities and countries).

A feature new to this edition is the discussion questions included throughout the volume. These questions are meant to foster discussion and further research into the topics related to the history, current issues, and future concerns related to physical, human, economic, and political geography.

Both a physical and a social science, geography is unique among social sciences in the demands it makes for visual support. For this reason, *World Geography* contains more than 100 maps, more than 1300 photographs, and scores of other graphical elements. In addition, essays are punctuated with more than 500 textual sidebars and tables, which amplify information in the essays and call attention to especially important or interesting points.

Both English and metric measures are used throughout this set. In most instances, English measures are given first, followed by their metric equivalents in parentheses. It should be noted that in cases of measures that are only estimates, such as the areas of deserts or average heights of mountain ranges, the metric figures are often rounded off to estimates that may not be exact equivalents of the English-measure estimates. In order to enhance clarity, units of measure are not abbreviated in the text, with these exceptions: kilometers are abbreviated as km. and square kilometers as sq. km. This exception has been made because of the frequency with which these measures appear.

Reference works such as this would be impossible without the expertise of a large team of contributing scholars. This project is no exception. Salem Press would like to thank the more than 175 people who wrote the signed essays and contributed entries to the gazetteers. A full list of contributors follows this note. We recognize the efforts of Dr. Ray Sumner, of California's Long Beach City College, for the expertise and insights that she brought to the previous editions of this book, and which have formed the strong foundation for this new edition. We also acknowledge the work of the editor of this current volume, Joe Castagno, Educational Reference Publishing, LLC.

INTRODUCTION

When Henry Morton Stanley of the *New York Herald* shook the hand of David Livingstone on the shore of Central Africa's Lake Tanganyika in 1871, the moment represented the high point of geography to many people throughout the world. A Scottish missionary and explorer, Livingstone had been out of contact with the outside world for nearly two years, and European and American newspapers had buzzed with speculation about his disappearance. At that time, so little was known about the geography of the interior of Africa that Stanley's finding Livingstone was acclaimed as a brilliant triumph of explorations.

The field of geography in Stanley and Livingstone's time was—and to a large extent still is—synonymous with explorations. Stories of epic journeys, both historic and contemporary, continue to exert a powerful attraction on readers. Mountains, deserts, forest, caves, and glaciers still draw intrepid explorers, while even more armchair travelers are thrilled by accounts and pictures of these exploits and discoveries. We all love to travel— to the beach, into the mountains, to our great national parks, and to foreign countries. In the need and desire to explore our surroundings, we are all geographers.

Numerous geographical societies welcome both professional geographers and the general public into their membership, as they promote a greater knowledge and understanding of the earth. The National Geographic Society, founded in 1888 "for the increase and diffusion of geographical knowledge," has awarded more than 11,000 grants for scientific exploration and research. Each year, the society invests millions of dollars in expeditions and fieldwork related to environmental concerns and global geographic issues. The findings are recorded in the pages of the familiar yellow-bordered *National Geographic* magazine, now produced in 40 local-language editions in many countries around the world, publishing around 6.8 million copies monthly, with some 60 million readers. The magazine, along with the National Geographic International television network, reaches more than 135 million readers and viewers worldwide and has more than 85 million subscribers.

An even older geographical association is Great Britain's' Royal Geographical Society, which grew out of the Geographical Society of London, founded in 1830 with the "sole object" of promoting "that most important and entertaining branch of knowledge—geography." Over the century that followed, the Royal Geographical Society focused on exploration of the continents of Africa and Antarctica. In the society's London headquarters adjacent to the Albert Hall, visitors can still view such historic artifacts as David Livingstone's cap and chair, as well as diaries, sketches, and maps covering the great period of the British Empire and beyond. Today the society assists more than five hundred field expeditions every year.

With the aid of satellites and remote-sensing instruments, we can now obtain images and data from almost anywhere on Earth. However, remote and inaccessible places still invite the intrepid to visit and explore them in person. Although the outlines of the continents have now been completed, and their interiors filled in with details of mountains and rivers, cities and political boundaries, remote places still exert a fascination on modern urbanites.

The enchantment of tales about strange sights and courageous journeys has been with us since the ancient voyages of Homer's *Ulysses*, Marco Polo's travels to China, and the nautical expeditions of Christopher Columbus, Ferdinand Magellan, and James Cook. While those great travelers are from the remote past, the age of exploration is far from over—a fact repeatedly demonstrated by the modern Norwegian navigator Thor Heyerdahl. Moreover, new journeys of discovery are still taking place. In 1993, after dragging a sled wearily across the frigid wastes of Antarctica for more than three months, Sir Ranulph Twisleton-Wykeham-Fiennes announced that the age of exploration is not dead. Six years later, in 1999, the long-missing body of British mountain climber George Mallory was found on the slopes of Mount Everest, near whose top he had mysteriously vanished in 1924. That discovery sparked a new wave of admiration and respect for explorers of such courage and endurance.

How many people have been enthralled by the bravery of Antarctic explorer Robert Falcon Scott and the noble sacrifice his injured colleague Lawrence Oates made in 1912, when he gave up his life in order not to slow down the rest of the expedition? There can be no doubt that the thrills and the dangers of exploring find resonance among many modern readers.

The struggle to survive in environments hostile to human beings reminds us of the power of our planet Earth. Significant books on this theme have included Jon Krakauer's *Into Thin Air* (1998), an account of a disastrous expedition climbing Mount Everest, and Sebastian Junger's *The Perfect Storm* (1997), the story of the worst gale of the twentieth century and its effect on a fishing fleet off the East Coast of North America. *Endurance* (1998), the epic of Sir Ernest Shackleton's survival and leadership for two years on the frozen Arctic, attracts the same people who avidly read *Undaunted Courage* (1996) the story of Meriwether Lewis and William Clark's epic exploration of the Louisiana Purchase territories in the early nineteenth century. In 1997 *Seven Years in Tibet* premiered, a popular film about the Austrian Heinrich Harrer, who lived in Tibet in the mid-twentieth century. The more urban people become, the greater their desire for adventurous, remote places, a least vicariously, to raise the human spirit.

There are, of course, also scientific achievements associated with modern exploration. In November 1999, the elevation of Mount Everest, the world tallest peak was raised by 7 feet (2.1 meters) to a new height of 29,035 feet (8,850 meters) above sea level; the previously accepted height had been based on surveys made during the 1950s. This new value was the result of Global Positioning System (GPS) technology enabling a more accurate measurement than had been possible with land-based earthbound surveying equipment. A team of climbers supported by the National Geographic Society and the Boston Museum of Science was equipped with GPS equipment, which enabled a fifty-minute re-

cording of data based on satellite signals. At the same time, the expedition was able to ascertain that Mount Everest is moving northeast, atop the Indo-Australian Plate, at a rate of approximately 2.4 inches (10 centimeters) per year.

In 2000, the International Hydrographic Organization named a "new" ocean, the Southern Ocean, which encompasses all the water surrounding Antarctica up to 60° south latitude. With an area of approximately 7.8 million square miles (20.3 million sq. km.), the Southern Ocean is about twice the size of the entire United States and ranks as the world's fourth largest ocean, after the Pacific, Atlantic, and Indian Oceans, but just ahead of the Arctic Ocean.

Despite the humanistic and scientific advantage of geographic knowledge, to many people today, geography is a subject where one merely memorized longs lists of facts dealing with "where" questions. (Where is Andorra? Where is Prince Edward Island? Where is Kalamazoo?) or "what" questions (What is the highest mountain in South America? What is the capital of Costa Rica?) This approach to the study of geography has been perpetuated by the annual National Geographic Bee, conducted in the United States each year for students in grades four through eight. Participants in the competition display an astonishing recall of facts but do not have the opportunity of showing any real geographic thought. To a geographer, such factual knowledge is simply a foundation for investigating and explaining the much more important questions dealing with "why"—"Why is the Sahara a desert?"

Geographers aim to understand why environments and societies occur where and as they do, and how they change. Geography must be seen as an integrative science; the collection of factual data and evidence, as in exploration, is the empirical foundation for deductive reasoning. This leads to the creation of a range of geographical methods, models, theories, and analytical approaches that serve to unify a very broad area of knowledge—the interaction between natural and human environments. Although geography as an academic discipline became established in nineteenth century Germany, there have always been geographers, in the sense of people curious about their world. Humans have always wanted to know about day and night, the shape of the earth, the nature of climates, differences in plants and animals, as well as what lies beyond the horizon. Today, as we hear about and actually experience the sweeping effects of globalization, we need more than ever to develop our geographic skills. Not only are we connected by economic ties to the countries of the world, but we must also appreciate the consequences of North America's high standard of living.

Political boundaries are artificial human inventions, but the natural world is one biosphere. As concern over global warming escalates, national leaders meet to seek a solution to the emission of greenhouse gases, rising ocean levels, and mass extinctions. Are we connected to our environment? At a time when the rate of species extinction is a hundred times above normal, and the human population is crowding in increasing numbers into huge urban centers, we have, nevertheless, taken time each year in April to celebrate Earth Day since 1970. We need now to realize that every day is Earth Day.

Geography languished in the United States in the 1960s, as social studies was taught with a history emphasis in schools. American students became alarmingly disadvantaged in geographic knowledge, compared with most other countries. Fortunately, members of the profession acted to restore geography to the curriculum. In 1984, the National Geographic Society undertook the challenge of restoring geography in the United States. The society turned to two organizations active in geographic education: The Association of American Geographers, the professional geographers" group with more than 10,000 members, mostly in higher education in the United States; and the National Council for Geographic Education that supports geography teaching at all levels—from kindergarten through university, with members that include U.S. and international teachers, professors, students, businesses, and others who support geography education. The council administers the Geographic Alliances, found in every state of the United Sates, with a national membership of about 120,000 schoolteachers. Together, they produced the "Guidelines in Geographic Education," which introduced the Five

Themes of Geography, to enhance the teaching of geography in schools. Using the themes of Location, Place, Human/Environment Interaction, Movement and Regions, teachers were able to plan and conduct lessons in which students encountered interesting real-world examples of the relevance and importance of geography. Continued research into geographic education led to the inclusion of geography in 1990 as one of the core subjects of the National Education Goals, along with English, mathematics, science, and history.

Another milestone was the publication in 1994 of "Geography for Life," the national Geography Standards. The earlier Five Themes were subsumed under the new Six Essential Elements: The World in Spatial Terms; Places and Regions; Physical Systems; Human Systems; Environment Systems; Environment and Society; and The Uses of Geography. Eighteen geography standards are included, describing what a geographically informed person knows and understands. States, schools, and individual teachers have welcomed the new prominence of geography, and enthusiastically adopted new approaches to introduce the geography standards to new learners. The rapid spread of computer technology, especially in the field of Geographical Information Science, has also meant a new importance for spatial analysis, a traditional area of geographical expertise. No longer is geography seen as an outdated mass of useless or arcane facts; instead, geography is now seen, again, to be an innovative an integrative science, which can contribute to solving complex problems associated with the human-environmental relationship in the twenty-first century.

Geographers may no longer travel across uncharted realms, but there is still much we long to explore, to learn, and seek to understand, even if it is only as "armchair" geographers. This reference work, *World Geography*, will help carry readers on their own journeys of exploration.

Ray Sumner
Long Beach City College

Joseph M. Castagno
Educational Reference Publishing, LLC

CONTRIBUTORS

Emily Alward
Henderson, Nevada Public Library

Earl P. Andresen
University of Texas at Arlington

Debra D. Andrist
St. Thomas University

Charles F. Bahmueller
Center for Civic Education

Timothy J. Bailey
Pittsburg State University

Irina Balakina
Writer/Editor, Educational Reference Publishing LLC

David Barratt
Nottingham, England

Maryanne Barsotti
Warren, Michigan

Thomas F. Baucom
Jacksonville State University

Michelle Behr
Western New Mexico University

Alvin K. Benson
Brigham Young University

Cynthia Breslin Beres
Glendale, California

Nicholas Birns
New School University

Olwyn Mary Blouet
Virginia State University

Margaret F. Boorstein
C.W. Post College of Long Island University

Fred Buchstein
John Carroll University

Joseph P. Byrne
Belmont University

Laura M. Calkins
Palm Beach Gardens, Florida

Gary A. Campbell
Michigan Technological University

Byron D. Cannon
University of Utah

Steven D. Carey
University of Mobile

Roger V. Carlson
Jet Propulsion Laboratory

Robert S. Carmichael
University of Iowa

Joseph M. Castagno
Principal, Educational Reference Publishing LLC

Habte Giorgis Churnet
University of Tennessee at Chattanooga

Richard A. Crooker
Kutztown University

William A. Dando
Indiana State University

Larry E. Davis
College of St. Benedict

Ronald W. Davis
Western Michigan University

Cyrus B. Dawsey
Auburn University

Frank Day
Clemson University

M. Casey Diana
University of Illinois at Urbana-Champaign

Stephen B. Dobrow
Farleigh Dickinson University

Steven L. Driever
University of Missouri, Kansas City

Sherry L. Eaton
San Diego City College

Femi Ferreira
Hutchinson Community College

Helen Finken
Iowa City High School

Eric J. Fournier
Samford University

Anne Galantowicz
El Camino College

Hari P. Garbharran
Middle Tennessee State University

Keith Garebian
Ontario, Canada

Laurie A. B. Garo
University of North Carolina, Charlotte

Jay D. Gatrell
Indiana State University

Carol Ann Gillespie
Grove City College

Nancy M. Gordon
Amherst, Massachusetts

Noreen A. Grice
Boston Museum of Science

Johnpeter Horst Grill
Mississippi State University

Charles F. Gritzner
South Dakota State University

C. James Haug
Mississippi State University

Douglas Heffington
Middle Tennessee State University

Thomas E. Hemmerly
Middle Tennessee State University

Jane F. Hill
Bethesda, Maryland

Carl W. Hoagstrom
Ohio Northern University

Catherine A. Hooey
Pittsburg State University

Robert M. Hordon
Rutgers University

Kelly Howard
La Jolla, California

Paul F. Hudson
University of Texas at Austin

Huia Richard Hutton
University of Hawaii/Kapiolani Community College

Raymond Pierre Hylton
Virginia Union University

Solomon A. Isiorho
Indiana University/Purdue University at Fort Wayne

Ronald A. Janke
Valparaiso University

Albert C. Jensen
Central Florida Community College

Jeffry Jensen
Altadena, California

Bruce E. Johansen
University of Nebraska at Omaha

Kenneth A. Johnson
State University of New York, Oneonta

Walter B. Jung
University of Central Oklahoma

James R. Keese
California Polytechnic State University, San Luis Obispo

Leigh Husband Kimmel
Indianapolis, Indiana

Denise Knotwell
Wayne, Nebraska

James Knotwell
Wayne State College

Grove Koger
Boise Idaho Public Library

Alvin S. Konigsberg
State University of New York at New Paltz

Doris Lechner
Principal, Educational Reference Publishing LLC

Steven Lehman
John Abbott College

Denyse Lemaire
Rowan University

Dale R. Lightfoot
Oklahoma State University

Jose Javier Lopez
Minnesota State University

James D. Lowry, Jr.
East Central University

Jinshuang Ma
Arnold Arboretum of Harvard University Herbaria

Dana P. McDermott
Chicago, Illinois

Thomas R. MacDonald
University of San Francisco

Robert R. McKay
Clarion University of Pennsylvania

Nancy Farm Männikkö
L'Anse, Michigan

Carl Henry Marcoux
University of California, Riverside

Christopher Marshall
Unity College

Rubén A. Mazariegos-Alfaro
University of Texas/Pan American

Christopher D. Merrett
Western Illinois University

John A. Milbauer
Northeastern State University

Randall L. Milstein
Oregon State University

Judith Mimbs
Loftis Middle School

Karen A. Mulcahy
East Carolina University

B. Keith Murphy
Fort Valley State University

M. Mustoe
Omak, Washington

Bryan Ness
Pacific Union College

Kikombo Ilunga Ngoy
Vassar College

Joseph R. Oppong
University of North Texas

Richard L. Orndorff
University of Nevada, Las Vegas

Bimal K. Paul
Kansas State University

Nis Petersen
New Jersey City University

Mark Anthony Phelps
Ozarks Technical Community College

John R. Phillips
Purdue University, Calumet

Alison Philpotts
Shippensburg University

Julio César Pino
Kent State University

Timothy C. Pitts
Morehead State University

Carolyn V. Prorok
Slippery Rock University

P. S. Ramsey
Highland Michigan

Robert M. Rauber
University of Illinois at Urbana-Champaign

Ronald J. Raven
State University of New York at Buffalo

Neil Reid
University of Toledo

Susan Pommering Reynolds
Southern Oregon University

Nathaniel Richmond
Utica College

Edward A. Riedinger
Ohio State University Libraries

Mika Roinila
West Virginia University

Thomas E. Rotnem
Brenau University

Joyce Sakkal-Gastinel
Marseille, France

Helen Salmon
University of Guelph

Elizabeth D. Schafer
Loachapoka, Alabama

Kathleen Valimont Schreiber
Millersville University of Pennsylvania

Ralph C. Scott
Towson University

Guofan Shao
Purdue University

Wendy Shaw
Southern Illinois University, Edwardsville

R. Baird Shuman
University of Illinois, Champaign-Urbana

Sherman E. Silverman
Prince George's Community College

Roger Smith
Portland, Oregon

Robert J. Stewart
California Maritime Academy

Toby R. Stewart
Alamosa, Colorado

Ray Sumner
Long Beach City College

Paul Charles Sutton
University of Denver

Glenn L. Swygart
Tennessee Temple University

Sue Tarjan
Santa Cruz, California

Robert J. Tata
Florida Atlantic University

John M. Theilmann
Converse College

Virginia Thompson
Towson University

Norman J. W. Thrower
University of California, Los Angeles

Paul B. Trescott
Southern Illinois University

Robert D. Ubriaco, Jr.
Illinois Wesleyan University

Mark M. Van Steeter
Western Oregon University

Johan C. Varekamp
Wesleyan University

Anthony J. Vega
Clarion University

William T. Walker
Chestnut Hill College

William D. Walters, Jr.
Illinois State University

Linda Qingling Wang
University of South Carolina, Aiken

Annita Marie Ward
Salem-Teikyo University

Kristopher D. White
University of Connecticut

P. Gary White
Western Carolina University

Thomas A. Wikle
Oklahoma State University

Rowena Wildin
Pasadena, California

Donald Andrew Wiley
Anne Arundel Community College

Kay R. S. Williams
Shippensburg University

Lisa A. Wroble
Redford Township District Library

Bin Zhou
Southern Illinois University, Edwardsville

REGIONS

OVERVIEW

THE HISTORY OF GEOGRAPHY

The moment that early humans first looked around their world with inquiring minds was the moment that geography was born. The history of geography is the history of human effort to understand the nature of the world. Through the centuries, people have asked of geography three basic questions: What is Earth like? Where are things located? How can one explain these observations?

Geography in the Ancient World

In the Western world, the Greeks and the Romans were among the first to write about and study geography. Eratosthenes, a Greek scholar who lived in the third century BCE, is often called the "father of geography and is credited with first using the word geography (from the Greek words *ge*, which means "earth," and *graphe*, which means "to describe"). The ancient Greeks had contact with many older civilizations and began to gather together information about the known world. Some, such as Hecataeus, described the multitude of places and peoples with which the Greeks had contact and wrote of the adventures of mythical characters in strange and exotic lands. However, the ancient Greek scholars went beyond just describing the world. They used their knowledge of mathematics to measure and locate. The Greek scholars also used their philosophical nature to theorize about Earth's place in the universe.

One Greek scholar who used mathematics in the study of geography was Anaximander, who lived from 610 to 547 BCE. Anaximander is credited with being the first person to draw a map of the world to scale. He also invented a sundial that could be used to calculate time and direction and to distinguish the seasons. Eratosthenes is also famous for his mathematical calculations, in particular of the circumference of Earth, using observations of the Sun. Hipparchus, who lived around 140 BCE, used his mathematical skills to solve geographic problems and was the first person to introduce the idea of a latitude and longitude grid system to locate places.

Such early Greek philosophers as Plato and Aristotle were also concerned with geography. They discussed such issues as whether Earth was flat or spherical and if it was the center of the universe, and debated the nature of Earth as the home of humankind.

Whereas the Greeks were great thinkers and introduced many new ideas into geography, the Roman contribution was to compile and gather available knowledge. Although this did not add much that was new to geography, it meant that the knowledge of the ancient world was available as a base to work from and was passed down across the centuries. Geogra-

---

CURIOSITY: THE ROOT OF GEOGRAPHY

The earliest human beings, as they hunted and gathered food and used primitive tools in order to survive, must have had detailed knowledge of the geography of their part of the world. The environment could be a hostile place, and knowledge of the world meant the difference between life and death. Human curiosity took them one step further. As they lived in an ancient world of ice and fire, human beings looked to the horizon for new worlds, crossing continents and spreading out to all areas of the globe. They learned not only to live as a part of their environment, but also to understand it, predict it, and adapt it to their needs.

phy in the ancient world is often said to have ended with the great work of Ptolemy (Claudius Ptolemaeus), who lived from 90 to 168 CE. Ptolemy is best known for his eight-volume *Guide to Geography*, which included a gazetteer of places located by latitude and longitude, and his world map.

Geography in China

The study of geography also was important in ancient China. Chinese scholars described their resources, climate, transportation routes, and travels, and were mapping their known world at the same time as were the great Western civilizations. The study of geography in China begins in the Warring States period (fifth century BCE). It expands its scope beyond the Chinese homeland with the growth of the Chinese Empire under the Han dynasty. It enters its golden age with the invention of the compass in the eleventh century CE (Song dynasty) and peaks with fifteenth century CE (Ming dynasty) Chinese exploration of the Pacific under admiral Zheng He during the treasure voyages.

Geography in the Middle Ages

With the collapse of the Roman Empire in the fifth century CE, Europe entered into what is commonly known as the Early Middle Ages. During this time, which lasted until the fifteenth century, the geographic knowledge of the ancient world was either lost or challenged as being counter to Christian teachings. For example, the early Greeks had theorized that Earth was a sphere, but this was rejected during the Middle Ages. Scholars of the Middle Ages believed that the world was a flat disk, with the holy city of Jerusalem at its center.

The knowledge and ideas of the ancient world might have been lost if they had not been preserved by Muslim scholars. In the Islamic countries of North Africa and the Middle East, some of the scholarship of the ancient world was sheltered in libraries and universities. This knowledge was extensively added to as Muslims traveled and traded across the known world, gathering their own information.

Among the most famous Muslim geographers were Ibn Battutah, al-Idrisi, and Ibn Khaldun. Ibn Battutah traveled east to India and China in the fourteenth century. Al-Idrisi, at the command of King Roger II of Sicily, wrote *Roger's Book*, which systematically described the world. Information from *Roger's Book* was engraved on a huge planisphere (disk), crafted in silver; this once was considered a wonder of the world, but it is thought to have been destroyed. Ibn Khaldun (1332-1406) is best known for his written world history, but he also was a pioneer in focusing on the relationship of human beings to their environment.

The Age of European Exploration

Beginning in the fifteenth century, the isolation of Europe came to an end, and Europeans turned their attention to exploration. The two major goals of this sudden surge in exploration were to spread the Christian faith and to obtain needed resources. In 1418 Prince Henry the Navigator established a school for navigators and began to gather the tools and knowledge needed for exploration. He was the first of many Europeans to travel beyond the limits of the known world, mapping, describing, and cataloging all that they saw.

The great wave of European exploration brought new interest in geography, and the monumental works of the Greeks and Romans—so carefully preserved by Muslim scholars—were rediscovered and translated into Latin. The maps produced in the Middle Ages were of little use to the explorers who were traveling to, and beyond, the limits of the known world. Christopher Columbus, for example, relied on Ptolemy's work during his voyages to the Americas, but soon newer, more accurate maps were drawn and, for the first time, globes were made. A particularly famous map, which is still used as a base map, is the Mercator projection. On the world map produced by Gerardus Mercator (born Geert de Kremer) in 1569, compass directions appear as straight lines, which was a great benefit on navigational charts.

When the age of European exploration began, even the best world maps crudely depicted only a few limited areas of the world. Explorers quickly began to gather huge quantities of information, making detailed charts of coastlines, discovering new continents, and eventually filling in the maps of those continents with information about both the natural and human features they encountered. This age of exploration is often said to have ended when Roald Amundsen planted the Norwegian flag at the South Pole in 1911. At that time, the world map became complete, and human beings had mapped and explored every part of the globe. However, the beginning of modern geography is usually associated with the work of two nineteenth century German geographers: Alexander von Humboldt and Carl Ritter.

The Beginning of Modern Geography

The writings of Alexander von Humboldt and Carl Ritter mark a leap into modern geography, because these writers took an important step beyond the work of previous scholars. The explorers of the previous centuries had focused on gathering information, describing the world, and filling in the world map with as much detail as possible. Humboldt and Ritter took a more scientific and systematic approach to geography. They began not only to compile descriptive information, but also to ask why: Humboldt spent his lifetime looking for relationships among such things as climate and topography (landscape), while Ritter was intrigued by the multitude of connections and relationships he observed within human geographic patterns. Both Humboldt and Ritter died in 1859, ending a period when information-gathering had been paramount. They brought geography into a new age in which synthesis, analysis, and theory-building became central.

European Geography

After the work of Humboldt and Ritter, geography became an accepted academic discipline in Europe, particularly in Germany, France, and Great Britain. Each of these countries emphasized different aspects of geographic study. German geographers continued the tradition of the scientific view, using observable data to answer geographic questions. They also introduced the concept that geography could take a chorological view, studying all aspects, physical and human, of a region and of the interrelationships involved.

The chorological view came to dominate French geography. Paul Vidal de la Blache (1845-1918) was

NATIONAL GEOGRAPHIC SOCIETY AND GEOGRAPHIC RESEARCH In 1888 the National Geographic Society was founded to support the "increase and diffusion of geographic knowledge" of the world. In its first 110 years, the society funded more than five thousand expeditions and research projects with more than 6,500 grants. By the 1990s it was the largest such foundation in the world, and the results of its funded projects are found on television programs, video discs, video cassettes, and books, as well as in the *National Geographic* magazine, established in 1888. Its productions are cutting-edge resources for information about archaeology, ethnology, biology, and both cultural and physical geography.

the most prominent French geographer. He advocated the study of small, distinct areas, and French geographers set about identifying the many regions of France. They described and analyzed the unique physical and human geographic complex that was to be found in each region. An important concept that emerged from French geography was "possibilism." German geographers had introduced the notion of environmental determinism—that human beings were largely shaped and controlled by their environments. Possibilism rejected the concept of environmental determinism, asserting that the relationship between human beings and the environment works in two directions: The environment creates both limits and opportunities for people, but people can react in different ways to a given environment, so they are not controlled by it.

British geographers, influenced by the French approach, conducted regional surveys. British regional studies were unique in their emphasis on planning and geography as an applied science. From this work came the concept of a functional region—an area that works together as a unit based on interaction and interdependence.

American Geography

Prior to World War II, only a small group of people in the United States called themselves geographers. They were mostly influenced by German ideas, but the nature of geography was hotly debated. Two schools of geographers were philosophical adversaries. The Midwestern School, led by Richard Hartshorne, believed that description of unique regions was the central task of geography.

The Western (or Berkeley) School of geography, led by Carl Sauer, agreed that regional study was important, but believed it was crucial to go beyond description. Sauer and his followers included genesis and process as important elements in any study. To understand a region and to know where it is going, they argued, one must look at its past and how it got to its present state.

In the 1930s, environmental determinism was introduced to U.S. geography but ultimately was rejected. Although geography in both Europe and the United States was essentially an all-male discipline, the United States produced the first famous woman geographer, Ellen Churchill Semple (1863-1932).

World War II illustrated the importance of geographic knowledge, and after the war came to an end in 1945, geographers began to come into their own in the United States. From the end of World War II to the early 1960s, U.S. geographers produced many descriptive regional studies.

In the early 1960s, what is often called the quantitative revolution occurred. The development of computers allowed complex mathematical analysis

to be performed on all kinds of geographic data, and geographers began to analyze a wide range of problems using statistics. There was great enthusiasm for this new approach to geography at first, but beginning in the 1970s, many people considered a purely mathematical approach to be somewhat sterile and thought it left out a valuable human element.

In the 1980s and 1990s, many new ways to look at geographic issues and problems were developed, including humanism, behaviorism, Marxism, feminism, realism, structuration, phenomenology, and postmodernism, all of which bring human beings back into focus within geographical studies.

Geography in the Twenty-first Century

Geography increasingly uses technology to analyze global space and answer a wide range of questions related to a host of concerns including issues related to the environment, climate change, population, rising sea levels, and pollution. The Geographic Information System (GIS), in particular, provides a powerful way for people trained in geography to understand geographic issues, solve geographic problems, and display geographic information. Geographers continue to adopt a wide variety of philosophies, approaches, and methods in their quest to answer questions concerning all things spatial.

Wendy Shaw

Mapmaking in History

Cartography is the science or art of making maps. Although workers in many fields have a concern with cartography and its history, it is most often associated with geography.

Maps of Preliterate Peoples

The history of cartography predates the written record, and most cultures show evidence of mapping skills. The earliest surviving maps are those carved in stone or painted on the walls of caves, but modern preliterate peoples still use a variety of materials to express themselves cartographically. For example, the Marshall Islanders use palm fronds, fiber from coconut husks (coir), and shells to make sea charts for their inter-island navigation. The Inuit use animal skins and driftwood, sometimes painted, in mapping. There is a growing interest in the cartography of early and preliterate peoples, but some of their maps do not fit readily into a more traditional concept of cartography.

Mapping in Antiquity

Early literate peoples, such as those of Egypt and Mesopotamia, displayed considerable variety in their maps and charts, as shown by the few maps from these civilizations that still exist. The early Egyptians painted maps on wooden coffin bases to assist the departed in finding their way in the afterlife; they also made practical route maps for their mining operations. It is thought that geometry developed from the Egyptians' riverine surveys. The Babylonians made maps of different scales, using clay tablets with cuneiform characters and stylized symbols, to create city plans, regional maps, and "world" maps. They also divided the circle in the sexigesimal system, an idea they may have obtained from India and that is commonly used in cartography to this day.

The Greeks inherited ideas from both the Egyptians and the Mesopotamians and made signal contributions to cartography themselves. No direct evidence of early Greek maps exists, but indirect evidence in texts provides information about their cosmological ideas, culminating in the concept of a perfectly spherical earth. This they attempted to measure and divide mathematically. The idea of climatic zones was proposed and possibly mapped, and the large known landmasses were divided into first two continents, then three.

Perhaps the greatest accomplishment of the early Greeks was the remarkably accurate measurement of the circumference of Earth by Eratosthenes (276-196 BCE). Serious study of map projections began at about this time. The gnomonic, orthographic, and stereographic projections were invented before the Christian era, but their use was confined to astronomy in this period. With the possible single exception of Aristarchus of Samos, the Greeks believed in a geocentric universe. They made globes (now lost) and regional maps on metal; a few map coins from this era have survived.

Later Greeks carried on these traditions and expanded upon them. Claudius Ptolemy invented two projections for his world maps in the second century CE. These were enormously important in the European Renaissance as they were modified in the light of new overseas discoveries. Ptolemy's work is known mainly through later translations and reconstructions, but he compiled maps from Greek and Phoenician travel accounts and proposed sectional maps of different scales in his *Geographia*. Ptolemy's prime meridian (0 degrees longitude) in the Canary Islands was generally accepted for a millennium and a half after his death.

Roman cartography was greatly influenced by later Greeks such as Ptolemy, but the Romans themselves improved upon route mapping and surveying. Much of the Roman Empire was subdivided by instruments into hundredths, of which there is a cartographic record in the form of marble tablets. In Rome, a small-scale map of the world known to the Romans was made on metal by Marcus Vipsanius Agrippa, the son-in-law of Augustus Caesar, and displayed publicly. This map no longer exists, however.

Cartography in Early East Asia

As these developments were taking place in the West, a rich cartographic tradition developed in Asia, particularly China. The earliest survey of China (Yu Kung) is approximately contemporaneous with the oldest reported mapmaking activity of the Greeks. Later, maps, charts, and plans accompanied Chinese texts on various geographical themes. Early rulers of China had a high regard for cartography—the science of princes. A rectangular grid was introduced by Chang Heng, a contemporary of Ptolemy, and the south-pointing needle was used for mapmaking in China from an early date.

These traditions culminated in Chinese cartographic primacy in several areas: the earliest printed maps (about 1155 CE), early printed atlases, and terrestrial globes (now lost). Chinese cartography greatly influenced that in other parts of Asia, particularly Korea and Japan, which fostered innovations of their own. It was only after the introduction of ideas from the West, in the Renaissance and later, that Asian cartographic advances were superseded.

Islamic Cartography

A link between China and the West was provided by the Arabs, particularly after the establishment of Islam. It was probably the Arabs who brought the magnetized needle to the Mediterranean, where it was developed into the magnetic compass.

Some scholars have argued that the Arabs were better astronomers than cartographers, but the Arabs did make several clear advances in mapmaking. Both fields of study were important in Muslim science, and the astrolabe, invented by the Greeks in antiquity but developed by the Arabs, was used in both their astronomical and terrestrial surveys. They made and used many maps, as indicated by the output of their most famous cartographer, al-Idrisi (who lived about 1100–1165). Some of his work still exists, including a zonal world map and detailed charts of the Mediterranean islands.

At about the same time, the magnetic compass was invented in the coastal cities of Italy, which gave rise to advanced navigational charts, including information on ports. These remarkably accurate charts were used for navigating in the Mediterranean Sea. They were superior to the European maps of the Middle Ages, which often were concerned with religious iconography, pilgrimage, and crusade. The scene was now set for the great overseas discoveries of the Europeans, which were initiated in Portugal and Spain in the fifteenth century.

In the next four centuries, most of the coasts of the world were visited and mapped. The early, projectionless navigational charts were no longer adequate, so new projections were invented to map the enlarged world as revealed by the European overseas explorations. The culmination of this activity was the development of the projection, in 1569, of Gerardus Mercator, which bears his name and is of special value in navigation.

Early Modern Mapmaking

Europeans began mapping their own countries with greater accuracy. New surveying instruments were invented for this purpose, and a great land-mapping activity was undertaken to match the worldwide coastal surveys. For about a century, the Low Countries of Belgium, Luxembourg, and the Netherlands dominated the map and chart trades, producing beautiful hand-colored engraved sheet wall maps and atlases.

France and England established new national observatories, and by the middle of the seventeenth century, the Low Countries had been eclipsed by France in surveying and making maps and charts. The French adopted the method of triangulation of Mercator's teacher, Gemma Frisius. Under four generations of the Cassini family, a topographic survey of France more comprehensive than any previous survey was completed. Rigorous coastal surveys were undertaken, as well as the precise measurement of latitude (parallels).

The invention of the marine chronometer by John Harrison made it possible for ships at sea to determine longitude. This led to the production of charts of all the oceans, with England's Greenwich eventually being adopted as the international prime meridian.

Quantitative, thematic mapping was advanced by astronomer Edmond Halley (1656–1742) who produced a map of the trade winds; the first published magnetic variation chart, using isolines; tidal charts; and the earliest map of an eclipse. The Venetian Vincenzo Coronelli made globes of greater beauty and accuracy than any previous ones. In the German lands, the study of map projections was vigorously pursued. Johann H. Lambert and others invented a number of equal-area projections that were still in use in the twentieth century.

Ideas developed in Europe were transmitted to colonial areas, and to countries such as China and Russia, where they were grafted onto existing cartographic traditions and methods. The oceanographic explorations of the British and the French built on the earlier charting of the Pacific Ocean and its islands by native navigators and the Iberians.

Nineteenth Century Cartography

Cartography was greatly diversified and developed in the nineteenth century. Quantitative, thematic mapping was expanded to include the social as well as the physical sciences. Alexander von Humboldt used isolines to show mean air temperature, a method that later was applied to other phenomena. Contour lines gradually replaced less quantitative methods of representing terrain on topographic maps. Such maps were made of many areas, for example India, which previously had been poorly mapped.

Extraterrestrial (especially lunar) mapping, had begun seriously in the preceding two centuries with the invention of the telescope. It was expanded in the nineteenth century. In the same period, regular national censuses provided a large body of data that could be mapped. Ingenious methods were created to express the distribution of population, diseases, social problems, and other data quantitatively, using uniform symbols.

Geological mapping began in the nineteenth century with the work of William Smith in England, but soon was adopted worldwide and systematized, notably in the United States. The same is true of transportation maps, as the steamship and the railway increased mobility for many people. Faster land travel in an east-west direction, as in the United States, led to the official adoption of Greenwich as the international prime meridian at a conference held in Washington, D.C., in 1884. Time zone maps were soon published and became a feature of the many world atlases then being published for use in schools, offices, and homes

A remarkable development in cartography in the nineteenth century was the surveying of areas newly occupied by Europeans. This occurred in such places as the South American republics, Australia, and Canada, but was most evident in the United States. The U.S. Public Land Survey covered all areas not previously subdivided for settlement. Property maps arising from surveys were widely available, and in many cases, the information was contained in county and township atlases and maps.

Modern Mapping and Imaging

Cartography was revolutionized in the twentieth century by aerial photography, sonic sounding, satellite imaging, and the computer. Before those developments, however, Albrecht Penck proposed an ambitious undertaking—an International Map of the World (IMW). Cartography historically had been a nationalistic enterprise, but Penck suggested a map of the world in multiple sheets produced cooperatively by all nations at the scale of 1:1,000,000 with uniform symbols. This was started in the first half of the twentieth century but was not completed, and was superseded by the World Aeronautical Chart (WAC) project, at the same scale, during and after World War II.

The WAC project owed its existence to flight information made available following the invention of the airplane. Both photography and balloons were developed before the twentieth century, but the new, heavier-than-air craft permitted overlapping aerial photographs to be taken, which greatly facilitated the mapping process. Aerial photography revolutionized land surveys—maps could be made at less cost, in less time, and with greater accuracy than by previous methods. Similarly, marine surveying was revolutionized by the advent of sonic sounding in the second half of the twentieth century. This enabled mapping of the floor of the oceans, essentially unknown before this time.

Satellite imaging, especially continuous surveillance by Landsat since 1972, allows temporal monitoring of Earth. The computer, through Geographical Information Systems (GIS) and other technologies, has greatly simplified and speeded up the mapping process. During the twentieth century, the most widely available cartographic product was the road map for travel by automobile.

Spatial information is typically accessed through apps on computers and mobile devices; traditional maps are becoming less common. The new media also facilitate animated presentations of geographical and extraterrestrial distributions. Cartographers remain responsive to the opportunities provided by new technologies, materials, and ideas.

Norman J. W. Thrower

MAPMAKING AND NEW TECHNOLOGIES

The field of geography is concerned primarily with the study of the curved surface of Earth. Earth is huge, however, with an equatorial radius of 3,963 miles (6,378 km.). How can one examine anything more than the small patch of earth that can be experienced at one time? Geographers do what scientists do all of the time: create models. The most common model of Earth is a globe—a spherical map that is usually about the size of a basketball.

A globe can show physical features such as rivers, oceans, the continents, and even the ocean floor. Political globes show the division of Earth into countries and states. Globes can even present views of the distant past of Earth, when the continents and oceans were very different than they are today. Globes are excellent for learning about the distributions, shapes, sizes, and relationships of features of Earth. However, there are limits to the use of globes.

How can the distribution of people over the entire world be described at one glance? On a globe, the human eye can see only half of Earth at one time. What if a city planner needs to map every street, building, fire hydrant, and streetlight in a town? To fit this much detail on a globe, the globe might have to be bigger than the town being mapped. Globes like these would be impossible to create and to carry around. Instead of having to hire a fleet of flatbed trucks to haul oversized globes, the curved surface of the globe can be transformed to a flat plane.

The method used to change from a curved globe surface to a flat map surface is called a map projection. There are hundreds of projections, from simple to extremely complex and dating from about two thousand years ago to projections being invented today. One of the oldest is the gnomonic projection. Imagine a clear globe with a light inside. Now imagine holding a piece of paper against the surface of the globe. The coastlines and parallels of latitude and meridians of longitude would show through the globe and be visible on the paper. Computers can do

the same thing because there are mathematical formulas for nearly all map projections.

Geometric Models for Map Projections

One way to organize map projections is to imagine what kind of geometric shape might be used to create a map. Like the paper (a plane surface) against the globe described above, other useful geometric shapes include a cone and a cylinder. When the rounded surface of any object, including Earth, is flattened there must be some stretching, or tearing. Map projections help to control the amount and kinds of distortion in maps. There are always a few exceptions that cannot be described in this way, but using geometric shapes helps to classify projections into groups and to organize the hundreds of projections.

Another way to describe a map projection is to consider what it might be good for. Some map projections show all of the continents and oceans at their proper sizes relative to one another. Another type of projection can show correct distances between certain points.

Map Projection Properties

When areas are retained in the proper size relationships to one another, the map is called an equal-area map, and the map projection is called an equal-area projection. Equal-area (also called equivalent or homolographic) maps are used to measure areas or view densities such as a population density.

If true angles are retained, the shapes of islands, continents, and oceans look more correct. Maps made in this way are called conformal maps or conformal map projections. They are used for navigation, topographic mapping, or in other cases when it is important to view features with a good representation of shape. It is impossible for a map to be both equal-area and conformal at the same time. One or the other must be selected based on the needs of the map user or mapmaker.

One special property—distance—can only be true on a few parts of a map at one time. To see how far it is between places hundreds or thousands of miles apart, an equidistant projection should be used. There will be several lines along which distance is true. The azimuthal equidistant projection shows true distances from the center of the map outward. Some map projections do not retain any of these properties but are useful for showing compromise views of the world.

Modern Mapmaking

Modern mapmaking is assisted from beginning to end by digital technologies. In the past, the paper map was both the primary means for communicating information about the world and the database used to store information. Today, the database is a digital database stored in computers, and cartographic visualizations have taken the place of the paper map. Visualizations may still take the form of paper maps, but they also can appear as flashes on computer screens, animations on local television news programs, and even on screens within vehicles to help drivers navigate. Communication of information is one of the primary purposes of making maps. Mapping helps people to explore and analyze the world.

Making maps has become much easier and the capability available to many people. Desktop mapping software and Internet mapping sites can make anyone with a computer an instant cartographer. The maps, or cartographic visualizations, might be quite basic but they are easy to make. The procedures that trained cartographers use to make map products vary in the choice of data, software, and hardware, but several basic design steps should always take place.

First, the purpose and audience for whom the map is being made must be clear. Is this to be a general reference map or a thematic map? What image should be created in the mind of the map reader? Who will use the map? Will it be used to teach young children the shapes of the continents and oceans, or to show scientists the results of advanced research? What form will the cartographic visualization take?

SLIDING ROCKS GET DIGITAL TREATMENT

Dr. Paula Messina studied the trails of rocks that slide across the surface of a flat playa in Death Valley, California. The sliding rocks have been studied in the past, but no one had been able to say for certain how or when the rocks moved. It was unclear whether the rocks were caught in ice floes during the winter, were blown by strong winds coming through the nearby mountains, or were moved by some other method.

Messina gave the mystery a totally digital treatment. She mapped the locations of the rocks and the rock trails using the global positioning system (GPS) and entered her rock trail data into a geographic information system (GIS) for analysis. She was able to determine that ice was not the moving agent by studying the pattern of the trails. She also used digital elevation models (DEM) and remotely sensed imagery to model the environment of the playa. She reported her results in the form of maps using GIS' cartographic output capabilities. While she did not solve completely the mystery of the sliding rocks, she was able to disprove that winter ice caused the rocks to slide along together in rafts and that there are wind gusts strong enough to move the biggest rock on the playa.

Will it be a paper map, a graphic file posted to the Internet, or a video?

The answers to these questions will guide the cartographer in the design process. The design process can be broken down into stages. In the first stage of map design, imagination rules. What map type, size and shape, basic layout, and data will be used? The second stage is more practical and consists of making a specific plan. Based on the decisions made in the first stage, the symbols, line weights, colors, and text for the map are chosen. By the end of this stage, there should be a fairly clear plan for the map. During the third stage, details and specifications are finalized to account for the production method to be used. The actual software, hardware, and methods to be used must all be taken into consideration.

What makes a good map? Working in a digital environment, a mapmaker can change and test various designs easily. The map is a good one when it communicates the intended information, is pleasing to look at, and encourages map readers to ask thoughtful questions.

New Technologies

Mapping technology has gone from manual to magnetic, then to mechanical, optical, photochemical, and electronic methods. All of these methods have overlapped one another and each may still be used in some map-making processes. There have been recent advances in magnetic, optical, and most of all, electronic technologies.

All components of mapping systems—data collection, hardware, software, data storage, analysis, and graphical output tools—have been changing rapidly. Collecting location data, like mapping in general, has been more accessible to more people. The development of the Global Positioning System (GPS), an array of satellites orbiting Earth, gives anyone with a GPS receiver access to location information, day or night, anywhere in the world. GPS receivers are also found in planes, passenger cars, and even in the backpacks of hikers.

Satellites also have helped people to collect data about the world from space. Orbiting satellites collect images using visible light, infrared energy, and other parts of the electromagnetic spectrum. Active sensing systems send out radar signals and create images based on the return of the signal. The entire world can be seen easily with weather satellites, and other specialized satellite imagery can be used to count the trees in a yard.

These great resources of data are all stored and maintained as binary, computer-readable information. Developments in laser technology provide large amounts of storage space on media such as optical disks and compact disks. Advances in magnetic technology also provide massive storage capability in the form of tape storage, hard drives, and cloud storage. This is especially important for saving the large databases used for mapping.

Computer hardware and software continue to become more powerful and less expensive. Software continues to be developed to serve the specialized needs that mapping requires. Just as word processing software can format a paper, check spelling and grammar, draw pictures and shapes, import tables and graphics, and perform dozens of other functions, specialized software executes maps. The most common software used for mapping is called Geographic Information System (GIS) software. These systems provide tools for data input and for analysis and modeling of real-world spatial data, and provide cartographic tools for designing and producing maps.

Karen A. Mulcahy

CHINA

A nation of immense size, physical and cultural diversity, and historical grandeur, China occupies much of East Asia. In Chinese writing, the character for the country means "central land or kingdom"—representing the belief of the early peoples that their country was not only the geographical center of the world but also the central, or true, civilization.

China has one of the world's earliest civilizations and a recorded history dating back 3,500 years. It is a nation of extremes and superlatives: the world's largest population, tremendous geographical diversity, the world's highest mountain but also a basin with an elevation below sea level, the most deaths in history from both earthquakes and floods, the world's longest artificial structure, and huge projects that required millions of workers—the Grand Canal, the Great Wall, and diking the Huang He—and now the world's largest construction project and largest dam.

Landforms

China's land area is 3.72 million square miles (9.63 million sq. km.), third in size after Russia and Canada and slightly larger than the United States. There is great diversity of landforms and landscapes, with the surface topography generally sloping downward from west to east in a four-step staircase. Mountains occupy 43 percent of the land surface, high plateaus 26 percent, basins 19 percent, and plains 12 percent. More than 3,000 islands are in its territorial seas, extending

China's first emperor, Qin Shi Huang, is famed for having united the Warring States' walls to form the Great Wall of China. Most of the present structure, however, dates to the Ming dynasty.

CHINA

12 nautical miles (22.2 km.) out from the coastline.

In the west, toward India and Nepal, lie the Tibetan Plateau and Himalaya Mountains. The plateau has an average elevation greater than 13,000 feet (4,000 meters) above sea level. Mountain ranges bordering it include the Himalayas on the south, Pamirs on the west, and Kunlun on the north. The Himalayas, shared with India, Nepal, and Pakistan, have eleven of the seventeen mountains in the world with elevations of at least 26,000 feet (8,000 meters). These peaks include Mt. Everest, the world's highest, on the Nepal-Tibet frontier.

The Tibetan Plateau and Himalayas started being folded, faulted, and uplifted about 40 million years ago as the Indian subcontinent, pushing northward, began colliding with Eurasia. This process continues, with the Himalayas still uplifting at the rate of about 3 inches (8 centimeters) per century.

Other major mountain ranges in western China—the Tien Shan, Kunlun, and Tsinling—(also spelled Chinling or Qinling)—were formed in an earlier episode of continental collision and mountain building that began about 340 million years ago and continued for 80 million years, as

Li River near Guilin, Guangxi.

part of the aggregation of the world's landmasses into the supercontinent of Pangaea.

A number of features, from the northwest and extending to the central and southern regions, have elevations of 3,300 to 6,600 feet (1,000 to 2,000 meters). The mountain ranges enclose a series of plateaus and basins, including the Tarim Basin in the northwest, with the Taklimakan Desert, Lop Nur salt lake, and Turfan Depression. The latter is as much as 505 feet (154 meters) below sea level; in its interior continental site, it has summer temperatures reach as high as 117° Fahrenheit (47° Celsius) and winter temperatures fall as low as -22° Fahrenheit (-30° Celsius).

The area also includes the loess plateau in north central China—extensive deposits of windblown silt from the Gobi Desert located to the northwest. The world's thickest silt deposits, tens of meters thick, are here along the Huang He River in north China, rivaled only by deposits of loess along North America's Missouri River. Farther to the south is the Yunnan Plateau, with one of the world's best-known example of karst terrain forming rugged scenic topography along the Li River near the city of Guilin (Kweilin). Karst forms when subsurface limestone rock is progressively dissolved by groundwater, developing sinkholes and caves. After uplift and erosion of the overlying rock layers, the honeycomb of rock pinnacles is exposed.

The coastal plain, extending on the east coast from Manchuria in the north to Hong Kong in the south, has elevations generally below 3,280 feet (1,000 meters). The offshore continental shelf, formed of sand and silt eroded off the mainland, is below sea level and extends out to water depths down to 650 feet (200 meters).

Rivers

China's rivers have been intimately involved in the rise of early agriculture and civilization, in transportation and trade, but also in immense destruction and human tragedy. Most of China's rivers flow from west to east into the East China Sea, with a few flowing to the far south. The two major rivers are the Chang Jiang ("long river"), often called the Yangtze, after its downstream portion, and the Huang He, or Yellow River.

The Yangtze runs 3,900 miles (6,300 km.), making it the world's third-longest. Originating in the Himalayas in the west, it has long been a transportation and trade artery through central China, past Shanghai, to the East China Sea. Cutting through the Wu Shan before traversing the broad, flatter coastal plain to the east, it creates the Three Gorges area—a stretch of deep scenic gorges with looming rock walls, narrows, and bends. While providing irrigation for agricultural lands, it has frequently flooded with devastating consequences.

The Yellow, or Huang He, River, the world's fourth-longest, wanders through the north central plain of China to the Yellow Sea. The river's yellowish color is caused by the heavy load of loess sediment it carries. Its channel is confined by natural levees of deposited silt and by dikes, which elevate the river level to as much as 16 feet (5 meters) above the surrounding floodplain. Habitation on the fertile floodplain dates back over 500,000 years. With the high population density, frequency of flooding, and vulnerability to broken levees, the river has claimed more lives—estimates range as high as 10 million—and caused more human suffering than any other single natural feature in the world. It is referred to ruefully as "China's Sorrow."

Earthquakes

China, a region of complex tectonic relationships, that is, mountain-building and crustal deformation, has a long history of earthquakes. Largely because of the collision by India and uplift of the Tibetan Plateau, a broad band of structural deformation trends easterly through central China. China has some of recorded history's greatest earthquakes and, because of its large population and long history of habitation, some of the most devastating.

Mineral and Energy Resources

China's geological diversity and variety of structures and terranes (crustal blocks with their own geological development) have resulted in a rich assembly of mineral deposits and energy resources. China has large reserves of coal, mostly in the northeast. Oil and gas are present in some basins of sedimentary rocks, especially in the Daqing (Taching) oilfield in the extreme northeast, where large-scale production and refining began in the early 1960s. In 2020, Daqing was the world's fourth most productive oilfield. Oil is also extracted from large oilfields in the South China Sea. Despite being one of the world's largest oil producers, China must still import half its oil to meet domestic needs.

China has developed ambitious programs to reduce the country's traditional dependence on coal and, more recently, oil. In the period from 2017 to 2020, China earmarked more than $367 billion for renewable energy generation: hydro, solar, wind, and nuclear. Hydroelectric power is produced from

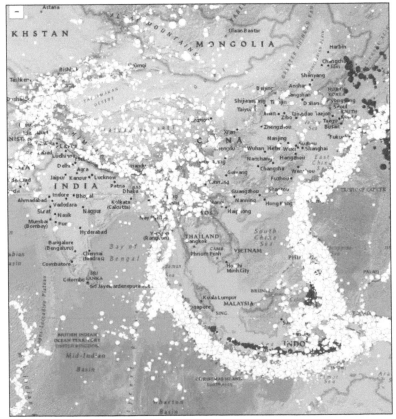

Earthquakes with a magnitude 4.5 and over (1900-2015). The yellow star is the 2008 Sichuan earthquake.

dams on several rivers, most notably the Gezhouba on the Yangtze and the controversial Three Gorges Dam—the world's largest-capacity hydroelectric power station. China has made substantial headway in developing solar projects, and now has some of the largest solar farms in the world. China is also the world's leader in wind-power generation, and ranks third in the world in nuclear energy capacity. Iron ore is abundant, and there are significant amounts of tungsten, tin, antimony, aluminum, zinc, molybdenum, lead, mercury, and rare-earth metals.

Climate and Biogeography

The latitude, range, and size of China are remarkably similar to those of the United States. China is largely in the temperate climate zone, but ranges from subarctic in the north to subtropical in the south. The eastern coastal region is warm and humid, the north central is cool and dry, and the western mountains are alpine and cold.

The wildlife is diverse, including species found only in China: giant pandas (found in remote mountain areas of central China), golden monkeys, white-lipped deer, Yangtze River dolphins, and Chinese alligators.

There are more than 11,000 species of woody plants, including some trees unique to China: metasequoia, China cypress, golden larch, (Cathay) silver fir, and the elmlike eucommia. Much of the natural vegetation was removed over many centuries of settlement and intensive cultivation, but some natural forests are being preserved in mountain regions.

Natural flora, roots, and animal parts have long been used as chemical medicines. As early as the sixteenth century, more than 1,800 kinds of herbal medicines were known.

Human Geography

China is divided into twenty-three provinces and five nominally autonomous regions (Xinjiang, Tibet since annexation in 1950, Inner Mongolia, Ningxia, and Guangxi), plus municipalities administered by the central government—Beijing (formerly Peking), Shanghai, Tianjin, and most recently, Hong Kong and Macao.

CHINA'S MOST DEVASTING EARTHQUAKES

Historians have found that records of more than ten thousand earthquakes in 3,000 years of Chinese history. More than five hundred of the earthquakes were of disaster proportions. Some of the most notable quakes:

1290: Earthquake in Chihli, in northeast China, kills 100,000 people.

1556: Earthquake in Shaanxi (Shensi), in central China, has world's highest death toll: 830,000 people living in nonreinforced earthen buildings and caves are killed when their homes collapse

1920: An estimated magnitude 8.5 earthquake in Gansu (Kansu), in central China, kills 180,000.

1927: An estimated magnitude 8.3 earthquake in Xining, Qinghai province, in west central China, kills 200,000.

1976: Magnitude 8.0 earthquake in Tangshan, northeast China, claims the world's highest twentieth century death toll in a natural disaster: 243,000 are killed; another 165,000 are seriously injured.

2008: Magnitude 7.9 earthquake in Wenchuan County, Sichuan is the deadliest earthquake in China since the 1976 Tangshan earthquake and the strongest since the 1950 Chayu earthquake: 87,587 are killed; 374,176 were reported injured, and 18,222 listed as missing as of July 2008.

China is the world's most populous nation. In 2018 its population was about 1.3 billion, about 18 percent of the entire world's population. Because of the burgeoning numbers and the strain on land and resources, China initiated national family planning in the 1970s. The birth rate dropped from 34.1 per 1,000 people in 1969 to 12.1 per 1,000 by 2018, and the rate of growth of the population overall was reduced by 2018 to 0.37 percent per year. China relaxed its strict one-child-per-family policy in 2016, although its legacy is evident in the country's rapidly aging population.

The cities, many densely populated, had about 60 percent of the total population in 2019. The rest

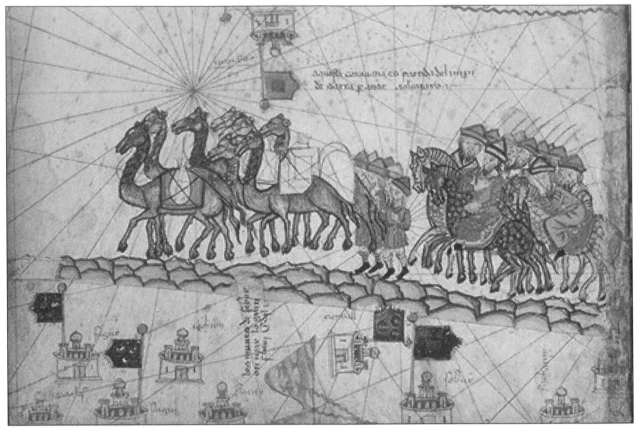

Caravan on the Silk Road, 1380.

of the people continue a largely rural, agrarian life, despite recent, rapid industrialization and development. Spontaneous migration from the countryside to the cities was discouraged or prohibited in the 1950s because of the lack of productive employment there, but the influx increased again in the 1980s. Large expanses of rugged terrain in the western regions are sparsely populated.

Metropolitan Shanghai is the largest city in China, with more than 26 million people in 2019. That same year, Beijing, the capital city for the last 800 years, had 20 million, Chongqing had 6.6 million, Guangzhou 14.9 million, Tianjin 13.4 million, and Shenzhen 10.3 million.

Ethnically, nearly 92 percent of the populace is of the Han culture group. There are fifty-six other nationalities, or ethnic groups, recognized by the Chinese government, most of whom inhabit the north and west regions.

Peopling and Settlement

More than a half-million years ago, the China region was home to primitive early humans—*Homo erectus* ("Peking Man"). The first dynasty is believed to have been the Xia, beginning about 1994 BCE. There was a transition from a slave society to a feudal society from the eighth to the third century BCE. A centralized, unified Chinese state was established with the Ch'in Dynasty—from which the name China derives—by emperor Qin Shi Huangdi. It is his funerary terra-cotta warriors that have been excavated near the Ch'in Dynasty capital of Xi'an (Sian) in central China.

China made many significant advances that placed it at the cultural forefront of the world, especially during the Han (202 BCE-220 CE) and Tang (618-907 CE) dynasties: metal smelting and casting (copper, bronze, iron), porcelain, textiles (silk and its culturing), invention of paper (105 CE), movable wood-block printing (about 1040 CE), navigation and the magnetic compass, and gunpowder and weaponry such as the crossbow. In the fifteenth century (Ming Dynasty), the explorer Zheng He led several voyages from China as far as Arabia and East Africa.

A major overland artery of cultural and economic exchange—for religious pilgrims, wandering armies, and caravans of traders—linked Europe and the Middle East to Xian, beginning about the second century BCE. Called the Jade or Emperor's Road, it later became known as the Great Silk Road.

In an early cultural exchange with the Western world, Marco Polo (and earlier his father) traveled to China overland from Italy. Arriving in 1275 in the Beijing area, he served the emperor Kublai Khan until leaving in 1292. His writings provided an influential travel window for Europeans into the geography and life of the Far East.

Historical Infrastructure

China has had three of the greatest, largest, and most labor-intensive construction projects of human civilization: the Great Wall, the Grand Canal, and the Three Gorges Dam.

The Great Wall of China is a sinuous fortification running along the northern frontier of China, from the eastern sea to the far interior. It was intended as protection from raids by nomadic peoples to the north. Largely constructed during the Qin (Ch'in) Dynasty (from 221 BCE) of earth and rock, it was re-paired and extended with brick and stonework in the Ming Dynasty (fourteenth to seventeenth centuries). Its official length is 13,171 miles (21,196 km.). Much of it had a height of about 25 feet (8 meters), with regular watchtowers and a roadway or stepped pathway on top that was several feet wide. Popularly called the Great Wall, it is called the 10,000-Li Wall by the Chinese. (The li equals about one-third of a mile.)

Grand Canal

The longest artificial waterway in the world, the Grand Canal was started in the fourth century BCE and rebuilt over the centuries. By the seventh century CE, it extended 1,100 miles (1,800 km.) from Beijing south to the coastal city of Hangzhou. It was used for water transport of grain and other goods and for irrigation. Much of the southern portion is still navigable.

Three Gorges Dam

The Three Gorges Dam is the largest hydroelectric dam engineering project in history. Begun in 1994, the immense dam complex reached its full capacity in 2012. With its series of dams on the Yangtze River and a 400-mile-long (645 km.) reservoir, the project produces an unprecedented 22,500 megawatts of electricity where it is needed most—in China's relatively underdeveloped interior provinces.

Grand Canal tour boats, Suzhou. The city was called Venice of the East by Marco Polo because of its canals and stone bridges.

The Three Gorges project has also had a number of controversial impacts. Some 150,000 acres (61,000 hectares) of land are now permanently under water. Prior to the project, this land supported 140 towns, 1,350 villages, 16 major archaeological sites, and some 2 million people, all of whom needed to be relocated. From an environmental perspective, the dam has had disastrous impacts. The construction and the flooding of hundreds of factories and waste dumps caused massive pollution. The banks along the reservoir and the river have been eroding, causing a number of catastrophic landslides. Dozens of aquatic species have been endangered by the dam, most notably the baiji, or Chinese river dolphin. The dam complex has also critically diminished water supply to downstream residential centers and ecosystems.

Economic Geography

Agriculture supports most of China's people. Only about 21.5 percent of the land is arable, most of it in eastern China, where nearly all such land is under cultivation. Half the arable land is irrigated—China has, in fact, more irrigated land than any other country in the world.

The northeast plain has fertile soil, with crops of wheat, maize (corn), millet, soybeans, and sugar beets. The north central plain also produces cotton and fruits. The Chang Jiang plain in east central China has rice and fish. Paddy rice, grown there and farther south, accounted for more than 40 percent of China's grain production in the 2010s; wheat comprised 20 percent. The south also has sweet potatoes, tea, and fish. The inland grasslands, mostly in the north and west, provide pasture for livestock—horses, cattle, sheep, and goats.

Industry and Trade

Today, China is the world's largest economy, a distinction it has held since 2017. This achievement took nearly 40 years to attain.

Prior to the late 1970s, China had the highly centralized and planned economy typical of the communist-governed countries of that era. Since then, there has been a gradual transition—some would say a dramatic and rapid shift, in view of China's

The Three Gorges Dam is the largest hydroelectric dam in the world.

A PLA air force Chengdu J-20 stealth fighter aircraft during the opening of Airshow China in Zhuhai.

past social and economic history—to a socialist market economy. This has increased productivity, economic growth and development, and modern industrialization. A prime aspect of this is that market forces help in the allocation of economic resources. Controlled encouragement is given for both privately owned and foreign enterprises. The latter typically include joint ventures in China with foreign firms or consortia in industry, trade and commerce, and tourism.

In 2020, along with being the world's largest economy, China is now the world's largest exporter of goods (since 2010) and the world's largest trad-ing nation. In the early 2020s, China's main exports are computers and computer-related machinery, telecommunications equipment, clothing, furniture, textiles, rice, and tea. Its major imports include integrated circuits and other computer components, medical and optical equipment, metals, cars, and soybeans. China's most recent five-year economic plan—its thirteenth—unveiled in 2016, seeks to accelerate growth by moving the economy from one that relies on exports to a more consumer-driven model.

Helen Finken and Robert Carmichael

Gate of Manifest Virtue, Forbidden City, Beijing.

JAPAN, KOREA, AND TAIWAN

Japan, Korea, and Taiwan are all located in East Asia. Mountains and highlands dominate at least 70 percent of their landscapes. Monsoon climates prevail, affecting the precipitation patterns in each country. Thanks to their close proximity to the Asian mainland, these nations share a strong heritage with Chinese civilization. Confucianism has shaped the social structure in these societies, and Buddhism, diffused by way of China, persists as a major religious element. Despite their unique cultural traditions and social norms, history and geography have created lasting links among them. Traces of Chinese civilization are still evident in the ideological framework of these cultures. Mandarin Chinese is the official language in Taiwan, and Chinese characters are part of the written language in Japan and Korea.

Since the mid-nineteenth century, Japan has been an important economic influence in East Asia. Beginning in 1868, the Meiji Restoration rapidly transformed Japan from an isolated nation to an industrial and military power. Japan then annexed Korea and colonized Taiwan. The subsequent development in Korea and Taiwan was tailored to providing resources and markets for the rapidly expanding Japanese industries. By World War II, Japan was competing against the Soviet Union (now Russia), the United States, and the Allied countries in Europe. The total defeat of Japan as a result of the war cost the country a great deal. Taiwan and Korea became independent, and the Japanese economy was severely crippled.

U.S. efforts under the Marshall Plan during 1947-1952, and the Korean War (1950-1953) helped the Japanese postwar recovery. By the 1980s, Japan had ascended to an economic power competing for world market share with the United States and Europe. Development in Taiwan and South Korea also gained momentum in the postwar era. During the Cold War, their strategic locations attracted generous aid from both Japan and the United States. Taiwan and South Korea, although trailing Japan economically, nevertheless have advanced to the head of the Asian economies. Today, Japan, South Korea, and Taiwan are among the most developed economies in East Asia.

Japan

Located between 24° and 46° north latitudes, Japan covers 145,914 square miles (377,915 sq. km.)—approximately the size of the U.S. states of California or Montana. Formed in the shape of a crescent off the eastern coast of Asia, Japan is one of the world's most extensive archipelagos. It comprises four principal islands and more than 3,200 adjacent small islands and islets.

Japan's four main islands—Hōkkaidō, Honshū, Shikōkū, and Kyūshū—extend from northeast to southwest for about 1,400 miles (2,250 km.). No point in Japan is more than 93 miles (150 km.) from the sea. Separated from the Asian mainland by the Sea of Japan (called the East in North and South Korea) in the west, Japan is about 124 miles (200 km.) east of the Korean peninsula at the nearest point of the Korea Strait. The southern coast of Japan on the Pacific is characterized by long, narrow, gradually shallowing inlets that provide numerous natural harbors. The northern coast of Japan, on the Pacific Ocean and the coast of the Sea of Japan, is generally smooth and unindented.

About 75 percent of Japan is mountainous. Mount Fuji in central Honshū is the highest peak, rising 12,388 feet (3,776 meters) above sea level. The Japanese Alps, formed by the convergence of three mountain ranges in the middle of Honshū Island—the Hida, the Kiso, and the Akaishi—divide Japan into eastern and western halves. Lacking extensive lowlands, small coastal plains are found in

JAPAN

the more rugged western half along the Pacific Ocean, including the Kantō, Nōbi, Kinki, and Sendai plains on Hōnshū Island and the Ishikari Plain on Hōkkaidō Island. The largest of these, the Kantō Plain, covers about 6,244 square miles (16,172 sq. km.). Rivers in Japan are numerous but generally short, with steep, swift courses. The Shinano River in central Hōnshū is the longest.

Climates in Japan, due to its elongated shape, vary significantly from north to south. Similar to the climate regime in the eastern United States from Maine to Florida, a humid continental climate—warm summers with long, cold winters—prevails in northern Japan; a humid subtropical climate—long, hot summers with short, mild winters—characterizes southern Japan.

Japan lacks most mineral resources and has only small deposits of chromite, magnesium, sulfur, lead, zinc, and copper. Coal deposits of moderate scale are located primarily on Hōkkaidō and Kyūshū islands. Japan depends heavily on imports of fossil fuels and industrial minerals to drive its economy.

Japan has nearly 127 million people. Over 50 percent of them are clustered in the core along the Pacific coast extending southwest from central Hōnshū to the northern Kyūshū islands for approximately 700 miles (1,130 km.). All the major cities are concentrated here. Surrounded by intensively farmed and highly productive rice paddies, the boundary of several primary urban industrial complexes is blurred: the Tokyo-Yokohama-Kawasaki metro area on the Kantō, Nagoya on the Nōbi Plain, and the Osaka-Kobe-Kyoto triangle in the Kinki District. The capital of Japan, Tokyo, has nearly 13.9 million people (2019), and the Tokyo-Osaka conurbation, also known as Tokaido, is one of the world's largest megalopolises.

Contrasted to the diverse ethnic population in the United States, the Japanese population is highly uniform, with a small Ainu minority on Hōkkaidō Island. Religious practices—reflecting the traditions of Shinto, Buddhism, Confucianism, Daoism, and Christianity—emphasize the maintenance of harmonious relations with each other and the fulfillment of social obligations as a group member.

Taiwan

Located between 21° 45′ and 25° 50′ north latitudes, Taiwan consists of Taiwan Island and nearby small island groups. The tobacco-leaf-shaped Tai-

Dabajian Mountain, Taiwan.

TAIWAN

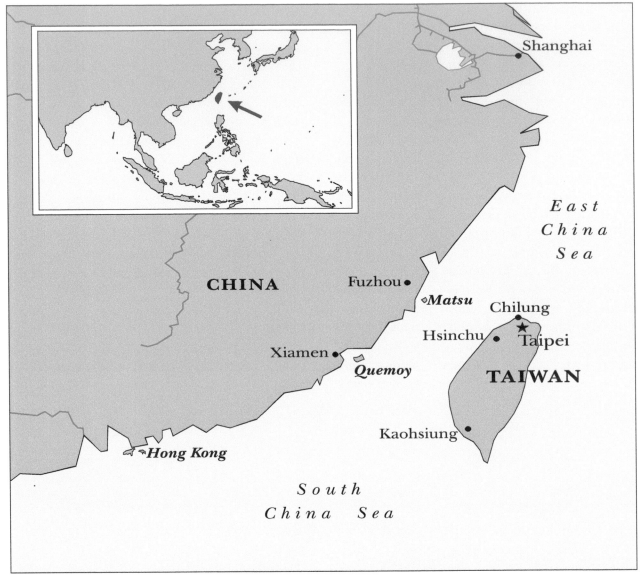

wan Island is about 100 miles (160 km.) east of China separated by the Taiwan Strait. The Tropic of Cancer crosses just south of the center of the island. With a total area of 13,892 square miles (35,980 sq. km.), Taiwan is about equivalent to the combined size of the U.S. states of Connecticut and Massachusetts. It is the most densely populated nation of this region, with 1,695 people per square mile.

Taiwan Island extends approximately 250 miles (402 km.) north-south, and 90 miles (145 km.) east-west. As with Japan, 75 percent of Taiwan Island is mountainous. The Chungyang Mountains, running north-south, dominate the eastern half of the island; Yu Shan, the highest peak, rises 13,110

feet (4,000 meters) above sea level. The gently rolling plains and scenic ports in western Taiwan contrast with rugged eastern Taiwan.

Rivers originating in the Chungyang Mountains are numerous but generally short, with swift courses. Taiwan's longest river, the Choshui River, is 114 miles (183 meters) long. The only navigable river on the island is the Tamsui River, which runs through the capital, Taipei. Monsoon climates dominate Taiwan. A wet season and a dry season alternate between northern and southern Taiwan. Long, hot, humid summers with short, mild winters characterize the entire island except the higher elevation of the Chungyang Mountains. A warm cli-

KOREA

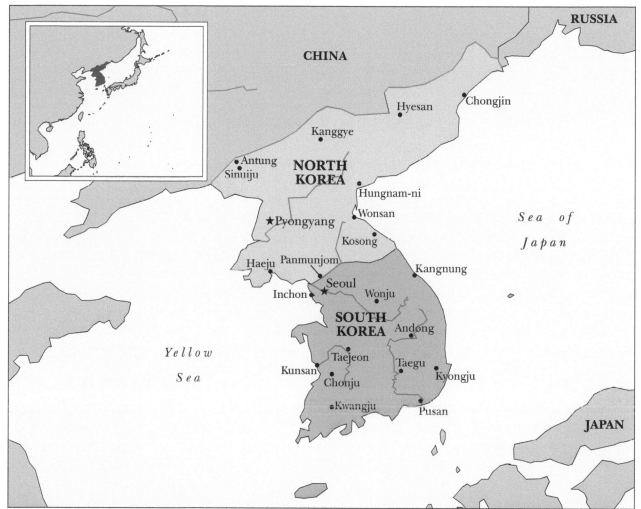

mate, copious rainfall, and fertile volcanic and alluvial soils enable multicrop farming. Rice and sugar are the most extensive crops. Although there are small deposits of more than twenty different minerals, Taiwan depends heavily on mineral and energy imports.

Approximately 23.5 million people live in Taiwan. More than 85 percent of them live along the alluvial plains on the west coast, from Taipei in the north to Kaohsiung in the south. The majority of the population are Taiwanese who immigrated centuries ago from Guangdong and Fujian provinces in southeast China. Aborigines and immigrants who fled mainland China in 1949 make up the rest. From 1895 to 1945, Taiwan was a Japanese colony as a result of the Sino-Japanese War and was transformed to a rice and sugar supplier for Japan. Independence came

after World War II and was followed by rapid urban and economic development. Taipei has grown to a major metropolitan area with more than 8.5 million people. Kaohsiung has emerged as a major port, handling international trade and commerce. Hsinchu, a satellite city southwest of Taipei, has been established as a technopole attracting international research and development projects.

Korea

The Korean Peninsula is located between 38° and 43° north latitude at the juncture of the northeast part of the Asian continent and the Japanese archipelago. It juts out southeast for approximately 620 miles (1,000 km.), with the East Sea (called the Sea of Japan by the Japanese) on the east and the Yellow Sea on the west. The Korea Strait separates South

Korea and Japan on the southeast. The northwestern land of the peninsula borders China and Russia.

The Koreans are an ethnically homogeneous people with no prominent minorities. Traces of Chinese civilization are evidenced in Korea's social ethics and its written language. Religious practices in Korea mirror a diverse heritage from Buddhism, Confucianism, Christianity, and derivatives of the traditional Shamanism. The Koreans were once the major transmitters of Buddhism and Confucianism from mainland Asia to the Japanese archipelago, but in 1910, Japan annexed Korea.

The Allied victory over Japan during World War II in 1945 ended the Japanese occupation of the Korean Peninsula. The Korean War broke out shortly thereafter, in 1950, consequent to the growing internal conflicts and the divided control of northern Korea by the Soviet Union and southern Korea by the United States. In 1953 the war ended with an armistice that created North Korea and South Korea. The Demilitarized Zone (DMZ) that runs along the thirty-eighth parallel is approximately 2.5 miles (4 km.) wide and 150 miles (241 km.) long.

North Korea makes up 55 percent of the Korean Peninsula, with an area of 46,540 square miles (120,538 sq. km.)—slightly larger than the U.S. state of Louisiana. As in Japan and Taiwan, mountains and highlands dominate the landscape, occupying 80 percent of the country. All the mountains above 6,562 feet (2,000 meters) on the peninsula are located in North Korea, and the northwestern borderland has the highest elevation. The three longest rivers in North Korea are the Yalu River, the longest and most navigable; the Tumen River in the northwest; and the Taedong River, which runs through Pyongyang, the capital.

A humid continental climate prevails. Winters are long, cold, and dry. Summers are short, hot, and rainy as a result of the summer monsoon. Compared with South Korea, North Korea has a much smaller population, 25.3 million people, but is better endowed with mineral deposits—especially

Mount Kumgang in Korea.

Seoul, South Korea.

An industrial plant in Hamhung, North K orea.

Incheon Grand Bridge, Incheon, South Korea.

coal, magnesite, lead, and zinc—and with large forest resources and significant hydroelectric potentials. The population in North Korea is clustered on the limited lowlands along the west coast and the thin strip along the east coast. Pyongyang is the nation's largest city, with a population of more than 2.8 million.

South Korea makes up the remaining 45 percent of the Korean peninsula with an area of 38,502 square miles (99,720 sq. km.)—a little larger than the U.S. state of Indiana. The rising eastern coast is smooth; the subsiding southern and western coasts are jagged and irregular. As in all the other nations in this region, 70 percent of South Korea is mountainous. The T'aebaek and Sobaek ranges are the most prominent. Offshore to the southwest of the peninsula lies the country's highest peak, Mount Halla, which rises to 6,400 feet (1,950 meters) above sea level on Cheju Island. Major rivers include South Korea's longest, the Naktong River, and the Han River, which runs through the capital, Seoul. Summers in South Korea are warmer and winters are milder than in North Korea. South Korea also has more arable land and a much larger population, 51.4 million (2018), but no significant mineral deposits except tungsten.

Although not as densely populated as Taiwan, South Korea is still one of the world's more densely populated countries, with 1,336 people per square mile. Seoul is the country's largest city and the world's fifth-largest city in 2020, with more than 25.6 million people in its metropolitan area. Pusan (Busan), South Korea's second-largest city, has a population of more than 8 million in its metro area. Imports of raw materials and semifinished components and exports of finished industrial products characterize the economy in South Korea.

Linda Q. Wang

DISCUSSION QUESTIONS: REGIONS

Q1. How have the landforms of mainland East Asia affected the settlement patterns within China and the Koreas? How has the island nature of Japan and Taiwan contributed to the development of those two countries?

Q2. What are the parallels between the industrialization of China and that of the rest of East Asia? In what ways are the economies of the four countries similar? How do they differ?

Q3. What is the Three Gorges Dam? Compare and contrast the positive and negative aspects of its construction and operation. In what ways did the creation of the Three Gorges Dam reflect the attitudes of the Chinese government?

SOUTHEAST ASIA

Southeast Asia comprises portions of the mainland of Asia and major island archipelagos. Mainland Southeast Asia includes the countries of Myanmar (formerly Burma), Thailand, Cambodia, Laos, Vietnam, and most of Malaysia. Two states of Malaysia, Sarawak and Sabah, are located on the island of Borneo, which Malaysia shares with Indonesia and the small Sultanate of Brunei. Southeast Asian states located on islands are Singapore, Indonesia, Timor-Leste, and the Philippines.

Almost all of Southeast Asia falls within the tropics. The equator crosses the middle of the large Indonesian island of Sumatra. The northernmost limits of Southeast Asia, marked by the upper regions of the Irrawaddy River in Myanmar, approach 30° north latitude. This would be equivalent to Midway Island in the central Pacific, or slightly farther north than the Hawaiian Islands. The climate of the region is, therefore, tropical in nature, although some higher mountain zones experience significant periods of cold weather. In general, areas subject to tropical climate conditions receive large amounts of precipitation in the rainy season.

Thailand and Myanmar

Although significant cultural distinctions exist between the two modern countries of Myanmar and Thailand, they share some geographical similarities. This is due in large part to the fact that mainland Southeast Asia, including the Malay Peninsula, is marked by a north-to-south extension of mountain chains that originate in China. In both Myanmar and Thailand, these north-to-south mountains provide rainfall drainage that feeds important river systems. The importance of these mountains for Thailand's lumber industry has been considerable, particularly given the dense concentration of teak forests, which have been heavily exploited.

In Myanmar, the principal river is the Irrawaddy, which runs from a point beyond 24° north latitude, well north of the northern city of Mandalay, southward to its delta west of the major capital city of Yangon (formerly Rangoon) at 16° north latitude. The delta is situated on the extension of flat land jutting out into the sea between the Bay of Bengal and the Andaman Sea. The broad peninsular effect is, in part, the product of deposits carried by the Irrawaddy and other north-to-south rivers in the same region. Along its way to the sea, the Irrawaddy takes in water from dozens of tributaries carrying

A woman dancing Legong Bapang Saba. *Balinese dances incorporate eye and facial expressions.*

SOUTHEAST ASIA

drainage from mountains to the west and east of Myanmar's central valley region.

Thailand's main rivers form a fanlike network that converges in the region surrounding the immense modern capital city and river port of Bangkok. Several rivers empty into the Gulf of Thailand in the baylike inlet that leads from their deltas to the main body of the Gulf of Thailand.

Malaysia and Singapore

The area known as Malaysia comprises the southernmost peninsular extension of mainland Southeast Asia (which it shares with Thailand and Myanmar) with Sarawak and Sabah, two northwestern states on the island of Borneo. Singapore, lo-

cated at the southern tip of the Malay Peninsula, briefly was integrated into Malaysia before becoming an independent island nation. Singapore plays an important role in Southeast Asian trade links with the West thanks to its position astride the shipping lanes from the Suez Canal to East Asia and destinations farther east across the Pacific Ocean.

The northernmost provinces of Malaysia have been linked periodically to provinces of southern Thailand. Possibly because of the importance of the Isthmus of Kra—the narrow neck of the Malay Peninsula that, since early historical times, invited land passage for trade from the Andaman Sea to the Gulf of Thailand—the area has always been in geopolitical contention. Both sides of the modern border share similar topographical features, namely narrow lowland plains that lend themselves ideally to rice-paddy cultivation, particularly in northern Malaysia's state of Kedah.

South from Kedah is the topographically more varied Malaysian state of Perak. Perak's landscape consists of impressive mountain outcroppings separated by broad valleys. Vegetation is luxuriant, with extensive coconut palm groves mixed with cultivated fields. This geographical location gained worldwide recognition in the later decades of the twentieth century when remarkably well-preserved human fossils, traces of what is now known as the Perak Man, were discovered in the region. For several decades, beginning in the late nineteenth century, the Perak region, together with the area around Malaysia's capital at Kuala Lumpur, contributed a major share of the world's tin production.

Malaysia's western coastal states are separated from less-populated and less-developed states on the eastern side of the peninsula (most notably the state of Terengganu) by a substantial north-to-south mountain system. The eastern coast is subject to seasonally strong winds from the South China Sea and heavy rainfall, particularly in the monsoon season.

Vietnam, Cambodia, and Laos

Before independence from France after World War II, Vietnam, Cambodia, and Laos were known collectively as Indochina. Topographically, the Annamite mountain range of Laos and Vietnam stands out as a continuation of the north-to-south ranges that originate in China. In this respect, the eastern quarter of the Southeast Asian landmass, with its coastline on the South China Sea, resembles the western areas of Myanmar and Thailand facing the Bay of Bengal and the Andaman Sea. Along the southern coast of Indochina, specifically where Cambodia borders on the Gulf of Thailand, the area's only northwest-to-southeast mountain mass, the Cardamom and Dâmrei (Elephant) ranges, is shared as a local subregion of Thailand and Cambodia.

The former Indochinese region shares another main geographical feature with the western parts of Southeast Asia: the long course of the Mekong River originating in the mountainous north and flowing southward through the plains area that is now central Cambodia until it enters the sea in southern Vietnam. The latter zone, dominated by Ho Chi Minh City (formerly Saigon, the former South Vietnamese capital), stands in contrast to the mountainous core areas of Vietnam. At their greatest height (at Ngoc Linh, midway on Vietnam's north-south range), these mountains rise more than 8,500 feet (2,600 meters). Laos, farther inland, is almost entirely mountainous.

The Mekong River marks Laos's mountainous border with Thailand before it flows into Cambodia. After this point, the topography of the Mekong is that of a plains system. The flatness of the land leading to the sea from Cambodia's capital at Phnom Penh (where the Mekong joins the Tonlé Sap River coming from the vast Tonlé Sap Lake in central Cambodia) to Vietnam's Mekong Delta is striking. In this area, extensive lowland irrigation potential makes it possible for Vietnam to grow the bulk of its essential annual rice crop.

The second-most-concentrated area of Vietnam's population and agricultural production is the coastal area next to the Gulf of Tonkin in the north. There, the northern city of Hanoi is served by the port of Haiphong. South from Haiphong is the marshy lowland delta area where the Red River flows into the Gulf of Tonkin.

The Indonesian Archipelago

Although the islands making up Indonesia are too numerous to describe in detail, certain features of the main islands stand out. The two best known large islands, Sumatra and Java, have striking characteristics.

Java is one of the most densely populated areas in the world. Several major cities of Indonesia are on Java, including the capital, Jakarta, and the country's second-biggest city, Surabaya. The central areas of the entire length of the island comprise a long chain of mountains of volcanic origin, some of which are still active. Java's coastal areas, mainly on the northern side of the island, are relatively flat. Most of Java's rice paddies are located on these

Subak irrigation system, Bali.

rain-fed plains, although paddies can be found interspersed between mountainous zones and along the terraced slopes of mountains throughout the island.

Sumatra's mountain chain extends along the entire length of the southwestern side of the island. The highest peak (relatively close to the coast of the Indian Ocean) is Mount Kerintji, nearly 13,000 feet (3,960 meters) in elevation. Several other mountains are between 10,000 and 12,000 feet (3,000 and 3,650 meters) high. This rugged, heavily forested habitat hosts rare animal species, including the famous but critically endangered Sumatran tiger and the equally menaced orangutan (from the Malay word meaning "man of the forest"). Sumatra's eastern half may be called a lowland plains area. The plains are fed by several major rivers running from the mountains toward the Straits of Malacca and the Java Sea. The main rivers are the Rokan, Kampar, Indragiri, and Musi. The island's principal large cities are found there. The modern cities of Medan and

Palembang dominate the northern and southern portions of the lowlands, respectively. In the coastal area of Palembang, and also the offshore islands of Bangka and Belitung, Indonesia's important petroleum reserves are located.

Beyond the well-known islands of Sumatra and Java, there are several other major Indonesian islands that have quite different characteristics. One of these is the large mass of Kalimantan, as the Indonesian portion of the island of Borneo is called, covering 242,312 square miles (627,600 sq. km.). Along the western side of Kalimantan are the Malaysian states of Sarawak and Sabah, as well as the independent Sultanate of Brunei. After Sumatra, Brunei is one of the richest petroleum-producing areas of Southeast Asia.

Like Kalimantan, the irregularly shaped island of Sulawesi (formerly Celebes) was heavily forested and represents major, although controversial, developmental and environmental prospects for Indonesia. In recent years, scientists have reported that more than 80 percent of Sulawesi's forest has been lost, and nearly all of its wetlands.

To the east of Java are several smaller islands that have played, and continue to play, an important role in Indonesia's cultural and political history. Some, like the Moluccas (traditionally known as the Spice Islands), were famous for their contribution to east-west trade in once-rare cinnamon and nutmeg. Best known is the small island of Bali, home to the only Hindu religious minority group that was able to maintain its separate cultural identity over the centuries following Indonesia's major conversion to Islam. Bali is world-famous for its exotic island culture and attracts large numbers of foreign tourists.

A multitude of lesser-known islands extend eastward up to the western half of New Guinea, which was joined to Indonesia in 1962 and renamed Irian Jaya. It was renamed again in 2002, this time as Pa-

pua. Smaller Indonesian islands include Lombok, Sumba, Sumbawa, and the western half of the former Portuguese colony of Timor.

Timor-Leste

The eastern half of Timor is home to tiny Timor-Leste (or East Timor), Asia's newest country, as well as one of its smallest and poorest countries. Once part of the Portuguese colony, Timor-Leste was annexed by Indonesia, which ruled it for more than two decades. In 2002, after a long and devastating struggle, East Timor gained full independence. Today, Timor-Leste derives more than 90 percent of its revenue from the exploitation of offshore oil and natural-gas deposits.

The Philippines

A second island archipelago located east of the South China Sea, not so complex as the island Re-

Filipinos planting rice. Agriculture employs 30 percent of the Filipino workforce as of 2014.

public of Indonesia, makes up the modern Republic of the Philippines. The Philippines is separated from the Asian mainland of Indochina by the South China Sea. The islands were named after the Spanish king Philip II, who claimed colonial control over them in the sixteenth century. The main islands of the Philippines, which extend between about 4° and 17° north latitude, are Luzon in the north, Mindoro, Panay, Negros, Cebu, Bohol, and Leyte in the central portion, and Mindanao in the south. The country's capital and only large city, Manila, is on Luzon. The lesser islands of Palawan and the chain of the Sulu Islands extend almost to the northern tip of Malaysia's Sabah state (northern Borneo).

Although heavily forested, the Philippines produces a variety of agricultural products for domestic consumption and world export. These include rice and corn as staples, and coconuts and coconut by-products (mainly copra) as major export commodities.

Byron Cannon

SOUTH ASIA

South Asia is traditionally regarded as comprising two Himalayan countries (Bhutan and Nepal), two island countries (Maldives and Sri Lanka), and the four continental countries of Afghanistan, Bangladesh, India, and Pakistan. However, there is debate about the inclusion of Afghanistan, which has transitional historic ties to both South Asia and the Middle East.

As defined here, the total land area of South Asia is 2.25 million square miles (5.83 million sq. km.), almost two-thirds the size of the United States. Nearly half the region is located farther south than the southernmost point of the United States. India dominates South Asia, in both area and population. The Maldives is the smallest country in area and population. All countries of South Asia were either British colonies or controlled by Britain sometime in the past.

Physical Geography

South Asia has diverse physical characteristics, but there are three broad physiographic divisions: the northern mountains, the southern peninsula, and the great Indo-Gangetic Plains. The first division forms an inverted "U" across the north of the region and has several mountain ranges, including the Sulaiman ranges of eastern Afghanistan and western Pakistan and the Kirthar ranges of Balochistan, Pakistan. These are relatively low mountains, but quite rugged.

The north face of Mt. Everest in the Himalayas from the Tibetan side of the China-Nepal border.

SOUTH ASIA

The Himalayas are the primary mountains of this physiographic division, extending about 1,500 miles (2,414 km.) from Kashmir to the northeast corner of India. The Himalayas contain many high peaks, including the world's tallest—Mount Everest at 29,028 feet (8,848 meters). Other ranges intersecting the Himalayas include the Hindu Kush of Afghanistan and Pakistan and the Karakoram of Kashmir. The Assam-Burma ranges of eastern India and western Myanmar are also part of this physiographic division and consist of several low hills. All ranges in this physiographic division are geologically young.

Located entirely within India, the southern peninsula has a triangular shape and consists mainly of a huge plateau called the Deccan. It is relatively low, with much of its surface only 1,000 to 2,000 feet (300 to 600 meters) above sea level, and tilts gently from west to east. Both the east and west perimeters of the plateau are defined by ranges of low mountains called *Ghats* (steps).

The Western Ghats, with an elevation of just over 8,000 feet (2,500 meters), border the plateau on the west; the Eastern Ghats, rising in places to 5,000 feet (1,500 meters), border the east. The Deccan plateau is composed of hard ancient rocks and thus is subject to erosion. It is not well-watered, either by plateau rivers or by rainfall. Agriculture is possible only with irrigation, but the Deccan plateau contains most of the mineral wealth of India: copper, iron, gold, lead, manganese, and coal. Peninsular India also has a coastal lowland zone of varying width, which contains fertile soil and receives sufficient rainfall for agriculture.

Covering more than 300,000 square miles (777,000 sq. km.), the Indo-Gangetic Plain lies between the northern mountains and the southern peninsular physiographic divisions. These plains range between 80 and 200 miles (130 and 325 km.) wide. They are composed of alluvial soils deposited by the Indus, Ganges, Brahmaputra, and many other rivers—most of which originate in the Himalayas.

The combination of adequate rainfall, river flow, fertile soils, and a long growing season makes these plains South Asia's most productive agricultural region as well as its most densely populated area. Leading urban centers such as Islamabad, New Delhi, and Dhaka are located in this physiographic division. The large Irrawaddy River basin of central Myanmar is located outside this zone, but possesses most of the characteristics of the Indo-Gangetic Plain. South Asia contains two deserts: the Thar in southwestern India and the desert of Balochistan in Pakistan.

Climatic conditions in South Asia demonstrate remarkable extremes: from perpetual snowfields in the Himalayas to the year-round tropical heat of the Deccan, and from some of the world's driest climates to some of its wettest. In general, winter temperatures decrease in a northwest direction: The highest temperatures are found in the Deccan and the lowest in northern Pakistan and northeastern India. Summer temperatures show a reverse trend:

The highest are in northern Pakistan and northwestern India and the lowest in the coastal region.

Average winter and summer temperatures are 60° and 80° Fahrenheit (15° and 27° Celsius), respectively. May is the hottest month in most places in the region, and temperature falls in early June as a result of monsoon rains. The extreme seasonality of rainfall is the main characteristic of the climate of South Asia. Over much of the region, 70 to 90 percent of the annual rainfall occurs in the summer (June to September), and the rest of the year is dry. Mean annual rainfall varies greatly, from near zero in the deserts of Pakistan and adjacent India, to more than 200 inches (5,080 millimeters) in parts of Western Ghats and in the Plateau of Meghalaya in India between Brahmaputra valley and the Bengal delta region.

The monsoon is the most important climate feature of South Asia. It affects Southeast and East Asia as well, but to a less dramatic degree. The term monsoon, believed to be of Arabic origin, describes a seasonal reversal in wind direction experienced in the Arabian Sea and along the East African coast. The two types of monsoons are the southwest (SW), or wet, monsoon, and the northeast (NE), or dry, monsoon.

The southwest monsoon reaches its height between June and September, when moisture-laden air moves in a southwest direction from the sea to the land. This is caused by the formation of low pressure over the land and high pressure in the Indian Ocean. It produces two areas of heavy orographic, or mountain-induced, rainfall: one along the Western Ghats and the other along the flanks of the eastern and central Himalayas. The southwest monsoon is important for agriculture throughout the region. During the winter, the pressure systems reverse and the winds blow from land to sea, bringing dry weather with occasional light rains. This northeast monsoon brings cooler weather to South Asia, especially in the north.

Forest areas differ widely in South Asian countries—from only 2.2 percent in Pakistan to 25 percent in Nepal to more than 85 percent in Bhutan, Myanmar, and Nepal. Tropical monsoon forests in this region are found in areas subject to a monsoon-type climate. Because of the two distinct seasons, many trees of this forest type lose part of their foliage in the dry months. Often cut for firewood, trees of monsoon forests make poor lumber. Monsoon forests support a rich collection of wildlife.

South Asia contains the world's largest mangrove forest, called the Sunderbans—the home of the Bengal tiger. It covers the southwestern coastal area of Bangladesh and southeastern coast of West Bengal, India. The Sunderbans Mangroves ecoregion is the habitat for hundreds of living organisms—from mammals, birds, fish, and crocodiles, to algae, fungi, and bacteria. Elephants, deer, and leopards are found, particularly in the forests of northeastern India. The Deccan of southern India and other dry land areas of South Asia are characterized by grasslands with scattered trees.

Human Geography

South Asia is an ancient land with an old civilization rich in traditions of literature, art, technology, and religion. The Indus Valley civilization developed in the Indus floodplain of Pakistan around 3500 BCE. This civilization is believed to have been developed by ancestors of the modern inhabitants of southern India, the Dravidians. A distinctive human group, known as Indo-Aryans, entered the region from the northwest mountain wall of Pakistan around 2000 BCE. The synthesis of these two groups produced the Hindu religion, which combines the beliefs and practices of Dravidians and Aryans.

Religions

Buddhism, a system of beliefs that originated in Hinduism, was born in Bihar, India, in the sixth century BCE by Gautama Buddha. This religion flourished during the time of the Indian monarch Asoka in the fourth century BCE. Later, several other religions developed in South Asia; notable among them is Sikhism, founded by Guru Nanak in 1499, which contains elements of both Hinduism and Islam.

Islam originated outside South Asia and entered the region from the Middle East around the eighth century CE. It challenged the dominance of Hinduism about 200 years later. Large-scale religious conversions took place in India during the Muslim rule, which lasted until 1757. The most intense Muslim influence was perhaps during the Mogul period (1526-1739). Islam was attractive to the Dalits (Untouchables) and other Hindus of lower castes.

Christianity was probably first introduced in Kerala, Goa, and other parts of the South Indian coast around 52 CE, but the number of Christians increased considerably during the British period (1758-1947). The effects of British colonial rule were far-reaching in many areas. For example, British rule influenced the organization of South Asia's production and trade infrastructure. By 1948 all countries of South Asia had won independence.

Followers of almost all the world's religions can be found in South Asia, with Hinduism, Islam, Buddhism, and Christianity the four major religions. Hinduism has the largest number of followers in the region. It is the dominant religion of India and Nepal, and has numerous adherents in Bangladesh, Bhutan, and Sri Lanka.

Muslims (followers of Islam) constitute the majority in Afghanistan, Bangladesh, Maldives, and Pakistan, but a minority in India, Sri Lanka, and Myanmar. Buddhism is the dominant religion of Bhutan, Myanmar, and Sri Lanka. Christians are minorities in almost all countries of South Asia. Sri Lanka has the highest religious diversity: 70.2 percent of its population are Buddhist, 12.6 percent Hindu, 7.4 percent Christian, and 9.7 percent Muslim. Afghanistan, the Maldives, and Pakistan are the three most religiously homogeneous countries of South Asia.

Religion is of great importance in South Asian life and has been the basis for many conflicts among its peoples. Religious differences were responsible for the 1947 partitioning of British India into Pakistan and India. Today, the impact of religion continues to shape the political geography of South Asia. India and Pakistan have fought several wars to control the disputed Kashmir region in the extreme north of the Indian subcontinent. During the partition of British India, the Hindu king of Muslim-dominated Kashmir joined with India. Immediately, Pakistan invaded parts of Kashmir, and India deployed its armed forces to halt the Pakistani aggression. A line of control was established, which, with minor modifications, continues today.

Kashmir Conflict

The tensions between Pakistan and India over the Kashmir issue provide just one example of religion-based conflicts in South Asia. For example, the Sikhs of the Punjab, India, have agitated for the creation of an independent Sikh state, Khalistan, although this effort has lost most of its public support. Similarly, the Tamils of Sri Lanka, mostly Hindus who live in the northeastern part of the country, had long demanded an independent Hindu state. For both the Sikhs and the Tamils, substantial migrations to other, mostly Western, countries diminished the intensity of the respective independence movements.

Languages

The languages of South Asia are diverse. Most of the region's languages show a high degree of regional concentration, making language a powerful symbol of regional and ethnic consciousness. Language has played a major role in the redrawing of India's state boundaries. Demand for a separate homeland by Sri Lankan Tamils, Sikh separatism in India's Punjab, the Sindhi movement in the Sind province of Pakistan, and the language-based cultural distinctiveness of Bangladesh all attest to the force of language in shaping the political landscape of South Asia. Language has played a major role in the domestic affairs of all but two South Asian countries: Bangladesh and the Maldives.

The languages of South Asia are derived primarily from two language families: the Indo-European, dominant in the central and northern parts of the region, and the Dravidian, dominant in southern India. Languages of the Sino-Tibetan family are spoken by people living along the southern slopes of the Himalayas in Nepal, Bhutan, the mountains of eastern India, and Myanmar. The Ladakhi language of Kashmir belongs to this family. Tribal people of the Chota Nagpur Plateau in central India speak Austroasiatic languages; people of the northwestern parts of Afghanistan speak languages of the Altaic family. Hindi, with more than 500 million speakers,

is the language spoken by the largest number of people in South Asia. A language of the Indo-European family, it is spoken by the people of northern India, but understood by many people of South Asia. In 2019, Hindi ranked fourth in the world (after Mandarin, Spanish, and English) with respect to number of speakers. Other South Asian languages ranked in the world's top twenty are Bengali (5), Punjabi (9), Marathi (10), Telugu (11), Tamil (18), and Urdu (20). Telugu and Tamil are languages of the Dravidian family, and the rest are Indo-European.

Economic Geography

Agriculture remains an important sector in all the countries of South Asia, although the industrial and service sectors are gaining prominence as well. The most industrialized country of South Asia is India; in 2017, it was the sixth most industrialized country in the world in terms of the relative output of the industrial sector compared to other sectors. Infrastructure built by the British and the availability of minerals such as coal, iron ore, and natural gas were significant in the industrialization of India.

At India's independence in 1947, only 2 percent of all Indian workers were engaged in the industrial sector; today, this sector employs more than 23 percent. India's main industries are textiles, telecommunications, chemicals, pharmaceuticals, biotechnology, and food processing. An important modern industry is thriving in Bangalore—India's Silicon Valley. Several thousand information-technology companies are based there, including many from the United States.

Pakistan has made some economic progress since 2000, primarily due to economic reforms, though conditions remain changeable. Textiles and apparel are its main exports. The country faces an enormous public debt and needs external assistance to stay solvent. Although Bangladesh is considered one of the world's fastest growing economies, it remains predominantly an agricultural country, with rice as its main product, although many industries have been established. Exports include jute and cotton textiles, clothing, frozen foods, and leather. The industrial sector contributed 29.3 percent of the gross domestic product in 2017.

Sri Lanka continues to struggle economically. The country relies on the traditional exports of rubber, tea, coconut-based products, gems, and inexpensive apparel. An upsurge in tourism has been a hopeful sign. The least-industrialized countries of South Asia are Bhutan, the Maldives, and Nepal.

Poverty

South Asia is the world's second-most poverty-afflicted region (after sub-Saharan Africa), with low average incomes, low levels of education, poor diets, and overall poor health. The region accounts for about one-third of the world's poorest populations. India alone has an estimated 285 million people out of more than 1.2 billion below the poverty line. Wealth is not equitably distributed, and most inhabitants of the region do not experience the benefits of economic development.

It is thought that at least 25 percent of the children in South Asia are malnourished and underweight. Reduction in poverty and hunger will not be possible without economic development. Important barriers for economic development in South Asian countries are lack of natural resources, including petroleum; overpopulation and a high rate of population growth; widespread illiteracy; political instability; corruption; recurrent natural hazards; and a lack of mutual cooperation among the countries of South Asia.

Bimal K. Paul

Mongolia and Asian Russia

The enormous dimensions of the combined Mongolia-Asian Russia region accommodate a great diversity of peoples and their languages and cultures, landforms, climates, flora, and fauna. Siberia is the Asian part of Russia, occupying more than 75 percent of the Russian Federation's territory. The image of Siberia that exists in many people's minds is of a frozen penal colony. And indeed, political prisoners were exiled there as early as 1593. Many religious dissenters, criminals, and rebellious peasants were sent there over the centuries, a practice that culminated in the notorious Gulag system exposed to the world in the works of Alexandr Solzhenitsyn. Just between 1934 and 1947, upwards of 10 million people were sentenced to the forced labor camps. The region has also been exploited for at least two centuries for its furs, a major reason for building the Trans-Siberian Railway. Under Soviet administration, Novosibirsk, Irkutsk, Omsk, and Yekaterinburg became important scientific and commercial centers, and Khabarovsk and the port of Vladivostok emerged as major cities of the Russian Far East. Mongolia was modernized in numerous ways under Soviet-style socialism. The country's transition to a market economy has been difficult.

Physiography

Mongolia and Asian Russia constitute a huge chunk of land. Mongolia, at about 600,000 square miles (1.55 million sq. km.), is more than twice the size of

Yinderitu Lake in the Badain Jaran Desert in Inner Mongolia.

MONGOLIA AND ASIAN RUSSIA

Texas. Virtually all of Asian Russia—that part of Russia lying east of the Ural Mountains—lies north of 55° north latitude, and covers about 6 million square miles (15.54 sq. km.) of some of the coldest, most inhospitable regions on Earth. Mongolia's northern border with Siberia runs for almost 1,700 miles (2,735 km.), and on its east, south, and west it shares nearly 3,000 miles (4,830 km.) of boundary with China.

Asian Russia can be divided into three regions. Western Siberia extends east to the Yenisei River and includes the vast Western Siberia Lowland; the Altai Mountain system, which also covers the northwestern third of Mongolia; and the Kuznetsk Basin with its bordering mountains. Eastern Siberia,

which contains the Central Siberian Plateau and reaches all the way to the Bering Strait, includes a long Arctic region, as well as Lake Baikal and the mountain ranges that border Mongolia. The Far East includes the Kamchatka Peninsula, the northwestern coast of the Sea of Okhotsk, and the Amur-Primoski region and Sakhalin Island, both of which lie just north-northwest of Japan.

The West Siberian Lowland is a large swamp, seldom higher than 200 to 300 feet (60 to 90 meters) and poorly drained by the Ob-Irtysh River system. The Central Siberian Plateau extends from the Yenisei River in the west to the Lena River in the east, and at its highest reaches 4,000 feet (1,200 meters). Siberia's largest mountain ranges lie in its formidable northeast regions north and west of the Sea of Okhotsk. These include the Yablonoi, Stanovoi, and Verkhoyansk Mountains in the southwest and west; the Chersky range, reaching more than 9,000 feet (2,700 meters), in the center; and the Kamchatka and Anadyr ranges in the east. Active volcanoes dot the Kamchatka Peninsula.

Except for the Amur, which flows into the Pacific just west of Sakhalin, Siberia's great rivers drain into the Arctic Ocean. The Lena River originates near Lake Baikal and flows north along the Verkhoyansk Mountains; to the east, the Indigirka, Kolyma, and Anadyr run, generally parallel, to the north. Lake Baikal, in south-central Siberia, is the world's deepest lake, with a maximum depth of 5,387 feet (1,642 meters). Mongolia's largest river, the Selenga, is joined by the Orhon and drains into Lake Baikal.

Mongolia's two largest rivers running eastward, the Onon and the Kerulen, eventually flow into the Pacific after crossing China. The Kobdo River rises in the Altai glaciers and drains into the dry southern region, as does the Dzavhan, which flows out of the southern reaches of the Khangai range. Mongolian rivers tend to be fast-running and steep; therefore, they are potential sources of electric power. Mongolia's lakes, more than 3,000 of them, are generally salty and ephemeral. The largest freshwater lake, Hövsgöl Nuur, lies in the far north, southwest of the Siberian city of Irkutsk. There are three large western lakes: the saline lakes Uvs Nuur and Hyargas Nuur, and the freshwater Har Us Nuur, which drains into the Hyargas.

In Mongolia, the highest peak in the Altai Range, Nayramadlin Orgil, rises to 14,350 feet (4,375 meters), while the Khangai Mountains in the northeast reach 13,000 feet (4,000 meters) at their highest. The Khangai range has many upland grazing slopes, and numerous extinct volcanoes punctuate its northern reaches as well as the country's easternmost tip. A smaller mountain range, the Khentei, runs along the Siberian border north of the capital city, Ulaanbaatar.

South of Eastern Siberia's Sayan Mountains and Lake Baikal, large basins cover much of Mongolia.

The Gobi Desert and grassland dominate the country's southeast.

Climate

The extreme cold of Siberian winters is caused by an area of high atmospheric pressure centered over Lake Baikal. Siberian winter temperatures have reached around -100° Fahrenheit (-73° Celsius) in some regions around Verkhoyansk, but as a result of the long hours of sunlight, summer temperatures in central Siberia frequently exceed 90° Fahrenheit (32° Celsius). Midsummer rain accounts for most of Siberia's precipitation, which is never heavy except on the Kamchatka Peninsula, where 40 inches (102 centimeters) is the norm. Kamchatka accumulates more than 50 inches (127 centimeters) of snow in the winter, and snow depths in northwest Siberia average 32 to 40 inches (81 to 102 centimeters). Snow cover in the north commonly lasts more than 250 days, although southern regions receive far less. Since the rivers run northward and are frozen up to six months in the south and eight months in the north, they tend to thaw and flood closer to their sources.

Mongolia's location in high latitudes (roughly between the fifty-second and forty-first parallels) means a continental climate with extremely cold winters, moderate summers, and low rainfall totals. Mean temperatures for Ulaanbaatar vary from -15° Fahrenheit (-26.1° Celsius) in January to 63° Fahrenheit (17.2° Celsius) in July. The corresponding figures for the Gobi region are 0° Fahrenheit (-17.8° Celsius) and 37° Fahrenheit (2.8° Celsius).

Southern Siberian steppe: windbreaker trees on the Russian steppe, Altai Krai, Russia. March 2004.

Annual rainfall can reach 14 inches (35.5 centimeters) in the northern mountains, but can be less than 4 inches (10 centimeters) in the south. Ulaanbaatar averages 9 inches (23 centimeters). Despite its susceptibility to blizzards and sandstorms, Mongolia averages about 250 days a year of bright sunshine.

Flora and Fauna

Asian Russia is a land of forests, with huge coniferous stands and the forested swamp of the taiga. Western Siberia abounds in fir, aspen, birch, alder, poplar, willow, and larch. Farther to the east, stone pine and spruce appear, and the Amur region includes these species as well as cedar, oak, maple, lime, and other hardwoods. In southwestern Siberia, the taiga yields to true steppe, with stipa grasses and steppe flowers.

North of the taiga, a belt of tundra 200 miles (320 km.) wide extends along the Arctic coast, producing little vegetation other than mosses, lichens, a few grasses and low bushes, willows, and dwarf birches. From north to south, Mongolia comprises five distinct vegetation types: alpine scrub plants found at elevations of 6,000 feet (1,830 meters); forest-steppe, with many coniferous forests; the grassy steppes; semidesert; and desert.

The northern forests have abundant stands of larch, cedar, spruce, pine, and fir, as well as the deciduous aspens, birches, and poplars. The steppes are blanketed by wormwood, various grasses, and numerous fodder species, and are brilliant in summertime with many flowers. The desert areas support only xerophytes, upon which camels, goats, and sheep manage to sustain themselves.

Siberia has always had many commercially valuable wild animals. Trapping has been a prominent occupation, and many people in the frigid climate have relied on fur clothing. Among the best-known animals are the squirrel, hare, wolf, brown bear, fox, elk, reindeer, polar bear, sable, ermine, lemming, and weasel. The Amur region in the far east hosts antelope, deer, panther, and the endangered Siberian tiger. The Kuril Islands north of Japan are a rich fishing ground, and the Arctic coast is home to walrus, seals, and whales. Mongolia is equally rich in animal species. The lynx, red deer, roe deer, musk deer, elk, brown bear, snow leopard, ermine, otter, wolverine, squirrel, sable, and large wild boar frequent the forested mountains.

The steppes are home to the tarbagan, a marmot-like creature with a valuable pelt. Besides the sheep, camels, and goats, the dry regions host the Gobi bear and domesticated cattle, highland yak, and Mongolian horses. Birdlife abounds in the steppes of both countries, including bustards, cranes, pheasants, partridges, larks, and falcons; ducks, geese, gulls, pelicans, cormorants, and swans inhabit the rivers and lakes. Snowy owls, condors, and golden eagles can be sighted in some areas, and the many fish species include salmon, pike, perch, and trout.

People and Cultures

In 2020, an estimated 40 million people lived in Siberia. Of these, approximately 10 percent are indigenous. The indigenous population includes a multitude of different groups. For the sake of convenience, anthropologists sometimes assign them to groups geographically, according to language. Four such language families have been identified: Uralic, Altaic, Yeniseian, and Paleosiberian.

The surviving indigenous groups, among others, include the Khant and the Mansi in western Siberia, and the Buriat, Yakut, Tuvintsi, Shor, Altain, Khakass, Nivkhi, Ulchi, and Orochi. The Chukchi of the far eastern Arctic region near and along the Bering Sea are thought to have a genetic relationship to Native American indigenous groups. The habits, culture, and belief systems of virtually every Siberian indigenous population were greatly diminished during the Soviet period. The rapid industrialization that occurred during those years caused great environmental damage that has yet to be remedied.

All Siberians are Russian citizens regardless of their ethnic background. Most of the population traces its ancestry to European Russia. Smaller Siberian populations are of Ukrainian, German, and Moldovan origin. Mongol, Turkic, and Tatar groups are also represented.

An image of an early 20th-century Oirat caravan, part of the Altaic group, traveling on horseback, possibly to trade goods.

In 2018, Mongolia's population reached 3.1 million. About 90 percent of the population are ethnic Mongols, who are divided into numerous groups, each with their own dialects, customs, and dress. The largest of these groups, the Khalkhas, constitute nearly 85 percent of the population and their dialect is the country's official language. The largest non-Mongol group are Kazaks (less than 4 percent). Other small indigenous groups make up the remainder. Religion, suppressed during the communist era, has reemerged, particularly Tibetan Buddhism, or Lamaism. The communist years also saw a general movement of people from the countryside into the cities. The Mongolian language is a cluster of closely related languages, customarily divided into Eastern Mongolian (Khalka, Buriat, and Chakhar) and Western Mongolian (Oirat and Kalmyk). They all belong to the larger Uralic-Altaic language family, named for the Ural and Altai mountain chains.

Frank Day

TRANSCAUCASIA AND THE CENTRAL ASIAN REPUBLICS

The primary factor in considering these areas as a single region lies in their common history as republics of the former Soviet Union. They share, for the most part, the same economic, developmental, and ecological circumstances.

Transcaucasia

Transcaucasia, also sometimes referred to as South Caucasus, includes the countries of Georgia, Armenia, and Azerbaijan. These countries, all formerly republics of the now-defunct Soviet Union, are nestled between the Black and Caspian Seas, between Russia to the north and Iran and Turkey to the south. The Caucasus Mountains are found in all three of these countries, reaching heights in excess of 16,400 feet (5,000 meters).

The mountains have dictated much about life in these lands. One important effect of the mountains is the presence of more than fifty distinct ethnic groups in a space roughly the size of California. The

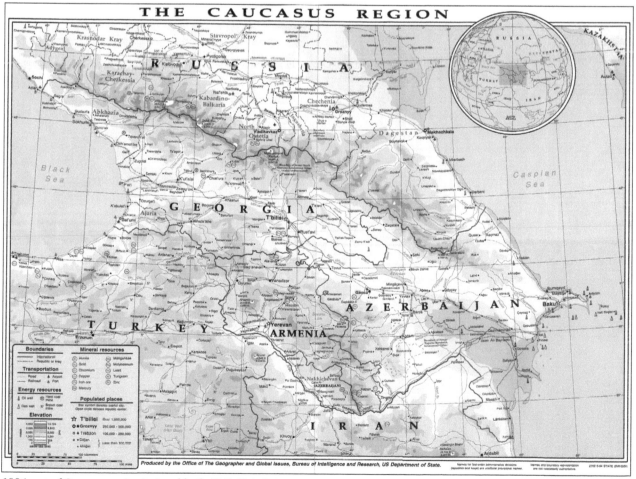

1994 map of Caucasus region prepared by the U.S. State Department.

View over the Old City at Sunset, Bukhara, Uzbekistan.

largest group is the Azerbaijanis, numbering over 9 million, while the smallest groups number in the hundreds. The name of each country denotes the dominant ethnic group within its boundaries. The great number of groups reflects two factors: the isolation that mountains provide, and the fact that this region, between the seas, was the route by which many nomadic herders moved between the Central Asian steppes and points to the south.

The mountainous terrain also dictates that agriculture, urbanization, and transportation be condensed into lowland valleys. The lower reaches of these valleys have a humid, subtropical climate where a broad range of crops can be grown.

Central Asian Republics

This subregion comprises the nations of Kazakhstan, Uzbekistan, Turkmenistan, Kyrgyzstan, and Tajikistan. The basic uniting features among these countries are that most of the people there speak Turkish languages, they share a common political history, and they adhere to Islam. The desert of the western lands (Kazakhstan, Turkmenistan, and Uzbekistan) gives way to a mountainous region that is reminiscent of Transcaucasia. The mountains are generally higher than 23,950 feet (7,300 meters) and include glaciers. As in Transcaucasia, the valleys of the region are fertile subtropical areas, producing a wide variety of agricultural exports.

Petroleum

The presence of oil in the Caspian Sea has had an enormous effect on both sides of the sea. Oil is produced on a large scale in Georgia, Armenia, Azerbaijan, Kazakhstan, and Turkmenistan, although every nation in the region produces some oil. With the exception of Georgia, which has Black Sea ports, these countries are landlocked. Under the Soviet Union, pipelines were constructed from the hub of the Caspian oil region, the Azerbaijani capital of Baku, to the Black Sea through the Russian seaport of Novorossiysk. Since the fall of the Soviet Union, alternative routes have been undertaken with the help of U.S., Japanese, British, and French oil companies.

In the first decade of the twenty-first century, more than a dozen pipelines were built using ports in Georgia, Turkey, and Iran. A pipeline from Kazakhstan to China became fully operational in 2014. The path of existing and proposed pipelines is the subject of international debate and intrigue, as the political and economic ramifications of these projects are enormous. All the nations are at the mercy of the country or countries through which their pipelines flow. Not only are the ports a point of potential dominance, but also the entirety of the line is subject to sabotage. Debate continues among

the oil-producing regions as to how best to export their product.

The Soviet Legacy

These countries have a shared history as republics of the former Soviet Union, and all continue to depend to some degree on Russia for defense. Their Soviet legacies also have left them with several internal threats of concern. The Soviets created a number of ethnic exclaves (ethnic regions that are placed within the boundaries of another ethnic group) when they drew the borders of the republics. This has led to years of discord between Azerbaijan and Armenia concerning the Armenian exclave of Nagorno-Karabakh, now referred to as Artsakh by its Armenian residents. The presence of a number of ethnic groups in Georgia has led to considerable political turmoil and violence. Uzbekis are distributed throughout the region, creating another po-

On the southern shore of Issyk Kul lake, Issyk Kul Region.

Kazakh man on a horse with golden eagle, image created between 1911 and 1914.

tential for conflict. Exclaves are found in all the countries of the region.

The Transcaucasian and Central Asian republics have faced challenges and hardships in their transition to freer, more market-based economies. Many of these difficulties stem from the policies and directives promulgated from Moscow during the Soviet period. Trade imbalances were maintained such that the republics ran deficits with the always dominant Russian republic. Upon independence in 1991, the new nations were therefore saddled with enormous debts to repay to Russia. This situation gave Russia an edge in trade agreements for decades to come.

The legacy of the Soviet Union is particularly evident in the infrastructure framework in both the Transcaucasian and Central Asian republics. Railroads were generally constructed on a north-south pattern, with Moscow as the core. From Central Asia, virtually all rail traffic to points west still must pass through Russia. Making matters more difficult, the gauge used differs from those used by railroads in Europe and China. Exportation of goods by rail to areas outside the former Soviet Union therefore requires laborious and time-consuming unloading and reloading. Ground transportation is hampered by a general lack of paved roads outside of cities, and those roadways that do exist are generally oriented toward Moscow. New and remedial road construction, when it occurs, is today largely funded through international grants. The energy infrastructure left behind by the Soviets, much of it developed in the 1960s and 1970s, is insufficient to meet the needs of populations in the 2020s.

Soviet rule also left environmental devastation. Shoddy equipment in the oil fields ringing the Caspian Sea has caused oil to drain into water sources. The water also is contaminated by industrial waste. The Soviet "virgin lands campaign" of the 1950s, in which marginal areas were plowed and planted with wheat, led to serious erosion in Kazakhstan. The Karakum Canal through the central part of Turkmenistan has caused the Aral Sea to dry up rapidly, and much of the lake's total volume has disappeared. The result is expanded desertification, as the sand and sediment exposed by the drying process have been blown throughout the region. The environmental disaster has prompted international efforts to restore the Aral Sea, with some positive results.

Also during the Soviet period, radioactive dumping grounds were located in the desert of Kazakhstan. Pesticides, primarily associated with the cotton industry, caused widespread water contamination in the region. Uzbekistan has the highest infant mortality and cancer rates of the region, both of which have been linked to pesticides and industrial pollution.

Mark Anthony Phelps

DISCUSSION QUESTIONS: REGIONS

Q1. What is meant by the term "Siberia"? Why does it carry onerous connotations? In what ways does Asian Russia differ from the rest of Asia?

Q2. How does the region's long domination by Russia and the Soviet Union manifest itself today? What positive affects arose from the association? In what ways are the countries of Transcaucasia and Central Asia still tied to Moscow?

Q3. What cultural factors are shared by the countries of Transcaucasia and those of the Central Asian Republics? In what ways do geographical features contribute to each area's regional and national identities?

THE MIDDLE EAST

In this volume, the term "Middle East" refers to the 2.4-million-square-mile (6.2-million-square-km) swath of Southwest Asia that includes the following 14 countries: Turkey, Syria, Iran, Iraq, Jordon, Lebanon, Israel, Saudi Arabia, Kuwait, Oman, Yemen, United Arab Emirates, Bahrain, and Qatar. (Egypt and Libya—African countries that share a common religion and language with much of the Asian Middle East—are examined in *World Geography: Africa*.)

Climates

The Middle East lies north of the equator and is centered at approximately 30° north latitude and 45° degrees east longitude. This location puts the Middle East within Earth's subtropical high-pressure belt, an area of descending air masses. These atmospheric dynamics bring about high temperatures and excessive evaporation at the surface. Hence, the region is exceedingly arid and dominated by a hot desert climate.

Although most desert areas receive rainfall occasionally, precipitation in much of the Middle East is so low as to be virtually unmeasurable. Temperatures remain relatively warm throughout the year, ranging from about 60° to more than 100° Fahrenheit (16° to 38° Celsius). In the desert, there is a greater difference between daytime and nighttime temperatures than between seasonal temperatures.

North and south of the hot desert zone lie narrow zones of hot steppe climate that receive more rainfall annually, 5 to 10 inches (130 to 250 millimeters), enough to marginally support grasses and small herbaceous vegetation. The hot steppe climate extends from the Mediterranean coast of southern Israel, into Jordan and Syria, through Mesopotamia, and around the Zagros and Elburz mountains of Iran. North of this zone is a narrow region of dry summer subtropical climate with more rainfall annually, 10 to 20 inches (250 to 500 milli-

meters), and a more distinct seasonal transition of temperatures between summer and winter.

This dry summer subtropical climate extends from Israel into Lebanon, Syria, and northern Iraq, and around the coasts of Turkey. The higher elevations in Turkey and Iran have cooler temperatures, more rainfall, and even occasional snow. At the southwestern corner of the Arabian peninsula, a small area of hot steppe encircles a small area of dry winter subtropical climate in the Asir Mountains.

Physiography

Plate-tectonic forces have created most of the physiography of the Middle East. The Arabian plate is moving away from Africa and colliding with the Turkish and the Iranian plates. These collisions have uplifted the surface to form the Zagros Mountains in Iran. In Turkey, the situation is more complex in terms of plate collisions that formed its rugged landscape. There, the mountain-building process is ongoing, which explains the frequent earthquakes occurring each year within the collision zones that are upthrusting mountains. The higher mountain masses, the Zagros, Asir, Elburz, and Turkish Mountains, mountains—range in elevation from 10,000 to 15,000 feet (3,000 to 4,600 meters), comparable to the Rocky Mountains of North America.

DEFINING THE MIDDLE EAST

The Middle East is defined here as the southwestern corner of Asia, including the peninsula of Arabia, Asia Minor (Turkey), and Iran. It forms the central part of a broader cultural and physical region—referred to as the Middle East, the Arab World, or even the World of Islam—stretching from Northern Africa into Central Asia. Another term often used for this region is "Near East," which is sometimes used to include Pakistan and India.

THE MIDDLE EAST

Most of the other mountainous areas are older, more eroded, and at lower elevations, ranging from 1,000 to 6,000 feet (300 to 1,800 meters), comparable to the Appalachians of North America. Behind the leading edge of plate collisions, lowlands or basins are formed, such as Mesopotamia, the Persian Gulf, and the basins of interior Iran. As the North African and Arabian tectonic plates pull apart, a great rift is being formed, extending from East Africa through the Red Sea, into the Dead Sea-Jordan Valley, and the Bekaa Valley in Lebanon.

These physiographic features offer advantages to humans. The highlands are a source of metallic minerals and building materials for settlements and towns and have traditionally provided a haven for persecuted minority groups. The basins and lowlands in between provide good agricultural soils, salts, phosphates, and potash, and sometimes petroleum and natural gas deposits.

The Middle East in History

The Middle East was the site of the world's oldest civilizations, which arose in 5,000 to 6,000 BCE in Mesopotamia and in the nearby Nile River Valley, in northeast Africa. These grew and commanded huge empires, based primarily on the development of agriculture and the regional trade linkages afforded by an advantageous location with respect to the rest of the ancient world. Since the Middle East is located near the center of the world's largest land-masses, the African and Eurasian continents, land trade between Africa, Europe, and Asia had to flow through it, making Middle Eastern merchants and rulers wealthy. This wealth and its control sparked numerous conflicts between contending empires, often crushing smaller kingdoms (such as ancient Israel) allied with the opposing empire or seeking their own small piece of the lucrative trade. When continental trade began to move by sea, the Middle East still had a locational advantage with its land-penetrating seas. The Mediterranean Sea, the Red Sea, and the Persian Gulf continued to direct seaborne trade to the ancient centers of civilization, Egypt and Mesopotamia.

The advent of European oceanic exploration and the redirection of global trade to bypass the Middle East ended the region's dominance and instituted a period of decline and stagnation. Later, the impact of European colonialism and the discovery of petroleum in the twentieth century awakened the people of the Middle East to a renewed effort toward national aspirations and development. This has brought a multitude of problems and challenges: declining traditional cultural and religious values; growing political experimentation; meeting educational aspirations; confronting poverty and disease; questioning the status of women in a traditional male society; and developing an industrial infrastructure. And still today, the Middle East seethes with opposing groups pushing their own agendas: traditionalists vs. modernists, Islamists vs. secularists, socialists vs. capitalists, democrats vs. authoritarians.

The history of the Middle East is not just one of political and economic struggle; it is also one of spiritual revolution. The region is the source of the world's four great monotheistic religions. In ancient Palestine, the Israelites formed the foundation of today's Jewish religion and its offspring, Christianity. In ancient Persia (now Iran), Zoroastrianism rose to prominence but declined under competition from Christianity and the region's newest religion, Islam. Islam grew from humble origins in the Arabian Desert (at Mecca and Medina) to become one of the largest and most rapidly expanding faiths in the world.

Population of the Middle East

In 2018, the total population of the Middle East was estimated to be about 332 million, with an average

Bactrian camels by sand dunes in Gobi Desert.

Jerusalem, Dome of the Rock.

Mecca in Saudi Arabia.

Baghdad, Iraq.

annual growth rate of 1.3 percent. The region's population doubling time is about fifty years—a rate less than half that of the span from 1950 to 2000.

More than 50 percent of the region's population live in urban areas, engaged primarily in trade and services. The other half are still employed in agriculture and living in small villages. The ancient nomadic way of life has almost disappeared through the attraction of urban jobs and efforts of governments to settle nomadic people. The region's largest cities are Tehran, Iran (15.2 million); Istanbul, Turkey (14.6 million); Baghdad, Iraq (8.5 million); and Riyadh, Saudi Arabia (6.5 million). Other notable urban centers include Dubai, Damascus, Tel Aviv, and Beirut.

Population concentrations in the Middle East are strongly associated with the presence of fresh water, either from large rivers such as those in Mesopotamia (the Tigris and Euphrates), or in areas of greater rainfall (dry subtropical climates and higher elevations). The Middle East's two largest countries in 2020 were Iran (estimated population of 83 million) and Turkey (estimated population of 81 million). Turkey's population is found mainly along its coastlines that lie within the dry summer subtropical climate. Iran's population is primarily focused in western Iran, in and around the Zagros Mountains, and along the northern coastline north of the Elburz Mountains. Other countries with sizeable populations include Iraq, Saudi Arabia, Yemen, and Syria.

As a cultural region, the Middle East is often defined in terms of a religion (Islam) and a language (Arabic), yet not everyone is a Muslim and not everyone speaks Arabic as a native tongue, especially in Turkey and Iran. Monotheism has long existed there, in the faiths of Judaism, Christianity, Islam, and Zoroastrianism. However, approximately 90 percent of the population is Muslim and, because Islam emphasizes learning the Quran (Koran) in Arabic, this language dominates the region.

Beirut, Lebanon.

Petroleum

Petroleum is the primary resource and producer of revenue in the Middle East. The region is estimated to have slightly less than half of the world's known supply of petroleum, with Saudi Arabia and Iran having the greatest reserves. In 2017 Saudi Arabia had 297.7 billion barrels of crude oil reserves, Iran had an estimated 211.6 billion barrels, Iraq 142.5 billion barrels, the United Arab Emirates 105.8 billion barrels, and Kuwait 101.5 billion barrels. In contrast, the United States is estimated to have reserves of 35 billion barrels of crude oil.

The Middle East's natural gas reserves also play an important role in meeting the world's energy needs. In 2018, Iran's known reserves of natural gas were estimated at 33.6 trillion cubic meters. Qatar had 25.6 trillion cubic meters, while Saudi Arabia had 7.1 trillion cubic meters. By comparison, the United States had an estimated 5.9 trillion cubic meters of natural gas reserves.

In 1960, the Organization of the Petroleum Exporting Countries (OPEC) was created by four Middle East countries (Iran, Iraq, Kuwait, and Saudi Arabia) plus Venezuela as a means to coordinate oil prices and stabilize the oil market. Nine more oil-producing countries (including several from outside the Middle East) soon joined the founding five. By the early 1970s, OPEC accounted for more than half the world's oil production.

In 1968, a subset of OPEC members formed the Organization of Arab Petroleum Exporting Countries (OAPEC). In 1973, OAPEC simultaneously instituted production cuts and ordered an oil embargo directed at the United States and those European countries that supported Israel in the Yom Kippur War. This marked the first time an oil embargo was used as an economic weapon for political gain. Oil prices spiked worldwide, and the revenue to the oil-producing countries skyrocketed. In the late 1970s, the Iranian Revolution and the Iran-Iraq War caused another disruption in supplies, and oil prices spiked again. But by then the U.S. and Europe had already begun examining ways to reduce their dependence on OPEC oil. As a result, by 1985 OPEC's market share had dropped to less than 30 percent.

In the decades since, although OPEC's market share has moved slightly up and down, the group has never regained its economic or political clout. Repeated attempts to limit the world supply of oil by imposing production quotas have been all but ignored by member countries seeking to increase revenue. And with non-OPEC oil production increasing almost daily, alternative-fuel use rising, and conservation and climate-friendly initiatives yielding results, demand for OPEC oil has continued to fall. As a result, in the first two decades of the twenty-first century, OPEC-initiated fluctuations in oil supply and oil prices have caused few disruptions to the world economy.

Other Industries

Outside the petroleum states, the economy of the Middle East is primarily based on agriculture, focused on the valleys of the Tigris and Euphrates Rivers, where food and fiber are grown. Where rainfall is adequate in the steppe climatic zone, grains (primarily barley and wheat) are grown, and sheep and goats are herded. In areas of dry summer subtropical climate, fruit and nut crops are favored, particularly grapes, figs, apricots, oranges, plums, pears, peaches, cherries, olives, pistachios, almonds, and walnuts. Industrial crops such as cotton, tobacco, sugar beets, sugarcane, linseed, and hemp are grown in Israel, Turkey, and Syria. Narcotic crops are found in particular areas of the Middle East: opium poppies in Turkey and *qat* in Yemen. Within the Middle East, Israel's agriculture is the most mechanized and foreign-market-orientated.

Islam is the largest religion in the Middle East. Here, Muslim men are prostrating during prayer in a mosque.

It's reasonable to state that, except for the petroleum industry, the Middle East is considered to be a part of the nonindustrialized world. The only truly industrial country in the region is Israel, although a few countries are developing their industrial base. Iran and Turkey are in the forefront of industrial development, and Syria, Iraq, and the Gulf States are expanding their manufacturing industries. Banking and other service industries are reviving in Lebanon. Turkey, Lebanon, and Israel have fairly well-developed tourism industries. Elsewhere in the Middle East, political turmoil and the strictly conservative atmospheres have discouraged potential tourists from visiting. Kuwait, Qatar, and the United Arab Emirates, partially in recognition of this perception, have developed state airlines to carry Western tourists to their many upscale seaside resorts. But in general, most non-petroleum industry in the Middle East is focused on the domestic markets, with food processing, beverages, construction materials, and clothing as the primary products. In the Gulf States, diversification has centered around the oil and natural gas industries in refining and the manufacturing of goods from petroleum or using petroleum in the production of products, such as aluminum.

Thomas Baucom

DISCUSSION QUESTIONS: REGIONS

Q1. What religions arose from the cultures of the Middle East? Which is considered the dominant faith in the Middle East? How does religion continue to shape the region?

Q2. How important is oil to the Middle East economies? When has oil been used as a political weapon? Define OPEC and its role in the region.

Q3. What legacies remain from the period when European countries controlled the Middle East? What positive impacts did the European domination produce? What is the role of Western countries in the Middle East today?

PHYSICAL GEOGRAPHY

OVERVIEW

CLIMATE AND HUMAN SETTLEMENT

"Everyone talks about the weather," goes an old saying, "but nobody does anything about it." If everyone talks about the weather, it is because it is important to them—to how they feel and to how their bodies and minds function. There is plenty they can do about it, from going to a different location to creating an artificial indoor environment.

Climate

The term "climate" refers to average weather conditions over a long period of time and to the variations around that average from day to day or month to month. Temperature, air pressure, humidity, wind conditions, sunshine, and rainfall—all are important elements of climate and differ systematically with location. Temperatures tend to be higher near the equator and are so low in the polar regions that very few people live there. In any given region, temperatures are lower at higher altitudes. Areas close to large bodies of water have more stable temperatures. Rainfall depends on topography: The Pacific Coast of the United States receives a great deal of rain, but the nearby mountains prevent it from moving very far inland. Seasonal variations in temperature are larger in temperate zones.

Throughout human history, climate has affected where and how people live. People in technologically primitive cultures, lacking much protective clothing or housing, needed to live in mild climates, in environments favorable to hunting and gathering. As agricultural cultivation developed, populations located where soil fertility, topography, and climate were favorable to growing crops and raising livestock. Areas in the Middle East and near the Mediterranean Sea flourished before 1000 BCE.

Many equatorial areas were too hot and humid for human and animal health and comfort, and too infested with insect pests and diseases.

Improvements in technology allowed settlement to range more widely north and south. Sturdy houses and stables, internal heating, and warm clothing enabled people to survive and be active in long cold winters. Some peoples developed nomadic patterns, moving with herds of animals to adapt to seasonal variations.

A major challenge in the evolution of settled agriculture was to adapt production to climate and soil conditions. In North America, such crops as cotton, tobacco, rice, and sugarcane have relatively restricted areas of cultivation. Wheat, corn, and soybeans are more widely grown, but usually further north. Winter wheat is an ingenious adaptation to climate. It is sown and germinates in autumn, then matures and is harvested the following spring. Rice, which generally grows in standing water, requires special environmental conditions.

Tropical Problems

Some scholars argue that tropical climates encourage life to flourish but do not promote quality of life. In hot climates, people do not need much caloric intake to maintain body heat. Clothing and housing do not need to protect people from the cold. Where temperatures never fall below freezing, crops can be grown all year round. Large numbers of people can survive even where productivity is not high. However, hot, humid conditions are not favorable to human exertion nor (it is claimed) to mental, spiritual, and artistic creativity. Some tropical areas, such as South India, Bangladesh, Indonesia, and Central Africa, have developed large populations living at relatively low levels of income.

Slavery

Efforts to develop tropical regions played an important part in the rise of the slave trade after 1500 CE. Black Africans were kidnapped and forceably transported to work in hot, humid regions. The West Indian islands became an important location for slave labor, particularly in sugar production. On the North American continent, slave labor was important for producing rice, indigo, and tobacco in colonial times. All these were eclipsed by the enormous growth of cotton production in the early years of U.S. independence. It has been estimated that the forced migration of Africans to the Americas involved about 1,800 Africans per year from 1450 to 1600, 13,400 per year in the seventeenth century, and 55,000 per year from 1701 to 1810. Estimates vary wildly, but at least 7.7 million Africans were forced to migrate in this process.

European Migration

Migration of European peoples also accelerated after the discovery of the New World. They settled mainly in temperate-zone regions, particularly North America. Although Great Britain gained colonial dominion over India, the Netherlands over present-day Indonesia, and Belgium over a vast part of central Africa, few Europeans went to those places to live. However, many Chinese migrated throughout the Nanyang (South Sea) region, becoming commercial leaders in present-day Malaysia, Thailand, Indonesia, and the Philippines, despite the heat and humidity. British emigrants settled in Australia and New Zealand, Spanish and Italians in Argentina, Dutch (Boers) in South Africa—all temperate regions.

Climate and Economics

Most of the economic progress of the world between 1492 and 2000 occurred in the temperate zones, primarily in Europe and North America. Climatic conditions favored agricultural productivity. Some scholars believe that these areas had climatic conditions that were stimulating to intellectual and tech-

---

IRELAND'S POTATO FAMINE AND EUROPEAN EMIGRATION

Mass migration from Europe to North America began in the 1840s after a serious blight destroyed a large part of the potato crop in Ireland and other parts of Northern Europe. The weather played a part in the famine; during the autumns of 1845 and 1846 climatic conditions were ideal for spreading the potato blight. The major cause, however, was the blight itself, and the impact was severe on low-income farmers for whom the potato was the major food.

The famine and related political disturbances led to mass emigration from Ireland and from Germany. By 1850 there were nearly a million Irish and more than half a million Germans in the United States. Combined, these two groups made up more than two-thirds of the foreign-born U.S. population of 1850. The settlement patterns of each group were very different. Most Irish were so poor they had to work for wages in cities or in construction of canals and railroads. Many Germans took up farming in areas similar in climate and soil conditions to their homelands, moving to Wisconsin, Minnesota, and the Dakotas.

---

nological development. They argue that people are invigorated by seasonal variation in temperature, sunshine, rain, and snow. Storms—particularly thunderstorms—can be especially stimulating, as many parents of young children have observed for themselves.

Climate has contributed to the great economic productivity of the United States. This productivity has attracted a flow of immigrants, which averaged about 1 million a year from 1905 to 1914. Immigration approached that level again in the 1990s, as large numbers of Mexicans crossed the southern border of the United States, often coming for jobs as agricultural laborers in the hot conditions of the Southwest—a climate that made such work unattractive to many others.

Unpredictable climate variability was important in the peopling of North America. During the 1870s and 1880s, unusually favorable weather encouraged a large flow of migration into the grain-producing areas just west of the one-hundredth meridian. Then came severe drought and much agrarian distress. Between 1880 and 1890, the combined population of Kansas and Nebraska increased by about a million, an increase of 72 percent. During the 1890s, however, their combined population was virtually constant, indicating that a large out-migration was offsetting the natural increase. Much of the area reverted to pasture, as climate and soil conditions could not sustain the grain production that had attracted so many earlier settlers.

Climate variability can be a serious hazard. Freezing temperatures for more than a few hours during spring can seriously damage fruits and vegetables. A few days of heavy rain can produce serious flooding.

Recreation and Retirement

Whenever people have been able to separate decisions about where to live from decisions about where to work, they have gravitated toward pleasant climatic conditions. Vacationers head for Caribbean islands, Hawaii, the Crimea, the Mediterranean Coast, even the Baltic coast. "The mountains" and "the seashore" are attractive the world over. Paradoxically, some of these areas (the Caribbean, for instance) have monotonous weather year-round and thus have not attracted large inflows of permanent residents. Winter sports have created popular resorts such as Vail and Aspen in Colorado, and numerous older counterparts in New England. Large numbers of Americans have retired to the warm climates in Florida, California, and Arizona. These areas then attract working-age adults who earn a living serving vacationers and retirees. Since these locations are uncomfortably hot in summer, their attractiveness for residence had to await the coming of air conditioning in the latter half of the twentieth century.

Human Impact on Climate

Climate interacts with pollution. Bad-smelling factories and refineries have long relied on the wind to disperse atmospheric pollutants. The city of Los Angeles, California, is uniquely vulnerable to atmospheric pollution because of its topography and wind currents. Government regulations of automobile emissions have had to be much more stringent there than in other areas to keep pollution under control.

Human activities have sometimes altered the climate. Development of a large city substitutes buildings and pavements for grass and trees, raising summer temperatures and changing patterns of water evaporation. Atmospheric pollutants have contributed to acid rain, which damages vegetation and pollutes water resources. Many observers have also blamed human activities for a trend toward global warming. Much of this has been blamed on carbon dioxide generated by combustion, particularly of fossil fuels. A widespread and continuing rise in temperatures is expected to raise water levels in the oceans as polar icecaps melt and change the relative attractiveness of many locations.

Paul B. Trescott

FLOOD CONTROL

Flood control presents one of the most daunting challenges humanity faces. The regions that human communities have generally found most desirable, for both agriculture and industry, have also been the lands at greatest risk of experiencing devastating floods. Early civilization developed along river valleys and in coastal floodplains because those lands contained the most fertile, most easily irrigated soils for agriculture, combined with the convenience of water transportation.

The Nile River in North Africa, the Ganges River on the Indian subcontinent, and the Yangtze River in China all witnessed the emergence of civilizations that relied on those rivers for their growth. People learned quickly that residing in such areas meant living with the regular occurrence of life-threatening floods.

Knowledge that floods would come did not lead immediately to attempts to prevent them. For thousands of years, attempts at flood control were rare. The people living along river valleys and in floodplains often developed elaborate systems of irrigation canals to take advantage of the available water for agriculture and became adept at using rivers for transportation, but they did not try to control the river itself. For millennia, people viewed periodic flooding as inevitable, a force of nature over which they had no control. In Egypt, for example, early people learned how far out over the riverbanks the annual flooding of the Nile River would spread and accommodated their society to the river's seasonal patterns. Villagers built their homes on the edge of the desert, beyond the reach of the flood waters, while the land between the towns and the river became the area where farmers planted crops or grazed livestock.

In other regions of the world, buildings were placed on high foundations or built with two stories on the assumption that the local rivers would regularly overflow their banks. In Southeast Asian countries such as Thailand and Vietnam, it is common to see houses constructed on high wooden posts above the rivers' edge. The inhabitants have learned to allow for the water levels' seasonal changes.

Flood Control Structures

Eventually, societies began to try to control floods rather than merely attempting to survive them. Levees and dikes—earthen embankments constructed to prevent water from flowing into low-lying areas—were built to force river waters to remain within their channels rather than spilling out over a floodplain. Flood channels or canals that fill with water only during times of flooding, diverting water

away from populated areas, are also a common component of flood control systems. Areas that are particularly susceptible to flash floods have constructed numerous flood channels to prevent flooding in the city. For example, for much of the year, Southern California's Los Angeles River is a small stream flowing down the middle of an enormous, 20- to 30-foot-deep (6–9 meters) concrete-lined channel, but winter rains can fill its bed from bank to bank. Flood channels prevent the river from washing out neighborhoods and freeways.

Engineers designed dams with reservoirs to prevent annual rains or snowmelt entering the river upstream from running into populated areas. By the end of the twentieth century, extremely complex flood control systems of dams, dikes, levees, and flood channels were common. Patterns of flooding that had existed for thousands of years ended as civil engineers attempted to dominate natural forces.

The annual inundation of the Egyptian delta by the flood waters of the Nile River ceased in 1968 following construction of the 365-foot-high (111 meters) Aswan High Dam. The reservoir behind the 3,280-foot-long (1,000-meter) dam forms a lake almost ten miles (16 km.) wide and almost 300 miles (480 km.) long. Flood waters are now trapped behind the dam and released gradually over a year's time.

Environmental Concerns

Such high dams are increasingly being questioned as a viable solution for flood control. As human understanding of both hydrology and ecology have improved, the disruptive effects of flood control projects such as high dams, levees, and other engineering projects are being examined more closely.

Hydrologists and other scientists who study the behavior of water in rivers and soils have long known that vegetation and soil types in watersheds can have a profound effect on downstream flooding. The removal of forest cover through logging or clearing for agriculture can lead to severe flooding in the future. Often that flooding will occur many miles downstream from the logging activity. Devastating floods in the South Asian country of Bangladesh, for example, have been blamed in part on clear-cutting of forested hillsides in the Himalaya Mountains in India and Nepal. Monsoon rains that once were absorbed or slowed by forests now run quickly off mountainsides, causing rivers to reach unprecedented flood levels. Concerns about cause-and-effect relationships between logging and flood control in the mountains of the United States were one reason for the creation of the U.S. Forest Service in the nineteenth century.

In populated areas, even seemingly trivial events such as the construction of a shopping center parking lot can affect flood runoff. When thousands of square feet of land are paved, all the water from rain runs into storm drains rather than being absorbed slowly into the soil and then filtered through the watertable. Engineers have learned to include catch basins, either hidden underground or openly visible but disguised as landscaping features such as ponds, when planning a large paving project.

Wetlands and Flooding

Less well known than the influence of watersheds on flooding is the impact of wetlands along rivers. Many river systems are bordered by long stretches of marsh and bog. In the past, flood control agencies often allowed farmers to drain these areas for use in agriculture and then built levees and dikes to hold the river within a narrow channel. Scientists now know that these wetlands actually serve as giant sponges in the flood cycle. Flood waters coming down a river would spread out into wetlands and be held there, much like water is trapped in a sponge.

Draining wetlands not only removes these natural flood control areas but worsens flooding problems by allowing floodwater to precede downstream faster. Even if life-threatening or property-damaging floods do not occur, faster-flowing water significantly changes the ecology of the river system. Waterborne silt and debris will be carried farther. Trying to control floods on the Mississippi River has

had the unintended consequence of causing waterborne silt to be carried farther out into the Gulf of Mexico by the river, rather than its being deposited in the delta region. This, in turn, has led to the loss of shore land as ocean wave actions washes soil away, but no new alluvial deposits arrive to replace it.

In any river system, some species of aquatic life will disappear and others replace them as the speed of flow of the water affects water temperature and the amount of dissolved oxygen available for fish. Warm-water fish such as bass will be replaced by cold-water fish such as trout, or vice versa. Biologists estimate that more than twenty species of freshwater mussels have vanished from the Tennessee River since construction of a series of flood control and hydroelectric power generation dams have turned a fast-moving river into a series of slow-moving reservoirs.

Future of Flood Control

By the end of the twentieth century, engineers increasingly recognized the limitations of human interventions in flood control. Following devastating floods in the early 1990s in the Mississippi River drainage, the U.S. Army Corps of Engineers recommended that many towns that had stood right at the river's edge be moved to higher ground. That is, rather than trying to prevent a future flood, the Corps advised citizens to recognize that one would inevitably occur, and that they should remove themselves from its path. In the United States and a number of other countries, land that has been zoned as floodplains can no longer be developed for residential use. While there are many things humanity can do to help prevent floods, such as maintaining well-forested watersheds and preserving wetlands, true flood control is probably impossible. Dams, levees, and dikes can slow the water down, but eventually, the water always wins.

Nancy Farm Männikkö

ATMOSPHERIC POLLUTION

Pollution of the Earth's atmosphere comes from many sources. Some forces are natural, such as volcanoes and lightning-caused forest fires, but most sources of pollution are byproducts of industrial society. Atmospheric pollution cannot be confined by national boundaries; pollution generated in one country often spills over into another country, as is the case for acid deposition, or acid rain, generated in the midwestern states of the United States that affects lakes in Canada.

Major Air Pollutants

Each of eight major forms of air pollution has an impact on the atmosphere. Often two or more forms of pollution have a combined impact that exceeds the impact of the two acting separately. These eight forms are:

1. Suspended particulate matter: This is a mixture of solid particles and aerosols suspended in the air. These particles can have a harmful impact on human respiratory functions.

2. Carbon monoxide (CO): An invisible, colorless gas that is highly poisonous to air-breathing animals.

3. Nitrogen oxides: These include several forms of nitrogen-oxygen compounds that are converted to nitric acid in the atmosphere and are a major source of acid deposition.

4. Sulfur oxides, mainly sulfur dioxide: This sulfur-oxygen compound is converted to sulfuric acid in the atmosphere and is another source of acid deposition.

5. Volatile organic compounds: These include such materials as gasoline and organic cleaning solvents, which evaporate and enter the air in a vapor state. VOCs are a major source of ozone formation in the lower atmosphere.

6. Ozone and other petrochemical oxidants: Ground-level ozone is highly toxic to animals and plants. Ozone in the upper atmosphere, however, helps to shield living creatures from ultraviolet radiation.

7. Lead and other heavy metals: Generated by various industrial processes, lead is harmful to human health even at very low concentrations.

8. Air toxics and radon: Examples include cancer-causing agents, radioactive materials, or asbestos. Radon is a radioactive gas produced by natural processes in the earth.

All eight forms of pollution can have adverse effects on human, animal, and plant life. Some, such as lead, can have a very harmful effect over a small range. Others, such as sulfur and nitrogen oxides, can cross national boundaries as they enter the atmosphere and are carried many miles by prevailing wind currents. For example, the radioactive discharge from the explosion of the Chernobyl nuclear plant in the former Soviet Union in 1986 had harmful impacts in many countries. Atmospheric radiation generated by the explosion rapidly spread over much of the Northern Hemisphere, especially the countries of northern Europe.

Impacts of Atmospheric Pollution

Atmospheric pollution not only has a direct impact on the health of humans, animals, and plants but also affects life in more subtle, often long-term, ways. It also affects the economic well-being of people and nations and complicates political life.

Atmospheric pollution can kill quickly, as was the case with the killer smog, brought about by a temperature inversion, that struck London in 1952 and led to more than 4,000 pollution-related deaths. In the late 1990s, the atmosphere of Mexico City was so polluted from automobile exhausts and industrial pollution that sidewalk stands selling pure oxygen to people with breathing problems became thriving businesses. Many of the heavy metals and organic constituents of air pollution can cause cancer when people are exposed to large doses or for long periods of time. Exposure to radioactivity in the atmosphere can also increase the likelihood of cancer.

In some parts of Germany and Scandinavia in the 1990s, as well as places in southern Canada and the southern Appalachians in the United States, certain types of trees began dying. There are several possible reasons for this die-off of forests, but one potential culprit is acid deposition. As noted above, one byproduct of burning fossil fuels (for example, in coal-fired electric power plants) is the sulfur and nitrous oxides emitted from the smokestacks. Once in the atmosphere, these gases can be carried for many miles and produce sulfuric and nitric acids.

These acids combine with rain and snow to produce acidic precipitation. Acid deposition harms crops and forests and can make a lake so acidic that aquatic life cannot exist in it. Forests stressed by contact with acid deposition can become more susceptible to damage by insects and other pathogens. Ozone generated from automobile emissions also kills many plants and causes human respiratory problems in urban areas.

Air pollution also has an impact on the quality of life. Acid pollutants have damaged many monuments and building facades in urban areas in Europe and the United States. By the late 1990s, the distance that people could see in some regions, such as the Appalachians, was reduced drastically because of air pollution.

The economic impact of air pollution may not be as readily apparent as dying trees or someone with a respiratory ailment, but it is just as real. Crop damage reduces agricultural yield and helps to drive up the cost of food. The costs of repairing buildings or monuments damaged by acid rain are substantial. Increased health-care claims resulting from exposure to air pollution are hard to measure but are a cost to society nevertheless.

It is impossible to predict the potential for harm from rapid global warming arising from greenhouse gases and the destruction of the ozone layer by chlorofluorocarbons (CFCs), but it could be cata-

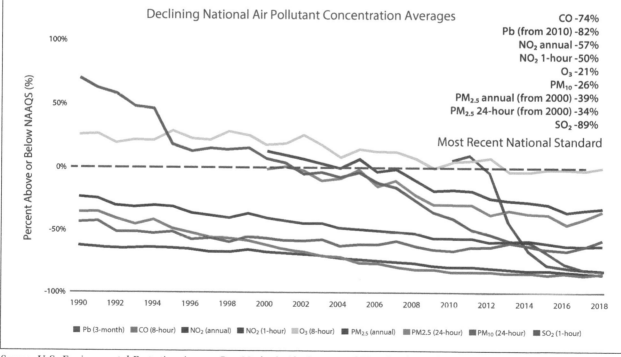

Air Quality Trends Show Clean Air Progress

While some pollutants continue to pose serious air quality problems in areas of the U.S., nationally, criteria air pollutant concentrations have dropped significantly since 1990 improving quality of life for many Americans. Air quality improves as America grows.

Declining National Air Pollutant Concentration Averages

CO -74%
Pb (from 2010) -82%
NO_2 annual -57%
NO_2 1-hour -50%
O_3 -21%
PM_{10} -26%
$PM_{2.5}$ annual (from 2000) -39%
$PM_{2.5}$ 24-hour (from 2000) -34%
SO_2 -89%

Most Recent National Standard

Pb (3-month) ■ CO (8-hour) ■ NO_2 (annual) ■ NO_2 (1-hour) ■ O_3 (8-hour) ■ $PM_{2.5}$ (annual) ■ PM2.5 (24-hour) ■ PM_{10} (24-hour) ■ SO_2 (1-hour)

Source: U.S. Environmental Protection Agency, Our Nation's Air, Status and Trends Through 2018.

strophic. Rapid global warming would cause the sea level to rise because of the melting of the polar ice caps. Low-lying coastal areas would be flooded, or, in the case of Bangladesh, much of the country. Global warming would also change crop patterns for much of the world.

Solutions for Atmospheric Pollution
Although there is still some debate, especially among political leaders, most scientists recognize that air pollution is a problem that affects both the industrialized and less-industrialized world. In their rush to industrialize, many nations begin generating substantial amounts of air pollution; China's extensive use of coal-fired power plants is just one example.

The major industrial nations are the primary contributors to atmospheric pollution. North America, Europe, and East Asia produce 60 percent of the world's air pollution and 60 percent of its food supply. Because of their role in supplying food for many other nations, anything that damages their ability to grow crops hurts the rest of the world. In 2018, about 76 million tons of pollution were emitted into the atmosphere in the United States. These emissions mostly contribute to the formation of ozone and particles, the deposition of acids, and visibility impairment.

Many industrialized nations are making efforts to control air pollution, for example, the Clean Air Act of 1970 in the United States or the international Montreal Accord to curtail CFC production. Progress is slow and the costs of reducing air pollution are often high. Worldwide, bad outdoor air caused an estimated 4.2 million premature deaths in 2016, about 90 percent of them in low- and middle-income countries, according to the World Health Organization. Indoor smoke is an ongoing health threat to the 3 billion people who cook and heat their homes by burning biomass, kerosene, and coal.

In the year 2019 the record of the nations of the world in dealing with air pollution was a mixed one. There were some signs of progress, such as reduced automobile emissions and sulfur and nitrous oxides in industrialized nations, but acid deposition remains a problem in some areas. CFC production has been halted, but the impact of CFCs on the ozone layer will continue for many years. However, more nations are becoming aware of the health and economic impact of air pollution and are working to keep the problem from getting worse.

John M. Theilmann

DISEASE AND CLIMATE

Climate influences the spread and persistence of many diseases, such as tuberculosis and influenza, which thrive in cold climates, and malaria and encephalitis, which are limited by the warmth and humidity that sustains the mosquitoes carrying them. Because the earth is warming as a result of the generation of carbon dioxide and other "greenhouse gases" from the burning of fossil fuels, there is intensified scientific concern that warm-weather diseases will reemerge as a major health threat in the near future.

Scientific Findings
The question of whether the earth is warming as a result of human activity was settled in scientific circles in 1995, when the Second Assessment Report of the Intergovernmental Panel on Climate Change, a worldwide group of about 2,500 experts, was issued. The panel concluded that the earth's temperature

had increased between 0.5 to 1.1 degrees Farenheit (0.3 to 0.6 degrees Celsius) since reliable worldwide records first became available in the late nineteenth century. Furthermore, the intensity of warming had increased over time. By the 1990s, the temperature was rising at the most rapid rate in at least 10,000 years.

The Intergovernmental Panel concluded that human activity—the increased generation of carbon dioxide and other "greenhouse gases"—is responsible for the accelerating rise in global temperatures. The amount of carbon dioxide in the atmosphere has risen nearly every year because of increased use of fossil fuels by ever-larger human populations experiencing higher living standards.

In 1998, Paul Epstein of the Harvard School of Public Health described the spread of malaria and dengue fever to higher altitudes in tropical areas of the earth as a result of warmer temperatures. Rising winter temperatures have allowed disease-bearing insects to survive in areas that could not support them previously. According to Epstein, frequent flooding, which is associated with warmer temperatures, also promotes the growth of fungus and provides excellent breeding grounds for large numbers of mosquitoes. Some experts cite the flooding caused by Hurricane Floyd and other storms in North Carolina during 1999 as an example of how global warming promotes conditions ideal for the spread of diseases imported from the Tropics.

Heat, Humidity, and Disease

During the middle 1990s, an explosion of termites, mosquitoes, and cockroaches hit New Orleans, following an unprecedented five years without frost. At the same time, dengue fever spread from Mexico across the border into Texas for the first time since records have been kept. Dengue fever, like malaria, is carried by a mosquito that is limited by temperature and humidity. Colombia was experiencing plagues of mosquitoes and outbreaks of the diseases they carry, including dengue fever and encephalitis, triggered by a record heat wave followed by heavy rains. In 1997 Italy also had an outbreak of malaria. An outbreak of zika in 2015–16, related to a virus spread by mosquitoes, raised concerns re-

garding the safety of athletes and spectators at the 2016 Summer Olympics in Rio de Janeiro and led to travel warnings and recommendations to delay getting pregnant for those living or traveling in areas where the mosquitoes are active.

The global temperature is undeniably rising. According to the National Oceanic and Atmospheric Administration, July 2019, was the hottest month since reliable worldwide records have been kept, or about 150 years. The previous record had been set in July 2017.

The rising incidence of some respiratory diseases may be related to a warmer, more humid environment. The American Lung Association reported that more than 5,600 people died of asthma in the United States during 1995, a 45.3 percent increase in mortality over ten years, and a 75 percent increase since 1980. Roughly a third of those cases occurred in children under the age of eighteen. Asthma is now one of the leading diseases among the young. Since 1980, there has been a 160 percent increase in asthma in children under the age of five.

Heat Waves and Health

A study by the Sierra Club found that air pollution, which will be enhanced by global warming, could be responsible for many human health problems, including respiratory diseases such as asthma, bronchitis, and pneumonia.

According to Joel Schwartz, an epidemiologist at Harvard University, air pollution concentrations in the late 1990s were responsible for 70,000 early deaths per year and more than 100,000 excess hospitalizations for heart and lung disease in the United States. Global warming could cause these numbers to increase 10 to 20 percent in the United States, with significantly greater increases in countries that are more polluted to begin with, according to Schwartz.

Studies indicate that global warming will directly kill hundreds of Americans from exposure to extreme heat during summer months. The U.S. Centers for Disease Control and Prevention have found that between 1979 and 2014, the death rate as a direct result of exposure to heat (underlying cause of death) generally hovered around 0.5 to 1 deaths per million people, with spikes in certain years). Overall, a total of more than 9,000 Americans have died from heat-related causes since 1979, according to death certificates. Heat waves can double or triple the overall death rates in large cities. The death toll in the United States from a heat wave during July 1999 surpassed 200 people. As many as 600 people died in Chicago alone during the 1990s due to heat waves. The elderly and very young have been most at risk.

Respiratory illness is only part of the picture. The Sierra Club study indicated that rising heat and humidity would broaden the range of tropical diseases, resulting in increasing illness and death from diseases such as malaria, cholera, and dengue fever, whose range will spread as mosquitoes and other disease vectors migrate.

The effects of El Niño in the 1990s indicate how sensitive diseases can be to changes in climate. A study conducted by Harvard University showed that warming waters in the Pacific Ocean likely contrib-

uted to the severe outbreak of cholera that led to thousands of deaths in Latin American countries. Since 1981, the number of cases of dengue fever has risen significantly in South America and has begun to spread into the United States. According to health experts cited by the Sierra Club study, the outbreak of dengue near Texas shows the risks that a warming climate might pose. Epstein and the Sierra Club study concur that if tropical weather expands, tropical diseases will expand.

In many regions of the world, malaria is already resistant to the least expensive, most widely distributed drugs. According to the World Health Organization (WHO), there were 219 million cases of malaria globally in 2017 and 435,000 malaria deaths, representing a decrease in malaria cases and deaths rates of 18 percent and 28 percent since 2010. Of those deaths, 403,000 (approximately 93 percent) were in the WHO African Region.

Bruce E. Johansen

PHYSIOGRAPHY

Asia is more a geographic term than a homogenous continent. The largest continent, Asia occupies the northern portion of the Eastern Hemisphere, extending northward beyond the Arctic Circle and southward nearly reaching the equator. It contains about one-third of the world's dry land, and one-twelfth of the whole surface of the globe. It has the greatest range of land elevation of all the continents and the longest coastline. Asia is also Earth's largest, youngest, and structurally most complicated continent. More than half of Asia remains seismically active, and new continental material is currently being produced in the island-arc systems that surround it to the east and southeast.

The Arctic Ocean bounds Asia on the north, the Pacific Ocean on the east, the Indian Ocean on the south, the Mediterranean and Black seas on the southwest, and Europe on the west. The boundary between Europe and Asia is a line that runs south from the Arctic Ocean along the eastern slope of the Ural Mountains and then turns southwest along the Zhem River to the northern shore of the Caspian Sea. West of the Caspian, the boundary follows the Kuma-Manych Depression to the Sea of Azov and the Kerch Strait.

The total area of Asia, including the Caucasian isthmus and excluding the island of New Guinea, amounts to about 17,159,955 square miles (44,444,100 sq. km.). The most distantly separated points of the Asian mainland are Cape Chelyuskin in north-central Siberia, Russia, to the north; the tip of the Malay Peninsula (Malaysia), Cape Piai, to the south; Cape Baba in Turkey to the west; and Cape Dezhnev (formerly known as East Cape) in

Linze, Zhangye, Gansu, China.

PHYSICAL GEOGRAPHY OF ASIA

northeastern Siberia, overlooking the Bering Strait, to the east.

Geological Units

Asia is made up of a large number of separate geological units that vary in age and origin. The Himalayan belt separates the main mass of central and northern Asia from the two separate blocks of peninsular India and Arabia. These southerly blocks originally were part of the ancient southern continent of Gondwanaland. They joined to the rest of Asia only in late Phanerozoic times (the last 100 million years), after the division of Gondwanaland. The northern part of Asia was part of the ancient continent of Laurasia.

Northern Asia is made up of ancient cratons—stable, broad, horizontal rock formations of Earth's crust. Mobile belts, which were active 500 million to 600 million years ago, separate these cratons. The Himalayan belt runs west-east from the Mediterranean through the Middle East to the Himalayas. From there, it turns south through Myanmar (formerly Burma), Malaysia, and Indonesia. The large, long-lived, mobile Himalayan belt has been active intermittently during the past 600 million years and is still active today.

Physiographic Divisions

Asia can be divided into five physiographic divisions. The first is North Asia, which includes the bulk of Siberia and the northeastern edges of the continent. East Asia includes the continental part of the Far East region of Siberia, the East Asian islands, Korea, and eastern and northeastern China. Central Asia includes the Plateau of Tibet, the Dzungarian and Tarim basins, Inner Mongolia, the Gobi Desert, the Sino-Tibetan mountain ranges, Middle Asia including the Turan Plain, the Pamirs, the Gissar-Alay Mountains, and the Tien Shan. South Asia includes the Philippine and Malay archipelagoes (Indonesia), Indochina (Vietnam, Laos, and Cambodia), peninsular India, Sri Lanka, the Indo-Gangetic Plain, and the Himalayas. West Asia includes the West Asian highlands (Asia Minor or Anatolia, Armenia, the Caucasus Mountains, Afghanistan, and Iran), the Levant (Israel, Lebanon, Syria, Jordan, and Iraq), and the Arabian Peninsula.

North Asia

Asiatic Laurasia is the vast continental area to the north of the Himalayas. This is a stable massif containing many ancient shield areas of Precambrian age (more than 570 million years ago). The Siberian platform is the largest of the shield areas and occupies the northern part of the continent.

Siberia extends from the Ural Mountains on the west to the Pacific Ocean on the east and southward from the Arctic Ocean to the hills of north central Kazakhstan and the borders of Mongolia and China. Its total area is about 5.2 million square miles (13.5 million sq. km.). Siberia falls into four major geographic regions, all of great size. In the west, the Ural Mountains are bordered by the huge West Siberian Plain, which is drained by the Ob and Yenisey rivers. East of the Yenisey River is Central Siberia, a vast area that consists mainly of plains and the Central Siberian Plateau. Farther east, the basin of the Lena River separates central Siberia from the complex series of mountain ranges, upland massifs, and intervening basins that makes up northeastern Siberia. The smallest of the four regions is the Baikal area, which is centered on Lake Baikal in the south central part of Siberia.

The 1991 eruption of Mount Pinatubo is the second largest volcanic eruption of the twentieth century.

In northeastern Siberia, the Verkhoyansk, Chersky, and Okhotsk-Chaun mountains are found. These mountains were formed by faults and folding of the land in the Mesozoic period (225 million to 65 million years ago) and are of moderate height. The Koryak mountain range, which is also found in this part of Siberia, is of a similar structure but was formed during the Cenozoic period (65 million years ago to 10,000 years ago). Volcanic activity was present in these areas during the Cenozoic era. Some plateaus are found in these areas of the ancient mountains, such as the Kolyma highlands.

The most significant feature of north central Siberia is the Central Siberian Plateau. This consists of a series of platform plateaus and stratified plains that were uplifted in the Cenozoic era. The plateaus are made of terraced and broken-up mesas with exposed horizontal volcanic intrusions. The intrusions were created by the flow of molten magma between layers of rock. The plains were formed from uplifted Precambrian blocks. The Putoran Mountains consist of a young uplifted mesa, broken up at the edges and partly covered with traprock.

The far north of Siberia has been subjected to several periods of glaciation, and erosive influences have dominated. In the south of the region, sedimentary features become more prevalent. South-north-flowing rivers such as the Ob, Lena, and Yenisey form the major transport arteries of the region, although they are frozen over for most of the year.

The Aldan Massif, in the southeast of the Siberian platform, contains very old early Precambrian high-grade metamorphic rocks (formed under heat and pressure), together with middle Precambrian sediments and granites. Gold occurs as lodes and placers in these rocks. In the southeast and southwest parts of the platform are large areas of late Precambrian rocks, representing mobile belts that were active 1,600 million to 1,800 million years ago.

To the southwest, the Siberian platform is cut by the Lake Baikal Rift Valley. A bow-shaped rift system of the late Tertiary age, the Baikal Rift Valley System runs for almost 1,242 miles (2,000 km.) along the southern edge of the Siberian craton. A series of fault troughs contains thick accumulations of sedimentary and volcanic rocks. Lake Baikal lies in the central part of the rift 1.8 miles (2.9 km.) below the blocks on each side of the Rift Valley.

East Asia

The Manchurian Plain is located in China's northeast, in the region formerly known as Manchuria. It is an undulating plain split into northern and southern halves by a low divide rising 500 to 850 feet (150 to 260 meters). The Sungari River drains the northern part, and the Liao River the southern part. Erosion rather than sedimentary deposits has shaped most of the surface of this area. The plain

A LAND OF EXTREMES

The Asian continent is home to the highest, lowest, wettest and coldest places on Earth. The world's highest peak is Mount Everest in Himalayan Tibet. Mount Everest is 29,035 feet (8,850 meters) high and still growing—about .16 inch (4 millimeters) per year—due to plate tectonics, the phenomenon that formed the Himalayas. The subduction of the Indian plate under the larger Asian plate pushes portions of the Asian plate upward. A similar phenomenon is occurring in Anatolia (Turkey) where the Arabian plate is sliding under the Anatolian plate.

The Dead Sea, between Israel and Jordan, is the lowest place on Earth's land surface. The Dead Sea located at 1,310 feet (400 meters) below mean sea level. Little or no life exists in the Dead Sea because of the water's high salt content.

Cherrapunji in the state of Assam in eastern India is the wettest place on Earth. The annual rainfall average is 450 inches (1,143 centimeters) per year. Much of the state is covered with dense tropical forests of bamboo and, at higher elevations, evergreens. Common animals of Assam include the elephant, tiger, leopard, rhinoceros, and bear.

Siberia is notorious for the length and severity of its almost snowless winters. The coldest place on Earth is Oimyakon in northeastern Siberia. There the temperature has been recorded to be as low as -90° Fahrenheit (-68° Celsius). The climate becomes increasingly harsh eastward, while precipitation also diminishes. The coldest recorded temperature for North America is -85° Fahrenheit (-65° Celsius) at Snag in the Yukon Territory of Canada.

The summit of Mount Fuji is the highest point in Japan.

has an area of about 135,000 square miles (350,000 sq. km.). The river valleys are wide and flat with a series of terraces formed by deposits of silt. The North China Plain is comparable in size to the Manchurian Plain. It lies at altitudes below 150 feet (45 meters) above sea level and is very flat. Sedimentary deposits brought down by the Yellow (Huang He) and Huai rivers from the Loess Plateau formed the plain. The Quaternary deposits (those from 10,000 to 1.8 million years old) reach thicknesses of 2,500 to 3,000 feet (750 to 915 meters).

Mesozoic (between 225 million and 65 million years ago) fold belts occur in northeast Asia, flanking the northeast margin of the Siberian platform. Thick layers of disintegrated sediments of Permian to Jurassic age are next to the platform. These sediments have been folded and intruded by Mesozoic granites containing gold, tin, and tungsten ores.

Farther east, toward the present continental edge, is a mobile belt active since early Phanerozoic times. Tertiary granites and Tertiary to Quaternary (more than 1 million years ago) andesite, dacite,

and basalt lavas are widespread. The island arcs of the present active oceanic margin, Kamchatka, the Kurils, Japan, and the Philippines fringe this belt. The northern region of East Asia includes the Khingan and Burein Mountains, the Zeya-Bureya Depression, and the Sikhote-Alin ranges.

Offshore Islands

The islands off the coast of East Asia and the Kamchatka Peninsula are related formations. The Ryukyu Islands, Japan, Sakhalin, and the Kuril Islands are fragments of the Ryuku-Korean, Hōnshū-Sakhalin, and Kuril-Kamchatka mountain-island arcs. These arcs date from the Mesozoic and Cenozoic periods and have knots at their junctions, the Japanese islands of Kyūshū and Hōkkaidō. The mountains are of low or moderate altitude and are made of folded and faulted blocks. There are some mountains of volcanic origin and small alluvial lowlands. Kamchatka is a mountainous peninsula that was formed by fragments of the Kamchatka-Koryak and Kuril-Kamchatka arcs. Cenozoic and current volcanoes exist in this area. The peninsula has a number of geysers and hot springs. There are vast plains made from alluvia and volcanic ashes.

Central Asia

This region is made up of mountains, plateaus, and tablelands formed from pieces of the ancient platforms. A folded area in the Paleozoic and Mesozoic periods eventually surrounded them. The Central Asian plains and tablelands include the Dzungarian Basin, the Tarim Basin, the Taklimakan Desert, and the Gobi and Ordos deserts. Features vary from surfaces leveled by erosion in the Mesozoic and Cenozoic periods to layered plateaus with low mountains, eroded plateaus on which loess has accumulated, and vast sandy deserts covered with windborne deposits.

The Tien Shan Mountains enclose the Dzungarian on the south. The Altai Mountains cut it off from the Mongolian People's Republic on the northeast. The larger portion of the Dzungarian lies at elevations between 1,000 and 1,500 feet (300 and 460 meters). In the lowest part, the elevation drops to 620 feet (190 meters). The Tarim Basin lies north of the Tibetan Plateau, about 3,000 feet (915 meters) above sea level. It is surrounded by great mountain ranges: the Tien Shan on the north, the Pamirs on the west, and the Kunlun Mountains on the south. The Taklimakan desert—one of the world's most barren—occupies the center of the basin. The area of the basin is about 215,000 square miles (556,000 sq. km.). It has elevations of 2,500 to 4,600 feet (760 to 1,400 meters) above sea level.

Alpine Asia, sometimes known as High Asia, includes the Pamir Mountains and the eastern Hindu Kush, the Kunlun Mountains, the Tien Shan, the Gissar and Alay ranges, the Tibetan Plateau, and the Karakoram Mountains. The Tien Shan consists of a system of ranges and depressions with two groups of ranges, northern and southern. A strip of depressions, which contain lower interior ranges of mountains, separates the northern and southern ranges.

Ancient metamorphic rock constitutes the larger portion of the mountains in the interior. Paleozoic

(from 245 million to 570 million years old) sedimentary and igneous sedimentary beds form its southern and northern chains. Mesozoic sandstone and conglomerates fill the depression in the interior area and make up the foothill ridges. The height of the Tien Shan is between 13,000 and 15,000 feet (4,000 and 4,600 meters). Individual peaks exceed 16,000 feet (4,900 meters) while the interior ranges reach 14,500 feet (4,400 meters). The Tibetan Plateau lies at elevations above 13,000 to 15,000 feet (4,000 to 4,600 meters), with its border ranges even higher. Individual peaks rise to heights of 23,000 to 26,000 feet (7,000 to 7,900 meters). The interior slopes of these border mountains are slight, while the exterior slopes are very steep.

South Asia

The Himalayas (in Sanskrit: *hima*, "snow," and *alaya*, "abode"), the loftiest mountain system in the world, form the northern limit of India. This great, geologically young mountain arc is about 1,550 miles (2,500 km.) long, stretching from the peak of Nanga Parbat in Indian-governed Jammu and Kashmir to the Namcha Barwa peak in the Tibet Autonomous Region of China. Between these extremes, the mountains fall across India, southern Tibet, Nepal, and Bhutan. The width of the system varies between 125 and 250 miles (200 and 400 km.).

Within India, the Himalayas are divided into three longitudinal belts—the Outer, Lesser, and Great Himalayas. At each extremity, there is a great bend in the system's alignment. From the bend, a number of lower mountain ranges and hills spread out. Those to the west lie wholly within Pakistan and Afghanistan. Those to the east straddle India's border with Myanmar. North of the Himalayas are the Plateau of Tibet and various trans-Himalayan ranges.

The mass of mountains of the Himalayas of South Asia and the Karakorams and the Hindu Kush of Central Asia gives rise to many of the major rivers of South Asia, namely the Indus, Ganges, Brahmaputra, Salween, Mekong, Chang Jiang, and Huang He. The Indo-Gangetic Plain is formed from the combined plains of the Indus, Ganges, and Brahmaputra rivers. From the same mass of mountains, ridge-like fingers point east and southeast into China and Southeast Asia. These fingers form the mountainous backbone of Indochina and peninsular Malaysia.

The Indo-Gangetic Plain lies between the Himalayas and the Deccan Plateau. The plain occupies the Himalayan front, formerly a seabed but now filled with river-borne alluvium to depths of up to 6,000 feet (1,830 meters). The plain stretches from the Pakistani provinces of Sind and Punjab in the west, where the Indus and its tributaries water it, eastward to the Brahmaputra valley in Assam. The Ganges basin (mainly in Uttar Pradesh and Bihar) forms the central and principal part of this plain. The eastern part is made up of the combined delta of the Ganges and Brahmaputra Rivers.

The Great Indian (or Thar) Desert forms an important southern extension of the Indo-Gangetic Plain. It is mostly in India, but also extends into Pakistan and is mainly an area of gently undulating terrain. Within it are several areas dominated by shifting sand dunes and numerous isolated hills.

The Ghats are mountains on the two sides of the Deccan Plateau. The Western Ghats, also called the Sahyadri, are a north-south chain of mountains that runs along the western edge of the Deccan Plateau. They rise abruptly from the coastal plain. The highest peak in the Western Ghats is Anai Mudi, at 8,842 feet (2,695 meters). The Eastern Ghats are a series of discontinuous low ranges running northeast to southwest along the coast of the Bay of Bengal. The highest peak is Arma Konda, at 5,512 feet (1,680 meters).

The Deccan Plateau or Indian craton is the triangular southern part of India and Sri Lanka lying between the Arabian Sea and the Bay of Bengal. This portion of Earth's crust is geologically quite different from the Himalayan mobile belt to the north. The southern part of India and Sri Lanka was originally joined to Africa as part of Gondwanaland. This land is the oldest and most stable in India. The plateau is mainly between 1,000 and 2,500 feet (305 and 760 meters) above sea level. Its general slope descends toward the east.

The Loess Plateau near Hunyuan in Datong, Shanxi Province.

Peninsular India and Sri Lanka are formed of platform plateaus and tablelands made from an ancient Precambrian crystalline basement rock. This was uplifted in the Mesozoic and Cenozoic periods and has been subject to humid-climate erosion ever since. The oldest Precambrian rocks are of early to middle Precambrian age and are found in the Dharwar belt of southwest India, the Arvalli region between Mumbai and Delhi, and the iron-ore belt of the Singbhum area.

The Dharwar belt is made up of an assemblage of gneisses, migmatites, and granites containing linear greenstone belts in which volcanic and sedimentary rocks, metamorphosed to a greater or lesser degree, are preserved. Some of the sediments contain rich hematite ores and manganese ores. Gold also occurs in association with basic volcanic rocks in the greenstone belt. A mobile belt of late Precambrian to early Paleozoic age (the Indian Ocean belt) crosses the southeast tip of India and runs through Sri Lanka and up the east coast as far as Madras. In Sri Lanka, a complex series of high-grade metamorphic rocks was formed during a sequence of metamorphic episodes, the final one occurring 500 million to 600 million years ago.

The Indochina peninsula consists of the western mountain area and the central and eastern mountains and plains. Although the structures in these mountain ranges are of Cretaceous to Quaternary age similar to those of the structures in the Himalayas, they do not seem to join the Himalayan ranges. The Malayan peninsula contains important granite intrusions from which valuable tin ores are derived.

The western mountain area of Myanmar is a zone of Cenozoic folding. Mountains of medium height are formed of folded blocks that decrease in size and height to the south. Mountains of low and moderate heights that have been slightly broken characterize the central and eastern regions of Thailand and Vietnam. The region has Mesozoic structures surrounding an ancient mass known as the Kontum block. Plateaus and lowlands, which have silt deposited by rivers, are associated with the Kontum block.

The Philippine and Malay archipelagoes that border the southeast margin of Asia are referred to as the island arc. Volcanism and coral reef building have produced these islands. They are bordered by deep ocean trenches and are characteristically unstable and mountainous. The Indian Ocean arcs —Sumatra and the Lesser Sunda Islands—consist of fragments of alpine folds formed from materials of different ages. Cenozoic era volcanoes and modern volcanic activity exist.

Volcanic mountains as well as alluvial lowlands are found in Sumatra and the Lesser Sundas. Borneo and the Malay Peninsula are formed from broken continental land. The mountains are made up of folded and faulted blocks. The lowlands are alluvial. The Pacific Ocean islands arcs, including the Celebes, the Molucca Islands, the Philippine Islands, and Taiwan, are fragments of folded mountain structures that were built up by volcanic action during the Cenozoic era. Volcanic activity and the building of coral reefs continue. Mountain areas of moderate height, volcanoes, alluvial lowlands, and coral reefs may all be found in this region.

West Asia

A series of massive and heavily eroded mountain ranges surrounds Iran's high interior basin. Most of the country is above 1,500 feet (460 meters), with one-sixth of it over 6,500 feet (2,000 meters). In the north, the 400-mile (645-km.)- long strip along the Caspian Sea, only 10 to 70 miles (16 to 110 km.) wide, falls sharply from 10,000-foot (3,000-meter) heights to the lake's edge, 90 feet (27 meters) below sea level. Along the southern coast, the land drops from a 2,000-foot (610-meter) plateau to meet the Persian Gulf and the Gulf of Oman.

The Iranian Highlands are a combination of mountain ranges. There are the Elburz and Turkmen-Khorasan Mountains, the Safid Kuh, and the western Hindu Kush in the north. In the south, there are the Zagros, Makran Soleyman, and Kirthar Mountains. There are also the plains of the interior, and the central Iranian, eastern Iranian, and central Afghanistan mountains. The Zagros range extends from the border with Armenia in the northwest to the Persian Gulf and eastward into Balochistan (Pakistan).

The southern portion of the Zagros broadens into a 125-mile (200-km.)-wide group of parallel ridges located between the plains of Mesopotamia (Iraq) and the great central plateau of Iran. The Elburz Mountains run along the south shore of the Caspian Sea to meet the border ranges of Khorasan to the east. Some of these mountains are active volcanoes; others were formed by volcanic activity but now are dormant. The highest of these peaks is snow-clad Qoliehye Damayand (Mount Demavend), which rises to somewhere between 18,400 and 19,000 feet (5,600 and 5,800 meters).

The interior plateau is arid and extends into Central Asia. It is cut by two smaller mountain ranges. The plateau's most remarkable features are the Dasht-e Kavir (Kavir Desert) and the Dasht-e Lut (Lut Desert). Anatolia (Turkey) is a predominantly mountainous area. True lowland is confined to the coastal areas. About one-fourth of the surface has an elevation greater than 4,000 feet (1,200 meters) and less than two-fifths lies below 1,500 feet (460 meters). Mountain heights exceed 7,500 feet (2,300 meters) in many areas, especially in the east. There, Anatolia's highest mountain, Mount Ararat, reaches 16,853 feet (5,137 meters), near the borders with Armenia and Iran.

Anatolia

The geology of Anatolia is complex. Sedimentary rocks range from the Paleozoic to Quaternary periods, with numerous intrusions and large areas of volcanic material. Four primary regions can be identified: the northern folded zone, the southern folded zone, the central massif, and the Arabian platform. The northern folded zone consists of a series of mountain ridges that increase in elevation toward the east. These ridges, called the Pontic Mountains, occupy an area about 90 to 125 miles (145 to 200 km.) wide immediately south of the Black Sea. The Dogukaradeniz Mountains in the west are the highest portion of the Pontic Mountains, rising to more than 10,000 feet (3,000 meters). The maximum elevation is 12,917 feet (3,937 meters) in the Kackar range.

The southern folded zone occupies the southern third of the country from the Aegean Sea to the Gulf of Iskenderun. From the gulf, it extends to the northeast and east around the northern side of the Arabian platform. Over most of its length, the Mediterranean coastal plain is narrow. The most prominent feature in this zone is the huge Taurus mountain system that runs parallel to the Mediterranean coast, extending along the southern border. The mountains' heights are often above 8,000 feet (2,500 meters), and several peaks are higher than 11,000 feet (3,350 meters).

The central massif is located in the western half of the country between the Pontic and Taurus systems. This elevated area is often referred to as the Anatolian Plateau. Southeastern Turkey between Gaziantep and the Tigris River rests on a stable massif called the Arabian platform. Relatively gentle relief characterizes it, with broad plateau surfaces descending to the south from about 2,500 feet (760 meters) at the mountain foot to 1,000 feet (300 meters) along the Syrian border.

Arabian Peninsula

The Arabian Peninsula is a tilted platform, highest along the Red Sea. The layered plains have undergone erosion due to dry conditions. Ancient marine sands and alluvial deposits from ancient seas take the form of vast, sandy deserts. The peninsula's highest peak, An-Nabi Shu'ayb, at 12,008 feet (3,660 meters) is located approximately 20 miles (32 km.) northwest of Sana, the capital of Yemen. Mesopotamia consists of the Tigris and Euphrates floodplains and of the deltas from Baghdad to the Persian Gulf. The original lowland is covered with late Cenozoic sedimentation. The elevated plain, however, has been cut by erosion during the late Cenozoic era. The regions of Israel, Lebanon, Syria, and Jordan are similar to the Mesopotamian plain, with the exception of the low mountains of Lebanon.

Dana P. McDermott

HYDROLOGY

The world's largest continent, Asia encompasses nearly one-third of the world's land surface. Therefore, it is not surprising that it has a vast, varied hydrological profile. Hydrology is concerned primarily with the cycle by which water from the surface escapes into the atmosphere and then returns to the surface in the form of precipitation. The rate at which water is distributed in a particular region and how it is returned are primarily related to temperature, one of the two factors that determine climate. (The other is precipitation.)

Asia can be divided into four regions by precipitation (and, to a lesser extent, climate) patterns: Monsoon Asia, which includes South Asia, Southeast Asia, and East Asia; Siberia; Central Asia; and Southwest Asia.

The distribution of water resources affects many aspects of life. For example, the environment is a factor in the type of livelihood in which one can engage. Likewise, human abuse of resources can cause serious repercussions to the environment. Asia is home to more than half the world's population, and many of its countries have only recently become industrialized. As a result, water demand and abuse on the continent are more acute in many areas than in most of the rest of the world.

Monsoon Asia

A wide band of the southern and eastern regions of Asia experiences monsoon climate. That consists of a wind reversal, in which winds bringing warm rains from around the equator blow toward the north about three months of the year, usually between June and October. From coastal India through southern coastal China, the effects of the season are felt full force. The farther one goes inland from the Indian or Pacific Ocean in these regions, the less the impact of the monsoons. The effects also are lessened as one goes farther north in the Pacific.

South Asia

The monsoon season in South Asia comes on the heels of successive dry and hot seasons. Two monsoon systems strike the region: A slightly later and cooler system that is formed in the Arabian Sea joins a system that sweeps up the Bay of Bengal. Life during the monsoon season, especially in regions closer to the coast, is dominated by flooding. During monsoon season, Bangladesh routinely has more than 90 inches (2,300 millimeters) of rain, and between 20 percent and 80 percent of the nation's land surface can be covered by water then. Meghalaya, the Indian state on the northern border of Bangladesh, has the world's highest average precipitation, with more than 430 inches (10,900 millimeters) per year. Monsoons impact the lower elevations of landlocked Nepal and Bhutan as well.

South Asia also has several massive rivers that originate in the valleys of the Himalaya and Hindu Kush mountain ranges, the highest such formations on the planet. These rivers are swollen with the midsummer volume of meltwater from the mountains' permanent snowcaps and glaciers. The peak of meltwater usually occurs just before monsoon season. Thus, the prime conduits of water removal for the monsoons are already swollen and the im-

---

MONSOON RAINS

The atmospheric effect of very high summertime temperatures throughout the Asian landmass is a massive zone of low pressure over the continent. This low-pressure zone attracts masses of air from areas of high pressure lying southeast of Asia over ocean waters. The high pressure pushes humid air over Southeast Asia, bringing vital and usually predictable monsoon rains to sustain agriculture throughout the region. The word "monsoon" comes from the Arabic *mawsim*, meaning "seasonal."

MAJOR WATERSHEDS OF ASIA

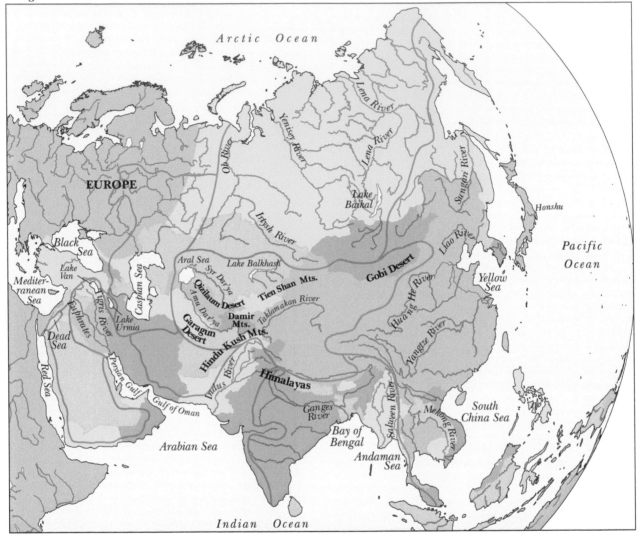

mediate ground areas already saturated. The volume of rainwater is especially problematic closer to the coast, particularly in Bangladesh. It is there that two of the three largest rivers of the region—the Ganges and the Brahmaputra—meet on their way to the Bay of Bengal to discharge their waters. Thus, most of the region's meltwater and monsoon water pass through the rivers of Bangladesh, a delta primarily formed from alluvial (waterborne) soil.

Around monsoon season, Bangladesh and the rest of the Bay of Bengal have violent thunderstorms that spawn lightning and tornadoes. The bay acts as a virtual funnel, bringing typhoons to Bangladesh. The result is usually a disaster, as flooding and violent storms claim the lives of thousands of inhabitants; in Bangladesh, floods of this

magnitude occur every four to five years. Villages and city blocks erode into the rivers. Diseases are spawned by water contaminated by decaying matter and raw sewage. Roads, bridges, and power grids disappear; small boats become the only means of transportation. Life can come to a standstill as a result of the volumes of water.

The benefit of the flooding is that the soil is reborn and the land made fertile by the annual layer of silt deposited. Bangladesh can nearly feed itself, despite its dense population, albeit at a low caloric level. Also, the flooding helps carry the water's contaminants of industrial waste and sewage to the open sea.

Water quality throughout most of South Asia is appalling. It is estimated that in India alone nearly 160,000 children age 0 to 6 die each year from vari-

ASIAN RIVERS MORE THAN 2,000 KILOMETERS (1,242 MILES) IN LENGTH

River	Country(ies)	Length Miles	Length Kilometers
Yangtze	China	3,915	6,300
Huang He	China	3,395	5,464
Mekong	China, Laos, Thailand, Cambodia, Vietnam	3,050	4,909
Lena	Russia	2,668	4,294
Ob'-Irtysh	China, Kazakhstan, Russia	2,640	4,248
Brahmaputra	China, India, Bangladesh	2,391	3,848
Indus	China, Pakistan*	2,243	3,610
Yenisei	Russia	2,167	3,487
Nizhnyaya Tunguska River	Russia	1,857	2,989
Yarlung Tsangpo River	Tibet, India, Bangladesh	1,765	2,840
Amur-Ussuri	China, Russia	1,755	2,824
Salween	China, Myanmar	1,740	2,800
Euphrates	Turkey, Syria, Iraq	1,715	2,760
Viljuj	Russia	1,647	2,650
Ganges	India	1,560	2,510
Amu Darya	China, Afghanistan, Tajikistan, Turkmenistan, Uzbekistan	1,553	2,500
Ishim	Kazakhstan, Russia	1,522	2,450
Ural	Russia, Kazakhstan	1,509	2,428
Aldan	Russia	1,412	2,273
Olenyok River	Russia	1,411	2,270
Syr Darya	Kyrgyzstan, Tajikistan, Uzbekistan, Kazakhstan	1,374	2,212
Irrawaddy	China, Myanmar	1,373	2,210
Xi Jiang	China	1,365	2,197
Kolyma River	Russia	1,323	2,129

Note: *Given the unresolved boundary between Pakistan and India, the Indus may well flow through Northern India as well.

eties of dysentery, resulting from exposure to contaminated water. As recently as 2015, more than 60 percent of India's nearly 1.3 billion inhabitants had no access to sewage treatment systems. There are few effective controls to keep industrial waste out of the waterways. It is ironic that the Hindu faith claims that the Ganges River is purer and holier than any other water on the planet, while worshippers make pilgrimages to walk into the sewage and industrial pollutants visible on its surface.

Southeast Asia

All of lowland Southeast Asia is heavily affected by the rains. Rice, a crop that requires flooded fields to grow, dominates the agricultural production of mainland Southeast Asia. Insular (island) Southeast Asia, on the whole, receives the most precipitation of any region on the planet. Padang, Indonesia, (on the island of Sumatra) normally records 170 inches (4,300 millimeters) of rain during the monsoon season. Parts of coastal Southeast Asia experience heavy rainfall in the monsoon's off-season, as winds sweeping down from coastal China bring moisture while the monsoon winds have reversed.

Two of the main rivers of Southeast Asia originate just north of the Himalayas: the Salween, which empties into the Andaman Sea from Myanmar, and the Mekong, which empties into the South China

Sea near Vietnam's Ho Chi Minh City. Major rivers that originate in lower ranges closer to the region include the Chao Praya, which flows into the Gulf of Thailand just south of Bangkok; the Red, which flows through Hanoi on its way to the Gulf of Tonkin; and the Irrawaddy, which flows through Yangon (Rangoon) on its way to the Andaman Sea. Like their South Asian counterparts, these rivers are flooded with meltwater just before monsoon season. These rivers often have basins or valleys that are irrigated by the river during drier seasons to produce a second growing season for rice.

Most of the region's lakes swell to several times their normal size during the monsoon season. The largest lake in the region, Tonle Sap in Cambodia, covers more than 9,270 square miles (24,000 sq. km.) during monsoon season, compared to its area of about 1,000 square miles (2,600 sq. km.) during the drier seasons.

East Asia

East Asia is affected perhaps even more dramatically by the monsoon than the other regions discussed, despite generally having less rainfall during the season. Southern China regularly has between 40 and 80 inches (roughly 1,000 and 2,000 millimeters) of rainfall, and rice dominates agricultural life. Rice can be double-cropped on the main Japanese island of Hōnshū, from roughly the 37th parallel and below. Rice also can be grown in the northeast-ern provinces of China (often referred to as Manchuria by non-Chinese peoples) and in South Korea, despite having latitudes north of Hōnshū. Clearly, the effects of the rain are felt there.

The interior of East Asia is primarily steppe and desert. North of the Qinling mountains, China has marginal rainfall. The farther one penetrates north and west into China, the less precipitation is found. All of western China is arid, because of its distance from the Pacific and the effect of the Himalaya and Hindu Kush ranges to the south.

Many large rivers originate in the mountain ranges of western China. The Yangtze, Yellow (Huang He), Mekong, Indus, Brahmaputra, Salween, Irtysh, Xi Jiang, and Tarim rivers all begin in western China. The Yangtze, Huang He, and Xi Jiang empty into the Pacific in China.

In China, the demands on the rivers have grown with industrialization and its accompanying urbanization. The building of the Three Gorges Dam and other dams around the country to generate hydro-electricity for industry has greatly lessened the volume of the flow of rivers. Many regions of China suffer from acute water shortages as the government diverts water from rural areas to cities.

Siberia

This area is found in the Asiatic portion of Russia. Precipitation is limited throughout the region, averaging between 0.6 and 0.8 inch (15 and 20 milli-

ASIA'S TEN LARGEST LAKES

| | | Area | |
Lake	Country(ies)	Square Miles	Square Kilometers
Lake Baikal	Russia	12,248	31,722
Lake Balkhash	Kazakhstan	6,562	16,996
Lake Issyk-Kul	Kyrgyzstan	2,408	6,236
Bratskoye Reservoir	Russia	2,115	5,478
Lake Urmia	Iran	2,008	5,200
Lake Taymyr	Cambodia	1,761	4,560
Poyang Hu	China	1,699	4,400
Lake Khanka	China, Russia	1,618	4,190
Quinghai Hu	China	1,616	4,185
Dongting Hu	China	1,089	2,820

April 18, 2014: The Hanhowuz (Khauzkhan) Reservoir is a splash of turquoise amidst desert browns. The reservoir was constructed in a natural depression to capture winter runoff and overflow from the canal for use later during the driest periods of summer. Phytoplankton thrive in the warm waters, as do many commercial fish—including Aral barbel, asp, and catfish.

meters) per year. Little precipitation comes inland from the Arctic Ocean. Likewise, the Pacific Coastal Range is high enough to produce a rain-shadow effect, in which rainwater is dumped on the windward side of this range, facing the coast. The opposite face of the mountains, and thus the interior of the land, receives virtually no rainfall. The distance of this region from the Indian Ocean would allow for little moisture, but even that tiny amount is blocked by the Himalaya, Hindu Kush, and Caucasus ranges to the south of Siberia.

The northern slopes of the ranges in Siberia are drained by several enormous rivers, forty of which are in excess of 600 miles (965 km.) in length. The Ob'-Irtysh, Lena, and Yenisey rivers are among the world's longest, each more than 2,500 miles (4,000 km.) long. These rivers are frozen for months each year. Spring and summer thaws send meltwater to the southern reaches of the river, while the northern part of the river is still frozen. When the water leaves the banks of the channel and the ground is frozen, the soil absorbs little of it. The Western Siberian Lowland along the Ob forms the world's largest swamp, the Vasyugan, because there is nowhere for the water to drain. Ten percent of Russia's land surface is swampland, all found in northern regions with similar hydrological challenges.

Lake Baikal contains nearly 20 percent of the world's lake fresh water and is the world's deepest lake, with a maximum depth of 5,315 feet (1,620 meters). The vast human-engineered reservoir, the Bratskoye Reservoir on the Angara River, is also located in this region.

Water quality in Siberia is dismal in populated industrial areas. A mere 18 percent of Siberian homes receive water of good quality. Much of the water in the region is polluted, with perhaps 50 percent absolutely unpotable. Chemical fertilizers have further damaged the water supply in less densely populated areas. Poorly designed water runoff systems have led to massive soil erosion in some areas. Dams on Siberian rivers have had an adverse effect on local forests, raising the local temperatures because water cools more slowly than land. This damaged environment produces acid rain, perpetuating the cycle of pollution, causing great environmental damage. Given that Siberian forests are the second-largest land-based source of oxygen on the planet (second only to the Amazon), degradation of the Siberian forest by this damaged hydrologic system could have an enormous impact on world climate.

Central Asia

A large region of Central Asia is steppe and desert. That band extends about 2,000 miles (3,200 km.) north and south, from western China through Mongolia, and about the same distance east to west, from central Kazakhstan to western China. The area is divided into eastern and western segments by the Tien Shan and Pamir mountain ranges.

This area lacks precipitation for several reasons. Its inland location prevents moisture from penetrating from the Pacific Ocean on the east. Little moisture is generated by the Arctic Ocean, with the northernmost reaches of the area receiving perhaps 0.8 inch (20 millimeters) of rainfall per year. Finally, the Himalayas and Hindu Kush, the highest mountains on Earth, block monsoon rains from the Indian Ocean from reaching the area. No area in the region (other than those adjacent to the highest ranges) exceeds 12 inches (300 millimeters) of rain per year. Snowmelt is a major water source during the summer months. The area primarily consists of alpine, desert, and steppe climates. Most of Turkmenistan lacks continuous surface-water flow.

The primary rivers flowing through the region that drain these ranges are the Amu-Darya and the Syr Darya. During the Soviet period, poorly

---

SYR AND AMU DARYA RIVERS

These two navigable Central Asian Rivers are unlike most others in Asia in that they drain into neither an ocean nor a sea with an outlet to the ocean, but empty instead into the landlocked Aral Sea. Most of the thirty-nine million people in Central Asia live near one of these rivers and use them to travel and move goods within the region. The 1,553-mile-long (2,500-kilometer-long) Amu Darya cuts through two deserts and forms a narrow strip of fertile valley in an otherwise uninhabitable land, much like the Nile River in Egypt. The 11,374-mile-long (2,212-kilometer-long) Syr Darya drains the large, fertile Fergana intermountain oasis where most other Central Asians live.

---

planned irrigation projects along these rivers led to the desiccation of the Aral Sea. Today, much of the South Aral Sea's total volume has disappeared. This has had disastrous results, as salt and dust storms formed from the former sea floor have seriously damaged local soils. Some estimates hold that residue from this process has raised the level of particulate matter in the atmosphere by 5 percent, affecting world climate adversely.

In addition, heavy use of fertilizers throughout the former Soviet republics has led to serious pollution of the rivers and underground water supplies. Industrial pollution and oil seepage from poorly constructed systems have exacerbated the problem. Pollution in an ecosystem as marginal as exists in much of Central Asia increases the rate of desertification. The Garagum and Qizilqum deserts of Turkmenistan are growing at a rate surpassed only by the Sahara. Furthermore, the deserts and steppe of Kazakhstan were used as dumping regions for radioactive materials by the former Soviet Union. The northeastern region of the country was also used by the Soviets for aboveground nuclear arms testing.

The high mountain terrain preserves some glaciers in the region. The world's longest glacier outside polar regions, the Fedchenko, is found in Tajikistan. Kyrgyzstan alone has more than 8,000 glaciers. Lakes are scattered throughout the region and generally have been polluted by overuse of fertilizers and pesticides.

Southwest Asia

This area consists of the region extending from Turkey in the west to Afghanistan in the east, and from Turkey in the north to the Arabian Peninsula in the south. Most of the land is steppe or desert. There are several exceptions: The Mediterranean coast can have between 4 and 50 inches (100 and 1,270 millimeters) of rain during the year, most of which falls between November and May. A belt north of the 37th parallel in Iraq receives 1 to 4 inches (25 to 100 millimeters) per year. Rainfall in excess of 4 inches (100 millimeters) occurs along the Caspian Sea coast in Iran. Black Sea coastal regions routinely have more than 40 inches (1,000 millimeters). Rain-fed agriculture occurs in each of these areas.

The rest of the region is essentially desert. Irrigation agriculture has been practiced along the Tigris and Euphrates rivers in Syria and Iraq for at least 6,000 years. Aside from the tributaries of these two rivers (most notably the Khabur, Great Zab, and Diyala), there are few sources for irrigation. The Orontes in Syria and the Jordan in Jordan and Israel are the two main exceptions. Salinization from poor drainage practices has doomed a number of traditional agricultural areas, such as in southern Iraq.

The region has many seasonal rivers, known as wadis. These channels flow during wet seasons, then disappear during the dry seasons. Even rivers that more or less flow perennially may disappear into the desert, forming underground rivers. The Barada River, the main source of water for the oasis that surrounds Damascus, is a prime example of this phenomenon.

A major source for water in the Arabian Peninsula is found in coastal desalinization plants. Kuwait derives virtually 100 percent of its water via desalinization, and in nearby Qatar and Bahrain, more than half of the water comes from this source. Saudi Arabia also produces water by this process, almost exclusively for agricultural usage. A deep and difficult-to-access aquifer under the peninsula, the Waisa, contains more water by volume

Mekong Delta.

than the Persian Gulf. In 2015, its rapid depletion of water for agricultural use led the Saudi government to restrict further drawing from it.

There are comparatively few large lakes in the region. The two largest are Lake Van in Turkey and Lake Urmia in Iran. The latter is too salty to support life.

The scant water resources in the area are a source of both international cooperation and conflict. Dams erected in Turkey and Syria have reduced the flow of water in the Euphrates dramatically downriver in Iraq (the Tigris and Euphrates systems account for over 90 percent of Iraq's water). Access to the waters of the Jordan presents a continual opportunity for bickering between Israel and Jordan, but also for cooperation in implementing new technology. Access to the water of the Sea of Galilee has been a sticking point in the return of the Golan Heights to Syria by Israel.

Mark Anthony Phelps

CLIMATOLOGY

In the first decades of the twenty-first century, the world is in a period of marked climatic change. Nowhere is the impact of climate change manifested more vividly than in Asia. Covering one-third of Earth's land surface and containing almost every known climate type, Asia is the largest landmass on Earth, and the most physically and climatically diverse continent. The highest point above sea level is found in Asia, as well as Earth's coldest inhabited place and its wettest site. Few common denominators unite the climates of this vast landmass.

Variety of Climates

Climates in Asia, subject to land influences as opposed to maritime influences, are noted for great seasonal variations in temperature and moisture. Climates differ greatly from region to region because of variations in the amount, intensity, and spatial distribution of solar energy, temperature, precipitation, atmospheric pressure and winds, and storms. If there is a unity in Asia's climatic diversity, it is provided by the monsoon effect and the impact of the Siberian high-pressure cell.

Asia's latitudes extend from well above the Arctic Circle to near the equator. Differences in heating and cooling between high and low latitudes and between snow-covered and vegetation-covered surfaces produce regional atmospheric pressure contrasts, causing wind. Seasonal land and sea winds generated by atmospheric pressure reversals produce the monsoon effect. Vast segments of Asia are thousands of miles from the ameliorating effects of warm ocean currents and from moist rain-bearing air masses. Other areas face the frozen Arctic or are located where mountain barriers inhibit advection of moisture.

Monsoon clouds in Madhya Pradesh.

CLIMATE REGIONS OF ASIA

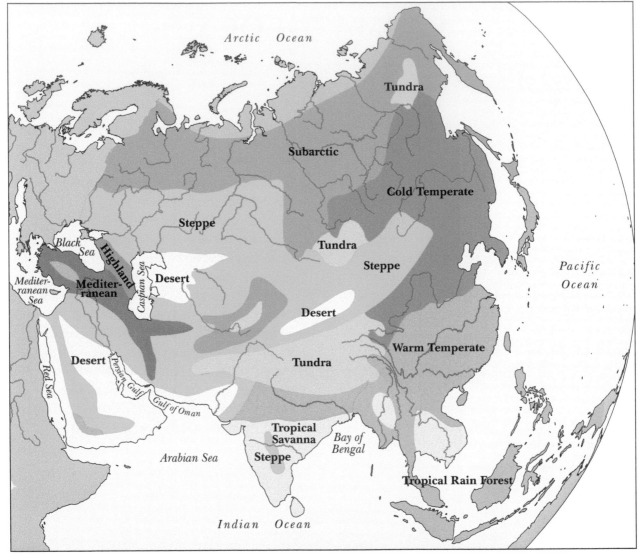

Using the Koeppen climatic classification, Asia can be divided into three major climate realms: Boreal Asia in the north, Desert Asia in the west and center, and Monsoon Asia in the south and east. Boreal Asia is the largest. A complex set of interactive climate controls and varied physiogeographic features influence local climates and give distinctive regional climatic character to places.

Climatic Controls

The unequal distribution of solar radiation over Asia is the primary factor in Asia's multifaceted weather and climate. The annual march of solar radiation is determined by the angle at which the Sun's rays strike the surface. In Asia's tropical belt,

the Sun remains high, with little seasonal variation, which accounts for continuous warm to hot year round temperatures. In Asia's midlatitudes, there are strong seasonal maximums and minimums in the amount of solar radiation received, which are reflected in greater seasonal variations in temperature than in the tropical belt. In Asia's high latitudes, there is a period in which limited or no solar radiation is received at the surface, resulting in a season with extremely low temperatures in the winter (or low-Sun) period. The greatest amount of solar radiation is received in the steppe and desert regions of southwest and west central Asia. Transformation of available solar radiation is an essential ingredient of the process that produces Asia's climate-particularly temperature ranges.

Moisture

Air masses from maritime tropical oceans and seas are the primary source of Asia's moisture. Asia is bordered on three sides by oceans: the Indian on the south, the Pacific on the east, and the Arctic on the north, along with the Red, Mediterranean, Black, and Caspian seas on the west. The rate of evaporation from water and land surfaces is dependent upon air temperature, dry air masses, and wind and surface conditions; therefore, the potential for evaporation declines from tropical latitudes toward polar latitudes. Average annual evaporation potential ranges from about 55 inches (1,400 millimeters) at 20° north latitude to 8 inches (200 millimeters) at 60° north latitude.

Specific humidity—usually measured in grams per kilogram—varies from 14 at 20° north latitude to 5 at 60° north latitude. Large amounts of evaporated water in the atmosphere and large amounts of latent and sensible heat along the east coast of Asia make this segment of Earth's surface an important breeding ground for tropical disturbances and storms. Moisture is an important climate-forming factor associated with precipitation, evaporation, cloudiness, fog, humidity, and continentality.

Basic Atmospheric Circulation

The basic atmospheric circulation features that give regional character to Asia's climates are the movement of air masses, transformation of air mass properties, and interactions between them along fronts. Asian air masses vary in temperature, moisture, and density. At any particular site, properties of air masses depend not only upon the nature of the source region but also on the modification the air mass experienced en route from the source region. These modifications are important in determining the nature of weather associated with an air mass.

Siberian High-pressure Cell

General atmospheric circulation over Asia is controlled by centers, or cells, of high or low pressure whose axes generally are east-to-west and whose

pressure centers vary dramatically from winter to summer. In winter, an extensive and well-developed high-pressure cell, centered over Mongolia, dominates the weather over most of East and South Asia. Triangular in shape, with the apex extending in the west to the Caspian Sea and the base anchored in the northeast near the Verkhoyansk Mountains of Siberia and in the southeast near the Chin Ling Mountains of east central China, the Siberian high effectively blocks penetration of moisture-bearing, moderating maritime air masses in winter.

Acting as a wedge forcing air masses to skirt northeasterly from the Black Sea across much of northern Siberia and to flow southerly along the Kamchatka Peninsula toward northern Japan, this intense high generates continental, land-trajectory, dry and cool, low-level air masses that surge from the north and northeast to the southwest across South and East Asia. The Siberian high, whose core and area of highest pressure is focused upon Lake Baikal, pulsates in intensity and breaks at times into smaller, less intense high-pressure cells.

February marks the height of the Siberian high's dominance of the winter circulation over Asia. The Siberian high weakens and shifts its center westward into a position over northeastern Central Asia in April, then dissipates in May. The weather map of Asia then begins to be dominated by an intensive, thermally induced low-pressure cell over southwestern Asia and focused upon the tip of the Arabian Peninsula, the Iranian Plateau, and the Thar Desert of Pakistan.

Southwest Asian Low

In summer, South Asia's weather is dominated by a large, deep, thermally induced low-pressure cell that extends from the Arabian Peninsula to central China and from central India to Central Asia. A complex cell, the Southwest Asian low experiences east-to-west locational oscillations and occasional intense pressure deepening. This low in summer interrupts the subtropical high-pressure system in the Northern Hemisphere by dividing the globe-girdling zonal band into two distinct large oceanic cells. One intense depression, the Southwest Asian low, causes a radical change in prevailing winds and storm tracts during the high-Sun period. Air masses from the stable eastern end of the Azores-Bermuda high-pressure cell skirt this low from a north to northwesterly direction across the eastern rim of the Arabian Peninsula; less intense air masses from the northern quadrant of the Azores-Bermuda high sweep eastward across Turkey and into the northern extremities of Central Asia.

Air masses and storms spawned under the western unstable quadrant of high pressure stationed over the Pacific Ocean sweep from the south and southeast in a northerly trajectory across Japan, extreme eastern China, and the Russian Maritime Provinces. The most constant and climatically significant air masses and storms are advected to this intense low-pressure cell in the form of southwesterly trade winds spawned from a semipermanent high-pressure cell in the Southern Hemisphere over the Indian Ocean. In India and East Asia, this modification of the general planetary wind system in summer constitutes the Asian monsoon.

Frontal Dynamics

Frontal zones and wind systems conform to the location and circulation of major air masses. In winter, four major zones of cyclonic activity are distinguished over Asia. One zone is located along the Asiatic Arctic front, well above the Arctic Circle in northern Siberia, extending along the shores of the continent. This front fluctuates greatly and Arctic air masses, at times, penetrate to the Black Sea, Southwest Asia, and east central Asia. The second zone is along the Southwest Asian polar front, which develops in winter over the Mediterranean Sea and extends to the Caspian Sea. The third zone is along the East Asian polar front, which aligns itself in a northeasterly path from extreme South Asia toward Japan. The final zone of cyclonic activity in winter is the east-west-aligned South Asian intertropical front located near 10° south latitude and traversing Java. Along these winter frontal zones and moving at various directions and speeds, depressions and anticyclones impart to the climates of Asia that special character by which one area differs from another.

During summer, three major zones of pronounced cyclonic activity can be identified. The first is the southward-displaced Asiatic Arctic front which at times extends east-west across northwestern Siberia along the 70° north parallel.

A second zone is the Asian polar front, which normally extends from the eastern tip of Lake Balkhash in Central Asia, over Mongolia, to the northernmost bend of the Amur River along the 50° north latitude. This oscillating and at times southward-dipping zone has been called the "barometric backbone of Asia" in summer or the "great ridge," because a large number of anticyclones are observed each year between 50° and 55° north latitude.

The third major zone is along the South Asian intertropical front. Extending from the south-central portion of the Arabian Peninsula, across Pakistan and northern India, and over China at approximately 25° north latitude, the South Asian intertropical front is well defined in some locations, but weak or absent in others. There, convergence of surface winds causes large-scale lifting of warm, humid, relatively unstable air, producing numerous weak, rain-generating disturbances.

Along major frontal zones are found pronounced horizontal variations in temperature, humidity, and stability and, because frontal zones are areas of steep horizontal temperature gradients, there are usually jet streams. Strong, narrow jet-stream currents, thousands of miles long, hundreds of miles wide, and several miles deep, are concentrated along a nearly horizontal axis in the upper troposphere or stratosphere, producing strong vertical and lateral shearing action.

Three distinct jet-stream systems—the polar-front jet stream, the subtropical jet stream, and the tropical easterly jet stream—have a major impact upon weather and climate in Asia. Latitudinal location of these jet systems, especially the polar jet, shifts considerably from day to day and from season to season, often following a meandering course. In general, however, the polar-front jet gives rise to storms and cyclones in the middle latitudes of Asia. The subtropical jet, noted for a predominant subsidence motion, gives rise to fair weather; and the tropical easterly jet is closely associated with the Indian monsoon.

Over most of Asia, the polar front jet and the subtropical jet are most intense on the eastern margin of the continent and are best developed during winter and early spring. Jet-stream wind speeds of up to 300 miles (480 km.) per hour have been encountered over Japan. At times in winter, the subtropical jet forms three planetary waves with ridges over eastern China and Japan. It frequently merges with the polar-front jet stream, producing excessively strong jet-stream wind speeds. During summer, the subtropical jet stream loses its intensity and appears only occasionally.

The Asian Monsoon

The Asian monsoon is a gigantic, complex weather system, composed of diverse heat and moisture cycles intimately related to local topography, modified air masses, quasi-stationary troughs and ridges, and jet streams. Weather over most of eastern and southern Asia during the winter is dominated by the

Siberian high-pressure cell and outflowing continental air masses. Low-level air flow, mainly from the north, is cold, dry, and stable. Successive waves of cold northeasterlies begin in late September and early October, progressing farther and farther south, reaching the south China coast by late November or early December. In the higher midlatitudes of Asia during the low-Sun period, a steep north-to-south pressure gradient persists.

Southwest Asia and the Indian subcontinent are protected from the cold Siberian air by blocking mountain ranges and highlands, largely because cold fronts extend from only 6,500 to 10,000 feet (2,000 to 3,000 meters) or so above sea level. During March and April, the Siberian high gradually weakens, and incursions of moist maritime tropical air from the south and east replace the cold-to-cool

The taiga in the river valley near Verkhoyansk, Russia, at 67°N, experiences the coldest winter temperatures in the northern hemisphere, but the extreme continentality of the climate gives an average daily high of 72° Fahrenheit (22° Celsius) in July.

City of Yawnghwe in the Inle Lake, Myanmar.

northeasterlies, producing widespread stratus clouds, fog, and drizzle that can persist for days.

In summer, a combination of the deep, elongated Southwest Asian low-pressure cell extending from the Arabian Peninsula to China, and the South Asian, reaching its maximum poleward displacement, sets the stage for a marked seasonal reversal of air flow. Warm, moist, and somewhat unstable southwesterly maritime air masses from relatively cooler oceanic source regions eventually overcome blocking atmospheric conditions. They then stream across India and flow northwestward over continental South Asia, China, and Japan. Considerable convective activity develops overland, and heavy showers and thunderstorms contribute significantly to the summer rainfall maximum of the region.

Characteristics, attributes, onset, and duration of the monsoon are site-specific. In all cases, low-level wind patterns and resultant weather in summer are complicated by topography. Distance from air mass source regions, moisture content of air masses, mountain barriers, and atmospheric

disturbances associated with cyclonic or inter-air mass convective activity determine the distribution and quantity of precipitation. The East Asian winter monsoon is much stronger than the South Asian winter monsoon, and the East Asian summer monsoon is much weaker than the South Asian summer monsoon.

Temperature

Northern Asia, specifically Siberia, is climatically isolated from moist tropical air masses. This region records the greatest mean annual temperature range on Earth. At Verkhoyansk in the valley of the Yana River, January temperatures average -56° Fahrenheit (-49° Celsius). July temperatures average +59° Fahrenheit (15° Celsius). In most of Siberia, the January mean annual temperature is below -13° Fahrenheit (-25° Celsius).

Much of southwestern Asia is subject to excessive heat. Summer temperatures in the interior lowlands of the Arabian Peninsula and Iran are so extreme that the region ranks among the hottest areas of the world. Mean daily temperatures in the Arabian Desert approach 96° Fahrenheit (36° Celsius) in July; in the Iranian valleys of Khuzestan and Luristan, daily maximums often exceed 111° Fahrenheit (44° Celsius), with temperatures in excess of 122° Fahrenheit (50° Celsius) reported for Abadan. However, Singapore, located about 90 miles (150 km.) north of the equator, has an annual mean temperature of 81.6° Fahrenheit (28° Celsius), with January averaging 78.8° Fahrenheit (26° Celsius) and July, 82.4° Fahrenheit (28° Celsius); there is little temperature change from one month to another, and the annual range is only 3.2 Fahrenheit degrees (2 Celsius degrees).

Asia's spatial temperature differences and ranges are greatest in winter, least in summer, and more extreme in all seasons within the landlocked core, rather than along the south and eastern maritime periphery. Winter isotherms reflect the influence of modifying ocean currents, mountain barriers, altitude above sea level, high-pressure cells, and solar radiation. Summer isotherms reflect solar radiation and latitude, along with altitude.

Precipitation

In general, the amounts of rain received increase from north to south and from the southwest to the southeast. In most of Siberia, annual average precipitation is scarcely 10 inches (250 millimeters); in Central Asia, the Tarim Basin, Gobi Desert, Plateau of Iran, and the Arabian Peninsula, scarcely 4 inches (100 millimeters). Some of the driest areas are in the southwest, with Eilat, Israel, averaging only about 1 inch (26 millimeters) of annual precipitation and Tabriz, Iran, only 0.3 inch (8 millimeters). On the other hand, the northeastern sector of the Indian subcontinent is considered by many scholars to be the wettest region on Earth. Here, the South Asian summer monsoon produces a veritable deluge. At Cherrapunji, the mean annual precipitation is 450 inches (1,143 centimeters), with a twenty-four-hour maximum of 38 inches (965 millimeters). Most of the precipitation occurs from May to September, and the monthly average is 0.2 inch (5 millimeters) in December and 115 inches (292 centimeters) in June.

Seasonal distribution of precipitation is important because the summer heat and continental character of Asia affect precipitation effectiveness and thermal efficiency. In the insular, island-studded South, Southeast, and East Asia, where there is limited or no frost, rainfall is relatively evenly distributed throughout the year. The interior and rain-shadow locations in South and East Asia experience a distinct dry season in the low-Sun period, or winter. Mediterranean Sea-influenced West Asia and the coastal regions of the Black and Caspian seas receive most of their precipitation in winter, at least three times as much precipitation in the wettest winter month versus the driest month of summer.

Precipitation in the dry realm of Southwest Asia, Central Asia, the Tarim Basin, and the Gobi Desert is minimal, erratic, and uncertain, but most of the precipitation comes in summer from violent convective showers or a combination of violent convective showers and frontal activity. In the subarctic and Arctic areas of Siberia, summer is the season of maximum temperatures, highest specific humidity, deepest penetration of maritime air masses under the influence of the summer monsoon, and maximum precipitation. Asia's central dry realm, aligned in a southwest-northeast orientation from the Red Sea coast of the Arabian Peninsula to the Plateau of Mongolia, separates Asia's southern and eastern warm-to-hot wet belt from Asia's north and northeastern cool-to-cold limited-precipitation belt.

Storms

Wind conveys heat, moisture, and lightweight materials from one location to another and, in part, determines the motions of minor atmospheric disturbances and cyclonic storms. Owing to the great variety in relief and exposure, Asia is subject to numerous local winds and their associated weather. Ubiquitous in Asia are the warm, dry, gusty, downslope foehn winds, usually generated by passing atmospheric disturbances in highland or mountainous areas.

Winds are given a medley of names: in Central Asia and Siberia, foehn-like winds experienced along the Caspian Sea are locally called *germich*; in Uzbekistan, *Afghanets*; in Tajikistan, *harmsil*; in and near the great Fergana Valley, *ursatevskiy* and *kastek*; along the Kazakh-Chinese border at the Dzhungarian Gate east of Lake Balkhash, *evgey*; in Iran, *samoon*; and on the islands of Indonesia, *bohorok* and *kumbang*. The largest and most pronounced winds are a major contributing factor to the broad aspects of Asia's climate.

Thunderstorms and Tornadoes

Thunderstorms reach their maximum development in Asia over lowlands within the equatorial region. Each year, areas of peninsular and insular (island) South Asia experience 180 days or more of such storms. Most thunderstorms are ordinary convective cumulonimbus clouds within maritime tropical air masses that produce localized precipitation. Air mass thunderstorms, randomly scattered, are primarily initiated by daytime solar heating of land surfaces; frontal and orographic (mountain-induced) thunderstorms have distinct patterns and movements, for they are triggered in a place or zone where unstable air is forced upward. Severe thunderstorms may produce hail, strong surface winds, and tornadoes. Convective thunderstorms develop most frequently because of strong insolation and low wind velocities. Clouds of convective thunderstorms reach elevations of about 40,000 feet (12,200 meters) or more, and rainfall associated with these cells is short in duration, intense, and localized.

Fewer than five thunderstorms per year occur over the tundra regions of Siberia, the hill and desert regions of Jordan, the southeastern quadrant of the Arabian Peninsula, southern Iran, Afghanistan, and Pakistan. Thunderstorm activity is a summer phenomenon in China, much of Central Asia, Singkiang, Mongolia, the Maritime Provinces of the Far East, and Siberia. Approximately twenty thunderstorms per year occur over Lebanon, Jordan, Israel, and Kuwait. They begin during the fall transitional period, are more prevalent in winter, and decline in number during the spring transitional period.

Spring is the season of maximum thunderstorm activity over most of the Indian subcontinent and the Arabian Peninsula. The most destructive spin-off of a thunderstorm is the tornado, an extremely violent rotating column of air that descends from a thunderstorm's cloud base and can cause great destruction along a narrow track. An Asian tornado travels in a straight track about 500 to 1,600 feet (150 to 500 meters) wide and several miles long, at speeds between 30 and 60 miles (50 and 100 km.) per hour. The tornado is one of the least common Asian storms, but it is the most violent. As the whirling mass of unstable air gains force, a rotating column of white condensation is formed at the base of the cloud. Dirt and debris sucked into this whirlwind darken the column of air as it reaches the ground.

Asia has relatively few tornadoes compared to central and eastern North America at similar latitudes: tornadoes occur about every three to five years in the northern Caspian Sea area and Central Asia during May, June, and July; once or twice each year in the northeastern part of the Indian subcontinent during March, April, and May; two to five or more annually in China and Japan during August and September; and at least one every four years in the Philippines, Indonesia, and Malaysia. And tornadoes can occur just about anywhere. In April 2019, a deadly tornado tore through the Bara District of Nepal. It was believed to be the first time that a tornado hit Nepal.

Tropical Cyclones

One of the most powerful and destructive types of cyclonic storms is the tropical cyclone. Referring loosely to any pressure depression originating above warm oceans in tropical regions, tropical cyclones are an important feature of the weather and climate of South and East Asia, particularly from July to October. Different names are used worldwide to describe this tropical storm: *typhoon* in the western Pacific, *baquios* or *baruio* in the Philippines, *tropical cyclone* in the Indian Ocean, *willey-willeys* in Australia, *hurricane* in the eastern Pacific and Atlantic, *cordanazo* in Mexico, and *taino* in Haiti.

Although the organization and development of tropical cyclones are not fully understood and are under intensive study, their formation is associated with warm ocean surfaces not less than 81° Fahren-

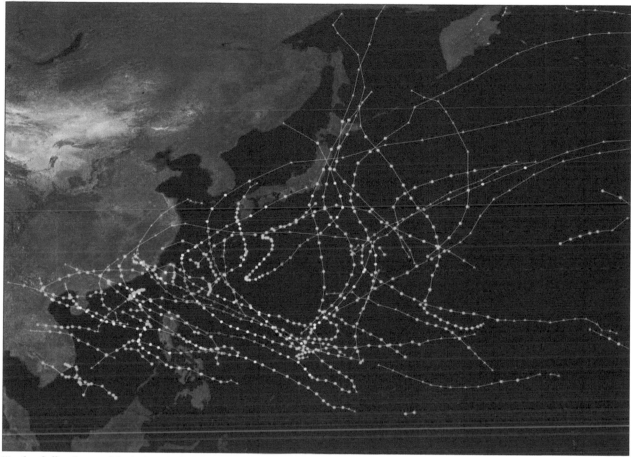

Tracks of all typhoons of the 2016 Pacific season.

heit (27° Celsius), located between 5° and 10° north latitude, with light to calm initial winds, and waves or troughs of low pressure deeply embedded in easterly wind streams converging into an unstable atmospheric zone. Large quantities of latent heat released through condensation converge and are transferred to higher levels, deepening the pressure center and intensifying the storm. An almost circular storm forms, with a center, or eye, of extremely low pressure. Winds spiral into the eye at great speeds.

Asian tropical cyclones travel slowly, at speeds of 10 to 30 miles (16 to 48 km.) per hour, cutting a destructive storm path 50 to 100 miles (80 to 160 km.) wide; winds in the wall cloud area can achieve speeds in excess of 120 miles (200 km.) per hour. Passage of a tropical cyclone over water and land is associated with strong winds and heavy rainfall. Storm tracks vary annually, and no two recorded tracks have been exactly the same.

Despite the irregularities, most tropical cyclones tend to move westward then poleward, finally turning eastward, toward higher latitudes under the influence of both internal circulation and external steering currents, penetrating into the belt of westerly winds. Tropical storms contribute between 25 and 50 percent of the annual precipitation received in many tropical weather stations. Flooding, destructive winds, and storm surges are responsible for much property damage and for human casualties.

Tropical cyclones occur in specific seasons, depending upon the geographical location of the storm-affected region. In Hong Kong, 83 percent of the annual recorded tropical cyclones have occurred between June and November. Approximately 50 percent of these severe tropical storms attain typhoon intensity. Storm frequency in the Bay of Bengal varies widely throughout the year, although the trend has been for the most severe cyclonic storms to occur in May and in October or November. An average of six to eight tropical cyclones is recorded there each year.

Frequency of tropical depressions and storms in the Arabian Sea is much less than in the Bay of Bengal. Nearly every summer, some part of East Asia is affected by severe tropical storms or tropical cyclones. From mid-November to April, few tropical storms pass over the coasts of China and Korea, but in the warm period of July through October, numerous tropical depressions, tropical storms, severe tropical storms, and tropical cyclones (typhoons) are experienced. From 1959 to 2015, 1,518 tropical cyclones formed in the northwest Pacific Ocean, with many affecting China, Korea, or Japan. Almost all occurred in the summer and fall seasons. Tropical cyclones usually weaken over land, and few penetrate and persist more than 300 miles (500 km.) inland. The rise in sea level, when combined with high tides, accounts for more damage and loss of life in Asia than the violent winds.

Artificial, local, microclimatic changes and long-term alteration of Asia's present macroclimates could plunge this part of the world into chaos. For the world, 2016 was the warmest year on record. For Asia, 2019 was the third-warmest year on record, according to the U.S. National Oceanic and Atmospheric Administration (NOAA). Only 2015 and 2017 were warmer. Asia's five warmest years have all occurred since 2007. For the period 1981 to 2019, the mean temperatures have increased by 0.63 Fahrenheit degrees (0.35 Celsius degrees) per decade, twice the previous amount. An annual change in temperature of only a few degrees could alter the climates of Asia sufficiently to render marginal agricultural regions unacceptable for food production and to wreak havoc on local food supplies. Human activities that modify climatic elements, combined with natural causes of climatic change, could lead to more frequent or more severe changes in Asia's climate regions. For Asians struggling to secure a meager existence from small plots of land, and urban dwellers seeking water for daily life, any climate change disrupts lifestyles—for humans and human institutions are adjusted to precisely the climate and weather that prevail.

William A. Dando

DISCUSSION QUESTIONS: CLIMATOLOGY

Q1. How have residents of the Middle East adapted to its extreme arid climates? What are the main sources of water in the region? How do crops survive?

Q2. How has the monsoon shaped the lives of people in Bangladesh and India? Do the benefits of the monsoon outweigh the drawbacks? Why or why not? Why do monsoons occur? How do the Asian monsoon and the South Asian monsoon differ?

Q3. Why is it so cold in Siberia? What influence does the Arctic Ocean have on Siberia's climate? Compare the Siberian climate to that of adjacent parts of East Asia. Why are the climates so different?

BIOGEOGRAPHY AND NATURAL RESOURCES

Overview

Minerals

Mineral resources make up all the nonliving matter found in the earth, its atmosphere, and its waters that are useful to humankind. The great ages of history are classified by the resources that were exploited. First came the Stone Age, when flint was used to make tools and weapons. The Bronze Age followed; it was a time when metals such as copper and tin began to be extracted and used. Finally came the Iron Age, the time of steel and other ferrous alloys that required higher temperatures and more sophisticated metallurgy.

Metals, however, are not the whole story—economic progress also requires fossil fuels such as coal, oil, natural gas, tar sands, or oil shale as energy sources. Beyond metals and fuels, there are a host of mineral resources that make modern life possible: building stone, salt, atmospheric gases (oxygen, nitrogen), fertilizer minerals (phosphates, nitrates, and potash), sulfur, quartz, clay, asbestos, and diamonds are some examples.

Mining and Prospecting

Exploitation of mineral resources begins with the discovery and recognition of the value of the deposits. To be economically viable, the mineral must be salable at a price greater than the cost of its extraction, and great care is taken to determine the probable size of a deposit and the labor involved in isolating it before operations begin. Iron, aluminum, copper, lead, and zinc occur as mineral ores that are mined, then subjected to chemical processes to separate the metal from the other elements (usually oxygen or sulfur) that are bonded to the metal in the ore.

Some deposits of gold or platinum are found in elemental (native) form as nuggets or powder and may be isolated by alluvial mining—using running water to wash away low-density impurities, leaving the dense metal behind. Most metal ores, however, are obtained only after extensive digging and blasting and the use of large-scale earthmoving equipment. Surface mining or strip mining is far simpler and safer than underground mining.

Safety and Environmental Considerations

Underground mines can extend as far as a mile into the earth and are subject to cave-ins, water leakage, and dangerous gases that can explode or suffocate miners. Safety is an overriding issue in deep mines, and there is legislation in many countries designed to regulate mine safety and to enforce practices that reduce hazards to the miners from breathing dust or gases.

In the past, mining often was conducted without regard to the effects on the environment. In economically advanced countries such as the United States, this is now seen as unacceptable. Mines are expected to be filled in, not just abandoned after they are worked out, and care must be taken that rivers and streams are not contaminated with mine wastes.

Iron, Steel, and Coal

Iron ore and coal are essential for the manufacture of steel, the most important structural metal. Both raw materials occur in many geographic regions. Before the mid-nineteenth century, iron was smelted in the eastern United States—New Jersey, New York, and Massachusetts—but then huge hematite deposits were discovered near Duluth, Minnesota, on Lake Superior. The ore traveled by ship to steel mills in northwest Indiana and northeast Illinois, and coal came from Illinois or Ohio. Steel also was made in Pittsburgh and Bethlehem in Pennsylvania, and in Birmingham, Alabama.

After World War II, the U.S. steel industry was slow to modernize its facilities, and after 1970 it had great difficulty producing steel at a price that could compete with imports from countries such as Japan, Korea, and Brazil. In Europe, the German steel industry centered in the Ruhr River valley in cities such as Essen and Düsseldorf. In Russia, iron ore is mined in the Urals, in the Crimea, and at Krivoi Rog in Ukraine. Elsewhere in Europe, the French "minette" ores of Alsace-Lorraine, the Swedish magnetite deposits near Kiruna, and the British hematite deposits in Lancashire are all significant. Hematite is also found in Labrador, Canada, near the Quebec border.

Coal is widely distributed on earth. In the United States, Kentucky, West Virginia, and Pennsylvania are known for their coal mines, but coal is also found in Illinois, Indiana, Ohio, Montana, and other states. Much of the anthracite (hard coal) is taken from underground mines, where networks of tunnels are dug through the coal seam, and the coal is loosened by blasting, use of digging machines, or human labor. A huge deposit of brown coal is mined at the Yallourn open pit mine west of Melbourne, Australia. In Germany, the mines are near Garsdorf in Nord-Rhein/Westfalen, and in the United Kingdom, coal is mined in Wales. South Africa has coal and is a leader in manufacture of liquid fuels from coal. There is coal in Antarctica, but it cannot yet be mined profitably. China and Japan both have coal mines, as does Russia.

Aluminum

Aluminum is the most important structural metal after iron. It is extremely abundant in the earth's crust, but the only readily extractable ore is bauxite, a hydrated oxide usually contaminated with iron and silica. Bauxite was originally found in France but also exists in many other places in Europe, as well as in Australia, India, China, the former Soviet Union, Indonesia, Malaysia, Suriname, and Jamaica.

Much of the bauxite in the United States comes from Arkansas. After purification, the bauxite is combined with the mineral cryolite at high temperature and subjected to electrolysis between carbon electrodes (the Hall-Héroult process), yielding pure aluminum. Because of the enormous electrical energy requirements of the Hall-Héroult method, aluminum can be made economically only where cheap power (preferably hydroelectric) is available. This means that the bauxite often must be shipped long distances—Jamaican bauxite comes to the United States for electrolysis, for example.

Copper, Silver, and Gold

These coinage metals have been known and used since antiquity. Copper came from Cyprus and takes its name from the name of the island. Copper ores include oxides or sulfides (cuprite, bornite, covellite, and others). Not enough native copper occurs to be commercially significant. Mines in Bingham, Utah, and Ely, Nevada, are major sources in the United States. The El Teniente mine in Chile is the world's largest underground copper mine, and major amounts of copper also come from Canada, the former Soviet Union, and the Katanga region mines in Congo-Kinshasa and Zambia.

Silver often occurs native, as well as in combination with other metals, including lead, copper, and gold. Famous silver mines in the United States include those near Virginia City (the Comstock lode) and Tonopah, Nevada, and Coeur d'Alene, Idaho. Silver has been mined in the past in Bolivia (Potosi mines), Peru (Cerro de Pasco mines), Mexico, and Ontario and British Columbia in Canada.

Gold occurs native as gold dust or nuggets, sometimes with silver as a natural alloy called electrum. Other gold minerals include selenides and tellurides. Small amounts of gold are present in sea water, but attempts to isolate gold economically from this source have so far failed. Famous gold rushes occurred in California and Colorado in the United States, Canada's Yukon, and Alaska's Klondike region. Major gold-producing countries include South Africa, Siberia, Ghana (once called the Gold Coast), the Philippines, Australia, and Canada.

The Exxon Valdez Oil Spill

On March 24, 1989, the tanker Exxon Valdez, with a cargo of 53 million gallons of crude oil, ran aground on Bligh Reef in Prince William Sound, Alaska. Approximately 11 million gallons of oil were released into the water, in the worst environmental disaster of this type recorded to date. Despite immediate and lengthy efforts to contain and clean up the spill, there was extensive damage to wildlife, including aquatic birds, seals, and fish. Lawsuits and calls for new regulatory legislation on tankers continued a decade later. Such regrettable incidents as these are the almost inevitable result of attempting to transport the huge oil supplies demanded in the industrialized world.

Petroleum and Natural Gas

Petroleum has been found on every continent except Antarctica, with 600,000 producing wells in 100 different countries. In the United States, petroleum was originally discovered in Pennsylvania, with more important discoveries being made later in west Texas, Oklahoma, California, and Alaska. New wells are often drilled offshore, for example in the Gulf of Mexico or the North Sea. The United States depends heavily on oil imported from Mexico, South America, Saudi Arabia and the Persian Gulf states, and Canada.

Over the years, the price of oil has varied dramatically, particularly due to the attempts of the Organization of Petroleum Exporting Countries (OPEC) to limit production and drive up prices. In Europe, oil is produced in Azerbaijan near the Caspian Sea, where a pipeline is planned to carry the crude to the Mediterranean port of Ceyhan, in Turkey. In Africa, there are oil wells in Gabon, Libya, and Nigeria; in the Persian Gulf region, oil is found in Kuwait, Qatar, Iran, and Iraq. Much crude oil travels in huge tankers to Europe, Japan, and the United States, but some supplies refineries in Saudi Arabia at Abadan. Tankers must exit the Persian Gulf through the narrow Gulf of Hormuz, which thus assumes great strategic importance.

After oil was discovered on the shores of the Beaufort Sea in northern Alaska (the so-called North Slope) in the 1960s, a pipeline was built across Alaska, ending at the port of Valdez. The pipeline is heated to keep the oil liquid in cold weather and elevated to prevent its melting through the permanently frozen ground (permafrost) that supports it. From Valdez, tankers reach Japan or California.

Drilling activities occasionally result in discovery of natural gas, which is valued as a low-pollution fuel. Vast fields of gas exist in Siberia, and gas is piped to Western Europe through a pipeline. Algerian gas is shipped in the liquid state in ships equipped with refrigeration equipment to maintain the low temperatures needed. Britain and Northern Europe benefit from gas produced in the North Sea, between Norway and Scotland.

Shale oil, a plentiful but difficult-to-exploit fossil fuel, exists in enormous amounts near Rifle, Colorado. A form of oil-bearing rock, the shale must be crushed and heated to recover the oil, a more expensive proposition than drilling conventional oil wells. In spite of ingenious schemes such as burning the shale oil in place, this resource is likely to remain largely unused until conventional petroleum is used up. A similar resource exists in Alberta, Canada, where the Athabasca tar sands are exploited for heavy oils.

John R. Phillips

RENEWABLE RESOURCES

Most renewable resources are living resources, such as plants, animals, and their products. With careful management, human societies can harvest such resources for their own use without imperiling future supplies. However, human history has seen many instances of resource mismanagement that has led to the virtual destruction of valuable resources.

Forests

Forests are large tracts of land supporting growths of trees and perhaps some underbrush or shrubs. Trees constitute probably the earth s most valuable, versatile, and easily grown renewable resource. When they are harvested intelligently, their natural environments continue to replace them. However, if a harvest is beyond the environment's ability to restore the resource that had been present, new and different plants and animals will take over the area. This phenomenon has been demonstrated many times in overused forests and grasslands that reverted to scrubby brushlands. In the worst cases, the abused lands degenerated into barren deserts.

The forest resources of the earth range from the tropical rain forests with their huge trees and broad diversity of species to the dry savannas featuring scattered trees separated by broad grasslands. Cold, subarctic lands support dense growths of spruces and firs, while moderate temperature regimes produce a variety of pines and hardwoods such as oak and ash. The forests of the world cover about 30 percent of the land surface, as compared with the oceans, which cover about 70 percent of the global surface.

Harvested wood, cut in the forest and hauled away to be processed, is termed roundwood. Globally, the cut of roundwood for all uses amounts to about 130.6 billion cubic feet (3.7 billion cubic meters). Slightly more than half of the harvested wood is used for fuel, including charcoal.

Roundwood that is not used for fuel is described as industrial wood and used to produce lumber, veneer for fine furniture, and pulp for paper prod-

ucts. Some industrial wood is chipped to produce such products as subflooring and sheathing board for home and other building construction. Most roundwood harvested in Africa, South America, and Asia is used for fuel. In contrast, roundwood harvested in North America, Europe, and the former Soviet Union generally is produced for industrial use.

It is easy to consider forests only in the sense of the useful wood they produce. However, many forests also yield valuable resources such as rubber, edible nuts, and what the U.S. Forest Service calls special forest products. These include ferns, mosses, and lichens for the florist trade, wild edible mushrooms such as morels and matsutakes for domestic markets and for export, and mistletoe and pine cones for Christmas decorations.

There is growing interest among the industrialized nations of the world in a unique group of forest products for use in the treatment of human disease. Most of them grow in the tropical rain forests. These medicinal plants have long been known and used by shamans (traditional healers). Hundreds of pharmaceutical drugs, first used by shamans, have been derived from plants, many gathered in tropical rain forests. The drugs include quinine, from the bark of the cinchona tree, long used to combat malaria, and the alkaloid drug reserpine. Reserpine, derived from the roots of a group of tropical trees and shrubs, is used to treat high blood pressure (hypertension) and as a mild tranquilizer. It has been estimated that 25 percent of all prescriptions dispensed in the United States contain ingredients derived from tropical rain forest plants. The value of the finished pharmaceuticals is estimated at US$6.25 billion per year.

Scientists screening tropical rain forest plants for additional useful medical compounds have drawn on the knowledge and experience of the shamans. In this way, the scientists seek to reduce the search time and costs involved in screening potentially useful plants. Researchers hope that somewhere in the dense tropical foliage are plant products that could treat, or perhaps cure, diseases such as cancer or AIDS.

Many as-yet undiscovered medicinal plants may be lost forever as a consequence of deforestation of large tracts of equatorial land. The trees are cut down or burned in place and the forest converted to grassland for raising cattle. The tropical soils cannot support grasses without the input of large amounts of fertilizer. The destruction of the forests also causes flooding, leaving standing pools of water and breeding areas for mosquitoes, which can spread disease.

Marine Resources

When renewable marine resources such as fish and shellfish are harvested or used, they continue to reproduce in their environment, as happens in forests and with other living natural resources. However, like overharvested forests, if the marine resource is overfished—that is, harvested beyond its ability to reproduce—new, perhaps undesirable, kinds of marine organisms will occupy the area. This has happened to a number of marine fishes, particularly the Atlantic cod.

When the first Europeans reached the shores of what is now New England in the early seventeenth century, they encountered vast schools of cod in the local ocean waters. The cod were so plentiful they could be caught in baskets lowered into the water from a boat.

At the height of the New England cod fishery, in the 1970s, efficient, motor-driven trawlers were able to catch about 32,000 tons. The catch began to decline that year, mostly as a result of the impact of fifteen different nations fishing on the cod stocks. As a result of overfishing, rough species such as dogfish and skates constitute 70 percent of the fish in the local waters. Experts on fisheries management decided that fishing for cod had to be stopped.

The decline of the cod was attributed to two causes: a worldwide demand for more fish as food and great changes in the technology of fishing. The technique of fishing progressed from a lone fisher with a baited hook and line, to small steam-powered boats towing large nets, to huge diesel-powered trawlers towing monster nets that could cover a football field. Some of the largest trawlers were floating factories. The cod could be skinned, the edible parts cut and quick-frozen for market ashore, and the skin, scales, and bones cooked and ground for animal feed and oil. A lone fisher was lucky to be able to catch 1,000 pounds (455 kilograms) in one day. In contrast, the largest trawlers were capable of catching and processing 200 tons per day.

In the 1990s, the world ocean population of swordfish had declined dramatically. With a worldwide distribution, these large members of the billfish family have been eagerly sought after as a food fish. Because swordfish have a habit of basking at the surface, fishermen learned to sneak up on the swordfish and harpoon them. Fishermen began to catch swordfish with fishing lines 25 to 40 miles (40 to 65 kilometers) long. Baited hooks hung at intervals on the main line successfully caught many swordfish, as well as tuna and large sharks. Whereas the harpoon fisher took only the largest (thus most valuable) swordfish, the longline gear was indiscriminate, catching and killing many swordfish too small for the market, as well as sea turtles and dolphins

As a result of the catching and killing of both sexually mature and immature swordfish, the reproductive capacity of the species was greatly reduced. Harpoons killed mostly the large, mature adults that had spawned several times. Longlines took all sizes of swordfish, including the small ones that had not yet reached sexual maturity and spawned. The decline of the swordfish population was quickly obvious in the reduced landings. But things have changed remarkably, thanks to a 1999 international plan that rebuilt this stock several years ahead of schedule. Today, North Atlantic swordfish is one of the most sustainable seafood choices.

Albert C. Jensen

NONRENEWABLE RESOURCES

Nonrenewable resources are useful raw materials that exist in fixed quantities in nature and cannot be replaced. They differ from renewable resources, such as trees and fish, which can be replaced if managed correctly. Most nonrenewable resources are minerals—inorganic and organic substances that exhibit consistent chemical composition and properties. Minerals are found naturally in the earth's crust or dissolved in seawater. Of roughly 2,000 different minerals, about 100 are sources of raw materials that are needed for human activities. Where useful minerals are found in sufficiently high concentrations—that is, as ores—they can be mined as profitable commercial products.

Economic nonrenewable resources can be divided into four general categories: metallic (hardrock) minerals, which are the source of metals such as iron, gold, and copper; fuel minerals, which include petroleum (oil), natural gas, coal, and uranium; industrial (soft rock) minerals, which provide materials like sulfur, talc, and potassium; and construction materials, such as sand and gravel.

Nonrenewable resources are required as direct or indirect parts of all the products that humans use. For example, metals are necessary in industrial sectors such as construction, transportation equipment, electrical equipment and electronics, and consumer durable goods—long-lasting products such as refrigerators and stoves. Fuel minerals provide energy for transportation, heating, and electrical power. Industrial minerals provide ingredients needed in products ranging from baby powder to fertilizer to the space shuttle. Construction materials are used in roads and buildings.

Location

When minerals have naturally combined together (aggregated) they are called rocks. The three general rock categories are igneous, sedimentary, and metamorphic. Igneous rocks are created by the cooling of molten material (magma). Sedimentary rocks are caused when weathering, erosion, transportation, and compaction or cementation act on existing rocks.

Metamorphic rocks are created when the other two types of rock are changed by heat and pressure. The availability of nonrenewable resources from these rocks varies greatly, because it depends not only on the natural distribution of the rocks but also on people's ability to discover and process them. It is difficult to find rock formations that are covered by the ocean, material left by glaciers, or a rain forest. As a result, nonrenewable resources are distributed unevenly throughout the world.

Some nonrenewable resources, such as construction materials, are found easily around the world and are available almost everywhere. Other nonrenewable resources can only be exploited profitably when the useful minerals have an unusually high concentration compared with their average concentration in the earth's crust. These high concentrations are caused by rare geological events and are difficult to find. For example, an exceptionally rare nonrenewable resource like platinum is produced in only a few limited areas.

No one country or region is self-sufficient in providing all the nonrenewable resources it needs, but some regions have many more nonrenewable resources than others. Minerals can be found in all types of rocks, but some types of rocks are more likely to have economic concentrations than others. Metallic minerals often are associated with shields (blocks) of old igneous (Precambrian) rocks. Important shield areas near the earth's surface are found in Canada, Siberia, Scandinavia, and Eastern Europe. Another important shield was split by the movement of the continents, and pieces of it can be found in Brazil, Africa, and Australia.

Similar rock types are in the mountain formations in Western Europe, Central Asia, the Pacific coast of the Americas, and Southeast Asia. Minerals for construction and industry are found in all three types of rocks and are widely and randomly distributed among the regions of the world.

The fuel minerals—petroleum and natural gas—are unique in that they occur in liquid and gaseous states in the rocks. These resources must be captured and collected within a rock site. Such a site needs source rock to provide the resource, a rock type that allows the resource to collect, and another surrounding rock type that traps the resource. Sedimentary rock basins are particularly good sites for fuel collection. Important fuel-producing regions are the Middle East, the Americas, and Asia.

Impact on Human Settlement

Nonrenewable resources have always provided raw materials for human economic development, from the flint used in early stone tools to the silicon used in the sophisticated chips in personal computers. Whole eras of human history and development have been linked with the nonrenewable resources that were key to the period and its events. For example, early human culture eras were called the Stone, Bronze, and Iron Ages.

Political conflicts and wars have occurred over who owns and controls nonrenewable resources and their trade. One example is the Persian Gulf War of 1991. Many nations, including the United States, fought against Iraq over control of petroleum production and reserves in the Middle East.

Since the actual production sites often are not attractive places for human settlement and the output is transportable, these sites are seldom important population centers. There are some exceptions, such as Johannesburg, South Africa, which grew up almost solely because of the gold found there. However, because it is necessary to protect and work the production sites, towns always spring up near the sites. Examples of such towns can be found near the quarries used to provide the material for the great monuments of ancient Egypt and in the Rocky Mountains of North America near gold and silver mines. These towns existed because of the nonrenewable resources nearby and the needs of the people exploiting them; once the resource was gone, the towns often were abandoned, creating "ghost towns," or had to find new purposes, such as tourism.

More important to human settlement is the control of the trade routes for nonrenewable resources. Such controlling sites often became regions of great wealth and political power as the residents taxed the products that passed through their community and provided the necessary services and protection for the traveling traders. Just one example of this type of development is the great cities of wealth and culture that arose along the trade routes of the Sahara Desert and West Africa like Timbuktu (in present-day Mali) and Kumasi (in present-day Ghana) based on the trade of resources like gold and salt.

Even with modern transportation systems, ownership of nonrenewable resources and control of their trade is still an important factor in generating national wealth and economic development. Modern examples include Saudi Arabia's oil resources, Egypt's control of the Suez Canal, South Africa's gold, Chile's copper, Turkey's control over the Bosporus Strait, Indonesia's metals and oil, and China's rare earth element.

Gary A. Campbell

NATURAL RESOURCES

Human existence requires consumption of a variety of natural resources. In a primitive society, consumption is more direct, with little artificial processing involved. In a highly advanced society, consumption becomes diverse in kind and sophisticated in quality. Although per capita consumption of natural resources is much higher in developed countries, it is increasing in developing countries.

Natural resources commonly are divided into two general categories: renewable and nonrenewable. Renewable resources are those that nature can replenish in time. Surface water, trees, and solar power are renewable. Nonrenewable resources can be used only once. Most mineral resources cannot be replenished once the reserve is depleted; therefore, humans are living mostly on finite resources.

However, throughout history, people have often found ways to substitute new resources for depleted ones. This practice has become critical to the preservation of the modern lifestyle.

Water

Although water is a renewable resource, it is not always available in sufficient quantity and of adequate quality. The amount of available water remains fairly fixed, while its usage expands as modernization and industrialization take place worldwide. Fresh surface water accounts for a mere 0.013 percent of Earth's total water reserve. Oceans have 96.5 percent, while ice caps and glaciers capture 3.47 percent. It is not the total amount that matters, but the amount of fresh water that is available for diverse human needs.

The Dhanbad mine complex in India.

SELECTED RESOURCES OF ASIA

Asia's water resource base is quite diverse, as it has both arid and semiarid regions and humid tropical regions where precipitation is high. Asia has some of the world's largest rivers in volume the Yangtze, Brahmaputra, Mekong, Ganges, Indus, and Yellow (Huang He) a number of large lakes, and extensive groundwater resources.

In spite of a comparatively rich endowment of water resources, the per-capita availability of water varies greatly throughout Asia. Papua New Guinea has the highest amount of fresh water available, while Singapore and the Arab states have the least. Water scarcity, a condition where water availability is below 1,000 cubic meters per capita per year, threatens many countries in Asia, including parts of India and China.

East Asia's water resource is fair to adequate at best. The region requires normal rainfall for its needs. Neither Japan nor Korea has a large river system left to be exploited, because most river resources are heavily used at the present. As water usage increases with rising living standards, Korea has been classified a region of water deficiency. One of the largest countries in territory in the world, China's regions vary in water supply and demand. In general, the interior has low rainfall and thus a poor water resource base. Many of the large river systems in China have been exploited; multipurpose facilities, including the Three Gorges Dam on the Yangtze River, have been built as the nation works to secure adequate water resources for the growing demands of agricultural, domestic, and in-

The Star of India is probably the largest such gem in the world. It is almost flawless and unusual in that it has stars on both sides of the stone. The greyish blue gem was mined in Sri Lanka.

dustrial use. Many of the dams also provide hydroelectric power generation.

Southeast Asia is a zone of tropical and subtropical climate. Under normal conditions, the region does not experience water shortages. In fact, it benefits from the heavy rainfall, which enables it to grow more than one crop in a year. Only the strong El Niño phenomenon tends to bring the region a serious condition of drought. The region's most important river system is the Mekong River; it originates in southern China and touches parts of China, Laos, Thailand, Cambodia, and Vietnam before reaching the South China Sea. Full development of the Mekong River would have significant impacts on the region as a whole.

The Indian subcontinent is a zone of monsoon climates, which bring heavy rainfall during the summer months. Rainfall is higher in the western part of the Indian peninsula than in the east. In most other seasons, the region receives little rain. Destructive tropical cyclones that affect the Bay of Bengal several times a year also occur during the summer months. Because of the heavy concentration of rainfall in the summer months, the region's most heavily populated areas depend greatly upon its major river systems, the Ganges and Indus, for their water. The basins of these river systems are important for both agriculture and domestic use. Both river systems flow into territories of two countries: the Indus River into India and Pakistan; the Ganges River into India and Bangladesh. More systematic exploitation of these rivers will require a greater spirit of cooperation and harmony in the region.

The Middle East

In Southwest Asia, an arid climate dominates except in a few areas on the Mediterranean coast. Not only is rainfall often marginal and irregular in most places, but it also comes during the winter. The water resources provided via snowmelt are vital for the region, providing relatively dependable surface water in Iraq, Syria, and Lebanon. But elsewhere in the Middle East, water scarcity has led to the construction of hundreds of desalination plants to produce water for drinking and irrigation. In 2019, 70 percent of Saudi Arabia's water needs were met by desalination. In Kuwait, the figure is 100 percent.

Southwest Asia's most important river systems are the Euphrates and Tigris. Through yearly flooding, these rivers created the fertile valley that gave rise to the Mesopotamian civilization. They still remain vital to Turkey, Syria, and Iraq. In a drive to increase the amount of irrigated land in its southern region, Turkey launched the Greater Anatolia Project, of which the Ataturk Dam on the Upper Euphrates, operational since 1992, was the first of nineteen dams to be built. By 2019, irrigation from the Ataturk Dam has more than tripled the harvest yield in the area.

In Israel, only the northern coastal zone receives adequate rainfall. Precipitation is lower in the south and inland. The Jordan River and Lake Tiberias are the only significant surface water sources in Israel, although some limited groundwater is available. Desalination plants now provide nearly half of Israel's drinking water.

Most of the Central Asian republics have arid or semiarid climates, as the region's major cities were based on oases along the famed Silk Road. Because the region is landlocked, all rivers except the Irtysh River end within the region. Central Asia's most important river system is that of the Amu Darya and Syr Darya, both of which empty into the Aral Sea. The Soviets diverted waters from these rivers to irrigate vast lands for cotton and rice farming, dramatically increasing the region's cotton and food production. However, it caused a disastrous impact on the region's ecology. As the water flows were re-

Vallur Thermal Power Station near Chennai, India.

The Kashiwazaki-Kariwa Nuclear Power Plant, a nuclear plant with seven units, is the largest single nuclear power station in the world.

duced drastically, the Aral Sea's surface shrunk greatly, exposing harmful salt particles to be blown away. The Aral Sea is one of the greatest human-made environmental disasters of the twentieth century. In recent years, mitigation projects initiated by UNESCO and the World Bank have had some modest success.

Forests

Forests are highly versatile natural resources. Not only are they raw materials for industrial and construction products, but they also are primary fuels in poor developing countries. About 27 percent of the world's wood consumption is for fuel. In addition to domestic and industrial uses, forests are Earth's lungs. When trees are destroyed for whatever reason, Earth's lung capacity is negatively affected. Forests' impact on the ecosystem is not local,

but global. Loss of woodlands also causes erosion and land degradation in general.

Trees are easily exploitable; for thousands of years, people have used wood for cooking and heating. Demands for lumber and wood products increase as living standards rise worldwide. The rapid decline of woodlands is taking place in most regions of the world, including Asia. The most worrisome development, which has attracted global concern, is the onslaught on the world's rain forests, a critical climatic regulator and the greatest reservoir of plant diversity. More than half the world's rain forests have been cleared, and the trend is accelerating. If the current rate continues, it is projected that the world's remaining rain forests will be gone by the year 2120. The forest resource base is fair to significant in East Asia, Southeast Asia, and South Asia. It is virtually nonexistent in Southwest Asia, where arid climates dominate. Forests that supply firewood and raw materials for export are located mainly in Thailand, Vietnam, Malaysia, Myanmar, Indonesia, and the Philippines. These forests face a dismal prospect of survival as the harvest of quality timber proceeds at a destructive rate. Domestic and foreign demands have led to uncontrolled or illegal logging in Myanmar, Cambodia, and Laos.

Throughout the region, forests have become casualties of persisting poverty and globalized commerce. Unless the harvest of tropical forests is restrained through rational logging practices, severe deforestation could usher in ecological disaster that will affect not only the region but the world. Although the region's hardwood sources are more widely spread from Japan to India, they do not meet the growing demands for timber. In East Asia, extensive wood shortages are being met by imports from North America, Russia, and elsewhere in Asia.

Although their commercial value is limited, forests in Korea and China show strong growth, assisted by government policies and increasing use of coal and natural gas as primary fuels for heating and cooking. In general, they are not fit for commercial harvest but are highly effective in erosion control and in providing wildlife habitats.

Nonrenewable Resources

Industrialized society needs many products to function properly and to satisfy consumers' sophisticated tastes. Every day, people use products made of mineral resources of which they are unaware. No one country dominates the mineral resource endowment there are too many different minerals involved, and the amount humans consume is too great for a single country to dominate. Only a few materials with rather highly restricted use are found mostly in a single country for example, helium in the United States and hafnium in France.

By definition, mineral resources are nonrenewable, but pure economics and technology make recycling common, and so some of them are considered renewable, at least in a practical sense. Another factor is that most mineral resources are substitutable. Only a few key mineral resources do not have adequate substitutes.

The nonrenewables that are most significant to the world's economy are energy sources. In 2015, oil was the most important, providing nearly 33 percent of the world's energy, followed by coal (29 percent), natural gas (23.4 percent), renewable energy sources (10 percent), and nuclear plants (4.4 percent).

Nonrenewable energy sources are found in quantity in a few countries. Asia as a region has vast energy resources, but petroleum is highly concentrated in the Middle East and Central Asia. Natural gas is somewhat more widespread, but this resource base is not as dominant as petroleum. Coal deposits appear in far more countries than do oil or natural gas.

East Asia

Japan is Asia's most resource-deficient nation. Although it is a major producer of some rare minerals, including bismuth, cadmium, indium, and tellurium, Japan lacks many critical raw materials. Its only significant energy resource is coal, which fills about 5 percent of the nation's demand, and is located mostly on the northern island of Hōkkaidō. Japan's rate of self-sufficiency for key industrial raw materials is the lowest among the world's most industrialized countries. Its import-dependency in energy, including coal, oil, and natural gas, is more than 80 percent. It imports nearly 100 percent of the petroleum it consumes. It imports all or nearly all domestically consumed iron ore, copper, tin, bauxite, nickel, and uranium. Japan's highly successful industrial economy is powered mostly by imported raw materials.

Japan's heavy dependence on foreign raw materials has resulted in two primary approaches to supplying energy needs. First, it has achieved a significant degree of diversification in suppliers. For example, Japan imports oil from Saudi Arabia, Indonesia, and Iran, supplemented by other sources with small quantities. Second, Japan has relied greatly on nuclear energy; by 2011, more than 30 percent of its energy supply came from this clean but controversial source. But following the 2011 earthquake and tsunami that damaged several nuclear reactors, the Japanese government closed down all the country's reactors. In 2015, several reactors reopened. The government has now proposed to revive the nuclear industry, with a target of nuclear providing 20 percent of Japan's energy needs by 2030.

Korea is even poorer in its endowment of key industrial raw materials. It is a major producer of only one mineral, magnesium ore. It depends extensively on foreign sources for petroleum, natural gas, pulp, and nonferrous metals. The nations' most abundant energy resource, coal, is mined in both South Korea and North Korea, although the North's deposits of coal and iron ore are much greater than the South's. Both Koreas have ambitious nuclear power programs to compensate for their poor energy-resource potential.

China's vast territory includes a mineral resource base that is rich in many categories. China is a leading producer of rare minerals; it leads in tin production and is a major producer of lead and zinc. China is well endowed with coal it is, in fact, the largest coal producer in the world. China's coal deposits are widely scattered, but heavy concentrations are found in the Loess Plateau in the north central region and the Sichuan Basin in the south central region. The northeast is another coal-producing region. In general, coal deposits are located far from major industrial centers, necessitating costs for extensive transportation. The heavy use of coal for cooking and heating produces greenhouse gases and has contributed to China's notorious air-pollution problems.

China is one of the world's major petroleum producers. Many of the oil deposits are located in the northeast, the southwest, and the far west in Xinjiang Province (also the country's largest natural-gas-producing area). Oil in the northeast region is centered in Daqing, which in 2015 was the world's fourth-largest oil producer. The Chinese also obtain substantial amounts of oil from offshore drilling. Despite these many sources, by 2018 China was importing more than 70 percent of its crude oil. As industrial development and living standards progress at a rapid rate, the demand for petroleum products will likely continue to rise.

China, as a major producer of uranium, also has rapidly developed its nuclear-energy capacity. To-day, nuclear energy supplies nearly 5 percent of China's electricity. As of March 2019, the country had 46 reactors operating, with nearly a dozen more being built.

Taiwan's raw material base is poor. Like those of Korea and Japan, Taiwan's industrial economy is based largely upon imported raw materials. This foreign dependency is manifested by the locations of the nation's major industrial centers. The coastal cities, mostly in the west, are convenient for unloading imported raw materials.

Southeast Asia

In Indochina, Vietnam has the most abundant resource base. It exports coal and crude oil, and also produces a variety of industrial minerals from its northwest region. Even more significant resources in Vietnam are petroleum and natural gas. Vietnam's oil is derived mostly from the Bach Ho field and adjacent fields on the continental shelf off the southeast coast. Production is expected to drop 10 percent per year at least through 2025, mostly due to overexploitation of the wells.

Cambodia has few mineral resources in quantity, except for abundant phosphates and gemstone deposits. It is pursuing oil and natural gas exploration. Likewise, Laos has few proven mineral resources except some tin deposits. Myanmar also has limited resources to export. Political stability and systematic economic development could change this, as these countries possess a few marketable mineral resources.

The Philippines has some gold deposits and a few industrial minerals, including nickel and copper. More importantly, the nation has modest petroleum and natural gas reserves. Its vast continental shelf may yield even more energy resources in the future. Thailand, once one of the world's premier tin producers, now ranks number 15, as its production continues to decline significantly. It also has some energy resources. Its natural gas reserve is much more substantial than its oil resources.

Indonesia and Malaysia have the most abundant resources in the region. They are not only important producers of tin, timber, and natural rubber, but also major producers of energy resources. Along with oil-rich Brunei, these two countries share about 2 percent of the world's total energy production. Indonesia was a member of the Organization of Petroleum Exporting Countries (OPEC) from 1962 to 2008 and then again from 2014 to 2016 This region's energy resources, which include more natural gas reserves than oil, are mainly exported to industrialized economies in East Asia.

South Asia

India is the region's largest and most populous nation. Its resource base is the most diverse, and its consumption the greatest. It produces industrial materials such as copper, tin, lead, zinc, nickel, manganese, aluminum, and uranium. India's most abundant energy resource is coal. High-quality coal deposits in the Chota-Nagpur Plateau have made the northeast region India's major steelmaking and metals center. Iron ore deposits in surrounding areas also have helped development. India ranked twenty-fifth among the world's oil-producing nations in 2019. Domestic production does not come close to meeting the demand, and dependency on foreign fuels remains a national concern.

In high-density Bangladesh, endowments of major industrial minerals are extremely scarce. In 2019, its natural gas reserves met 56 percent of the country's energy needs. Fertile land and its agricultural potential continue to be the nation's foremost resources, and its abundant labor force the most important industrial resource.

In Sri Lanka, rubber is the most prominent raw material produced and exported, a legacy of the colonial plantation economy. Sri Lanka is also known for its variety of gemstones, including rubies and sapphires. Pakistan's vast Balochistan region has significant mineral potential, including oil and natural gas deposits. Gypsum, chromite, rock salt, and other minerals contribute to the country's chemical industries.

Southwest Asia

A few oil-rich countries dominate the resource economy of Southwest Asia. Because the region possesses more than 60 percent of the world's known oil reserves, it is important to the world and to the West's industrial economy. Most oil-rich countries in the region are members of the OPEC cartel. A few countries in the region have minimal or no oil reserves.

Saudi Arabia is the largest, most important oil producer in the region. Having 297 billion barrels of oil in reserve, Saudi Arabia expects its oil to last at least until the year 2100. As in most Gulf states, Saudi Arabia's oil is located mainly in its eastern regions along the Persian Gulf. It is in the process of developing systematic petrochemical industries in the Gulf and Red Sea areas, using its abundant natural gas.

The Persian Gulf states of Bahrain, Qatar, and the United Arab Emirates are also rich in oil reserves, but both their physical size and their oil reserves are much smaller than those of Saudi Arabia. Bahrain's oil is expected to last only two decades. In spite of its small physical size, Kuwait is one of the major oil producers of the region, with about 5.7 percent of the world's proven oil reserves. It has the world's seventh-largest oil reserves, which may last more than fifty years at current production levels. Kuwait is also rich in natural gas, which its growing petrochemical industry uses as raw materials. Oil

OPEC and OAPEC

OPEC, the Organization of Petroleum Exporting Countries, was formed in 1960 by five founding members: Iran, Iraq, Kuwait, Saudi Arabia, and Venezuela. However, it was not well known until 1973, when it demonstrated its capacity to influence world oil prices through production control after a successful oil embargo. Designed primarily to maximize member countries' oil revenues, the organization achieves this goal through brokered allocation of production quotas. OPEC has the ability to adjust the level of oil supply in the world market, thus indirectly controlling the oil price. As of 2020, there are eight additional OPEC members—Algeria, Angola, Congo, Equatorial Guinea, Gabon, Libya, Nigeria, and United Arab Emirates—for a total of thirteen countries.

A later organization, OAPEC, the Organization of Arab Petroleum Exporting Countries, was created as a political force after the Arab-Israeli War of 1967 and includes Algeria, Bahrain, Egypt, Iraq, Kuwait, Libya, Qatar, Saudi Arabia, Syria, and United Arab Emirates.

reserves in both Oman and Yemen are limited and insignificant.

Iraq has the world's fifth-largest oil reserves about 8 percent of the world's proven oil reserves. Iraq's oil exports have been subject to United Nations' sanctions, which limit Iraq's oil exports to that necessary for the purchase of food and medicine. Iraq's production capacity in the year 2000 was far greater than its production and export level. Wars, sanctions, military occupation, and civil unrest have added a degree of uncertainty to the accuracy of official statistics regarding the Iraqi oil industry. Other Arab countries in the region—Jordan, Lebanon, and Syria—have very limited oil reserves in their territories. Only Syria produces some oil, about 0.05 percent of the world's total. Oil nevertheless remains the main source of its export earnings. Jordan's principal exportable mineral is phosphate, for which prices tend to fluctuate in the world market. Jordan has the world's fifth-largest deposit of oil shale, which it will begin exploiting in the very near future with the completion of the Attarat Power Plant in 2020.

Iran is the region's other major oil producer, with the world's third-largest oil reserves. The country estimates it has eighty years of production left. Its oil fields are located along the Zagros Mountains in the southwestern region. Iran also produces coal, iron ores, and copper.

Turkey is a non-Arab Islamic nation that does not share the region's predominant energy resource, oil. It produces a relatively small amount of oil in its southern and southeastern regions. It also produces coal and iron ore, which support iron and steel mills in the northern region, as well as chromium and other minerals.

Afghanistan is more important for its strategic location than for what it produces. Repeated warfare and civil wars have prevented the country from exploring its natural resources fully. It extracts small quantities of natural gas, coal, and other minerals, but its full potential for mineral resources is unknown.

In resource endowment, Israel suffers greatly from its small physical size. Key materials, such as metal-bearing ores and petroleum, are virtually nonexistent. It extracts potash, bromine, and other materials from the Dead Sea floor for its chemical industries. Most industrial materials needed for its growing manufacturing sector are imported from abroad.

Central Asia

Central Asia's five Islamic countries were mainly raw-material sources for the Soviet industrial machine under Moscow rule. They are rich with natural gas, oil, coal, and iron ore, and produce a variety of minerals. It is likely that the region will produce far more variety and quantity of minerals as systematic exploration takes place in its vast plains and mountain zones.

Kazakhstan, the region's largest state, is also its richest in natural resources. Kazakhstan has the world's twelfth-largest proven reserves of oil and natural gas. But with only three refineries within its borders, most of the oil is exported to Russia. It has large reserves of coal at Karaganda in the central region, and various Soviet-era smelting operations in the region. Central Kazakhstan has one of the largest copper reserves in the world. Other raw materials with which the region is richly endowed include copper, lead, zinc, chromium, and nickel. Along with Uzbekistan, Kazakhstan is a leading producer of uranium, the source of energy used to generate electric energy at nuclear power plants.

Turkmenistan also has significant oil reserves, as it shares the shore of the oil-rich Caspian Sea, but it is even richer in natural gas reserves. It has the sixth-largest gas reserves in the world, which it exports to Europe, earning precious foreign currency. Other Central Asian republics have vital resources, but at a much lower level. Natural gas is found in Uzbekistan, and a relatively small quantity of oil is produced in Kyrgyzstan. Mountainous Tajikistan is poorest in natural resources, as its only significant energy resource is some coal deposits.

Like the Persian Gulf region, the Caspian Sea basin is rich in energy resources. The Caspian Sea has oil on its eastern shore, and its western shore and inland regions are rich in oil and natural gas. The traditional oil-producing region of Baku in Azerbaijan, Georgia, Chechnya, and the foothills of the Russian Caucasus are all significant producers of oil and natural gas.

It is speculated that the Caspian basin could someday become a major energy-producing region, even rivaling the Persian Gulf region. Today, more than a dozen pipelines are operational, moving oil throughout the region and even to Europe, China, and beyond. Nevertheless, it could be some years before the conditions are met to fully exploit the Caspian's rich endowment of energy resources.

Walter B. Jung

DISCUSSION QUESTIONS: NATURAL RESOURCES

Q1. How has the abundance of oil and natural gas deposits in and around the Caspian Sea transformed the Transcaucasian Republics? In what ways has the Soviet-era infrastructure helped or hindered the energy industries in these countries? Through what means are the Transcauasian Republics able to export their oil and natural gas to foreign markets?

Q2. Why is China unable to provide all of its citizens with fresh water? What is the Chinese government doing to remedy the problem? What roles—positive and negative—has the Three Gorges Dam played in China's water situation?

Q3. Why have China, Japan, and a number of other Asian countries turned to nuclear power as an energy source? Why did Japan temporarily close all of its nuclear power plants? Why is Japan allowing some to reopen? What important lesson has Japan learned about nuclear-energy safety?

FLORA

Asia has the richest flora of Earth's seven continents. Because Asia is the largest continent, it is not surprising that 100,000 different kinds of plants grow within its different climate zones, ranging from tropical to Arctic. These plants which include ferns, conifers, and flowering plants make up 40 percent of Earth's plant species. The endemic plant species come from more than forty plant families and 1,500 genera.

Asia is divided into five major vegetation regions based on the richness and types of each region's flora: tropical rain forests in Southeast Asia, temperate mixed forests in East Asia, tropical rain/dry forests in South Asia, desert and steppe in Central and West Asia, and taiga and tundra in North Asia.

Tropical Rain Forests

The Asian regions richest in flora—tropical rain forests—are found in the island nations of Southeast Asia, which extend from Kinabalu in the north to Java in the south, and from New Guinea in the east to Sumatra in the west. In this vast archipelago running between Asia and Australia, 35,000 to 40,000 vascular plant species can be found. Tropical rain forests grow there all year round because of the region's warm temperatures and plentiful rainfall. The forests contain great varieties of tall trees, some towering 148 feet (45 meters) high. Within any 1-square-mile (2.5-square-kilometer) area, one can see as many as 100 tree species with no single species dominant.

The rain forests have mostly broad-leafed evergreens, with some palm trees and tree ferns. The uppermost branches of the trees form canopies that cover and protect the surface below. Because little sunlight penetrates the dense canopies, few shrubs and herbs grow in the rain forests. Instead, many vines, lianas, epiphytes, and parasites are twined on tree branches and trunks. Fringing the tropical rain

forests along the coasts, these plant species are replaced by mangrove.

Temperate Mixed Forests

Second in flora richness, East Asia's temperate mixed forests contain 30,000 to 35,000 plant species. This region ranges from Japan in the east to the Himalayan nations (Bhutan and Nepal) and northeastern India in the west, and from Russia's Amur River Valley in the north to China's Hainan Island in the south. East Asia's temperate weather is similar to the climate of North America, with hot summers and cool winters. From south to north or from the east coasts to lower elevations in mountainous areas in

Dates on date palm.

the west, the vegetation changes from evergreen to deciduous broad-leafed forests, with dense shrubs, bamboo, and herbs in different layers beneath the forest canopy. The major tree species are from the magnolia, oak, tea, laurel, spurge, azalea, and maple families. Herb species include members of the primrose, gentian, pea, carrot, foxglove, composite, buttercup, and rose families.

The Himalayan range is the point where the regions of South Asia, East Asia, and Central and West Asia join. From the Qinghai-Tibet Plateau in southwest China to the lower areas of the Himalayas, elevation usually is between about 5,000 and 13,000 feet (1,500 and 4,000 meters). Countless mountains with deep valleys showcase complex, multiple vegetation types from mixed forests and dense shrubs to alpine meadows in mountain plains. Many primary seed plants (that is, gymnosperms and flowering plants) grow there.

Untouched native vegetation in East Asia is usually found only in mountainous or remote areas. On most mountains at high elevations, a point exists where the temperatures are so cold that trees cannot grow. These points together form what is called the tree line. Near the tree line, only plants related to coniferous plants and alpine species grow. Above about 13,000 feet (4,000 meters) in high mountain areas, no vegetation grows. Instead, snowcaps or icebergs exist year round.

Tropical Rain/Dry Mixed Forests

The third-richest region, tropical rain/dry mixed forests, is found in South Asia, which reaches from the Philippines in the east to Pakistan in the west, and from the Himalayas in the north to Thailand in the south. Some 25,000 to 30,000 species grow there. This region has both tropical rain forest and tropical seasonal dry forests. The tropical rain forest is mainly found in the region's lowlands, and the seasonal dry forests in the highlands or mountainous areas. More often, these two types of forests are combined.

Tukulan sandy area in the taiga of the Central Yakutian Lowland.

The tropical seasonal dry forests usually grow in a climate with wet and dry seasons or under a somewhat cooler climate than the tropical rain forests. The canopy, formed from primarily deciduous broad-leafed species, is much thinner than the canopy in the tropical rain forest, so more sunlight reaches plants below. Many different plant species live together, forming tropical jungles. Tall, thick-trunked trees, colorful orchids, plentiful ferns, dense mosses, and twined vines and lianas dominate this vast region. The major components of these kinds of forests are members of the dipterocarpus, sweetsop, laurel, piper, fig, dissotis, akee, gardenia, periwinkle, milkweed, African violet, palm, and aroid families. In central and southern India and in some areas of Pakistan, tropical grasslands, called savanna, flourish. Because of the savanna's hot, dry weather, mainly coarse grasses grow there.

Desert and Steppe

The desert and steppe region in Central and West Asia has 20,000 to 25,000 species. This region stretches from north and northwest China and Mongolia in the east to Turkey in the west, and from Kazakhstan in the north to the Arabian Peninsula in the south. This region's vegetation transitions from semidesert or desert to the temperate grassland called the steppe.

Central and West Asia contains the largest desert-steppe landscape in the Northern Hemisphere. Few plant species grow in the steppe and nearly none in the desert. The herbs and few woody plants that grow in these dry areas are members of the grass, pink, mustard, pea, saxifrage, stonecrops, lignum vitae, forget-me-not, and lily families. Because the desert environment is so dry, plant species must be able to survive in the arid weather for long periods of time. Central and West Asia with its steppe between the desert in the south and coniferous forests in the north forms one of the world's largest foraging areas, providing food resources for both wild and domestic animals, such as camels, sheep, goats, cows, and horses.

Iris oxypetala (Iris lactea var. lactea). *From the collection of the Botanical Garden of Moscow State University (main area on the Sparrow Hills).*

Taiga and Tundra

The poorest region in flora richness, with only about 5,000 vascular plant species, is North Asia. This region is primarily Siberia, the eastern part of Russia, reaching from the Ural Mountains in the west to the Bering Strait in the east and from the Arctic Circle in the north to Mongolia and Kazakhstan in the south. The region's weather is temperate, with short, mild summers and long, cold winters. Few shrubs or herbs grow there. The predominant vegetation in North Asia is coniferous (boreal) forest. This region, called the taiga, contains mainly pines, spruces, firs, larches, and some species in the birch, aspen, and willow families. Because the trees there are straight and tall, the taiga provides timber for Russia's forestry industry. Small perennial herbs and a few shrubs grow in the taiga's swamps or marshes.

Farther north is the cooler Arctic area called the tundra. Plants that grow in the tundra are resistant to the cold climate. They have a short life during the summer and complete their life cycle quickly before winter comes. Tundra plant species are members of such common families as composites, peas, grasses, and reeds. Far beyond the tundra is Arctic ice.

Asia's native plant species provide shelter and food for animals. For example, arrow bamboo and umbrella bamboo, found in the forests of central to southwest China, are the main food of the giant panda. Many plants in Asia also provide food, ornaments, or medicine for humans.

Food Crops

Rice is the main food for humans in Asia, especially in the tropical areas. In temperate Asia, wheat, one of the world's main foods, joins rice as a primary food source. Various beans and peas provide plant protein in the human diet and are eaten with vegetables and grains. Asia also has many tropical fruit plants, such as the mango, banana, lychee, citrus fruits, and breadfruit. Pears, apples, grapes, peaches, and strawberries are temperate fruits. The kiwi, one of the most nutrient-rich fruits, originally came from central China. The Chinese not only eat kiwi but also make kiwi wine. Palm dates are another important fruit in West and Southwest Asia (that is, the Arabian Peninsula). Vegetables grown in Asia include various cabbages, lettuce, onions, garlic, celery, carrots, soybeans, cucumbers, and squash. Ginger also originally came from Asia.

Soybean oil is the major cooking oil in Asia. Although soybeans are native to Asia, they are now the number-two crop in the United States (after corn). Another oil plant, the sunflower, is grown in temperate Asia. In tropical Asia, people use mustard oil, palm oil, cotton oil, and peanut oil. In Central and West Asia, the most popular oil is olive oil. Many other foods people enjoy throughout the

Bamboo is used for mussels breeding and propagation (Abucay, Bataan, Philippines).

world are native Asian plants, for example, tea and coconuts. Black pepper and sugarcane also are grown in tropical Asia.

Ornamental and Medicinal Plants

Many of Asia's plant species have great ornamental value. Azaleas, dogwood, primroses, camellias, peonies, roses, lotuses, daisies, cherries, and begonias are used frequently in gardens and yards. Ornamental conifers from Asia include pines, spruces, cedars, junipers, umbrella pines, and yews. Besides these garden flowers and trees, there are thousands of Asian wildflowers, such as poppies, snapdragons, slippers, columbine, trillium, marigolds, buttercups, gentian, lilies, bluebells, and violets. Europeans who explored Asia centuries ago brought back ornamental plants to their home countries. As a result, many of these plants are now cultivated throughout the world. The world's largest flower, rafflesia, grows in the tropical rain forests of Sumatra. In full bloom, the flower's diameter is about 3 feet (1 meter).

Plants make up a large part of traditional Chinese medicine, which has existed for thousands of years. Today, some of these plants are used in alternative medicine. They include ephederas, eucommia, cinnamons, ginseng, sanqi, and ginkgo.

Scientific Value

Botanists view the region ranging from central China to the Himalayas to the northern part of South Asia as a key area for botanical research, especially for studying the origin of flowering plants. Native plant species in Asia are numerous and play a key role in the modern study of botany. Botanists also study Asian flora because some plants are relics of ancient times, from millions of years ago. Others have already become fossils.

Ancient species include such gymnosperms as the dawn redwood of central China. Dawn redwood is similar to California's redwood and giant sequoia. Another fossil-like tree species is East Asia's ginkgo. This species not only has great ornamental value, but also has great commercial value in alternative medicine. Ginkgo trees are also dust-resistant, which makes them a favorite in urban landscaping

Yachounomori Garden, Tatebayashi, Gunma, Japan.

and the ornamental industry. Other Asian plants such as the magnolia and its allied families may represent the most primitive flowering plants.

Introduced Plants

Asian flora today also contains introduced plant species from other parts of the world that play important roles in peoples' lives. For example, the rubber tree of South America is cultivated in tropical Asia. This tree produces raw material for the natural rubber industry, for which Asia is the leader.

Cacao, a tree species that provides the basis of chocolate, was also introduced from tropical America. Maize, (corn) one of the most common crops in Asia, was introduced from America (via Europe) several hundred years ago. Several vegetables, including the tomato, potato, eggplant, green pepper, hot pepper, and chili (all from the nightshade family), were introduced to Asia long ago. Peanuts, originally from Brazil, are also cultivated in Asia.

Coffee, an increasingly popular beverage in Asia, came originally from Africa. Vietnam and Indonesia are now major coffee producers. An introduced fruit tree is the pineapple, which came from tropical America but now is popular in tropical Asia. The Philippines, Thailand, India, and Indonesia are important producers. The sweet potato, from Central America, is also cultivated in China and Indonesia. Tobacco is another crop introduced to and cultivated in Asia. This controversial plant originally came from tropical America; today, China is by far the world's leading producer of tobacco.

Impact of Human Activity

Asia's highly diversified flora has contributed positively to the daily lives of people around the world. Asia's dense and rapidly growing population demands more vegetation as food sources; at the same time, vegetated land is being cleared for economic development. Deforestation, overgrazing, and urbanization have become major reasons for heavy losses of Asian flora, especially in South and East Asia. In Asia, more than 250 plant species are classified as "critically endangered." Among them are the Chinese silver fir, dawn redwood, and ginseng. These plants only grow in several isolated locations, and are rarely found in their original sites. As natural vegetation is cut for farming, grazing, or simply for cooking and heating fuel, fewer plants remain. Although scientists from around the world have worked on this problem for decades, the situation has not improved. The protection of Asian flora, along with the animals that depend on these flora, guarantees better lives for future generations.

Guofan Shao
Jinshuang Ma

Fauna

Asia is a vast and diverse continent with many different types of climate, temperature extremes, and large and small human populations. Many portions of Asia, such as Siberia, are almost empty of people. Some areas are quite densely populated, such as Java in Indonesia, Japan, India, and eastern China. Unfortunately for the fauna of Asia, in those areas where the human population is thin, the extremes of climate do not allow for an abundance of wildlife.

Many species are near extinction, endangered, or at risk because those areas of Asia that have a climate good for fauna are also densely populated. Some species, however, may have become too specialized for survival. The giant panda, for example, eats only bamboo and has an inefficient digestive system. This requires a giant panda to eat large quantities of bamboo. The panda also reproduces slowly. This characteristic may have evolved because the

Habitats and Selected Vertebrates of Asia

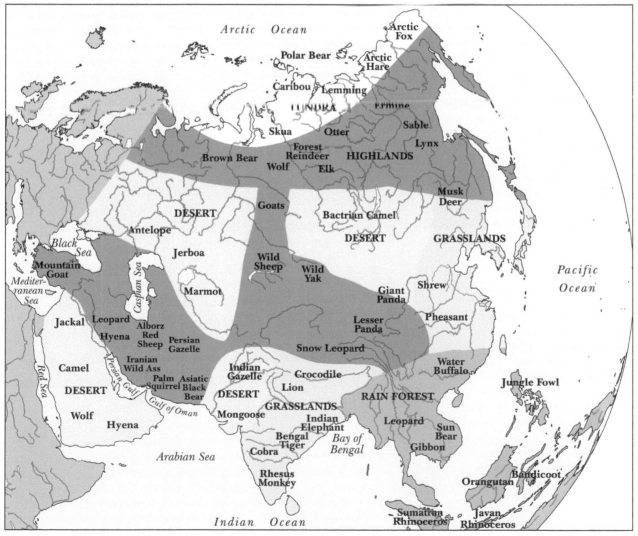

panda has few or no natural enemies except humans. This type of adaptation can prove to be fatal for a species when factors such as climate, habitat, and predators change.

Much of Asian fauna is under pressure as a result of habitat loss and overhunting. Most primates worldwide live only in tropical rain forests. The tropical rain forest is being rapidly cleared for timber, firewood, and land for agriculture. Species such as the rhinoceros are disappearing because they are hunted down just for their horns, which have been used for dagger handles in Yemen and for medicinal purposes. Many animals of all species are slaughtered because they are considered to be pests or simply in the way of the ever-growing human population.

The Tundra, Taiga, and Steppes
Northern Siberia can be characterized as tundra. Because the tundra is partly free from snow only during the short summer, conditions for life are poor. The principal animals of the tundra are the reindeer, Arctic hare, Arctic fox, wolf, and lemming. With the exception of the lemming, they live in the tundra in the summer only and migrate in autumn. Birds can be found in the tundra, but with the ex-

Endangered Mammals of Asia

Common Name	Scientific Name	Range	Status
Bear, Tibetan blue	*Ursus arctos pruinosus*	China (Tibet)	Trading restricted in US due to CITES*
Camel, wild bactrian	*Camelus ferus*	Southern Mongolia, northern China	Critically endangered
Deer, black musk	*Moschus spp.*	Central and eastern Asia	Endangered
Elephant, Asian	*Elephas maximus*	South-central and southeastern Asia	Endangered
Gazelle, Arabian	*Gazella*	Arabian Peninsula, Palestine, Sinai	Vulnerable
Gibbons	*Hylobates spp.*	China, India, Southeast Asia	Vulnerable
Langu, pig-tailed	*Simias concolor*	Islands off the coast of Sumatra, in Indonesia	Critically endangered
Leopard, snow	*Panthera uncia*	Central Asia	Vulnerable
Lion, Asiatic	*Panthera leo persica*	Turkey to India	Endangered
Orangutan, Sumatra	*Pongo abelii*	Sumatra	Critically endangered
Panda, giant	*Ailuropoda melanoleuca*	China	Vulnerable
Rhinoceros, Javan	*Rhinoceros sondaicus*	Indonesia, Indochina, Burma, Thailand, Sikkim, Bangladesh, Malaysia	Critically endangered
Tiger, South China	*Panthera tigris tigris*	Southern China	Critically endangered (possibly extinct in the wild) (60 living in zoos)
Yak, wild	*Bos mutus*	China (Tibet), India	Vulnerable

*Note: *CITES (Convention on International Trade in Endangered Species of Wild Fauna and Flora, also known as the Washington Convention) is a multilateral treaty to protect endangered plants and animals.*
Source: U.S. Fish and Wildlife Service, U.S. Department of the Interior.

ceptions of the willow grouse and ptarmigan, they also flee the tundra in winter. Many species of waders, the gray plover, and several kinds of sandpipers migrate to the tundra and breed there in the summer. The snow bunting and the Lapland bunting are also found there. Gyrfalcons (a type of large Arctic falcon), buzzards, and skuas feed on these smaller birds and lemmings.

The taiga takes in much of the rest of Siberia and is forested mostly by pine trees. Fauna is richer and more diverse in the taiga than in the tundra, because the greater degree of vegetation provides more food and cover. Mammals found there include the brown bear, wolf, glutton (a kind of wolverine), otter, ermine, sable, lynx, elk, and forest reindeer. The rivers of northern Asia have many species of freshwater fish and several types of sturgeon.

The steppes are the southern edges of Siberia, portions of Kazakhstan, western China, and northern Tibet. Those areas are relatively treeless and similar to the northern area of the U.S. Great Plains. The steppes were a place of origin for cattle, the horse, and the Bactrian (two-humped) camel. The animal life of the steppes includes burrowing rodents, jerboas, marmots, and piping hares, and larger animals such as diverse types of antelope. Wild sheep and goats live in the mountains and the plateau areas north of the Himalayas. Tibet is the home of the wild yak, long classified an endangered species, but since 2008 reclassified as "vulnerable."

East and Southwest Asia

Northeastern and eastern China, Korea, and Japan have several native species of deer. The giant panda lives in the lower mountain area of China near Tibet. In 2016, the giant panda's status was upgraded from "endangered" to "vulnerable." The red panda, now classified in its own distinct family, is native to the Himalayas. In 2008, the red panda was listed as an endangered species.

A rare animal in East Asia is the Siberian tiger, an endangered species that feeds on elk and inhabits a corner of the Russian Far East and possibly small

A giant panda, China's most famous endangered and endemic species, at the Chengdu Research Base of Giant Panda Breeding in Sichuan.

portions of China and North Korea. As of 2018, there were estimated to be 480 to 540 Siberian tigers in the wild. Their numbers have risen thanks to strident conservation efforts. A critically endangered animal is the Chinese alligator. The remaining populations of these timid reptiles, approximately 300 in the wild as of 2017, are limited to just a half-dozen regions within heavily populated southeastern Anhui Province. One of just two alligator species in the world, Chinese alligators are believed to have diverged from their American counterparts at least 20 million years ago. They reach lengths of about 5 to 7 feet (1.5 to 2 meters), which is only half the size of American alligators.

The great rivers of China support a rich variety of fish. In the Yangtze and Yellow (Huang He) rivers, the paddlefish is found. Its only close relative is the paddlefish of North America. The giant salamander, which can grow to a length of 4.7 feet (1.4 meters) or more, is found in Japanese waters. Most members of the carp family are in Southeast Asia and southern China.

The Bali myna is found only on Bali and is critically endangered.

FASHION VICTIM: THE TIBETAN ANTELOPE

Known by its elegant, lyrate horns and the striking black markings on its face and legs, the male Tibetan antelope, or chiru, is considered one of the world's most beautiful mammals. It inhabits the windswept Tibetan steppe. At the end of the twentieth century, the population of the chiru was down to perhaps 75,000 animals, from an estimated several million earlier in the century. By 2015, however, the population appeared to be increasing to around 150,000 animals in the wild. The chiru's wool, called shatoosh, is considered to be the finest in the world and a growing status symbol in Western fashion. The Tibetan antelopes must be slaughtered to harvest the wool, and it takes an average of three to four antelopes to make a single scarf. Ironically, sale of shatoosh has been illegal under the Convention on International Trade in Endangered Species (CITES) since 1979. However, in 1994, $100,000 worth of shahtoosh shawls were sold illegally at a U.S. charity auction to raise money for cancer patients—leading to the country's first criminal prosecution for shahtoosh sales. And as late as 1998, fashion magazines touted the virtues of shatoosh scarves, often priced in the thousands of dollars.

Japan's fauna includes the bear, wild boar, fox, deer, and antelope. Some of these species are very different from those on the Asian mainland. The Japanese macaque inhabits many areas. Those at the northern tip of Hōnshū form the northern limit of primate habitat in the world. There are eagles, hawks, falcons, pheasants, and more than 150 species of songbirds. Waterbirds include gulls, auks, grebes, and albatrosses. Reptiles include sea turtles, freshwater tortoises, sea snakes, and two species of poisonous snakes.

The Philippines has about 220 species of mammals, including as many as fifty-six species of bats. There are more than 500 species of birds, including jungle fowl (related to the chicken). The critically endangered Philippine (or monkey-eating) eagle is found in a few locales on Mindanao and Luzon. Fossils show that elephants once existed in the Philippines. It is possible that the climate of the Philippines was much drier in the somewhat distant past.

The population of Asian elephants in Thailand's wild has fallen to an estimated 2,000–3,000.

If so, there may have been savanna-like grassland that would have favored elephants. As the climate subsequently became wetter, tropical rain forest grew, which would have reduced the elephants' habitat and thus their population.

West Asia

Fauna in Iran includes the leopard, bear, hyena, wild boar, ibex, and gazelle, all of which inhabit the forested mountains. Seagulls, ducks, and geese line the shores of the Caspian Sea and the Persian Gulf. Buzzards nest in the desert. Deer, hedgehogs, foxes, and more than twenty species of rodents live in the semidesert high-altitude regions. Palm squirrels, Asiatic black bears, and perhaps a few lions inhabit Balochistan in the southeast. Amphibians and reptiles include frogs, salamanders, boas, racers, rat snakes, cat snakes, and vipers. More than 200 varieties of fish are found in the Persian Gulf, along with shrimp, lobster, and turtles. Sturgeon is found in the Caspian Sea.

The Arabian Peninsula has camels, both wild and domesticated, sheep, goats, and the Arabian horse (now rare there). Gazelles, oryx, and the ibex are becoming rare. Other wild animals are the hyena, wolf, and jackal. The baboon, fox, ratel, rabbit, hedgehog, and jerboa are among the smaller animals. Reptiles include the horned viper, a species of cobra; striped sea snakes; and the large desert monitor. Common birds include eagles, vultures, owls, and the lesser bustard. Flamingos, pelicans, and egrets live on the coasts.

Common insects include locusts, which can descend on fields like a biblical plague. Turkey has the wolf, fox, boar, wildcat, beaver, marten, jackal, hyena, bear, deer, gazelle, and mountain goat. Game birds are the partridge, wild goose, quail, and bustard. The rest of Southwest Asia—Israel, Syria, Lebanon, Jordan, and Iraq—have fauna that are a mix of those found in Saudi Arabia and Turkey.

The Oriental Region

Zoogeographers use the term "Oriental Region" to describe an area that includes India and extends eastward from India over the mainland and much of insular Southeast Asia. A major portion of the Oriental Region is tropical. Such a climate supports malaria-bearing mosquitoes and water-borne flukes carrying the schistosomiasis-causing parasite. These two problems are present in much of tropical Asia as well as in Africa.

In tropical rain-forest areas, monkeys are common. Larger primates are found only in tropical rain forests because the kind of cover and food supply that they seem to need exists only there. Gibbons are found in Assam in northeast India; Myanmar; Indochinese Thailand, Laos, Cambodia, and Vietnam; and the Greater Sunda Islands, Java, Sumatra, and Borneo. The orangutan is found only in Sumatra and Borneo. Indonesia is home to the world's two most critically endangered rhinoceros species, the Javan and Sumatran. In 2018, fewer than eighty Sumatran rhinos were living in the wild.

The largest group of Javan rhinoceros is in Ujung Kulon National Park in West Java, but the number of rhinoceros there was fewer than seventy in 2019. The only known wild population of Javan rhinos outside of Indonesia in the Cat Loc Nature Reserve in Vietnam, became extinct in 2010.

India

India is an important part of the Oriental zoogeographic region. Almost all orders of mammals are found in India. The primates there include diverse types of monkeys, including the rhesus monkey and the Hanuman langur. Wild herds of Indian elephants can be found in several areas such as the Periyar Lake National Park in Kerala and Bandipur National Park in Karnataka. The Indian rhinoceros is protected at Kaziranga National Park and Manas Wildlife Sanctuary in Assam. There are also four species of large cats: the leopard, snow leopard, Bengal tiger, and Asiatic lion. The Asiatic lion, classified as "endangered" is now found only in the Gir National Park in the Kathiawar Peninsula of

Chiang Dao Wildlife Sanctuary, north of the Thanon Thong Chai Range, in Chiang Mai Province, northern Thailand.

Young orangutan at Bukit Lawang, Sumatra.

Gujarat. Bengal tigers, also considered endangered, are found in the forests of the Tarai region of Uttar Pradesh, Bihar, and Assam; the Ganges delta in West Bengal; the Eastern Ghats; Madhya Pradesh; and eastern Rajasthan. The snow leopard, classified as "vulnerable," is found only in the Himalayan regions.

More than 1,200 species and perhaps 2,000 subspecies of birds are found in India. Herons, storks, ibises, and flamingos are well represented, and many of these are found in the Keoladeo Ghana National Park in Rajasthan. The Rann of Kach forms the nesting ground for one of the world's largest breeding colonies of flamingos.

Crocodiles are found in India's rivers, swamps, and lakes. The estuarine crocodile, which can grow as large as 30 feet (9 meters), feeds on the fish, birds, and crabs of muddy delta areas. The long-snouted gavial or gharial, which is similar to the crocodile, is found in several large rivers, including the Ganges

and Brahmaputra. There are almost 400 species of snakes. One-fifth of these are poisonous, including kraits and cobras. The Indian python inhabits marshy areas and grasslands. More than 2,000 species of fish are found in India, 20 percent of which are freshwater species. Commercially valuable insects include the silkworm, bee, and lac insect. The lac insect secretes a sticky, resinous material called "lac," from which shellac and a red dye are made.

Southeast Asia

Southeast Asia is located where two important divisions of the world's fauna come together. It constitutes the eastern half of the Oriental zoogeographic region. Bordering on the south and east is the Australian zoogeographic region. The eastern part of the Southeast Asian islands—Sulawesi (Celebes), the Moluccas, and the Lesser Sunda Islands (Bali, Sumba, Flores, and Timor)—forms an area of transition between these two faunal regions. Southeast Asia thus has a considerable diversity of wildlife throughout the region. The region has placental mammals as opposed to the marsupials of Australia, but also has hybrid species such as the bandicoot of eastern Indonesia. Small mammals such as monkeys and shrews are the most common. Larger mammals have been pushed into remote areas and national preserves.

Indonesia is located in the transitional zone between the Oriental and Australian faunal regions. The so-called boundary between these two zones is known as Wallace's Line. The line runs between

Japanese macaques at Jigokudani hot spring.

Siberian Ibex (Capra sibirica).

Borneo and Sulawesi in the north and Bali and Lombok in the south. A unique species of proboscis monkey lives only in Kalimantan (southern Borneo). The babirusa (hoglike animal with curved tusks) and anoa (a small, wild ox with straight horns) are found only in Sulawesi. A giant lizard, the Komodo dragon, occurs only on two small islands, Rinca and Komodo. It is classified as a "vulnerable" species, with approximately 3,000 individuals counted in a 2015 census. Insect life in Indonesia includes walkingsticks, large atlas beetles, luna moths, and bird-wing swallowtails.

DISCUSSION QUESTIONS: FLORA AND FAUNA

Q1. What important crops in Asia are derived from plants not native to the region? Where did these plants originate? How and when did they get to Asia?

Q2. Where do rain forests occur in Asia? What is the magnitude of their destruction, and why is it happening? How would losing the Asian rain forests affect Earth's atmosphere? How does the Asian rain-forest situation compare to what's happening in the Amazon?

Q3. Name several Asian animal species that have been classified as endangered or critically endangered. What are the factors surrounding the species' declining numbers? How can the trends be reversed?

Mammals in Vietnam include elephants, tapirs, tigers, leopards, rhinoceros, wild oxen such as gaurs and koupreys, black bears, sun bears, and several species of deer such as the small musk deer and barking deer. In Cambodia, small populations of elephants, wild oxen, rhinoceros, and several deer species can still be found, along with tigers, leopards, and bears. Snakes abound, with the four most dangerous species being the Indian cobra, the king cobra, the banded krait, and Russell's viper. The fauna of Myanmar and Thailand are similar to those found in Cambodia.

Dana P. McDermott

HUMAN GEOGRAPHY

OVERVIEW

THE HUMAN ENVIRONMENT

No person lives in a vacuum. Every human being and community is surrounded by a world of external influences with which it interacts and by which it is affected. In turn, humans influence and change their environments: sometimes intentionally, sometimes not, and sometimes with effects that are harmful to these environments, and, in turn, to humans themselves. Humans have always shaped the world in which they live, but developments over the past few centuries have greatly enhanced this capacity.

Many people feel a sense of alarm about the consequences of widespread adoption of modern technology, including artificial intelligence (AI) and accelerating human population growth in the world. Travel and transportation among the world's regions have been made surer, safer, and faster, and global communication is virtually instantaneous. The human environment is no longer a matter of local physical, biological, or social conditions, or even of merely national or regional concerns—the postmodern world has become a true global community.

Students of human geography divide the human environment into three broad areas: the physical, biological, and social environments. The study of ecology describes and analyzes the interactions of biological forms (mainly plants and animals) and seeks to uncover the optimal means of species cooperation, or symbiosis. Everything that humans do affects life and the physical world around them, and this world provides potentials for and constraints on how humans can live.

As people acquired and shared ever-more knowledge about the world, their abilities to alter and shape it increased. Humans have always had a direct impact on Earth. Even 10,000 years ago, Neolithic people cut down trees, scratched the earth's surface with simple plows, and replaced diverse plant forms with single crops. From this basic agricultural technology grew more complex human communities, and people were freed from the need to hunt and gather. The alteration of the local ecosystems could have deleterious effects, however, as gardens turned eventually to deserts in places like North Africa and what later became Iraq. Those who kept herds of animals grazed them in areas rich in grasses, and animal fertilizer helped keep them rich. If the area was overgrazed, however, destroying important ground cover, the herders moved on, leaving a perfect setup for erosion and even desertification. Today, people have an even greater ability to alter their environments than did Neolithic people, and ecologists and other scientists as well as citizens and politicians are increasingly concerned about the negative effects of modern alterations.

The Physical Environment

The earth's biosphere is made up of the atmosphere—the mass of air surrounding the earth; the hydrosphere—bodies of water; and the lithosphere—the outer portion of the earth's crust. Each of these, alone and working together, affect human life and human communities.

Climate and weather at their most extreme can make human habitation impossible, or at least extremely uncomfortable. Desert and polar climates do not have the liquid water, vegetation, and animal life necessary to sustain human existence. Humans can adapt to a range of climates, however. Mild variations can be addressed simply, with clothing and shelter. Local droughts, tornadoes, hurricanes, heavy winds, lightning, and hail can have devastating effects even in the most comfortable of climates. Excess rain can be drained away to make habitable land, and arid areas can be irrigated. Most people live in temperate zones where weather extremes are rare or dealt with by technological adaptation. Heating and, more recently, air conditioning can create healthy microclimates, whatever the external conditions. Food can be grown and then transported across long distances to supply populations throughout the year.

The hydrosphere affects the atmosphere in countless ways, and provides the water necessary for human and other life. Bodies of water provide plants and animals for food, transportation routes, and aesthetic pleasure to people, and often serve to flush away waste products. People locate near water sources for all of these reasons, but sometimes suffer from sudden shifts in the water level, as in tidal waves (tsunamis) or flooding. Encroachment of salt water into freshwater bodies (salination) is a problem that can have natural or human causes.

The lithosphere provides the solid, generally dry surface on which people usually live. It has been shaped by the atmosphere (especially wind and rain that erode rocks into soil) and the hydrosphere (for example, alluvial deposits and beach erosion). It serves as the base for much plant life and for most agriculture. People have tapped its mineral deposits and reshaped it in many places; it also reshapes itself through, for example, earthquakes and volcanic eruption. Its great variations—including vegetation—draw or repel people, who exploit or enjoy them for reasons as varied as recreation, military defense, or farming.

The Biological Environment

Humans share the earth with over 8 million different species of plants, animals, and microorganisms—of which only about 2 million have been identified and named. As part of the natural food chain, people rely upon other life-forms for nourishment.

Through perhaps the first 99 percent of human history, people harvested the bounty of nature in its native setting, by hunting and gathering. Domestication of plants and animals, beginning about 10,000 years ago, provided humans a more stable and reliable food supply, revolutionizing human communities. Being omnivores, people can use a wide variety of plants and animals for food, and they have come to control or manage most important food sources through herding, agriculture, or mechanized harvesting. Which plants and animals are chosen as food, and thus which are cultivated, bred, or exploited, are matters of human culture, not, at least in the modern world, of necessity.

Huge increases in human population worldwide have, however, put tremendous strains on provision of adequate nourishment. Areas poorly endowed with foodstuffs or that suffer disastrous droughts or blights may benefit from the importation of food in the short run, but cannot sustain high populations fostered by medical advances and cultural considerations.

Human beings themselves are also hosts to myriad organisms, such as fungi, viruses, bacteria, eyelash mites, worms, and lice. While people usually can coexist with these organisms, at times they are destructive and even fatal to the human organism. Public health and medical efforts have eradicated some of humankind's biological enemies, but others remain, or are evolving, and continue to baffle modern science.

The presence of these enemies to health once played a major role in locating human habitations to avoid so-called "bad air" (*mal-aria*) and the breeding grounds of tsetse flies or other pests. The use of pesticides and draining of marshy grounds have alleviated a good deal of human suffering. Human efforts can also control or eliminate biological threats to the plants and animals used for food, clothing, and other purposes.

Social Environments

Human reproduction and the nurturing of young require cooperation among people. Over time, people gathered in groups that were diverse in age if not in other qualities, and the development of towns and cities eventually created an environment in which otherwise unrelated people interacted on intimate and constructive levels. Specialization, or division of labor, created a higher level of material wealth and culture and ensured interpersonal reliance.

The pooling of labor—both voluntary and forced—allowed for the creation of artificial living environments that defied the elements and met human needs for sustenance. Some seemingly basic human drives of exclusivity and territoriality may be responsible for interpersonal friction, violence and, at the extreme, war. Physical differences, such as size, skin, or hair color, and cultural differences, including language, religion, and customs, have often divided humans or communities. Even within close quarters such as cities, people often separate themselves along lines of perceived differences. Human social identity comes from shared characteristics, but which things are seen as shared, and which as differentiating, is arbitrary.

People can affect their social environment for good and ill through trade and war, cooperation and bigotry, altruism and greed. While people still are somewhat at the mercy of the biological and physical environments, technological developments have balanced the human relationship with these. Negative effects of human interaction, however, often offset the positive gains. People can seed clouds for rain, but also pollute the atmosphere around large cities, create acid rain, and perhaps contribute to global warming.

Human actions can direct water to where it is needed, but people also drain freshwater bodies and increase salination, pollute streams, lakes, and oceans, and encourage flooding by modifying riverbeds. People have terraced mountainsides and irrigated them to create gardens in mountains and deserts, but also lose about 24 billion metric tons of soil to erosion and 30 million acres (12 million hectares) of grazing land to desertification each year. These negative effects not only jeopardize other species of terrestrial life, but also humans' ability to live comfortably, or perhaps at all.

Globalization

Humankind's ability to affect its natural environments has increased enormously in the wake of the Industrial Revolution. The harnessing of steam, chemical, electrical, and atomic energy has enabled people to transform life on a global scale. Economically, the Western world still to dominates global markets despite effort of China to capture the crown, and computer and satellite technology have made even remote parts of the globe reliant on Western information and products. Efficient transportation of goods and people over huge distances has eliminated physical barriers to travel and commerce. The power and influence of multinational corporations and national corporations in international markets continues to grow.

Human environmental problems also have a global scope: Extreme weather, changes in ocean temperatures and sea level rise, global warming, and the spread of disease by travelers have become planetary concerns. International agencies seek to deal with such matters, and also social and political concerns once left to nations or colonial powers, such as population growth, the provision of justice, or environmental destruction within a country. Pessimists warn of horrendous trends in population and ecological damage, and further deterioration of human life and its environments. Optimists dismiss negative reports as exaggerated and alarmist, or expect further technological advances to mitigate the negative effects of human action.

Joseph P. Byrne

POPULATION GROWTH AND DISTRIBUTION

The population of the world has been growing steadily for thousands of years and has grown more in some places than in others. On November 2019, the total population of the earth had reached 7.7 billion people. The population of the United States in August 2019 was approximately 329.45 million. India's population in November 2019 was 1.37 billion, making it the world's second most populous country. China's population was about 1.45 billion—about 1 in 5 people on the planet.

How Populations Are Counted

The U.S. Constitution requires that a census, or enumeration, of the population of the United States be conducted every ten years. The U.S. Census Bureau mails out millions of census forms and pays thousands of people (enumerators) to count people that did not fill out their census forms. This task cost about US$5.6 billion in the year 2010, and estimates for the 2020 census have risen to over US$15 billion. Despite this great effort, millions of people are probably not counted in every U.S. census. Moreover, many countries have much less money to spend on censuses and more people to count. Therefore, information about the population of many poor or less-developed countries is even less accurate than that for the population of the United States.

Counting how many people were alive a hundred, a thousand, or hundreds of thousands of years ago is even more difficult. Estimates are made from archaeological findings, which include human skeletons, ruins of ancient buildings, and evidence of ancient agricultural practices. Historical records of births, deaths, taxes paid, and other information are also used. Although it is not possible to estimate the global population 1,000 years ago with great accuracy, it is a fascinating topic, and many people have participated in estimating the total population of the planet through the ages.

History of Human Population Growth

Ancient ancestors of humans, known as hominids, were alive in Africa and Europe around 1 million years ago. It is believed that modern humans (*Homo sapiens sapiens*) coexisted with the Neanderthals (*Homo sapiens neandertalensis*) about 100,000 years ago. By 8000 BCE (10,000 years ago) fully modern humans numbered around 8 million. If the presence of archaic *Homo sapiens* is accepted as the beginning of the human population 1 million years ago, then the first 990,000 years of human existence are characterized by a very low population growth rate (15 persons per million per year).

Around 10,000 years ago, humans began a practice that dramatically changed their growth rate: planting food crops. This shift in human history, called the Agricultural Revolution, paved the way for the development of cities, government, and civilizations. Before the Agricultural Revolution, there were no governments to count people. The earliest censuses were conducted less than 10,000 years ago in the ancient civilizations of Egypt, Babylon, China, Palestine, and Rome. For this reason, historical estimates of the earth's total population are difficult to make. However, there is no argument that human numbers have increased dramatically in the past 10,000 years. The dramatic changes in the growth rates of the human population are typically attributed to three significant epochs of human cultural evolution: the Agricultural, Industrial, and Green Revolutions.

Before the Agricultural Revolution, the size of the human population was probably fewer than 10 million people, who survived primarily by hunting and gathering. After plant and animal species were domesticated, the human population increased its growth rate. By about 5000 BCE, gains in food production caused by the Agricultural Revolution meant that the planet could support about 50 million people. For the next several thousand years, the human population continued to grow at a rate of about 0.03 percent per year. By the first year of the common era, the planet's population numbered about 300 million.

At the end of the Middle Ages, the human population numbered about 400 million. As people lived in densely populated cities, the effects of disease increased. Starting in 1348 and continuing to 1650, the human population was subjected to massive declines caused by the bubonic plague—the Black Death. At its peak in about 1400, the Black Death may have killed 25 percent of Europe's population in just over fifty years. By the end of the last great plague in 1650, the human population numbered 600 million.

The Industrial Revolution began between 1650 and 1750. Since then, the growth of the human population has increased greatly. In just under 300 years, the earth's population went from 0.5 billion to 7.7 billion people, and the annual rate of increase went from 0.1 percent to 1.1 percent. This population growth was not because people were having more babies, but because more babies lived to become adults and the average adult lived a longer life.

The Green Revolution occurred in the 1960s. The development of various vaccines and antibiotics in the twentieth century and the spread of their use to most of the world after World War II caused big drops in the death rate, increasing population growth rates. Feeding this growing population has presented a challenge. This third revolution is called the Green Revolution because of the technology used to increase the amount of food produced by farms. However, the Green Revolution was really a combination of improvements in health care, medicine, and sanitation, in addition to an increase in food production.

Geography of Human Population Growth

The present-day human race traces its lineage to Africa. Humans migrated from Africa to the Middle East, Europe, Asia, and eventually to Australia, North and South America, and the Pacific Islands. It is believed that during the last Ice Age, the world's sea levels were lower because much of the world's water was trapped in ice sheets. This lower sea level created land bridges that facilitated many of the major human migrations across the world.

Patterns of human settlement are not random. People generally avoid living in deserts because they lack water. Few humans are found above the Arctic Circle because of that region's severely cold climate. Environmental factors, such as the availability of water and food and the livability of climate, influence where humans choose to live. How much these factors influence the evolution and development of human societies is a subject of debate.

The domestication of plants and animals that resulted from the Agricultural Revolution did not take place everywhere on the earth. In many parts of the world, humans remained as hunter-gatherers while agriculture developed in other parts of the world. Eventually, the agriculturalists outbred the hunter-gatherers, and few hunter-gatherers remain in the twenty-first century. Early agricultural sites have been found in many places, including Central and South America, Southeast Asia and China, and along the Tigris and Euphrates Rivers in what is now Iraq. The practice of agriculture spread from these areas throughout most of the world.

By the time Christopher Columbus reached the Americas in the late fifteenth century, there were millions of Native Americans living in towns and villages and practicing agriculture. Most of them died from diseases that were brought by European colonists. Colonization, disease, and war are major mechanisms that have changed the composition and distribution of the world's population in the last 300 years.

The last few centuries also produced another change in the geography of the human population. During this period, the concentration of industry in urban areas and the efficiency gains of modern agricultural machinery caused large numbers of people to move from rural areas to cities to find jobs. From 1900 to 2020 the percentage of people living in cities went from 14 percent to just about 55 percent. Demographers estimate that by the year 2025, more than 68 percent of the earth's population will live in cities. Scientists estimate that the human population will continue to increase until the year

2050, at which time it will level out at between 8 and 15 billion.

Earth's Carrying Capacity

Many people are concerned that the earth cannot grow enough food or provide enough other resources to support 15 billion people. There is great debate about the concept of the earth's carrying capacity—the maximum human population that the earth can support indefinitely. Answers to questions about the earth's carrying capacity must account for variations in human behavior. For example, the earth could support more bicycle-riding vegetarians than car-driving carnivores. Questions about carrying capacity and the environmental impacts of the human race on the planet are fundamental to the United Nations' goals of sustainable development. Dealing with these questions will be one of the major challenges of the twenty-first century.

Paul C. Sutton

GLOBAL URBANIZATION

Urbanization is the process of building and living in cities. Although the human impulse to live in groups, sharing a "home base" probably dates back to cave-dweller times or before. The creation of towns and cities with a few hundred to many thousands to millions of inhabitants required several other developments.

Foremost of these was the invention of agriculture. Tilling crops requires a permanent living place near the cultivated land. The first agricultural villages were small. Jarmo, a village site from c. 7000 BCE, located in the Zagros Mountains of present-day Iran, appears to have had only twenty to twenty-five houses. Still, farmers' crops and livestock provided a food surplus that could be stored in the village or traded for other goods. Surplus food also meant surplus time, enabling some people to specialize in producing other useful items, or to engage in less tangible things like religious rituals or recordkeeping.

Given these conditions, it took people with foresight and political talents to lead the process of city formation. Once in cities, however, the inhabitants found many benefits. Walls and guards provided more security than the open country. Cities had regular markets where local craftspeople and traveling merchants displayed a variety of goods. City governments often provided amenities like primitive street lighting and sanitary facilities. The faster pace of life, and the exchange of ideas from diverse people interacting, made city life more interesting and speeded up the processes of social change and invention. Writing, law, and money all evolved in the earliest cities.

Ancient and Medieval Cities

Cities seem to have appeared almost simultaneously, around 3500 BCE, in three separate regions. In the Fertile Crescent, a wide curve of land stretching from the Persian gulf to the northwest Mediterranean Sea, the cities of Ur, Akkad, and Babylon rose, flourished, and succeeded one another. In Egypt, a connected chain of cities grew, soon unified by a ruler using Memphis, just south of the Nile River's delta, as his strategic and ceremonial base. On the Indian subcontinent, Mohenjo-Daro and Harappa oversaw about a hundred smaller towns in the Indus River valley. Similar developments took place about a thousand years later in northern China.

These first city sites were in the valleys of great river systems, where rich alluvial soil boosted large-scale food production. The rivers served as a "water highway" for ships carrying commodities and luxury items to and from the cities. They also furnished water for drinking, irrigation, and waste disposal. Even the rivers' rampages promoted civilization, as making flood control and irrigation systems required practical engineering, an organized workforce, and ongoing political authority to direct them.

Eurasia was still full of peoples who were not urbanized, however, and who lived by herding, pirating, or raiding. Early cities declined or disappeared, in some cases destroyed by invasions from such forces around 1200 BCE. Afterward, the cities of Greece became newly important. Their surrounding land was poor, but their access to the sea was an advantage. Greek cities prospered from fishing and trade. They also developed a new idea, the city-state, run by and for its citizens.

Rome, the Greek cities' successor to power, reached a new level of urbanization. Its rise owed more to historical accident and its citizens' political and military talents than to location, but some geographical features are salient. In some ways, the fertile coastal plain of Latium was an ideal site for a great city, central to both the Italian peninsula and the Mediterranean Sea. There, the Tiber River becomes navigible and crossable.

In other ways, Rome's site was far from ideal. Its lower areas were swampy and mosquito-ridden. The seven hills, with their sacred sites later filled with public buildings and luxury houses, imposed a crazy-quilt pattern on the city's growth. Romans built cities with a simple rectangular plan all over Europe and the Middle East, but their home city grew in a less rational way.

At its peak, Rome had a million residents, a population no other city reached before nineteenth century London. It provided facilities found in modern cities: a piped water supply, a sewage disposal system, a police force, public buildings, entertainment districts, shops, inns, restaurants, and taverns. The streets were crowded and noisy; to control traffic, wheeled wagons could make deliveries only at night. Fire and building collapse were constant risks in the cheaply built apartment structures that housed the city's poorer residents. Still, few wanted to live anywhere but in Rome, their world's preeminent city.

In the Early Middle Ages after the western Roman Empire collapsed, feudalism, based on land holdings, eclipsed urban life. Cities never disappeared, but their populations and services declined drastically. Urban life still flourished for another millenium in the eastern capital of Constantinople. When Islam spread across the Middle East, it caused the growth of new cities, centered around a mosque and a marketplace.

In the twelfth and thirteenth centuries, life revived in Western Europe. As in the Islamic cities, the driving forces were both religious—the building of cathedrals—and commercial—merchants and artisans expanding the reach of their activities. Medieval cities were usually walled, with narrow, twisting streets and a lack of basic sanitary measures, but they drew ambitious people and innovative forces together. Italy's cities revived the concept of the city-state with its outward reach. Venice sent its merchant fleet all over the known world. Farther north, Paris and Bologna hosted the first universities. The feudal system slowly gave way to nation-states ruled by one king.

Modern Cities

Modern cities differ from earlier ones because of changes wrought by technology, but most of today's cities arose before the Industrial Revolution. Until the early nineteenth century, travel within a city was by foot or on horse, which limited street widths and city sizes. The first effect of railroads was to shorten travel time between cities. This helped country residents moving to the cities, and speeded raw materials going into and manufactured goods coming out of the factories that increasingly dotted urban areas. Rail transit soon caused the growth of a suburban ring. Prosperous city workers could live in more spacious homes outside the city and ride rail lines to work every day. This pattern was common in London and New York City.

Factories, the lifeblood of the Industrial Revolution, were built in pockets of existing cities. Smaller cities like Glasgow, Scotland, and Pittsburgh, Pennsylvania, grew as ironworking industries, using nearby or easily transported coal and ore resources, built large foundries there. Neither industrialists nor city authorities worried about where the people working there would live. Workers took whatever housing they could find in tenements or subdivided old mansions.

Beginning in the 1880s, metal-framed construction made taller buildings possible. These skyscrapers towered over stately three- to eight-story structures of an earlier period. Because this technology enabled expensive central-city ground space to house many profitable office suites, up through the 1930s, city cores became quite compacted. Many people believed such skyward growth was the wave of the future and warned that city streets were becoming sunless, dangerous canyons.

Automobiles kept these predictions from fully coming true. As car ownership became widespread, more roads were built or widened to carry the traffic. Urban areas began to decentralize. The car, like rail transit before it, allowed people to flee the urban core for suburban living. Because roads could be built almost anywhere, built-up areas around cities came to resemble large patches filling a circle, rather than the spokes-of-a-wheel pattern introduced by rail lines. Cities born during the automotive age tend to have an indistinct city center, surrounded by large areas of diffuse urban development. The prime example is Los Angeles: It has a small downtown area, but a consolidated metropolitan area of about 34,000 square miles (88,000 sq. km.).

Almost everywhere, urban sprawl has created satellite cities with major manufacturing, office, and shopping nodes. These cause an increasing portion of daily travel within metropolitan areas to be between one edge city and another, rather than to and from downtown. Since these journeys have an almost limitless variety of start points and destinations within the urban region, mass transit is only a partial solution to highway crowding and air pollution problems.

The above trends typify the so-called developed world, especially the United States. Many cities in poor nations have grown even more rapidly but with a different mix of patterns and problems. However, the basic pattern can be detected around the globe, as urban dwellers seek to better their own cir-

URBANIZATION AND DEVELOPING NATIONS

The urban population, or number of people living in cities, in North America accounts for about 75 percent of its total population. In Europe, about 90 percent of the population lives in cities. In developing countries, the urban population is often less than 30 percent. The term "urbanization" refers to the rate of population growth of cities. Urbanization mainly results from people moving to cities from elsewhere. In developing countries, the urbanization rate is very high compared to those of North America or Europe. The high rate of urbanization of these countries makes it difficult for their governments to provide housing, water, sewers, jobs, schools, and other services for their fast-growing urban populations.

cumstances. Today, 55 percent of the world's population lives in urban areas, and that percentage is expected to rise to 68 percent by 2050. Projections show that urbanization combined with the overall growth of the world's population could add another 2.5 billion people to urban areas by 2050, with close to 90 percent of this increase taking place in Asia and Africa, according to a United Nations data set published in May 2018.

Megacities and the Future

In the year 2019 the world had thirty-three megacities, defined as urban areas with a population of 10 million or more. The largest was Tokyo, with an estimated 37.5 million people in 2018, predicted to grow to around 37 million by 2030. Second-largest was Delhi, with more than 28.5 million in 2018 and predicted to grow to around 38.94 million by 2030. Megacities in the United States include New York-Newark with a population of 18.8 million and Los Angeles at 12.5.

Megacities profoundly affect the air, weather, and terrain of their surrounding territory. Smog is a feature of urban life almost everywhere, but is worse where the exhaust from millions of cars mixes with industrial pollution. Some megacities have slowed the problem by regulating combustion technology; none have solved it. Huge expanses of soil pre-

URBAN HEAT ISLANDS

Large cities have distinctly different climates from the rural areas that surround them. The most important climatic characteristic of a city is the urban heat island, a concentration of relatively warmer temperatures, especially at nighttime. Large cities are frequently at least 11 degrees Fahrenheit (6 degrees Celsius) warmer than the surrounding countryside.

The urban heat island results from several factors. Primary among these are human activities, such as heating homes and operating factories and vehicles, that produce and release large quantities of energy to the atmosphere. Most of these activities involve the burning of fossil fuels such as oil, gas, and coal. A second factor is the abundance of heat-absorbing urban materials, such as brick, concrete, and asphalt. A third factor is the surface dryness of a city. Urban surface materials normally absorb little water and therefore quickly dry out after a storm. In contrast, the evaporation of moisture from wet soil and vegetation in rural areas uses a large quantity of solar energy—often more than is converted directly to heat—resulting in cooler air temperatures and higher relative humidities.

empted by buildings and pavements can turn heavy rains into floods almost instantly, and the ambient heat in large cities stays several degrees higher than in comparable rural areas. Recent engineering studies suggest that megacities create instability in the ground beneath, compressing and undermining it.

How will cities evolve? Barring an unforeseen technological or social breakthrough, the current growth and problems will probably continue. The process of megapolis—metropolitan areas blending together along the corridors between them—is well underway in many areas. Predictions that the computer will so change the nature of work as to cause massive population shifts away from cities have not been proven correct. Despite its drawbacks, increasing numbers of people are drawn to urban life, seeking the economic opportunities and wider social world that cities offer.

Emily Alward

PEOPLE

The first anatomically and behaviorally modern human beings to settle in the Middle East and then spread throughout Asia did so approximately 50,000 years ago. They originated in Africa and migrated first to Asia and then onward in search of more reliable and plentiful food supplies.

Earliest Humans

The distant ancestors of these modern humans (or hominids, as scientists call them) inhabited Earth approximately 6 million years ago and journeyed into Eurasia more than 1 million years ago. Hunting during the Stone Age or Paleolithic period (about 2.5 million years ago) could not support a growing population. Only two people could survive on the vegetation and wildlife available in an area of 1 square mile (2.6 sq. km.) on the African grassland, scientists calculate. These early humans—and the people who came later—migrated over great distances just to survive.

Homo erectus, the first primate to look like modern humans and the first to use fire and manufacture relatively sophisticated tools, was also the first human species to migrate from Africa into Asia. These earlier humans primarily settled in areas with benign climates and on savanna lands that were similar to those where they had evolved. These regions included the Middle East, the Indian subcontinent, Southeast Asia, and parts of China. Archaeological evidence suggests that the first would-be settlers found interior Eurasia virtually uninhabitable due to the harsh environment and the limited food resources available.

The early hominids in Africa, the Middle East, and coastal Eurasia relied on a diet of plants and animals. Living in colder climates required a greater reliance on animal-based foods. When they ventured into colder climates, these hominids may have used fire and scavenged animal carcasses for skins or furs to keep warm. *Homo erectus* likely also lacked the social and survival skills needed to settle in interior Eurasia.

The First Migrants

During the Upper Paleolithic period, which began approximately 40,000 years ago, a revolution occurred. Neanderthal humans disappeared, leaving a world inhabited only by *Homo sapiens*. From eastern Africa, *Homo sapiens* emigrated eastward into Asia, not northward through the eastern Mediterranean as previously thought. These first emigrants may have numbered as few as 2,000 individuals—all in search of food. Many geographers and other scientists believe the constant search for food created the foundation for settlement of the world. *Homo sapiens* migrated toward Southeast Asia and Australia, and eventually entered Europe from Asia.

Without certain biological and cultural changes or mutations, however, humans might not have settled the world. For instance, the development of sophisticated language required a fully modern vocal apparatus. Chimpanzees, the primates that are the closest to humans, use tools, have relatively complex social lives, and show signs of self-awareness. But they lack spoken language skills and the ability to manipulate symbols and conceptualize things in remote time and space. Many scientists believe language is essential to humankind's capacity for self-awareness and conscious actions. The ability to form and maintain groups larger than the family unit—such as clans, tribes, villages, cities, kingdoms, and empires—was a prerequisite for large-scale settlements and ultimately, conquest of other settlements. The first thoroughly modern humans lived in family groups that divided labor within the family by sex.

According to the out-of-Africa theory, modern humans migrated from Africa only after they acquired the ability to communicate ideas, to build and navigate boats or rafts, and to manufacture relatively sophisticated stone implements. Some geographers suggest that without environmental changes such as global warming, agrarian and industrial civilizations would not have been possible.

Asian Challenges

Humans faced several challenges in migrating from Africa's warm savanna lands to the colder and drier regions of the world, such as Eurasia. The harsh climate of northern Asia apparently was too great an obstacle until about 40,000 years ago, when innovations in winter clothing and housing made survival in the region possible. Systematic fur trapping of wolves or Arctic foxes provided clothing for the harsh winters. Early humans manufactured bone and stone tools and learned to sew jackets, pants, and boots to protect themselves from the wintry weather in Asia's colder regions, including its barren deserts, high snow-covered mountains, forests, and steppes.

Using these new tools and skills, early humans migrated into the Arctic and subarctic latitudes between 35,000 and 20,000 years ago, and later crossed the Bering gateway to North America. They survived because of advances in housing and the management of fire. The period from 40,000 to 10,000 years ago (the Upper Paleolithic period) saw the spread of relatively sophisticated tools across the southern, more temperate margins of Siberia. The Upper Paleolithic people also penetrated southward to Mongolia, northern China, Korea, and Japan. The people who migrated to Asia settled in one of two distinct regions on the Eurasian landmass: Inner Asia and Outer Asia.

Inner Asia

Inner Asia is a vast landlocked region, encompassing approximately 8 million square miles (20 million sq. km.)—about one-seventh of the world's land. Dominating Inner Asia's landscape is a vast, arid plain, the largest plain or steppe in the world, which includes tundra (frozen desert), forests, grasslands or prairies that constitute the steppes, and the Gobi and other deserts. The geography of the region provided few natural barriers such as mountains and large lakes to protect inhabitants from aggressive marauders and armies. Geography is one factor that contributed to the development of the world's largest land-based empire, the Mongol Empire. The most important gateways into Inner Asia were through the northern and northwestern borders of China; across the Central Asian borders with Iran and Afghanistan, and through the passes of the Caucasus; and the passage between the Black Sea and the Carpathian Mountains that leads from the Balkans.

Inner Asia includes most of the former Soviet Union, Mongolia, and parts of China. Modern humans settled Inner Asia only after they developed warmer clothes, more reliable and less risky hunting skills and technology, and housing. Despite these adaptations, Inner Asia's harsh environment contributed to a low population density. Because of its central location, however, many of Asia's major trade routes passed through it. Agricultural and manufactured products, technologies, religions, and armies journeyed along these trade routes.

Outer Asia

Outer Asia includes well-watered subcontinents that lie in a great arc from Europe, to the Middle East, to India, and to South and East Asia. Population centers and the world's first urban, literate civilizations developed in Outer Asia. The major movements of people, trade goods, and cultural and technical innovations were along land routes such as the Silk Road. Lack of water in the region's sprawling deserts and steppes makes rain-dependent farming and extensive irrigation impractical or impossible.

Before the migration into Eurasia could expand very far geographically, humans had to master the art of hunting. Hunting is more dangerous, difficult, and unreliable than gathering plants. Animals can elude or even attack their pursuers. Successful hunters knew about their prey's grazing habits and regions and the best times to hunt certain animals—mammoths and other fur-bearing animals in the winter and reindeer in the spring. Hunters

Traditional Korean dance (Jinju geommu).

often cooperated with each other to kill their prey in large herds.

Another key to humankind's survival in Asia was their increased ability to communicate and to exchange technological, ecological, and geographical information. Sophisticated language was also central to planning and cooperation. The sophistication and specialization of hunting equipment increased as humans began to narrow their targets to particular animals, such as birds and sea mammals. Advances in weapons, such as in bows and arrows, spear-throwers, and knives, were especially important to survival. Projectile weapons such as the spear-thrower enabled the hunter to throw a spear harder, longer, and faster than it could be thrown with the arm alone. Hunters also needed to learn how to preserve and store their food to keep it from spoiling too quickly.

Significant geographic expansion throughout Asia occurred only as people developed sustainable irrigation systems to feed themselves. Great river systems made great civilizations possible and sustainable. Much of the world's population still lives in the cities that were established near major lakes, rivers, and oceans when the urban way of life first evolved in Southwest Asia. The world's first complex societies evolved in the alluvial plains of the Tigris and Euphrates rivers and their tributaries. Major waterways provided relatively reliable and sustainable supplies of water and fish as a food source. Large-scale irrigation systems such as those found in China and the Middle East permitted the development of empires. Such irrigation systems were first successful in the region from Egypt's Nile Valley through the Tigris and Euphrates valley to the head of the Persian Gulf. Irrigation technology then spread to other hearth areas (points of origin where civilizations began).

Agricultural Advances

During the first agricultural revolution, between 9000 and 7000 BCE, many humans went from a hunting and gathering way of life to growing plants. Advances at this time included the development of seed agriculture, including wheat and rice, the use of plow and draft animals such as sheep and goats, and the harnessing of water supplies for human benefit. Seed agriculture occurred across Southwest Asia from Greece to present-day Turkey and the western part of Iran, northern China, northeast India, and East Africa. The slash-and-burn system of cultivation was also developed. Under that system, plants such as wild cereals are cut close to the ground, left to dry for a time, and ignited. Burning enriches the soil with nutrients from the plants.

Agricultural breakthroughs were possible only where natural plant foods and animals were already plentiful, and the soils were naturally well irrigated and easy to till. People settled in places where water was available and the land suitable for cultivation. The invention of the wheel enabled development of large-scale irrigation systems in the Middle East and China. Agriculture could not have occurred without irrigation and other technological advances in the Fertile Crescent and other hearth areas. Tools included the sickle for harvesting wheat; the plow for working the soil; the yoke for hitching oxen to the plow; and the wheel, used first for grinding wheat, then for drawing water, and finally for transportation.

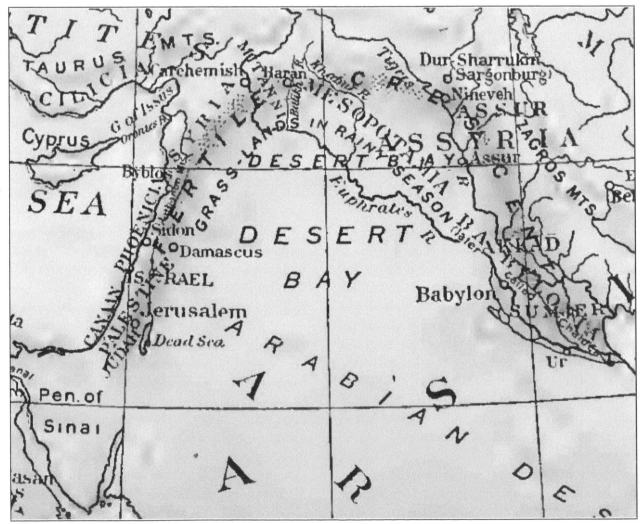

1916 map of the Fertile Crescent by James Henry Breasted, who popularised usage of the phrase.

Domesticated animals enabled humans to increase food supplies by increasing the amount of energy applied to producing food. Humans domesticated cattle and sheep and learned how to hunt rather than rely on gathering for food, clothing, and protection from the cold and harsh environment. The domestication of the horse was pivotal in extending the mobility of nomads in Asia, and the invention of wagons and the domestication of camels after 3000 BCE allowed for the colonization of most of Inner Asia. Portable dwellings made of felt, called yurts, provided shelter in harsh environments.

Hearth Areas

Geographers have identified three regions (hearth areas) in Asia where people learned how to improve their crops and farming techniques and to domesticate animals. Ideas, innovations, ideologies, language, and religion had their sources in cultural hearth areas. The hearth areas provided all the basic ingredients for successful agricultural development—the perfect balance of water, mild climate, plants, animals, and other natural resources. The Indo-European language family, which includes French, Italian, and Spanish, had its origins in Asia in the region bordered by the Tigris-Euphrates valley, the Black Sea, and the Caspian Sea, before spreading west, east, and south, developing into the languages of today.

The hearth areas that played a significant role in the settlement of Asia are in the Middle East, in the Tigris-Euphrates valley around the foothills of the ZagrosZagros Mountains;hearth area Mountains in present-day Iran and Iraq, around the Dead Sea Valley in Jordan and Israel, and on the Anatolian Plateau of Turkey. In South Asia, the hearth area is along the floodplains of the Ganges, Brahmaputra, Indus, and Irawaddy rivers in Assam, Bangladesh, Myanmar, and northern India. In China, the hearth area is along the floodplain of the Yuan River.

The Neolithic people who lived in these hearth areas learned how to breed plants and animals to produce desired genetic characteristics, including disease resistance. Farming supported larger populations than hunting and gathering. The people

freed from the constant need to hunt and gather their food later developed new skills, including governing large groups of people, making jewelry and pottery, weaving, building trade relations with other regions, and, eventually, leading large armies of conquest.

The domestication of plants and animals allowed for the development of settlements and complex societies in the floodplains along the Tigris and Euphrates rivers, especially during the first agricultural revolution. The earliest urban areas developed independently in the various hearth areas of the first agricultural revolution. Urbanization required leaders who could exact tributes, impose taxes, and control labor through either theological persuasion or military might.

The first states and cities emerged in southern Iraq and Iran, in the lower elevations of the great mountains of the Fertile Crescent. These regions offered natural habitats where wild wheat and barley grew naturally, sheep and goats flourished, and rainfall was relatively reliable. The first of the world's cities developed in these hearth areas. Military and economic needs and access to significant water supplies played a role in their development. In the Middle East, cities developed in the valleys of the Tigris and Euphrates rivers by approximately 3500 BCE. By 2500 BCE, cities had developed in the Indus Valley, and by 1800 BCE, in northern China.

Breakthroughs in the hearth areas permitted the transformation from loose communities to a proliferation of settled villages and kinship systems that regulated rights over land and other resources. The early empires of Egypt and China developed as towns and cities became centers of administration, military garrisons, and religious centers. Kings, khans, and other rulers used military power and theological authority to build and hold together their empires. Geographical expansion occurred as productive resources within the empire declined. A second agricultural revolution occurred when technologies permitted dramatic increases in crop and livestock yields, improved yokes for oxen or the substitution of horses for oxen, and the application of fertilizers and improved drainage systems.

Asia and Native Americans

A preponderance of scientific evidence indicates that Native Americans are descended from inhabitants of northeastern Asia who crossed a land bridge that linked northeast Asia to Alaska during glacial periods of lowered sea level. This colonization occurred only after humans developed the housing, clothing, and other required cultural capabilities, approximately 12,000 years ago.

The settlement of Asia that began approximately 50,000 years ago occurred relatively rapidly, due in part to humankind's ability to manipulate and advance its most significant competitive advantages—culture, language, and technology. The ability to exchange geographical and technological knowledge with other groups of humans over great distances accelerated the colonization of previously inhospitable regions. Most scientific evidence supports the origin of modern humankind in Africa and then their subsequent migration to Asia.

Fred Buchstein

POPULATION DISTRIBUTION

The largest continent in the world, Asia supports 4.5 billion of the approximately 7.5 billion people of the world. This translates to an average of 265 persons living in each square mile (100 persons per sq. km.), making Asia the most densely populated continent in the world. Europe, the second-most densely populated continent, has only about 188 persons per square mile (72 per sq. km.). As on other continents, population is not evenly distributed across Asia.

Early History

Asia has been the world's most populated continent for thousands of years. Although *Homo sapiens* (modern humans) originated in Africa, fossils found in China and Java, Indonesia, suggest that they appeared very early in Asia as well. The population of Asia originated in and migrated from a series of core areas over a period of more than 4,000 years. The three major clusters of population in Asia around 3000 BCE were in the North China Plain, the Indus Valley of Pakistan, and the Tigris and Euphrates valley of the Middle East (Mesopotamia). Productive agriculture in these cores permitted population growth and stimulated expansion.

Later, secondary and tertiary population cores developed in different parts of Asia, including southern Korea, the Red River basin of northern Vietnam, Java, the upper Ganges Valley in northern India, and the upper Mekong delta in Southeast Asia. New population cores also developed during the European colonial period, the period when the Asian population increased significantly for the first time.

European technology, coastally oriented trade, and the imposition of Western political structures were the principal causes of population increase and the change in its distribution. Most of Asia's dominant cities—Tokyo, Hong Kong, Shanghai, Kolkata (formerly Calcutta), Mumbai (formerly Bombay), Bangkok, Singapore, and Jakarta—grew dramatically during that period. Asia as a whole experienced rapid population growth after World War II.

Distribution of People

In 2020, the world population was concentrated in four major and several minor clusters. Two of the major clusters were East Asia and South Asia. Nearly 4 billion people—more than half the world's inhabitants—live in those two clusters. East Asia includes China (the world's most populous country), Japan, South and North Korea, and Taiwan; South Asia comprises India (the second-most populous country), Pakistan, Bangladesh, Sri Lanka, Nepal, Bhutan, and the Maldives. These clusters share several characteristics. Population density is highest in the coastal areas and along the fertile inland river valleys. Large cities, such as Mumbai, New Delhi, Dhaka, Beijing, and Shanghai, also are located in the coastal areas or in inland river valleys. In these metropolitan areas, population density often exceeds 1,500 persons per square mile (575 per sq. km.).

Elsewhere in East and South Asia, populations are moderately or even sparsely distributed, for example, in the mountains, rugged highlands, and arid lands. The great Himalayan mountain complex in South Asia is largely uninhabited. This zone includes the northern parts of Pakistan, India, Nepal, and Bhutan as well as the Assam-Burma ranges in eastern India. Other sparsely populated areas of these two regions include the deserts of Baluchistan (Pakistan) and Rajasthan (Thar); east-central parts of the Deccan Plateau in India; the western two-thirds of China, where the Gobi Desert is located; and the rugged mountains of central Japan.

POPULATION DENSITIES OF ASIAN COUNTRIES

(BASED ON CIA WORLD FACTBOOK DATA AS OF JANUARY 1, 2019)

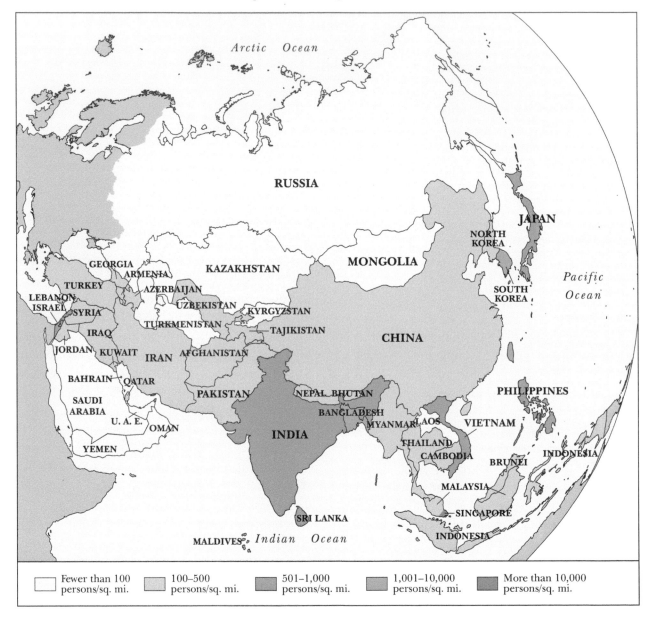

Fewer than 100 persons/sq. mi.   100–500 persons/sq. mi.   501–1,000 persons/sq. mi.   1,001–10,000 persons/sq. mi.   More than 10,000 persons/sq. mi.

Southeast Asia, which includes all Asian countries located east of South Asia and south and southwest of East Asia, has significant population concentrations, particularly in Indonesia, Vietnam, and the Philippines. This region also encompasses the mainland states of Myanmar, Thailand, Laos, and Cambodia, and the maritime states of Malaysia, Singapore, Brunei, and Timor-Leste.

Indonesia, with more than 260 million people, has the largest population of any nation in Southeast Asia and is the fourth-most populated country in the world. Java, an island of Indonesia, is considered a minor world population cluster. Parts of Java have densities of more than 3,000 people per square mile (1,160 per sq. km.). The government of Indonesia has attempted—with little success—to relocate some of Java's inhabitants to less-populated islands within the country.

The population in Southeast Asia is concentrated in the coastal plains and along the river valleys. High population densities are found in the lower valley and delta of the Irrawaddy River in

Myanmar; the deltas of the Chao Phraya in Thailand, and the Mekong and Red rivers in Vietnam; coastal lowlands of Vietnam; and major lowlands of the Philippines. Densities are lower than in the coastal areas and river valleys of East and South Asia, however. Although there is no arid area in Southeast Asia, the tropical rain forest areas of the region and mountains of Myanmar support few people.

Water and Population

Availability of water is the most important determinant of population concentration in the Middle East, which is characterized by arid and semiarid climates. The landscape of this region is dominated by deserts and mountains. The Arabian and Iranian deserts receive less than 5 inches (130 millimeters) of rainfall per year and are therefore not suitable for human settlement or agriculture. About 90 percent of the land area in the Middle East is either too rugged or too dry to be cultivated.

In the Middle East, populations are concentrated along coastal areas, near river valleys, and at oases. The eastern coast of the Mediterranean Sea, the coasts of Turkey, and the river valleys of the Euphrates and Tigris in Iraq are relatively densely populated. The mountain ranges of Iran, Turkey, and Lebanon support considerable human settlement. One main reason for this is the rainfall. Windward sides of the mountains in these countries receive relatively large amounts of rainfall.

The city-state of Singapore has the highest population density in Asia, with 22,290 persons per square mile (8,602 per sq. km.). If just non-city-states are considered, Bahrain, with 5,044 persons per square mile (1,947 per sq. km.), has the highest population density in Asia, followed by the Maldives (3,413 persons per square mile; 1,317 per sq. km.) and Bangladesh (3,021 persons per square mile; 1,107 per sq. km.). Bahrain and the Maldives are small countries, both in area and total population.

Taiwan, South Korea, and India have population densities greater than 1,000 persons per square mile. Five Asian countries—Mongolia, Saudi Arabia, Oman, Bhutan, and Laos—have population densities lower than 100 persons per square mile.

The lowest population density in Asia is found in Mongolia, with an average of 5.1 persons per square mile (2 per sq. km.). Most Asian countries with low population densities are located in the drier part of the continent. Climate and topographic conditions appear to be the two most important determinants of population distribution in Asia.

Within each country, population density differs markedly. For example, 94 percent of all Chinese reside in the humid eastern third of that country. Population densities are greatest in the Yangtze River Valley, Sichuan Basin, North China Plain, Pearl River delta, around Beijing, and around the northeastern city of Shenyang. Population is sparsely concentrated in the arid and mountainous western China.

In Bangladesh, the level, fertile floodplains of the Padma, Jamuna, and Meghna rivers, and the southern coastal area, support population densities that sometimes exceed 3,000 persons per square mile (1,160 per sq. km.). In contrast, the Chittagong Hill Tracts region in southeastern Bangladesh averages fewer than 200 persons per square mile (77 per sq. km.). The effective density is much higher in many small countries, because population is concentrated in small, highly productive zones. For example, in Abu Dhabi, a part of the United Arab Emirates, approximately three-fourths of the total population is concentrated in one large urban area of the same name.

Reasons for Distribution Patterns

Population distributions throughout the world reflect the interaction of complex physical, cultural, and historical forces. Parts of Asia with favorable climates, soils, and topography have supported high population densities for thousands of years. Population concentrations generally follow the suitability of land for agriculture; areas with low agricultural productivity usually have low population densities. The vast majority of the people in Asia are farmers and live in rural areas on floodplains, coastal areas, and along waterways. Exceptions to these generalizations are Japan, Singapore, South Korea, Taiwan, and parts of China, which are highly urbanized and industrialized.

The influence of various factors on population concentration differs within major regions of Asia. In East, South, and Southeast Asia, a realm often referred to as Monsoon Asia, the relationship between dense human populations and intensive agriculture is pronounced. There, rice is the principal food crop and is ideally suited to the climatic and topographic conditions of those regions. Rice cultivation requires large amounts of rainfall; correspondingly, population density is extremely high in the portions of East, South, and Southeast Asia that receive enough rainfall to allow rice cultivation. In contrast, most of the Middle East receives little rainfall, and people are therefore concentrated near sources of water.

People also have a tendency to live where other people are already established and opportunities are perceived to exist. Thus, densely populated areas most likely will continue to be densely populated if no explicit attempt is made to relocate current inhabitants. Indonesia, Malaysia, and Sri Lanka are trying to change their population distributions through programs that encourage migration from densely populated regions into sparsely populated areas.

Urbanization

Approximately 50 percent of Asians lived in urban centers in 2018. Only Africa has a lower level of urbanization than Asia. Some of the first cities in the world are believed to have developed in the lower Tigris and Euphrates valley (Mesopotamia) more than 6,000 years ago. City-based early civilizations or ancient cultural hearth areas emerged in the Middle East, and urbanization spread from there to Europe and elsewhere. Two other city-based civilizations developed in Asia as early as 3000 BCE: one in the Huang He Valley in northern China, the other one in the Indus Valley in Pakistan.

Of the four major regions of Asia, the Middle East is the most urbanized, with more than 60 percent of the people living in cities. Oil-rich countries of this region, particularly those with relatively small populations, have higher urbanization rates

than the non-oil-producing countries. Their urban residents enjoy a standard of living equaling that of affluent Western cities. Urban centers of the Middle East have grown rapidly, and cities such as Istanbul, Turkey, rank among the world's largest urban centers. Rapid urban growth there is the result of high rates of natural increase of population and massive rural-to-urban migration. Each contributes about half of the total growth of urban population. Rural-to-urban migration puts enormous pressure on the service sector in the major cities of the Middle East and has caused the development of squatter settlements on the outskirts of the urban areas. There, recent rural migrants often live without electricity, tap water, sewerage, or health-care facilities.

Traditional cities of the Middle East have two distinct sections: the older city core and the newer outer fringe. The old central core is densely populated; the streets are narrow and congested, because they were intended for foot traffic and small, animal-drawn carts, not for motorized vehicles. Newer, more spacious urban districts with wider streets and larger, more modern housing developed away from the older core. Several modern cities were founded virtually overnight on oil wealth, and contrast markedly with the region's colorful, complex ancient cities.

Although East Asia includes Japan, South Korea, and Taiwan, three highly urbanized countries, only 40 percent of the total population of East Asia resides in cities. This is because China, also part of East Asia, has a relatively low level of urbanization. When the Communist Party came into power in 1949, the country was only 9.1 percent urban; in 2020, the figure stood at 60 percent. The communist government limited rural-to-urban migration in the early 1950s by rationing food and restricting urban industrial development. The government viewed cities with contempt and distrust, because cities were inhabited by more educated and commerce-oriented people. Cities were also viewed as being nonproductive because the urban elites were perceived to be living off the products of rural labor. A bold and disruptive policy expressing this antiurban philosophy materialized during the Cul-

MAJOR URBAN CENTERS IN ASIA

tural Revolution (1966-1969), which was marked by a period of counterurbanization.

The Chinese government later changed its policy and allowed people to migrate from rural areas to cities. Between 1978 and 2020, the proportion of China's population living in urban areas more than quintupled. At least 844 million Chinese now live in urban centers, and China has more large cities than any country in the world.

East Asian cities have another source of urban growth: accretion of adjacent rural settlement. For example, part of Tokyo's growth is through boundary expansions that incorporate eighty-seven surrounding towns and cities. In East Asia, urbanization does not always equate to better living

standards. For example, North Korea is 62 percent urban but has a low standard of living.

The culturally diverse region of Southeast Asia displays remarkable variation in levels of urbanization. This region includes the only country in the world that is 100 percent urban: the city-state of Singapore. Brunei and Malaysia, which are more than 50 percent urbanized, experienced rapid growth in urban population when the oil industry expanded. Other countries in Southeast Asia have low levels of urbanization. This region shares most of the urban characteristics of the Middle East and East Asia. At least fifty cities in Southeast Asia have more than 1 million people.

South Asia is the least urbanized region of Asia, with slightly less than 35 percent of the total population living in cities. Urbanization has been slow but continual, and some cities experienced rapid growth. For example, the population of Mumbai, India, increased from about 1 million in 1891 to nearly 13 million by 1991 to more than 20 million in 2020. The functions of South Asian cities have remained essentially the same over the past two centuries. They were administrative centers of government, centers of commerce and manufacturing, and wellsprings of intellectual and religious activities.

South Asia's modern urbanization has its roots in the colonial period, when the British selected Kolkata, Mumbai, and Chennai (formerly Madras) as regional trading centers and coastal focal points for colonial exports and imports. Later, the British established cities in the interior parts of the region. Although not even 35 percent of the total Indian population lived in cities in 2020, this country had

the world's fourth-largest total urban population and had forty-seven cities with more than 1 million people.

As in other countries of South Asia, a high natural increase in the native urban population, rural-to-urban migration, territorial expansion of existing cities, and emergence of new cities are associated with urban growth in India. The contribution of these factors of growth differs from country to country. Most countries in South Asia are experiencing rapid urban growth, often exceeding 3 percent annually.

The most urbanized countries in South Asia are Bhutan (with 41.6 percent of its people living in cities), followed by the Maldives (40.2 percent) and Bangladesh (39.4 percent). Close behind are Pakistan (nearly 37 percent) and India (34.5 percent). Nepal and Sri Lanka come in at 20 and 18.6 percent, respectively. In many countries, urbanization

Traffic in Beijing.

is seen as an inevitable consequence of the modernization process, but in most South Asian countries, it has been more the result of increased rural poverty, which compelled rural people to migrate to the cities. Environmental hazards also encourage rural-to-urban migration. Floods and coastal cyclones are recurrent natural phenomena, especially in eastern India and Bangladesh. After a major natural disaster, many people in the affected areas migrate to urban areas in search of employment and financial security. An established, informal kinship network also affects migration flow by aiding migrants once they arrive in the city.

The cities of South Asia, especially the large ones, face problems in providing employment, shelter, and basic services to their residents. Most South Asian cities are densely populated, and congestion, noise, pollution, crime, and severe shortages of essential utilities such as water, sewers, and electricity characterize urban life.

About a quarter of the urban population of South Asia lives in poverty, many in slums and shantytowns. These squatter settlements are also found in the large cities of the Middle East, but are more prevalent in South Asian cities. Squatter settlements in this region usually are located in unhealthy, hazard-prone areas characterized by unpaved and narrow roads, poor or nonexistent street lighting, and irregular waste disposal. They develop on public land, along railroad lines or canals, on wetlands, and on other vacant lands.

The large or capital cities of Asia are mostly primate cities, meaning that they are much larger in size than the second-largest cities of their respective countries. A primate city has a relatively large proportion of the national urban population and is the

URBAN POPULATIONS AS PERCENT OF TOTAL POPULATIONS FOR SELECTED COUNTRIES, 2018

Country	Total Population (in Millions)	Urban Population as Percentage of Total
Bangladesh	159.0	36
Cambodia	16.4	24
People's Republic of China	1,384.7	55
Indonesia	262.8	55
India	1,330.0	34
Japan	126.0	92
North Korea	25.5	62
South Korea	51.2	81
Malaysia	32.4	75
Mongolia	3.2	68
Nepal	29.7	38
Pakistan	207.9	37
Myanmar (Burma)	53.7	31
Philippines	105.9	51
Singapore	5.9	100
Sri Lanka	22.6	18
Thailand	68.6	50
Vietnam	97.0	36

center of the economic, political, and cultural structure of the country. Primate cities also are found in some Asian countries at a subnational level. Shanghai, for example, is not a primate city at the national level because there are several similar-sized cities in China. However, it functions as a primate city in the east central region of China. Kolkata and Chennai function as primate cities in the eastern and southern regions of India, respectively.

Future Projections

Today, the population in Asia is growing at an annual rate of 1.1 percent. This is only slightly higher than the world population growth rate of 1.08 percent, equal to the corresponding growth rate of South America (also 1.1 percent), but lower than Africa's (2.6 percent). At that rate of growth, Asia adds about 50 million people each year to its population base. In 2020, the growth rate differs markedly among regions as well as among countries in Asia. The rate of natural increase in Asia is highest in the Middle East (1.3 percent), followed by South Asia (1.2 percent), Southeast Asia (1.0 percent), and East Asia (0.33 percent). In Asia, Japan and Georgia have the lowest (-0.2 percent) and Oman(4.08 percent) and Bahrain (4.26 percent) the highest population growth rate. The corresponding rate for the United States is 0.8 percent.

Variations in regional and national population growth rates are associated with the extent of economic development and the adoption of population-control measures. Most countries of East Asia are more developed and experience a lower level of population growth. By contrast, the oil-rich and predominantly Muslim countries of the Middle East have experienced high population growth rates because of religious opposition to family-planning programs. India, Sri Lanka, and Thailand are not as developed as the countries of East Asia, but they have moderate to low population growth because family-planning programs are strongly established there and face little religious opposition. India was the first developing country to introduce family planning as an official national program, which it did in 1951. China's controversial one-child policy, initiated in the 1970s, drastically reduced the population growth rate to 0.37 percent by 2018. The policy was relaxed in 2016, although its legacy persists in the country's aging population.

The United Nations estimates that Asia's population will increase from 4.5 billion in 2020, to 5.3 billion in 2050. The world population for 2050 is expected to reach 9.7 billion. Asia's share of the world population will remain the same, but its population size will increase. This increase will differ regionally as well as nationally within the continent. The population of South Asia is projected to surpass that of East Asia by the mid-twenty-first century, to become the world's largest population cluster. In 2020, India and China both have populations close to 1.4 billion. Most experts believe that India will soon surpass China as the most populous country.

Future population growth will create developmental challenges for some regions and countries of Asia. Most Asian countries could face severe problems coping with increased population. The growth of food production remains uncertain, particularly in South Asia.

Bimal K. Paul

CULTURE REGIONS

The massive continent of Asia generally is understood to have four cultural realms: East Asia, South Asia, Southeast Asia, and the Middle East (comprising North Africa and Southwest Asia), each of which can be broken into additional cultural regions. This list of Asia's culture realms does not include Northern Asia or Central Asia, because of the political history of Russia and the former Soviet Union. Russians are considered to be Europeans, but they conquered a huge area of land eastward to the Pacific Ocean, and the former Soviet Union included several republics in Central Asia. The Russians refer to their land east of the Ural Mountains as Asian, even though many of the people who live there are ethnic Russians, while the peoples around the Caspian Sea are mostly Turkic Muslims. Northern and Central Asia are often considered culturally as a part of Russia or the former Soviet Union, not as a separate Asian cultural realm.

The terms "Asia" and "Europe" have their roots in the language of the ancient Assyrians. Those ancient people, whose descendants now live in countries such as Syria and Iraq, used the word *asu* for the direction and land of the rising Sun, and the word *ereb* for the direction and land of the setting Sun. When the Greek Empire expanded under Alexander the Great, the Assyrian terms were used by the ancient Greeks. The ancient Greeks were not sure how far the land stretched to the east and west, but they labeled their maps with these terms. Much later, after the Crusades and by the Age of Exploration, the western Europeans referred to the great expanse of land to the east as Asia. Asia was not a homogenous region, however; the people there spoke many different languages, practiced many different religions, and lived in numerous kingdoms and empires unconnected to each other.

East Asia

The East Asia culture realm extends from Japan in the east to Taiwan (the Republic of China) in the southeast, Tibet in the west, and Mongolia in the north. It also includes the People's Republic of China, the Republic of Korea (South Korea), and the Democratic People's Republic of Korea (North Korea). This massive land area is tied together in

THE CHINESE DIASPORA

More people consider themselves to be Chinese or to be of Chinese ethnic heritage than any other group of people in the world. For millennia, people of Chinese ethnic heritage have migrated away from their cultural heartland. During the nineteenth and twentieth centuries, Chinese people moved in large numbers from China to every part of the world. The British brought thousands of Chinese people to their colonies in Southeast Asia to work in tin mines and to open small businesses for essential services. Today, Singapore is overwhelmingly Chinese, and other countries of Southeast Asia, such as Malaysia, have large Chinese minorities. Many Chinese people came to North America in the nineteenth century to work on the railroads and in the mines of western Canada and the United States.

Today, many Chinese people come to the United States and Canada as professionals, business people, or unskilled laborers, looking for a new life. In the 2017-18 academic year, there were 360,000 Chinese students enrolled in the U.S. Vancouver, Canada, is sometimes referred to as "Little Hong Kong" because of its large Hong Kong-born population. Every country in the world now has at least a small Chinese population, and Chinese restaurants are found in every major city in the world. Often, Chinese restaurant owners adapt their cuisine to the taste preferences of the local populations.

large part because of the strong historic influence of Chinese language, arts, and philosophy.

Since ancient times, Chinese civilization has had a profound impact on all of the peoples of this area, despite the fact that the Koreans, Japanese, Tibetans, and Mongolians speak languages that are unrelated to Chinese languages. The spread of Confucianism (a Chinese moral philosophy that is more than 2,000 years old) and Buddhism (a religious philosophy originating in India) are two other characteristics that tie this realm together.

Despite these strong unifying elements, East Asia can be divided into a number of regions based upon historical, economic, political, and cultural characteristics. The first region is the political unit of the People's Republic of China, which covers the largest part of East Asia. Due to its size, the centralizing force of its government, and the long history of its great civilization, China is often studied as a separate region. The Tibetan area has important differences in language and culture, but is included with the study of China because China invaded it in 1950 and made part of it a province of China in 1965.

The majority of people in Taiwan (the Republic of China) are ethnic Chinese, but are often studied with the Japanese and Koreans because they all share the experience of rapid economic development in the last half of the twentieth century. Although North Korea is one of the world's most isolated and least developed countries, it should nevertheless be included with South Korea because of the cultural and historical ties the Korean people share.

Finally, while a part of Mongolia is a province of China, the country of Mongolia remains an interesting area on its own because of its different culture, its historic ties to China (Genghis Khan and Kublai Khan were Mongolians who ruled the greatest Chinese empire), and its modern-day political alliance with the Russians through the former Soviet Union. In 1921 the Mongolians asked the Soviets to help them repel a Chinese effort to absorb Mongolia into what was then the Republic of China. The Soviets obliged, but indirectly took over the country themselves. Mongolia remained a satellite nation of the former Soviet Union until 1991. For

this reason, Mongolia is often studied with Asian Russia, but it would be more appropriate to think of Mongolia as a transition zone between China and Russia.

South Asia

South Asia is a substantial, extremely diverse culture realm, traditionally associated with the Indian subcontinent. The Indian subcontinent is a large landmass that extends southward from the great arc of mountain ranges known as the Hindu Kush (to the northwest), the Karakorams (north), and the Himalayas (north and northeast). This realm in-

THE INDIAN DIASPORA

People have been migrating from the Indian subcontinent to other parts of the world for thousands of years. Early on, they migrated into eastern Africa and the Southeast Asian islands and peninsulas. In both areas, they had an extraordinary impact on the language, religion, and arts of the local people. For example, Buddhism continues to thrive in Southeast Asia, and Sanskrit, the religious language of Hinduism, has influenced many of the writing systems and vocabularies of Southeast Asian languages. Indian traders are primarily responsible for bringing Islam to Southeast Asia. In the late nineteenth and early twentieth centuries, after the abolition of African slavery, the British transported hundreds of thousands of people from the Indian subcontinent to other parts of their far-flung empire to serve as cheap labor on British sugarcane, rubber, and tea plantations.

According to the United Nations Department of Economic and Social Affairs, Indians comprised the world's largest migrant diaspora populations in the world in 2019 with over 17.5 million (6.4% of global migrants or 0.4% of India's population) Indians out of total 272 million migrants worldwide. People of Indian heritage make up 35 percent of the population in Fiji, Guyana, Trinidad and Tobago, and other countries. Since the 1960s, people from the Indian subcontinent have migrated to Western countries in large numbers as professionals and to conduct business enterprises. They have also gone to the oil-rich countries as laborers. Finally, many Indians living in Eastern Africa have migrated to Western countries over the last quarter of the twentieth century. They are called "twice-migrants."

cludes India, Nepal, Bhutan, Bangladesh, Sri Lanka, Pakistan, and sometimes Afghanistan. Because of the height and extent of the magnificent mountains in the north, and the broad expanses of water to the west (the Arabian Sea), the south (the Indian Ocean), and the east (the Bay of Bengal), South Asia seems to be set apart from the rest of Asia. Only broad generalizations can be used to tie the area together, or to define contiguous culture regions within the realm.

South Asia has a long, shared cultural history reaching back to the ancient Harappan civilization of the Indus River valley (now in Pakistan). It also has a long history of invasions, particularly from the northwest through the mountain passes, such as the Khyber Pass, which helped to push Harappan culture traits farther across the subcontinent. Through these invasions, the peoples of South Asia were exposed to new linguistic influences (for example, Eu-

ropean languages and Persian), religious practices (Islam, Christianity, and Zoroastrianism, the traditional Persian religion), and political systems (British colonialism), which they reacted to and absorbed. Although no invader ever controlled the entire subcontinent, the region was profoundly influenced by all of them. Still, one can recognize culture regions within South Asia based upon the distribution of language families and religions practiced there.

With some exceptions, the realm can be divided generally into north and south regions, based on broad linguistic and religious differences. The majority of people from Afghanistan to Pakistan, through northern and central India, Nepal, and Bangladesh, speak a language belonging to the Indo-European language family. This means that most of the languages in this realm are related to the languages of Europe. This pattern exists be-

The Leshan Giant Buddha, 71 meters (233 ft) high; begun in 713, completed in 803.

165

A procession of Akharas marching over a makeshift bridge over the Ganges River. Kumbh Mela at Allahabad, 2001.

cause of historic invasions of people from the northwest that go back more than 3,000 years.

Islam also has had a strong influence across this northern area, except in Nepal and Bhutan. The people in the southern area (southern India and northeastern Sri Lanka) speak a language belonging to the Dravidian family. This language family originated in the Harappan civilization of the Indus Valley, but migrated southward; only a small minority of people still speak it in the north. Muslims and Christians can be found in the southern part of South Asia, although the Hindu religion dominates this area. Thus, travelers in South Asia find distinctive differences between the north and south.

Diversity exists within the broad categories just identified. Each country, as well as areas within countries, has unique features. For example, most Sri Lankans are Buddhist and speak Sinhalese, which is related to the languages of the north. This resulted from migrations that occurred about 2,000 years ago. In addition, Sri Lanka has absorbed different people who have arrived on its shores over the centuries.

Arab traders began visiting Sri Lanka in the tenth century, and some settled there. They called the island "Serendip," which means place of good fortune in Arabic. The British morphed the word into "serendipity," which is used today to mean good luck.

Far away, in the northern mountains, Bhutan is also a predominantly Buddhist country, but its people speak a language related to Tibetan. Nepal has a large Buddhist population, but the majority of its people call themselves Hindu. While the Nepalese language is related to the north Indian languages, many people speak languages related to Tibetan.

India, the largest nation in the region, can be studied on its own because of its size and diversity. While the majority of people are Hindu throughout the country, the many religious minorities account for more than 20 percent of the population.

Buddhist monks collecting alms in Luang Prabang, north Laos.

Pakistan, Bangladesh, and Afghanistan are Muslim countries. At one time, Pakistan and Bangladesh were a single country, but in 1971, Bangladesh separated from Pakistan. Besides being separated by more than 1,000 miles (1,600 km.), Pakistan and Bangladesh have different cultural traditions, and it was difficult for them to remain united.

Afghanistan is a Muslim nation, whose national boundary was essentially created by default. In the nineteenth century, the British and Russian Empires had both expanded into this area. To prevent conflict between themselves, they treated the area now called Afghanistan as a buffer zone. To this day, Afghanistan does not fit neatly into any culture region around it.

Southeast Asia

The realm of Southeast Asia is a tropical one. It extends from Myanmar (formerly Burma) in the northwest to Indonesia and Timor-Leste in the southeast. It is treated as a separate realm, in part, because of its location south and east of two other

great realms, South Asia and East Asia. No unifying themes hold this realm together outside of its relative location. Thus, geographers tend to divide Southeast Asia into two clearly defined regions: mainland Southeast Asia and insular Southeast Asia. The two parts of Southeast Asia are quite different from one another. The mainland, directly connected to the rest of Asia, is unified by its Buddhist traditions but not its languages. Insular Southeast Asia, however, is unified by its linguistic traditions but not by religion.

Mainland Southeast Asia includes Myanmar, Thailand, Laos, Vietnam, and Cambodia. All these countries share a strong tradition of practicing Buddhism, but the languages the people speak in the various countries do not appear to be related to one another. Myanmar is somewhat of a cultural transition zone between South Asia and Southeast Asia, because it has cultural traits that connect it to both areas. Thailand and Cambodia are famous for their great historic civilizations of Siam and Khmer, respectively. Laos is a mountainous country with many different cultural groups, of which the Lao are dominant. Vietnam is a long, narrow coastal nation, historically influenced by China.

Insular Southeast Asia comprises the long, narrow Malay Peninsula and the thousands of islands that extend south and eastward from the peninsula. It includes Malaysia, Singapore, Brunei, Indonesia, Timor-Leste, and the Philippines. In this region, the majority of people speak a language belonging to the Malay (sometimes called the Indonesian) language family. In fact, the standard languages of Malaysia and Indonesia (Bahasa Malay and Bahasa Indonesian) are mutually understandable in the same way that standard American English and standard British English are.

On the other hand, there is a great diversity of religions in this region. Malaysia, Indonesia, and Brunei are predominantly Muslim countries. Malaysia is notable for its large minority populations of Hindus, Buddhists, Christians, and people who follow traditional Chinese practices (ancestor worship, Taoism, and Confucianism). The Philippines and Timor-Leste are predominantly Roman Catholic, with Muslim minorities. Singapore has people who are Christian, Muslim, Hindu, and Buddhist, as well as those who follow the Chinese complex of religious beliefs.

The Middle East

The Middle East is sometimes described by the longer term "Southwest Asia and North Africa." Previously, it was commonly called the Near East. The terms "Middle East" and "Near East" come directly from the European point of view. When viewed on a map, this culture realm is east of Europe and is nearer to Europe than is the rest of Asia. It can also be considered to be in the middle between Europe and the bulk of Asia. Because of shared cultural traditions and history, parts of North Africa also belong to this cultural realm. Culture realms do not always remain neatly on the same continent, or even in the same country. The countries in this realm extend from Iran in the east to Morocco in the west.

The Dakhshinewar Temple was founded by Rani (Queen) of Janbaazar Rashmoni in 1855 on the east bank of the Ganges River.

Shintō purification rite after a ceremonial children's sumo tournament at the Kamigamo Jinja in Kyoto.

Southwest Asia and North Africa are tied together by a number of factors. First, there is a long shared history of great civilizations that reaches back thousands of years. These include the Egyptian civilization in the Nile Valley; the Sumerian, Babylonian, and Assyrian civilizations in the Tigris and Euphrates valleys; and the Persian civilization centered in Iran.

The second unifying factor is religion. All the countries of the Middle East, except Israel, are predominantly Muslim. The rise of three great, modern-world religions—Judaism, Christianity, and Islam—in this realm gives the region a religious importance that is felt throughout the world. Israel is the only country in the world that is predominantly Jewish; Christian minorities can be found in almost every country in the area. The third unifying factor is linguistic. This results from the importance of the Arabic language in Islam.

The holy book of Muslims, the Qur'an (Koran), was originally written in Arabic, and people had to learn Arabic to read the book and say their prayers.

As a result, many people became Arabic speakers. Even so, other people kept their original languages. For example, in Turkey people speak Turkish, which is unrelated to Arabic; in Iran, most people speak Farsi (Persian), which is related to European languages.

While the people of Southwest Asia and North Africa have a shared history and are predominantly Arabic-speaking Muslims, there are cultural differences that define regions within the realm. Tunisia, Algeria, and Morocco are often studied together as a region called the Maghreb. Israel, Lebanon, Syria, and Jordan are often studied together as a region called the Levant. Saudi Arabia, Kuwait, North Yemen, South Yemen, Oman, and the United Arab Emirates are studied together as an Arabian region. Turkey, Egypt, Israel, Iraq, and Iran can each be studied on its own because of unique culture histories. Some people, such as the Kurds, do not have their own nation. Their culture region spans four different countries: Turkey, Syria, Iraq, and Iran.

Carolyn V. Prorok

169

EXPLORATION

Explorers from many lands have trekked through Asia's 17 million square miles (44 million sq. km.). Many were lured by a sense of adventure and the desire for diplomatic advantage and military conquest. Others were attracted by the possibility of fame, missionary zeal, the prospect of wealth, and scientific curiosity about unknown cultures and histories.

Many explorers died or were killed chasing their dreams of finding lost cities buried beneath desert sands. Those who survived and succeeded in reaching their destinations opened Asia to further exploration and economic development. They rediscovered the glories and knowledge of lost civilizations; sometimes they provided vital military intelligence.

The opening of Asia also brought with it new markets and products, ideas, technologies, and religions—including Buddhism, Christianity, and Islam. Geologists and petroleum companies represent a more contemporary type of explorer—one who digs beneath the Caspian Sea and Taklimakan Desert in search of oil pools that could rival those of the Middle East.

As early as 50,000 years ago, prehistoric people first began exploring Asia in search of food. Written accounts of exploration dating to the fourth millennium BCE can be found in Egyptian hieroglyphics. During the second century CE, the Greek geographer and astronomer Ptolemy created a new vision of the world in his *Guide to Geography*. This work

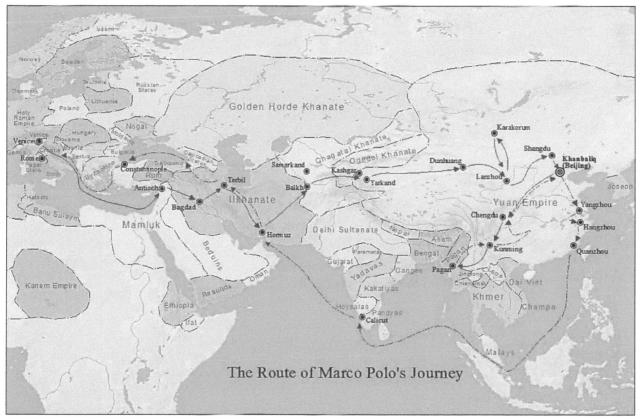

Map of Marco Polo's travels.

later convinced the Italian explorer Christopher Columbus that Asia could be reached from Europe by traveling westward across the Atlantic Ocean. Other European explorers used ancient geographies to help guide them in their explorations of Asia.

The Great Silk Road

The discovery and exploration of Asia owe much to the development of the northern and southern Silk Roads. These land routes promoted conquest, learning, the spread of new religions, and trade. Prior to the Silk Roads, only such powerful empires as the Chinese, Mongol, Persian, and Roman could guarantee safe travel over long distances. Conflict and chaos deterred explorers. The Great Silk Road reached its peak traffic around 750 CE. Seagoing trade eclipsed the land routes by 1200.

Sometime between the seventh and fourth centuries BCE, the followers of Buddha traveled throughout India and nearby regions in the first large-scale missionary effort in the history of the world's religions. Buddhist missionaries helped open and develop the land routes of Asia called the Great Silk Road. They entered China sometime before the common, or Christian, era and established religious communities.

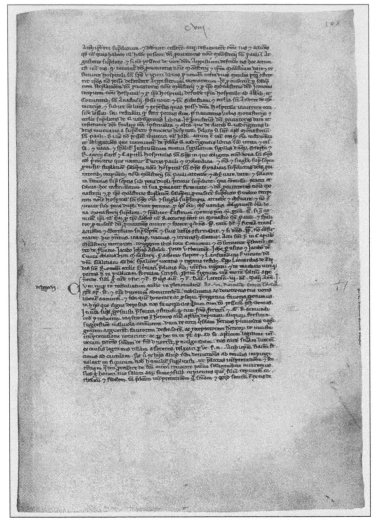

Text of the letter of Pope Innocent IV "to the ruler and people of the Tartars," brought to Güyüg Khan by John de Carpini, 1245.

Some Chinese converts to Buddhism journeyed to India to explore the roots of their faith firsthand. Buddhism's spread into China not only gave the Chinese a new religion, but also gave the world an entirely new art style—Serindian, a word formed by the combination of Seres (China) and India. China was isolated from India by the Himalayas and neither culture was familiar with the other. The art form, revolutionary in that it was the first time artists depicted Buddha in a human form, developed in the Peshawar Valley region in what is now northwestern Pakistan.

Lost Cities

Following a violent sandstorm, the great seventh century Buddhist explorer and pilgrim Hsuan-tsang found a lost town he called Ho-lo-lo Kia, believed to be one of the approximately 300 lost cities beneath the Taklimakan Desert.

In 139 BCE, the Han emperor of China, Wu-ti, sent his commander of the guards at the imperial palace gates, Zhang Qian (also known as Chang-ch'ien), to the western lands to negotiate a military agreement with a nomadic tribe, the Yueh-chih. The emperor needed the alliance to conquer the enemies of the Chinese, the Xiongnu (also known as Hsiung-nu). Zhang failed but through his thirteen years of travel learned of new opportunities for trade, especially for horses raised in the Ferghana Valley in what later became Uzbekistan. His journey led to China's discovering Europe and the birth of the Silk Road. Zhang told the Han emperor and his court of rich, unknown cities such as Ferghana and Samarkand. He may have explored parts of Persia and Rome. The Great Traveler, as Zhang Qian is called, is the father of the Silk Road. Traders and other explorers followed in his footsteps.

Military campaigns also spurred exploration for new markets and products. By the mid-eighth century CE, Muslim armies ruled much of the Silk Road and surrounding territories. Military domination gave them control over most of the trans-Asian trade. Charismatic Muslim preachers spread their religion, which led to mass religious conversions and the Islamization of the oasis-centered Silk Road and much of Central Asia. The Muslim involvement in long-distance trade was the main reason for the Islamization of the Silk Road.

Asians also pursued exploration. In 1405 the Chinese launched a series of voyages into the Indian Ocean. Under the command of a powerful Ming Dynasty court eunuch, Zheng He (also known as Cheng Ho), the Chinese visited the Maldive Islands, Calicut (now Kozhikode), Hormuz, and the east coast of Africa in search of trade connections.

Early Travelers

Perhaps the greatest Chinese traveler of the fifth century CE was Faxian (also known as Fa-hsien). In 399, when he was sixty-five years old, he practically walked from central China across the Taklimakan Desert, over the Pamir Plateau, and through India down to the Indian Ocean. He returned to China in 413 by sea, sailing via Ceylon and Sumatra, across the Indian Ocean and the China Sea. He succeeded in bringing back Buddhist religious texts and images of Buddhist deities. The seventh century Buddhist pilgrim, Xuanzang (also known as Hsuan-Tsang), traveled overland via the Silk Road west toward India. His account of his travels provided later explorers a travelogue for the Silk Road.

Nineteenth century explorers rediscovered the landlocked trade routes that ran from China to the Mediterranean Sea. They went in search of such legendary cities as Samarkand, Kashgar, Kotan, and Yarkand, and their historical treasures. The barriers to their explorations were formidable: the Gobi Desert, the Taklimakan Desert (considered by many explorers to be the harshest and most hostile desert in the world), and the six mountain chains between the Indian subcontinent and Central Asia.

The region of the Western Himalayas was so remote that it was not until the 1860s that travelers brought back news of cities such as Kotan, acclaimed in ancient times for its gold and jade. Few travelogues and maps existed to guide explorers. Not until the 1840s did the public learn of the travels of Chinese Buddhist monks who, from the fifth to seventh centuries CE, went to India in search of

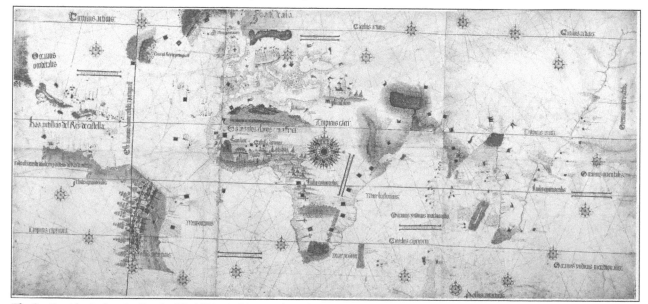

The Cantino planisphere (or Cantino World Map) of 1502 is the earliest surviving map showing Portuguese discoveries in the east and west.

their religious heritage. While the monks told strange tales of dragons and haunted waters, they provided few directions to help explorers retrace their travels.

Nineteenth century explorers almost all owned an account of Marco Polo's travelogue. Polo crossed the Pamirs from Balkh to Kashgar in 1274. Unfortunately, he devoted only three to four pages in his journal to describing his route. Later explorers had to retrace his footsteps or uncover new ways to go where they wanted. Only Jesuit missionaries penetrated the remote western Himalayas between 1600 and 1800. Their journals and maps guided future explorers. People who explored the lands between India and Russia and India and China literally had to fill in the blank spots on maps themselves.

Political considerations in England, Pakistan, Afghanistan, Russia, China, and India dictated the direction and pace of exploration in the 1840s. Powerful nations struggled over which would control Central Asia. The British, for example, secretly em-

Ferdinand von Richthofen.

ployed Indian nationals to map British India and beyond to assess the threat of military expansion posed by czarist Russia. The British needed to know how an invading army might attack in order to devise a defensive strategy, especially in Tibet.

Sea Routes

Two Portuguese maritime explorers, Bartolomeu Dias and Vasco da Gama, opened the sea route to Asia. Dias circumnavigated the Cape of Good Hope at the southern tip of Africa in 1488 and then of necessity returned to Portugal. Da Gama reached the tip of the Indian Ocean in 1497, giving the Portuguese a sea route to the Asian markets. Da Gama's achievement meant European nations alone controlled the seas and would until the end of the nineteenth century, when naval power shifted to the United States and Japan. The third Portuguese expedition to India in 1502 was meant to give the Portuguese military control of the Eastern trade. In the first naval battle off the Malabar Coast of India, the Portuguese, under the command of da Gama, outgunned and defeated an enemy Muslim flotilla.

Christian missionaries, especially the Jesuits, quickly followed Christian conquerors. Francis Xavier, a disciple and friend of the founder of the Jesuit order, Ignatius Loyola, was among the first missionaries to arrive in Goa on the west coast of India. In 1549 Xavier sailed to Kyūshū, Japan, to spread Christianity. He mastered Mandarin Chinese and spent years in China, providing Europeans with an unprecedented knowledge of Chinese culture and geography.

During the mid-1500s the English were anxious to find a northern route to Cathay (China) that would enable them to bypass the Muslim world and avoid confrontation with the Portuguese or Spanish. An English merchant, Anthony Jenkinson, pioneered the northern route via the Caspian Sea to Persia.

By the seventeenth century, the Dutch and English had displaced the Portuguese as the dominant commercial powers in the Far East. The Jesuits also expanded their efforts to verify rumors of lost Christian communities in Tibet and elsewhere in the Himalayas. A Jesuit priest, Bento de Goes, ven-

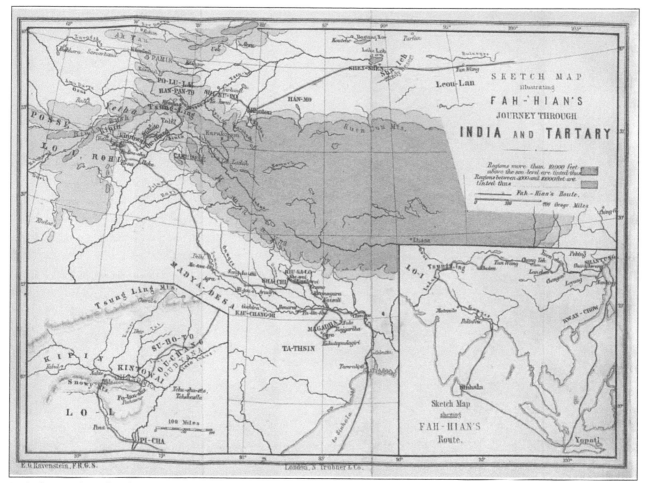

Faxian's route through India.

tured into the Asian interior to investigate whether Tibetan Buddhist lamas might actually be Christian priests and whether Marco Polo's Cathay was China. The first European to cross the Himalayas was the Jesuit priest António de Andrade, who succeeded in establishing a small mission.

Central Asia

Central Asia was the center of much exploration in the nineteenth century. European nations such as Russia and England wanted the upper hand in the struggle for empire called the Great Game. In 1864 the first European, William Johnson, reached the Taklimakan and became convinced that ancient cities long thought to be legends and myths actually existed underneath the desert sands. He returned to India with evidence that an old city existed near Urankash.

Exploration of the Silk Road took off with the expedition in 1895 of Sven Hedin, a Swedish cartographer and linguist who was to become famous for his travels across Central Asia and Tibet. One of his professors in Germany was Baron Frederick von Richthofen, the geographer who coined the term "Silk Road" to describe the ancient trade route. Hedin's treks included crossing the mountainous region, the Pamirs, which has many peaks more than 20,000 feet (6,000 meters) high, to the town of Kashgar. During his later travels, he discovered several ruined cities, including Loulan, in the Taklimakan Desert.

Hedin's exploration of Chinese (or eastern Turkmenistan), Tibet, and Mongolia provided Europeans with knowledge of regions they knew virtually nothing about. In fact, he claimed that of the 6,500 miles (10,600 km.) he traveled, 2,000 miles (3,200 km.) of it had never before been visited by a

175

European. He explored and mapped the area north of the Himalaya Mountains and surveyed the sources of the Indus, Sutlej, and Tsangpo-Brahmaputra rivers.

An archaeological frenzy to find other lost cities and their treasures around the Taklimakan, especially ancient manuscripts, ensued. Hungarian-born British archaeologist Sir Aurel Stein is among the best-known explorers. During his 1900, 1906, and 1913 expeditions, Stein retraced the routes taken by caravans on the Silk Road, discovering long-lost cities and languages no longer spoken. He spent much of his life trying to retrace the routes and battlefields of Alexander the Great. Stein happened on the great temples Chinese and western Buddhists had carved into the cliffsides at Dunhuang (also known as Tun-huang). He purchased a large cache of ancient manuscripts. By the mid-1920s Chinese leaders halted all exploration in their country in order to stop the plundering by European explorers of the archaeological treasure troves in Central Asian regions such as the Tarim Basin.

Scientists today are using new technologies to explore Asia. Geologists go in search of new petroleum reserves in areas such as the Caspian Sea. Geneticists trace humankind's migration from Africa to Asia. Previously unreadable manuscripts have been deciphered, shedding new light on the history of Asia. Satellite photography, radar, and other imaging technologies are being used to uncover further secrets of ancient Asian civilizations. Explorers continue their quest to know even more about the history, geography, people, and resources of Asia. The explorers and their instruments are different now, but their desire for discovery remains just as intense.

Fred Buchstein

URBANIZATION

Asia's population is larger than that of any other region in the world. According to the United Nations, Asia will have 45 percent of the world's total population by the year 2050. The recent and projected growth of the population is concentrated in Asia's cities. In 2020, there were already more than 512 cities in the world with populations exceeding 1 million, many of these in the developing world, which includes Asia, Africa, and South America. In Asia, the large cities are growing more quickly than are those in any other region.

As more people dwell in urban areas, competition for living space, working space, and so-called green spaces such as parks increases significantly, often driving up the costs of housing and rents on shops and other business properties. In addition to rising costs, problems created by increasing numbers of people in urban areas include air, water, and noise pollution; shortages of such basic necessities as food and electricity; and a wide range of social menaces such as disease, vandalism, and violent crime. Efforts to relieve population driven urban congestion and its associated social and environmental consequences are vital to Asia's economic and political future.

Patterns of Urbanization

Until the middle of the twentieth century, Asia's populations were heavily rural, depending upon agricultural work, small handicrafts, and local trade

Guangzhou, China.

Chongqing, China.

for their livelihoods. The Industrial Revolution that took place in Europe and North America in the eighteenth and nineteenth centuries did not develop in Asia, so the concentration of industry and workers in urban areas so familiar in the West did not occur. Instead, Asian cities reflected the region's role in the global economy as sources of raw materials for the West, importers of Western manufactured goods, and administrative centers running the affairs of outlying towns and villages.

In the twentieth and early twenty-first centuries, however, import and export patterns became more complex, and Asia's urban areas grew, as did their populations. Today, Asian cities have become as economically diversified as were those in the West. Millions of new jobs are being created in domestic and international business, and in government, medicine, education, and other social services.

In Asia, as in Africa and South America, this general pattern of economic development was accompanied by a rapid growth in the populations of urban centers. This urbanization of the population has taken many routes. In cities where better health care, more and better foods, and safer work environments have been available, higher birth rates and lower mortality have contributed to population growth. The common belief among rural residents that cities offer readily available employment, higher wages, and better lifestyles has attracted, or "pulled," labor from agricultural employment in rural areas toward urban settings.

Finally, and most commonly, rural populations with historically high mortality rates and the uncertainty associated with weather cycles and harvests have been "pushed" toward cities by overcrowding, food shortages, and rural poverty. One key "push" factor arising from Asia's traditional agricultural economies is the primogeniture system of landholding, in which only the eldest son can inherit the family's farmland. Younger sons must find another means of making their livelihood, and often go to a city to find it.

Rural dwellers moving to cities usually do so without already having secured housing or employment there. Those relocated people have tended to congregate where living costs are lowest. In Asia, where local government authority is often weak, the result has been expanding poverty zones, poorly served by city services and characterized by unemployment, crime, and even substance abuse. The fabric of the city itself also registers the impact of these new residents, in deteriorating roads, inadequate public transportation, lack of basic utilities, shortages of food and water supplies, and degradation of the local environment. As the cost of maintaining these urban goods and services rises, municipal authorities must either defer payments until some future date, creating new debt for the cities, or simply stop providing services altogether.

Although internal population shifts in Asia are overwhelmingly rural-to-urban, some countries have experienced historical migrations from cities to the countryside. In the 1960s, the People's Republic of China deliberately evacuated many cities and moved industries into China's rural interior. Chinese leaders during this crucial period of the Cold War expected that the United States might bomb China's cities, as Japan had done during its occupation of coastal China in World War II.

At about the same time, hundreds of thousands of Chinese intellectuals, professionals, and scholars were punished for political offenses by being sent from cities to the countryside, where they worked on farms and dam projects as a means of learning about the hard life of China's rural peasants.

Urban Pollution

The quality of life in urban environments is related to many factors: congestion, housing, the management of wastes, and the use of resources. As a city's population rises, the importance of each of these is-

Lotte World Tower, September 18, 2016.

A view of the Hozomon gate and the Nakamise beyond it as well as one of the main mikoshi from the top step of Sensō-ji during Sanja Matsuri.

sues grows as well. But in many Asian cities, the rate of urban population growth has greatly outpaced the ability of municipal authorities to manage its negative impacts. Air quality and air pollution have become particularly vexing problems in Asia, where cities strive to combine rapid economic and industrial growth with pollution-prevention measures. The results have been decidedly mixed.

In 2018, a study conducted by the World Health Association (WHO) found Asian countries dominating the top nine spots in the list of the world's most polluted countries. Bangladesh suffered from the worst air pollution. Other countries where the air was judged as unhealthy were Pakistan, India, Afghanistan, Bahrain, Mongolia, and Kuwait. Nepal and the United Arab Emirates ranked slightly better—eighth and ninth—with their air judged as being unhealthy for sensitive groups.

In India, air pollution has emerged as a very serious health issue. The WHO study indicated that twenty-two of the thirty most polluted cities in the world are in India. WHO named Gurugram, a suburb of New Delhi, as the world's most polluted city.

The city's air quality is so poor that the U.S. Environmental Protection Agency (EPA) has warned that anyone exposed to Gurugram's level of air pollution "may experience serious health effects."

China's cities have long been notorious for their high pollution levels and hazardous smog concentrations. This has been vividly illustrated by the dense haze that has enveloped the Chinese capital of Beijing for weeks at a time.

In 2014, the government enacted strict antipollution measures. Polluting industries nearby the many cities were either closed or relocated. Fuel and emission standards for motor vehicles were strengthened substantially. Overall coal consumption was reduced. Thanks to these policies, Beijing was able to cut its pollution levels by one-third just between 2018 and 2019. In the early 2020s, environmental experts predict that Beijing will no longer be on the list of the world's 200 most polluted cities.

Today, the most-polluted city in China, according to a Greenpeace report, is Henan, in the far west of the country. But in the city of Baoding, about 90 miles (145 km.) southwest of Beijing, things are looking brighter. In 2015, Baoding was near the top of the list of most polluted Chinese cities. Residents complained of smog so dense that nearby buildings were completely obscured. Three years later, anti-pollution policies have helped Baoding clean up its air, improving the city's pollution ranking significantly.

The news is not so positive for other Asian countries. Indonesia, South Korea, and Vietnam, for example, have seen their air-pollution levels spike dangerously. In Bangkok, Thailand, the city even resorted to cloud-seeding in the hope that the induced rainfall would help cleanse the atmosphere of the dense smog.

In every urban center in Asia, the main contributor to pollution is the burning of fossil fuels. Experts worry that climate change, by altering atmospheric dynamics, is further exacerbating the problem. Bearing these factors in mind, Asian countries are being urged to develop programs and otherwise take steps to resolve their air-pollution problems as soon as possible.

Urban Resource Conservation

As urban populations grow, greater demands are placed on basic municipal assets such as roads, gas and water pipelines, waste disposal systems, and electricity-distribution grids. These demands weigh heavily upon Asian cities, which have historically been underdeveloped relative to European or North American cities of the same size. In most cases, Asian cities lack the central municipal governments, modern equipment, and adequate financing that have enabled major cities in the West to install, maintain, and update complex infrastructures.

Growing urban populations in Asia and the lack of basic municipal facilities have produced acute shortages of basic resources, such as water and electricity, in many places. While Asian countries struggle to increase the supplies available, their cities are also encouraging urban residents to bear in mind the imperatives of conservation in their patterns of consumption.

Conservation has become an especially important strategy for managing urban electricity sup-

Natural color satellite image of a smog event in the heart of northern China.

plies. In India, rapid urban population growth has combined with the development of new energy-intensive industries to put huge new demands on the country's electrical grids. Although many new generating plants are under construction, India's cities have long been emphasizing voluntary conservation to eliminate the need for brownouts and electricity rationing.

Another consideration is waste management: solid-waste collection and sorting systems in Asia are not nearly as highly developed as they are in North America and Europe. Nevertheless, across the continent, officials have launched awareness campaigns to educate people on the links between poor waste management and the spread of dangerous human and animal diseases. New programs encourage manufacturers to use biodegradable packaging, and many individual households compost their own vegetable wastes. Garbage-collection services have become the norm in most cities, and recycling is encouraged through the availability of sorting bins for such materials as plastics and glass.

New laws and treaties have been enacted to confront a parallel problem: the importation to China and other Asian countries of plastics waste and hazardous waste products from abroad. For many years, countries in East and Southeast Asia bought huge amounts of plastics waste and recycled it into new products. China alone was buying upwards of half the industrial world's plastics waste: soda bottles, grocery sacks, and hundreds of other plastic items. In 2017, China imported a record 6.4 million tons. In many cases, the plastics were accompanied by unrecyclable contaminated waste. The Asian recipients of the garbage had little choice but to burn the hazardous materials or dump them in landfills and waterways. This created a whole new set of health and environmental challenges. Now, thanks to updated provisions of the 1989 Basel Convention, exportation of plastics wastes and hazardous wastes from one country to another will be forbidden beginning in 2021 without express written documentation. And some countries, including the Philippines and Malaysia, are blocking the importation of any and all plastics waste.

Urban Poor and Urban Youth

Middle Eastern and Asian cities, like many urban centers in Africa, have experienced rapid growth rates in the poorest segments of the urban population. One key factor in this pattern is the mass migrations to cities of the rural poor, especially children and teens. These young people are driven by a variety of factors, including lack of food, the disintegration of families through death, disease, and poverty, displacement due to war and terrorism, and the lure of better jobs in the cities. The impact of the arrival of poor rural youths in urban areas has taken a toll. Throughout the Middle East and Asia, cities lack the housing space and employment opportunities needed to accommodate these domestic migrants. As a result, huge shantytowns and slums have developed on the edges of urban centers. These congested and often dangerous areas have above-average rates of contagious disease, violent crime, prostitution, and child abuse.

The new arrivals face a host of challenges. Many come to the city seeking opportunity or escaping from their overcrowded homes in the countryside. Clean water, adequate food, and proper clothing are virtually unknown. Ethnic, linguistic, and religious differences in the population are evident in the slums, and give rise to the formation of rival gangs. Girls are especially susceptible to becoming victims of crime, and in some places their low social status makes it difficult for them to find legal employment. Officials in many cities have addressed these problems by providing wells and water pumps in the slums and by encouraging self-help options like better food selections, personal hygiene, and savings programs. New credit programs aimed at single and young women are designed to help them establish small businesses such as food stalls or clothing repair, in hopes that by earning their own money, they can raise their social status and reduce their vulnerability to crime.

The problem of poor urban youths is most acute in East and Southeast Asia. In Vietnam, 40 percent of the population is under twenty-four years of age. This is the group that is rapidly migrating from agricultural areas to cities. The cities of Hanoi, Ho Chi

Minh City (Saigon), and Hue have more squatters, homelessness, and gang violence than ever before.

In Thailand, the movement of youths to the capital has weakened traditional patterns of guidance by families and village leaders and left young people in Bangkok more susceptible to prostitution rackets, drug use, and violent crime. In China, the introduction of free markets has spurred huge migrations to urban areas, especially by young people. The imbalance between urban and rural incomes sharpened rapidly with the rollback of government controls on employment, and residency fueled the push toward the cities. Young people arriving in China's cities are typically homeless for several months, or longer. Lacking funds, many quickly become involved in drug use and crime.

Planned Cities

In some Asian countries, government officials are planning ahead for the predicted explosion of urban populations by designing completely new cities to be built on undeveloped lands known as greenfield sites. Although environmental organizations generally criticize this approach to managing urban growth, many municipal authorities are convinced of the importance of creating new planned cities. They maintain that urban centers designed from the ground up will function more efficiently, with fewer wasted resources, less pollution, and less social friction than in established cities.

A particularly ambitious undertaking was the creation in India of Navi Mumbai (New Bombay) a completely new city covering 250 square miles (650 sq. km.) on the outskirts of Mumbai (Bombay) itself. The new town features high-density housing, public transportation links to central Mumbai, ultramodern hospitals, schools and colleges, and cable and satellite communications systems in a fully landscaped environment. Navi Mumbai is designed to give residents convenient access both to the commercial and cultural assets of Mumbai and to parks and recreation facilities near their homes. In 2019, the population exceeded 1.1 million.

China has also made a huge commitment to planning new cities. Its old urban centers on the eastern and southern coasts have been rapidly en-

circled by planned urban, industrial, and manufacturing centers to accommodate China's growing population and rapid economic development. Millions of farmers and thousands of villages have been removed from China's coastal plains to make room for new urban complexes and expanding megacities.

Zhuhai, on the southeastern coast, has become a model for this type of urban development in China. Begun in 1979, Zhuhai took shape on land previously used to cultivate rice. The city council set out to create a modern seaside garden city with a manufacturing-based economy tied to international trade. By 2010 Zhuhai was a 665-square-mile (1,723-sq.-km.) city with hundreds of miles of highways, two harbors, an airport, a large electricity plant, water and waste treatment facilities, dozens of small and medium-sized factories, a reforestation program, and air, noise, and water quality monitoring systems. Zhuhai is often promoted by government officials as a model for new urban development in China. In 2014, Zhuhai was named the most livable city in China.

Elsewhere in China, however, other patterns of urban planning are observed. A number of major cities have been declared open to international trade and investment, and each has created new industrial and urban zones nearby to accommodate modern manufacturing and high-technology industries. These new development zones, with their employment opportunities and new residential facilities, are magnets for rural youths.

Qingdao, located on the eastern coast of the Shandong Peninsula overlooking the Korean Peninsula, is one of the open coastal cities undergoing rapid expansion. In 1992 approval was obtained for creation of the Qingdao Economic Development Area, the Qingdao Technological Development Zone, and six other special manufacturing and investment areas surrounding the ancient city of Qingdao. Existing agricultural populations were removed from these areas, and new housing, industrial, and transportation facilities designed and built. The area now has ports, airports, manufacturing plants, oil and chemical processing facilities, new roads and rail systems, schools, colleges, hospitals, laboratories, and huge residential developments, as well as a seaside holiday and resort area. In 2014 the Qingdao area had a population of more than 9 million people.

In Qingdao and other coastal cities, comprehensive development planning is radically changing traditional relationships between city and countryside and accelerating urban growth at unprecedented speeds. In 2012, China's urban population exceeded its number of rural residents for the first time. By 2019, nearly 60 percent of China's people lived in cities. This remarkable growth is indicative of the prevailing urbanization patterns in Asia.

Laura M. Calkins

DISCUSSION QUESTIONS:
HUMAN GEOGRAPHY—PEOPLE

Q1. When did the first modern human beings arrive in Asia? From where did they come? What path of settlement did they follow? Where did the earliest great civilizations of Asia develop?

Q2. What relationship is thought to exist between the indigenous people of Siberia and Native Americans? Does scientific proof confirm this relationship? How are these peoples the similar of different today?

Q3. Define the term "primate city." To which Asian cities might that term apply? Which countries have more than one primate city? How can that be so?

POLITICAL GEOGRAPHY

The forty-eight independent states of Asia vary greatly in size, population, and level of political and economic development. The most common regional classification of Asia divides the continent into five major regions:

East Asia—Japan, South Korea, North Korea, China, Taiwan, and Mongolia;

Southeast Asia—Vietnam, Laos, Cambodia, Thailand, Myanmar, Malaysia, Singapore, Brunei, Philippines, Indonesia, and Timor-Leste;

South Asia—Afghanistan, Bhutan, Bangladesh, India, Nepal, Sri Lanka, Maldives, and Pakistan;

Central Asia—Kyrgyzstan, Tajikistan, Uzbekistan, Kazakhstan, and Turkmenistan;

Southwest (or Western) Asia—Iran, Iraq, Kuwait, the Asian part of Turkey, Syria, Lebanon, Israel, the State of Palestine (a non-member observer state of the United Nations), Jordan, Saudi Arabia, Yemen, Oman, United Arab Emirates, Qatar, Bahrain, Armenia, Georgia, and Azerbaijan.

While Southwest Asia is the largest part of the *Middle East*, the names of the two regions are not quite synonymous. The Middle East is a transcontinental region that by most modern definitions includes Southwest Asia, all of Turkey, and Egypt (which is located in North Africa except for the Sinai Peninsula). In broader usage, the term "Middle East" is applied to an even larger region that includes Southwest Asia and all of the Arab countries of North Africa, and occasionally also Afghanistan and Pakistan.

The Asian part of Russia is sometimes classified as its own Asian region—North Asia, although Russia as a political entity is commonly classified as a European country since the large majority of its population and major cities are located in Europe. Turkey, Azerbaijan, Georgia, and Kazakhstan are also transcontinental countries with the territory on both sides of the geographical boundary between Europe and Asia, but their European parts are relatively small and these states are commonly included in the list of countries of Asia. Located on the dividing line between Europe and Asia—the watershed along the Greater Caucasus mountain range—the states of Georgia, Azerbaijan, and Armenia are also classified as Transcaucasian countries.

Colonial Heritage of Asia

Most Asian countries have been adversely affected by a history of colonial exploitation. In the modern era, much of the responsibility for Asian colonial subjugation lay with European powers. Indeed, most major European powers had one or more colonial enclaves in Asia.

European contact with Asia began during the medieval period, when merchants and explorers like Niccolò Polo, his brother Maffeo, and his son Marco, traveled across Central Asia on the famed Silk Road to trade with the courts of China and Japan. Such contact remained quite limited until the development of large, seaworthy ships and other technologies ushered in the era of maritime exploration in the late fifteenth and sixteenth centuries.

The first European maritime explorer to round the Cape of Good Hope and travel across the Arabian Sea to India was Vasco da Gama, who landed on India's coast in May 1498. Thereafter, European exploration of Asia expanded rapidly as explorers from Portugal, Spain, the Netherlands, France, and Britain sought gold and glory. By the nineteenth century, European governments had carved up most of Asia into spheres of influence.

The most successful colonizers were the British: Great Britain held colonial possessions all across Asia. From Hong Kong and the Malay peninsula to South Asia and much of Southwest Asia, Britain enforced its exploitative rule upon its new subjects. By the late nineteenth century, France also had established colonies in Indochina (the continental part of Southeast Asia comprising present-day Vietnam, Cambodia, and Laos), the Indian subcontinent, China, Lebanon, and Syria.

The islands that now collectively make up Indonesia were colonized by the Dutch and the Portuguese beginning in the fifteenth and sixteenth centuries. Spain held sway over those islands that would become the Philippines. The Russian Empire (and its successor state, the Soviet Union) extended its influence into Central Asia, Transcaucasia, and southern Siberia (North Asia) beginning in the mid-seventeenth century. Little of Asia remained outside the direct or indirect purview of European governments bent on building empires.

Modern colonizers of Asia were not solely European; during the late nineteenth and early twentieth centuries, Japan aggressively pursued colonial undertakings in East Asia. The militaristic governments of Japan during the Meiji (1868-1912), Taisho (1912-1926), and the early Showa (1926-1945) periods expanded into Korea and northern China (Manchuria). During World War II, the Japanese government sought to create an Asia-wide empire, the Greater East Asia Co-Prosperity Sphere. It was only with Japan's defeat in 1945 that its overseas colonies were liberated.

The United States acquired its sole Asian colony in 1898 when Spain ceded its longstanding colony of the Philippines to the United States after Spain's defeat in the Spanish-American War.

Decolonization

In the immediate post-World War II period that, the drive toward independence swept much of the rest of Asia; within several decades, most colonies had become fully independent. However, the independence movement continued beyond the immediate postwar era: Two European outposts in China were returned to China's sovereignty at the turn of the third millennium. In July 1997, Hong Kong was handed over to China by the United Kingdom amid tremendous national celebrations in Beijing, and tiny Portuguese-held Macao was transferred to China in December 1999.

The last former colony in Asia to become a sovereign state was East Timor (Timor-Leste). Colonized by Portugal in the sixteenth century, it declared independence in November 1975 only to be invaded by the Indonesian military days later and declared a province of Indonesia. In 1999, after decades of the East Timorese independence struggle and under international pressure, Indonesia relinquished control over East Timor. In 2002, the Democratic Republic of Timor-Leste became the first new independent nation of the twenty-first century.

Chiang Kai-shek, leader of the Kuomintang from 1925 until his death in 1975.

General Douglas MacArthur landing ashore during the Battle of Leyte in the Philippines on October 20, 1944.

Despite political independence, the effects of colonialism in Asia remained quite severe. Besides siphoning off natural resource wealth, the European colonizers also reoriented indigenous economies from self-sufficiency toward the cash- crop export markets, leaving those economies with severe developmental barriers in the wake of decolonization. Indigenous social and political systems were often torn asunder and replaced by European institutions that were ill-suited to local cultures. In carving up the Asian landmass, Europeans unknowingly contributed to future intraregional political instability, as borders were drawn by colonial powers to suit their own interests rather than developing in an evolutionary and rational process of state-building.

Modern Political Borders

Many of Asia's modern international borders are the result of colonial machinations and postindependence warfare. After vanquishing the Turkish Ottoman Empire in World War I, Britain and France created artificial borders for their newly obtained colonies in Southwest Asia. A glance at the borders of modern-day Jordan, Syria, Iraq, or Kuwait provides ample evidence for this. These borders continue to be the cause of much of the conflict in the Middle East.

Another example of the colonists' influence over the creation of borders can be found in the five Central Asian states, former republics of the Soviet Union. The borders of Uzbekistan, Kazakhstan, Kyrgyzstan, Tajikistan, and Turkmenistan were drawn artificially by the government of Joseph Stalin to enable the shrewd Soviet dictator to more easily divide and conquer the Central Asian peoples inhabiting this region. To this day, these now-independent countries continue to claim parcels of each other's territory.

HUMAN GEOGRAPHY

The U.S. Role in the Korean War

The war between North and South Korea involved more than nineteen countries, including the United States, Great Britain, Canada, and communist China. The United States took advantage of the Soviet Union's boycott of the United Nations' Security Council and used its considerable influence in that body to secure a United Nations resolution in support of repelling North Korean aggression. In July 1953, after three years of fighting and more than 2.5 million casualties, an armistice line was agreed upon by the parties to the conflict; this Demilitarized Zone (DMZ) at the thirty-eighth parallel created a border that was not very different from that which had existed at the war's outset. To date, there has not been an official peace treaty signed, and in 2020, 23,000 U.S. troops remain stationed in South Korea.

In addition, international borders in Asia were often the result of internecine war following independence. For example, after Britain gave up its colonies in South Asia in 1947, a bloody partition struggle raged for several months. In the end, the countries of India and Pakistan emerged from the strife, but not until more than 1 million people had died and millions were displaced. As a result of still another war, East Pakistan later seceded from Pakistan to become the sovereign country of Bangladesh.

Likewise, the Korean peninsula was thrown into tumult after Japan's defeat in 1945. Upon the Japanese surrender, Korea was partitioned into two countries at the thirty-eighth parallel of latitude, with a Soviet-backed communist government taking control in the north and a U.S.-backed authoritarian regime gaining power in the south. When Kim Il Sung's North Korean troops poured across the border on June 25, 1950, a war broke out between North Korea and South Korea—one of the deadliest wars in modern history. The Korean conflict remains ongoing to this day, as no peace treaty was ever signed between the two Korean states and neither accept the border between them as permanent.

Topography did play some role in determining Asia's present-day international borders. A variety of mountain ranges and river basins have served as natural boundaries between some Asian states. The world's highest mountain range, the Himalayas, separates India from China; the Hindu Kush mountains are a natural border between China and Pakistan; the Tian Shan and Pamir mountain ranges separate China from Tajikistan and Kyrgyzstan, respectively. Much of China's northern border with the Russian Federation follows the courses of three major rivers: the Argun, Amur, and Ussuri. In Southeast Asia, the Mekong River basin forms a partial boundary between Laos and its two western neighbors, Myanmar and Thailand. On the other side of Asia, the Jordan River and the Red Sea form part of the border between Israel and the Kingdom of Jordan.

Territorial Disputes

Issues of border delimitation remain a constant source of political instability in the Asian region. Besides the ongoing animosity between North and South Korea, many other countries in the region continue to be plagued by conflicts over borders. Hindu-majority India and Muslim-majority Pakistan both claim Jammu and Kashmir, a mountainous region in the western Himalayas populated mostly by Muslims. Started after the partition of India in 1947, this enduring conflict has spawned three major wars between these two countries; armed skirmishes and cross-border shelling are common. The danger of this territorial dispute increased after both India and Pakistan acquired nuclear weapons. On top of that, a part of Kashmir called Ladakh has been contested between India and China, which fought a war over this disputed territory (the 1962 Sino-Indian War). There is still no official border between these two countries.

In Southwest Asia, a longstanding territorial dispute between Iraq and Iran led to an eight-year war that lasted from 1980 to 1988. The Iran-Iraq War resulted in over 500,000 casualties but no territorial changes. Other unresolved territorial issues in Asia include the dispute between Japan and Russia over the Kuril Islands, the dispute between Malaysia and the Philippines over the northern part of the island of Borneo, and the South China

Sea disputes involving China, Indonesia, Malaysia, the Philippines, Vietnam, and Brunei.

Self-Proclaimed States and Ethnic Conflicts

With its numerous large and small ethnic groups, the legacy of arbitrary colonial borders, and huge gaps in socioeconomic development between different parts of many countries, Asia has been a cauldron of ethnic conflicts and separatist movements. In Southwest Asia, there are three breakaway, separatist states: Nagorno-Karabakh (Artsakh), Abkhazia, and South Ossetia (Alania)—all in Transcaucasia.

The self-declared Republic of Artsakh controls the Nagorno-Karabakh region of Transcaucasia, which is populated mostly by Armenians. However, it was designated part of Azerbaijan in the 1920s by Stalin's government of the Soviet Union, to which Armenia and Azerbaijan belonged at the time. As soon as the Soviet Union began to dissolve in 1991, the region declared independence and a war between Armenia and Azerbaijan erupted. The 1994 cease-fire agreement, which remains in place to this day, has been repeatedly broken.

The secession of South Ossetia (after the 1991-1992 war) and Abkhazia (after the 1992-1993 war) from Georgia is the latest episode in the centuries-long independence struggle of these Transcaucasian ethnic enclaves. Both are now *de facto* sovereign states with close ties to Russia, while most United Nations member states consider them legally parts of Georgia.

Other major ethnic, religious, and cultural conflicts and separatist movements in different parts of Asia include, among others:

in China—the struggle of ethnic Tibetans, Uyghurs, and other ethnic minorities against the repression of their religious freedoms, linguistic rights, and cultural practices;

in the Greater Kurdistan region—Kurdish nationalist movement aimed at creating a sovereign state of Kurdistan out of those parts of Turkey, Iraq, Iran, and Syria within the majority Kurdish population;

in India and Pakistan—indigenous insurgency movements in the Indian states of Assam, Tamil Nadu, and Nagaland; an ongoing conflict between the Naxalite Maoist groups and the Indian government; the Kashmiri separatism; the Khalistan movement seeking to create a Sikh state from parts of the Punjab region of India and Pakistan;

in Indonesia—numerous indigenous separatist movements in Aceh, Bali, Papua, and other parts of the country;

in Japan—discrimination and social marginalization of the indigenous Ainu and Okinawan peoples and the Burakumin outcast group;

in Lebanon—a history of sectarian violence among Sunnis, Alawites, and Christians; Lebanon is also the home base of Hezbollah, a Shia Islamist militant group active throughout the Middle East;

in Myanmar—ongoing violent persecution of the Muslim Rohingya people and armed insurrections by the Rohingya;

in Nepal—a history of violent Maoist insurgencies;

in Sri Lanka—a past history of militant Tamil nationalism and later large-scale displacement of Sri Lankan Tamils; and

in Yemen—the Houthi insurgency that has escalated into an ongoing civil war.

Geopolitical Rivalries

The Korean War was the first Asian theater of the Cold War. It was waged between North Korea, supported by communist China and the Soviet Union, against South Korea, supported by the United States and its allies In the 1960s and 1970s, Asia was again the scene of intense geopolitical rivalry between the two superpowers: the Soviet Union and the United States. The Vietnam War pitted U.S. forces against Soviet- and Chinese-backed insurgents in a bitter, eleven-year struggle.

In the aftermath of the downfall of the Soviet Union, its successor state, the Russian Federation, soon positioned itself as a powerful regional force alongside the rapidly developing China. The continued support by China and Russia of the dictatorial regime of North Korea has emboldened that country to pursue nuclear armament plans. Started in 2011, the Syrian Civil War once again heightened the conflicting ambitions of the United States and Russia in Asia, as they supported opposing sides of the war. China's increased military activity in the South China Sea has led to the United States' strengthening its naval presence in that region. However, in the increasingly multifaceted and interconnected world of the twenty-first century, no one country or couple of countries can influence global or even regional developments quite the way the superpowers did in the twentieth century.

The Status of Taiwan, Hong Kong, and Macao

With the dramatic communist takeover of mainland China in 1949, the Nationalist Chinese government of Chiang Kai-shek was forced to evacuate the mainland, establishing rule later that year on the island of Taiwan and retaining the name that the nationalist state had on the mainland—the Republic of China (ROC). In the succeeding years, the communist leadership in Beijing continuously demanded the unification of Taiwan with mainland China, the People's Republic of China (PRC). Early attempts at unification on the part of the PRC were met with hostility from both Chiang's military forces and the U.S. Seventh Fleet in the Taiwan Strait. As a founding member of the United Nations, the ROC continued to represent China in the UN for over twenty years. The United States and other non-communist countries did not maintain diplomatic relations with the communist PRC and referred to Taiwan as "China."

Throughout the 1950s and 1960s, however, more and more countries came to recognize Beijing as the legitimate Chinese government. In 1971, the United Nations changed China's representation in the UN from the ROC to the PRC. In 1979, the United States withdrew official recognition from the Republic of China, although maintaining the close military ties that had existed since its founding. Officially, the United States does not support the independence of Taiwan and opposes unilateral

The Great Hall of the People where the National People's Congress convenes.

changes to the status quo by either side. However, stability in the region has been periodically threatened in the recent past both by pressure for unification from the PRC and by internal political struggles within Taiwan between parties favoring gradual unification with the PRC and those in favor of declaring full sovereignty. The status of Taiwan remains uncertain as it maintains official diplomatic relations with several countries and holds separate membership in some international organizations, while the PRC only recognizes Taiwan as its province.

After Hong Kong and Macao were transferred to China's sovereignty, they officially became Special Administrative Regions of the PRC, and their governance has been based on China's constitutional principle of "One country, two systems." Under this principle, each region can maintain its own system of government and conduct its domestic and economic affairs, including immigration and international trade relations, independently of mainland China. While the population of Macao has been largely supportive of the implementation of the "One country, two systems" principle, several decisions of China's national legislature and Hong Kong's government resulted in large-scale protests in 2003, 2014, and 2019-2020. The protesters view these decisions as China's encroachment on Hong Kong's self-government and political freedoms.

The Middle East
Southwest Asia has been a focus of superpower and interstate rivalry since the founding of Israel in May 1948. Since then, several major wars have erupted between Israel and its Arab neighbors (Egypt, Jordan, and Syria). The 1948 Arab-Israeli War resulted in a fateful demographic change in the Middle East—hundreds of thousands of Palestinians were displaced from the territories where they lived and became Palestinian refugees. The 1967 Six-Day War ended in a stunning Israeli victory, as the Jewish state expanded its territory to include Egypt's Sinai Peninsula and Gaza Strip, Syria's Syria's Golan Heights, and Jordan's West Bank territory, and East Jerusalem—a part of the city of Jerusalem

that had been occupied by Jordan during the 1948 War.

The Yom Kippur War of 1973 ended favorably for Israel, as it seized territory west of the Suez Canal from Egypt. Less than six years later, Egypt's president Anwar Sadat made peace with Israel in return for an Israeli withdrawal from the Sinai Peninsula and U.S. security and financial concessions. It took another fifteen years before a formal peace agreement was reached between Israel and its eastern neighbor, Jordan. Efforts to bridge the chasm between Syria and Israel have been less fruitful. The two countries technically remain at war, and the so-called Purple Line—the ceasefire line established after the Six-Day War—still serves as the *de facto* border between the two countries.

The geographical area long described as "Palestinian territories" or "Occupied Palestinian Territory" is now referred to by the United Nations as the State of Palestine. It includes the West Bank, the Gaza Strip, and East Jerusalem—the designated capital of the State of Palestine. Since 2012, it is a *de jure* sovereign state with the status of a non-member observer in the UN.

Until September 11, 2001, the United States' involvement in the Middle East was focused on backing Israel and preventing Soviet influence in Arab countries. On the day that became known as 9/11, four mass-casualty attacks in the U.S. were perpetrated by terrorists affiliated with the Islamic militant group al-Qaeda, based in Afghanistan. In retaliation, the United States invaded Afghanistan, aiming to destroy al-Qaeda and drive from power its local ally, the Taliban, an Islamic fundamentalist military-political organization that controlled the country. The War in Afghanistan, the longest war in U.S. history, helped marginalize al-Qaeda but failed to stabilize and democratize Afghanistan. The Taliban eventually reemerged as a powerful force and reacquired control over much of the country.

Similarly, the 2003 invasion of Iraq by the United States, deeply rooted in America's anti-authoritarian ideology, helped remove that country's dictator Saddam Hussein but fueled sectarian violence and political instability in that country. It also furthered the rise of still another Islamist militant group, known as the Islamic State in Iraq and Syria (ISIS). For a time, ISIS was able to recruit anti-American followers from around the world and even maintain a quasi-state in the region. It took years of international effort to gain back territories captured by ISIS.

The still ongoing Syrian Civil War broke out in 2011. It is a multi-sided conflict between the Syrian government, supported by various allies (notably, Russia, Iran, and the Lebanese Hezbollah) on one side and various anti-government groups and Kurdish forces on the other side; ISIS militants have been also involved. Some of these groups also fight each other and are supported or opposed, in varying combinations, by the United States, Turkey, Saudi Arabia, Israel, and other countries. The fallout of the war has spilled over many borders. It led to serious tensions between Iran and Turkey,

stoked a resurgence of violence in Lebanon, exacerbated sectarian conflicts in other Middle Eastern states, and once again pitted the United States and Russia against each other, this time in a proxy war. The huge influx of Syrian refugees into Turkey, Jordan, Lebanon, and—by sea—to countries of the European Union became a humanitarian crisis of global proportions. It also heightened political divisions between and within the EU member countries.

Asian Political Systems

Of the dozens of countries in Asia, only a few can be considered fully consolidated democratic systems, including Japan, India, South Korea, and Israel. Elsewhere in Asia, the political systems fall somewhere between authoritarianism and democracy; there are also several one-party-states, theocracies, and monarchies.

2019–2020 Hong Kong protests.

In Central Asia, the former Soviet republics of Uzbekistan, Turkmenistan, Kyrgyzstan, Kazakhstan, and Tajikistan are nominal democracies but, in essence, are authoritarian regimes in which former Soviet communist party leaders and their designated successors have continued to rule in the postcommunist era.

In East Asia, several one-party communist regimes remain in existence. In the 1990s the communist states of China, Vietnam, and Laos started pursuing economic reforms under a moderated form of socialism that includes some characteristics of market economy, while militaristic North Korea remained wedded to Stalinist orthodoxy.

In Southwest Asia, Islamic fundamentalist regimes came to power in Iran and Afghanistan in 1979 and 1996, respectively. Ultimate political power in the Islamic Republic of Iran lies with a theocratic elite: an unelected group of Islamic ayatollahs and mullahs. In Afghanistan, the ultraconservative Taliban was ousted during the initial phase of the Afghanistan War but regained much of its power after the United States curtailed its military involvement there.

In addition to the authoritarian political systems already discussed, several sultanistic regimes exist on the Arabian Peninsula in Southwest Asia, where an emir, a sheik, or a ruling family holds political power based upon heredity. Such is the situation in Saudi Arabia, Qatar, the United Arab Emirates, and Oman. These countries are considered absolute monarchies while Jordan, Kuwait, and Bahrain are constitutional monarchies, in which the authority of the ruler is exercised within certain, constitutionally defined, limits.

In other parts of Asia, Nepal made a transition from monarchy to a republic in 2008, Brunei remains an absolute monarchy, and Bhutan and Cambodia are constitutional ones. Japan is nominally also a constitutional monarchy, but the emperor of Japan is largely a ceremonial figurehead, much like the modern monarchs of the United Kingdom.

Thomas E. Rotnem

DISCUSSION QUESTIONS: POLITICAL GEOGRAPHY

Q1. How has Asia's colonial legacy shaped the outlook of its peoples? How did the economic systems of the colonized countries change to suit the needs of the European governments? What parallels can be drawn between Britain's colonial empire and the territories that once belonged to the Soviet Union?

Q2. What are some of the factors that gave rise to the territorial disputes in the Middle East and the Transcaucasian republics? Why have some of these disputes evolved into separatist movements, conflicts, and, ultimately, outright wars?

Q3. How many different forms of government are there in Asia? Which is the most authoritarian? Which is the most democratic? How do Asian monarchies differ from those in Western Europe? What is meant by the term "theocracy"?

ECONOMIC GEOGRAPHY

OVERVIEW

TRADITIONAL AGRICULTURE

Two agricultural practices that are widespread among the world's traditional cultures, slash-and-burn and nomadism, share several common features. Both are ancient forms of agriculture, both involve farmers not remaining in a fixed location, and both can pose serious environmental threats if practiced in a nonsustainable fashion. The most significant difference between the two forms is that slash-and-burn generally is associated with raising field crops, while nomadism as a rule involves herding livestock.

Slash-and-Burn Agriculture

Farmers have practiced slash-and-burn agriculture, which is also referrred to as shifting cultivation or swidden agriculture, in almost every region of the world where the climate makes farming possible. Humans have practiced this method for about 12,000 years, ever since the Neolithic Revolution. Swidden agriculture once dominated agriculture in more temperate regions, such as northern Europe. It was, in fact, common in Finland and northern Russia well into the early decades of the twentieth century. Today, between 200 and 500 million people use slash-and-burn agriculture, roughly 7 percent of the world's population. It is most commonly practiced in areas where open land for farming is not readily available because of dense vegetation. These regions include central Africa, northern South America, and Southeast Asia

Slash-and-burn acquired its name from the practice of farmers who cleared land for planting crops by cutting down the trees or brush on the land and then burning the fallen timber on the site. The farmers literally slash and burn. The ashes of the burnt wood add minerals to the soil, which temporarily improves its fertility. Crops the first year following clearing and burning are generally the best crops the site will provide. Each year after that, the yield diminishes slightly as the fertility of the soil is depleted.

Farmers who practice swidden cultivation do not attempt to improve fertility by adding fertilizers such as animal manures but instead rely on the soil to replenish itself over time. When the yield from one site drops below acceptable levels, the farmers then clear another piece of land, burn the brush and other vegetation, and cultivate that site while leaving their previous field to lie fallow and its natural vegetation to return. This cycle will be repeated over and over, with some sites being allowed to lie fallow indefinitely while others may be revisited and farmed again in five, ten, or twenty years.

Farmers who practice shifting cultivation do not necessarily move their dwelling places as they change the fields they cultivate. In some geographic regions, farmers live in a central village and farm cooperatively, with the fields being alternately allowed to remain fallow, and the fields being farmed making a gradual circuit around the central village. In other cases, the village itself may move as new fields are cultivated. Anthropologists studying indigenous peoples in Amazonia, discovered that village garden sites were on a hundred-year cycle. Villagers farmed cooperatively, with the entire village working together to clear a garden site. That garden would be used for about five years, then a new site was cleared. When the garden moved an inconvenient distance from the village, about once every twenty years, the entire village would move to be closer to the new garden. Over a period of approximately 100 years, a village would make a circle through the forest, eventually ending up close to where it had been located long before any of the present villagers had been born.

In more temperate climates, individual farmers often owned and lived on the land on which they practiced swidden agriculture. Farmers in Finland, for example, would clear a portion of their land, burn the brush and other covering vegetation, grow grains for several years, and then allow that land to remain fallow for from five to twenty years. The individual farmer rotated cultivation around the land in a fashion similar to that practiced by whole villages in other areas, but did so as an individual rather than as part of a communal society.

Although slash-and-burn is frequently denounced as a cause of environmental degradation in tropical areas, the problem with shifting cultivation is not the practice itself but the length of the cycle. If the cycle of shifting cultivation is long enough, forests will grow back, the soil will regain its fertility, and minimal adverse effects will occur. In some regions, a piece of land may require as little as five years to regain its maximum fertility; in others, it may take 100 years. Problems arise when growing populations put pressure on traditional farmers to return to fallow land too soon. Crops are smaller than needed, leading to a vicious cycle in which the next strip of land is also farmed too soon, and each site yields less and less. As a result, more and more land must be cleared.

Nomadism

Nomadic peoples have no permanent homes. They earn their livings by raising herd animals, such as sheep, cattle, or horses, and they spend their lives following their herds from pasture to pasture with the seasons. Most nomadic animals tend to be hardy breeds of goats, sheep, or cattle that can withstand hardship and live on marginal lands. Traditional nomads rely on natural pasturage to support their herds and grow no grains or hay for themselves. If a drought occurs or a traditional pasturing site is unavailable, they can lose most of their herds to starvation.

THE HERITAGE SEED MOVEMENT

Modern hybrid seeds have increased yields and enabled the tremendous productivity of the modern mechanized farm. However, the widespread use of a few hybrid varieties has meant that almost all plants of a given species in a wide area are almost identical genetically. This loss of biodiversity, or, the range of genetic difference in a given species, means that a blight could wipe out an entire season's crop. Historical examples of blight include the nineteenth century Great Potato Famine of Ireland and the 1971 corn blight in the United States.

In response to the concern for biodiversity, there has been a movement in North America to preserve older forms of crops with different genes that would otherwise be lost to the gene pool. Nostalgia also motivates many people to keep alive the varieties of fruits and vegetables that their grandparents raised. Many older recipes do not taste the same with modern varieties of vegetables that have been optimized for commercial considerations such as transportability. Thus, raising heritage varieties also can be a way of continuing to enjoy the foods one's ancestors ate.

In many nomadic societies, the herd animal is almost the entire basis for sustaining the people. The animals are slaughtered for food, clothing is woven from the fibers of their hair, and cheese and yogurt may be made from milk. The animals may also be used for sustenance without being slaughtered. Nomads in Mongolia, for example, occasionally drink horses' blood, removing only a cup or two at a time from the animal. Nomads go where there is sufficient vegetation to feed their animals.

In mountainous regions, nomads often spend the summers high up on mountain meadows, returning to lower altitudes in the autumn when snow begins to fall. In desert regions, they move from oasis to oasis, going to the places where sufficient natural water exists to allow brush and grass to grow, allowing their animals to graze for a few days, weeks, or months, then moving on. In some cases, the pressure to move on comes not from the depletion of food for the animals but from the depletion of a water source, such as a spring or well. At many natural desert oases, a natural water seep or spring provides only enough water to support a nomadic group for a few days at a time.

In addition to true nomads—people who never live in one place permanently—a number of cultures have practiced seminomadic farming: The temperate months of the year, spring through fall, are spent following the herds on a long loop, sometimes hundreds of miles long, through traditional grazing areas; then the winter is spent in a permanent village.

Nomadism has been practiced for millennia, but there is strong pressure from several sources to eliminate it. Pressures generated by industrialized society are increasingly threatening the traditional cultures of nomadic societies, such as the Bedouin of the Arabian Peninsula. Traditional grazing areas are being fenced off or developed for other purposes. Environmentalists are also concerned about the ecological damage caused by nomadism.

Nomads generally measure their wealth by the number of animals they own and so will try to develop their herds to be as large as possible, well beyond the numbers required for simple sustainability. The herd animals eat increasingly large amounts of vegetation, which then has no opportunity to regenerate, and desertification may occur. Nomadism based on herding goats and sheep,

---

DESERTIFICATION

Desertification is the extension of desert conditions into new areas. Typically, this term refers to the expansion of deserts into adjacent nondesert areas, but it can also refer to the creation of a new desert. Land that is susceptible to prolonged drought is always in danger of losing its vegetative ground cover, thereby exposing its soil to wind. The wind carries away the smaller silt particles and leaves behind the larger sand particles, stripping the land of its fertility. This naturally occurring process is assisted in many areas by overgrazing.

In the African Sahel, south of the Sahara, the impact of desertification is acute. Recurring drought has reduced the vegetation available for cattle, but the need for cattle remains high to feed populations that continue to grow. The cattle eat the grass, the soil is exposed, and the area becomes less fertile and less able to support the population. The desert slowly encroaches, and the people must either move or die.

---

for example, has been blamed for the expansion of the Sahara Desert in Africa. For this reason, many environmental policymakers have been attempting to persuade nomads to give up their roaming lifestyle and become sedentary farmers.

Nancy Farm Männikkö

COMMERCIAL AGRICULTURE

Commercial farmers are those who sell substantial portions of their output of crops, livestock, and dairy products for cash. In some regions, commercial agriculture is as old as recorded history, but only in the twentieth century did the majority of farmers come to participate in it. For individual farmers, this has offered the prospect of larger income and the opportunity to buy a wider range of products. For society, commercial agriculture has been associated with specialization and increased productivity.

Commercial agriculture has enabled world food production to increase more rapidly than world population, improving nutrition levels for millions of people.

Steps in Commercial Agriculture
In order for commercial agriculture to exist, products must move from farmer to ultimate consumer, usually through six stages:

1. Processing, packaging, and preserving to protect the products and reduce their bulk to facilitate shipping.

2. Transport to specialized processing facilities and to final consumers.

3. Networks of merchant middlemen who buy products in bulk from farmers and processors and sell them to final consumers.

4. Specialized suppliers of inputs to farmers, such as seed, livestock feed, chemical inputs (fertilizers, insecticides, pesticides, soil conditioners), and equipment.

5. A market for land, so that farmers can buy or lease the land they need.

6. Specialized financial services, especially loans to enable farmers to buy land and other inputs before they receive sales revenues.

Improvements in agricultural science and technology have resulted from extensive research programs by government, business firms, and universities.

International Trade
Products such as grain, olive oil, and wine moved by ship across the Mediterranean Sea in ancient times. Trade in spices, tea, coffee, and cocoa provided powerful stimulus for exploration and colonization around 1500 CE. The coming of steam locomotives and steamships in the nineteenth century greatly aided in the shipment of farm products and spurred the spread of population into potentially productive farmland all over the world. Beginning with Great Britain in the 1840s, countries were willing to relinquish agricultural self-sufficiency to obtain cheap imported food, paid for by exporting manufactured goods.

Most of the leaders in agricultural trade were highly developed countries, which typically had large amounts of both imports and exports. These countries are highly productive both in agriculture and in other commercial activities. Much of their trade is in high-value packaged and processed goods. Although the vast majority of China's labor force works in agriculture, their average productivity is low and the country showed an import surplus in agricultural products. The same was true for Rus-

sia. India, similar to China in size, development, and population, had relatively little agricultural trade. Australia and Argentina are examples of countries with large export surpluses, while Japan and South Korea had large import surpluses. Judged by volume, trade is dominated by grains, sugar, and soybeans. In contrast, meat, tobacco, cotton, and coffee reflect much higher values per unit of weight.

The United States
Blessed with advantageous soil, topography, and climate, the United States has become one of the most productive agricultural countries in the world. Technological advances have enabled the United States to feed its own residents and export substantial quantities with only 3 percent of its labor force engaged directly in farming. In the 2020s there are about 2 million farms cultivating about 1 billion acres. They produced about US$133 billion worth of products. After expenses, this yielded about US$92.5 billion of net farm income—an average of only about US$25,000 per farm. However, most farm families derive substantial income from nonfarm employment.

There is a great deal of agricultural specialization by region. Corn, soybeans, and wheat are grown in many parts of the United States (outside New England). Some other crops have much more limited growing areas. Cotton, rice, and sugarcane require warmer temperatures. Significant production of cotton occurred in seventeen states, rice in six, and sugarcane in four. In 2018, the top 10 agricultural producing states in terms of cash receipts were (in descending order): California, Iowa, Texas, Nebraska, Minnesota, Illinois, Kansas, North Carolina, Wisconsin, and Indiana Typically the top two states in a category account for about 30 percent of sales. Fruits and vegetables are the main exception; the great size, diversity, and mild climate of California gives it a dominant 45 percent.

Socialist Experiments
Under the dictatorship of Joseph Stalin, the communist government of the Soviet Union established a program of compulsory collectivized agriculture

in 1929. Private ownership of land, buildings, and other assets was abolished. There were some state farms, "factories in the fields," operated on a large scale with many hired workers. Most, however, were collective farms, theoretically run as cooperative ventures of all residents of a village, but in practice directed by government functionaries. The arrangements had disastrous effects on productivity and kept the rural residents in poverty. Nevertheless, similar arrangements were established in China in 1950 under the rule of Mao Zedong. A restoration of commercial agriculture after Mao's death in 1976 enabled China to achieve greater farm output and farm incomes.

Most Western countries, including the United States, subsidize agriculture and restrict imports of competing farm products. Objectives are to support farm incomes, reduce rural discontent, and slow the downward trend in the number of farmers. Farmers in the European Union will see aid shrink in the 2021–2027 period to 365 billion euros (US$438 billion), down 5 percent from the current Common Agricultural Policy (CAP). Japan's Ministry of Agriculture, Forestry and Fisheries (MAFF) has requested 2.65 trillion yen (roughly US$24 billion) for the Japan Fiscal Year (JFY) 2018 budget, a 15 percent increase over last year. The budget request eliminates the direct payment subsidy for table rice production, but requests significant funding for a new income insurance program, agricultural export promotion, and underwriting goals to expand domestic potato production. In 2019, trade wars with China and punishing tariffs have led to increased subsidies by the U.S. government, totaling US$10 billion dollars in 2018 and US$14.5 billion dollars in 2019.

Problems for Farmers

Farmers in a system of commercial agriculture are vulnerable to changes in market prices as well as the universal problems of fluctuating weather. Congress tried to reduce farm subsidies through the Freedom to Farm Act of 1996, but serious price declines in 1997-1999 led to backtracking. Efforts to increase productivity by genetic alterations, radiation, and feeding synthetic hormones to livestock have drawn critical responses from some consumer groups. Environmentalists have been concerned about soil depletion and water pollution resulting from chemical inputs.

Productivity and World Hunger

Despite advances in agricultural production, the problem of world hunger persists. Even in countries that store surpluses of farm commodities, there are still people who go hungry. In less-developed countries, the prices of imported food from the West are too low for local producers to compete and too high for the poor to buy them.

Paul B. Trescott

MODERN AGRICULTURAL PROBLEMS

Ever since human societies started to grow their own food, there have been problems to solve. Much of the work of nature was disrupted by the work of agriculture as many as 10,000 years ago. Nature took care of the land and made it productive in its own intricate way, through its own web of interdependent systems. Agriculture disrupts those systems with the hope of making the land even more productive, growing even more food to feed even more people. Since the first spade of soil was turned over and the first plants domesticated, farmers have been trying to figure out how to care for the land as well as nature did before.

Many modern problems in agriculture are not really modern at all. Erosion and pollution, for example, have been around as long as agriculture.

However, agriculture has changed drastically within those 10,000 years, especially since the dawn of the Industrial Revolution in the seventeenth century. Erosion and pollution are now bigger problems than before and have been joined by a host of others that are equally critical—not all related to physical deterioration. Modern farmers use many more machines than did farmers of old, and modern machines require advanced sources of energy to unleash their power. The machines do more work than could be accomplished before, so fewer farmers are needed, which causes economic problems.

Cities continue to grow bigger as land—usually the best farmland around—is converted to homes and parking lots for shopping centers. The farmers that remain on the land, needing to grow ever more food, turn to the research and engineering industries to improve their seeds. These industries have responded with recombinant technologies that move genes from one species to another; for example, genes cut from peanuts may be spliced into chickens. This creates another set of cultural problems, which are even more difficult to solve because most are still "potential"—their impact is not yet known.

Erosion

Soil loss from erosion continues to be a huge problem all over the world. As agriculture struggles to feed more millions of people, more land is plowed. The newly plowed lands usually are considered more marginal, meaning they are either too steep, too thin, or too sandy; are subject to too much rain; or suffer some other deficiency. Natural vegetative cover blankets these soils and protects them from whatever erosive agents are active in their regions: water, wind, ice, or gravity. Plant cover also increases the amount of rain that seeps downward into the soil rather than running off into rivers. The more marginal land that is turned over for crops, the faster the erosive agents will act and the more erosion will occur.

Expansion of land under cultivation is not the only factor contributing to erosion. Fragile grasslands in dry areas also are being used more inten-

sively. Grazing more livestock than these pastures can handle decreases the amount of grass in the pasture and exposes more of the soil to wind—the primary erosive agent in dry regions.

Overgrazing can affect pastureland in tropical regions too. Thousands of acres of tropical forest have been cleared to establish cattle-grazing ranges in Latin America. Tropical soils, although thick, are not very fertile. Fertility comes from organic waste in the surface layers of the soil. Tropical soils form under constantly high temperatures and receive much more rain than soils in moderate, midlatitude climates; thus, tropical organic waste materials rot so fast they are not worked into the soil at all. After one or two growing seasons, crops grown in these soils will yield substantially less than before.

Tropical fields require fallow periods of about ten years to restore themselves after they are depleted. That is why tropical cultures using slash-and-burn methods of agriculture move to new fields every other year in a cycle that returns them to the same place about every ten years, or however long it takes those particular lands to regenerate. The heavy forest cover protects these soils from exposure to the massive amounts of rainfall and provides enough organic material for crops—as long as the forest remains in place. When the forest is cleared, however, the resulting grassland cannot provide the adequate protection, and erosion accelerates. Grasslands that are heavily grazed provide even less protection from heavy rains, and erosion accelerates even more.

The use of machines also promotes erosion, and modern agriculture relies on machinery: tractors, harvesters, trucks, balers, ditchers, and so on. In the United States, Canada, Europe, Russia, Brazil, South Africa, and other industrialized areas, machinery use is intense. Machinery use is also on the rise in countries such as India, China, Mexico, and Indonesia, where traditional nonmechanized methods are practiced widely. Farming machines, in gaining traction, loosen the topsoil and inhibit vegetative cover growth, especially when they pull behind them any of the various farm implements designed to rid the soil of weeds, that is, all vegetation except the desired crop. This leaves the soil more exposed to erosive weather, so more soil is carried away in the runoff of water to streams.

Eco-fallow farming has become more popular in the United States and Europe as a solution to reducing erosion. This method of agriculture, which leaves the crop residue in place over the fallow (nongrowing) season, does not root the soil in place, however. Dead plants do not "grab" the soil like live plants that need to extract from it the nutrients they need to live, so erosion continues, even though it is at a slower rate. Eco-fallow methods also require heavier use of chemicals, such as herbicides, to "burn down" weed growth at the start of the growing season, which contributes to accelerated erosion and increases pollution.

Pollution

Pollution, besides being a problem in general, continues to grow as an agricultural problem. With the onset of the Green Revolution, the use of herbicides, insecticides, and pesticides has increased dramatically all over the world. These chemicals are not used up completely in the growth of the crop, so the leftovers (residue) wash into, and contaminate, surface and groundwater supplies. These supplies then must be treated to become useful for other purposes, a job nature used to do on its own. Agricultural chemicals reduce nature's ability to act as a filter by inhibiting the growth of the kinds of plant life that perform that function in aquatic environments. The chemical residues that are not washed into surface supplies contaminate wells.

As chemical use increases, contamination accumulates in the soil and fertility decreases. The microorganisms and animal life in the soil, which had facilitated the breakdown of soil minerals into usable plant products, are no longer nourished because the crop residue on which they feed is depleted, or they are killed by the active ingredients in the chemical. As a result, soil fertility must be restored to maintain yield. Chemical replacement is usually the method of choice, and increased applications of chemical fertilizers intensify the toxicity of this cyclical chemical dependency.

Chemicals, although problematic, are not as difficult to contend with as the increasingly heavy silt load choking the life out of streams and rivers. Accelerated erosion from water runoff carries silt particles into streams, where they remain suspended and inhibit the growth of many beneficial forms of plant and animal life. The silt load in U.S. streams has become so heavy that the Mississippi River Delta is growing faster than it used to. The heavy silt load, combined with the increased load of chemical residues, is seriously taxing the capabilities of the ecosystems around the delta that filter out sediments, absorb nutrients, and stabilize salinity levels for ocean life, creating an expanding dead zone.

This general phenomenon is not limited to the Mississippi Delta—it is widespread. Its impact on people is high, because most of the world's population lives in coastal zones and comes in direct contact with the sea. Additionally, eighty percent of the world's fish catch comes from the coastal waters over continental shelves that are most susceptible to this form of pollution.

Monoculture

Modern agriculture emphasizes crop specialization. Farmers, especially in industrialized regions, often grow a single crop on most of their land, perhaps rotating it with a second crop in successive years: corn one year, for example, then soybeans, then back to corn. Such a strategy allows the farmer to reduce costs, but it also makes the crop, and, thus, the farmer and community, susceptible to widespread crop failure. When the crop is infested by any of an ever-changing number and variety of pests—worms, molds, bacteria, fungi, insects, or other diseases—the whole crop is likely to die quickly, unless an appropriate antidote is immediately applied. Chemical antidotes can do the job but increase pollution. Maintaining species diversity—growing several different crops instead of one or two—allows for crop failures without jeopardizing the entire income for a farm or region that specializes in a particular monoculture, such as tobacco, coffee, or bananas.

Chemicals are not the only modern methods of preventing crop loss. Genetically engineered seeds are one attempt at replacing post-infestation chemical treatments. For example, splicing genes into varieties of rice or potatoes from wholly unrelated species—say, hypothetically, a grasshopper—to prevent common forms of blight is occurring more often. Even if the new genes make the crop more resistant, however, they could trigger unknown side effects that have more serious long-term environmental and economic consequences than the problem they were used to solve. Genetically altered crops are essentially new life-forms being introduced into nature with no observable precedents to watch beforehand for clues as to what might happen.

Urban Sprawl

As more farms become mechanized, the need for farmers is being drastically reduced. There were more farmers in the United States in 1860 than there were in the year 2000. From a peak in 1935 of about 6.8 million farmers farming 1.1 billion acres, the United States at the end of the twentieth century counted fewer than 2.1 million farmers farming 950 million acres. As fewer people care for land, the potential for erosion and pollution to accelerate is likely to increase, causing land quality to decline.

As farmers are displaced and move into towns, the cities take up more space. The resulting urban sprawl converts a tremendous amount of cropland into parking lots, malls, industrial parks, or suburban neighborhoods. If cities were located in marginal areas, then the concern over the loss of farmland to commercial development would be nominal. However, the cities attracting the greatest numbers of people have too often replaced the best cropland. Taking the best cropland out of primary production imposes a severe economic penalty.

James Knotwell and Denise Knotwell

WORLD FOOD SUPPLIES

All living things need food to begin the life process and to live, grow, work, and survive. Almost all foods that humans consume come from plants and animals. Not all of Earth's people eat the same foods, however, nor do they require the same caloric intakes. The types, combinations, and amounts of food consumed by different peoples depend upon historic, socioeconomic, and environmental factors.

The History of Food Consumption

Early in human history, people ate what they could gather or scavenge. Later, people ate what they could plant and harvest and what animals they could domesticate and raise. Modern people eat what they can grow, raise, or purchase. Their diets or food composition are determined by income, local customs, religion or food biases, and advertising. There is a global food market, and many people can select what they want to eat and when they eat it according to the prices they can pay and what is available.

Historically, in places where food was plentiful, accessible, and inexpensive, humans devoted less time to basic survival needs and more time to activities that led to human progress and enjoyment of leisure. Despite a modern global food system, instant telecommunications, the United Nations, and food surpluses at places, however, the problem of providing food for everyone on Earth has not been solved.

According to the United Nations Sustainable Development Goals that were adopted by all Member States in 2015, an estimated 821 million people were undernourished in 2017. In developing countries, 12.9 per cent of the population is undernourished. Sub-Saharan Africa has the highest prevalence of hunger; the number of undernourished people increased from 195 million in 2014 to 237 million in 2017. Poor nutrition causes nearly half (45 per cent) of deaths in children under five—3.1 million children each year. As of 2018, 22 per cent of the global under-5 population were still chronically undernourished in 2018. To meet challenge of Goal 2: Zero Hunger, significant changes both in terms of agriculture and conservation as well as in financing and social equality will be required to nourish the 821 million people who are hungry today and the additional 2 billion people expected to be undernourished by 2050.

World Food Source Regions
Agriculture and related primary food production activities, such as fishing, hunting, and gathering, continue to employ more than one-third of the world's labor force. Agriculture's relative importance in the world economic system has declined with urbanization and industrialization, but it still plays a vital role in human survival and general economic growth. Agriculture in the third millennium must supply food to an increasing world population of nonfood producers. It must also produce food and nonfood crude materials for industry, accumulate capital needed for further economic growth, and allow workers from rural areas to industrial, construction, and expanding intraurban service functions.

Soil types, topography, weather, climate, socioeconomic history, location, population pressures, dietary preferences, stages in modern agricultural development, and governmental policies combine to give a distinctive personality to regional agricultural characteristics. Two of the most productive food-producing regions of the world are North America and Asia. Countries in these regions export large amounts of food to other parts of the world.

Foods from Plants
Most basic staple foods come from a small number of plants and animals. Ranked by tonnage produced, the most important food plants throughout the world are corn (maize), wheats, rice, potatoes, cassava (manioc), barley, soybeans, sorghums and millets, beans, peas and chickpeas, and peanuts (groundnuts).

More than one-third of the world's cultivated land is planted with wheat and rice. Wheat is the dominant food staple in North America, Western and Eastern Europe, northern China, and the Middle East and North Africa. Rice is the dominant food staple in southern and eastern Asia. Corn, used primarily as animal food in developed nations, is a staple food in Latin America and Southeast Africa. Potatoes are a basic food in the highlands of South America and in Central and Eastern Europe. Cassava (manioc) is a tropical starch-producing root crop of special dietary importance in portions of lowland South America, the west coast countries of Africa, and sections of South Asia. Barley is an important component of diets in North African, Middle Eastern, and Eastern European countries. Soybeans are an integral part of the diets of those who live in eastern, southeastern, and southern Asia. Sorghums and millets are staple subsistence foods in the savanna regions of Africa and south Asia, while peanuts are a facet of dietary mixes in tropical Africa, Southeast Asia, and South America.

Food from Animals
Animals have been used as food by humans from the time the earliest people learned to hunt, trap, and fish. However, humans have domesticated only a few varieties of animals. Ranked by tonnage of meat produced, the most commonly eaten animals are cattle, pigs, chickens and turkeys, sheep, goats, water buffalo, camels, rabbits and guinea pigs, yaks, and llamas and alpacas.

Cattle, which produce milk and meat, are important food sources in North America, Western Europe, Eastern Europe, Australia and New Zealand, Argentina, and Uruguay. Pigs are bred and reared for food on a massive scale in southern and eastern Asia, North America, Western Europe, and Eastern Europe. Chickens are the most important domesticated fowl used as a human food source and are a part of the diets of most of the world's people.

Sheep and goats, as a source of meat and milk, are especially important to the diets of those who live in the Middle East and North Africa, Eastern Europe, Western Europe, and Australia and New Zealand.

Water buffalo, camels, rabbits, guinea pigs, yaks, llamas, and alpacas are food sources in regions of the world where there is low consumption of meat for religious, cultural, or socioeconomic reasons. Fish is an inexpensive and wholesome source of food. Seafood is an important component to the diets of those who live in southern and eastern Asia, Western Europe, and North America.

The World's Growing Population

The problem of feeding the world is compounded by the fact that population was increasing at a rate of nearly 82 million persons per year at the end of second decade of the twenty-first century. That rate of increase is roughly equivalent to adding a country the size of Germany to the world every single year.

Also compounding the problem of feeding the world are population redistribution patterns and changing food consumption standards. In the year 2050, the world population is projected to reach approximately 10 billion—4 billion people more than were on the earth in 2000. Most of the increase in world population is expected to occur within the developing nations.

Urbanization

Along with an increase in population in developing nations is massive urbanization. City dwellers are food consumers, not food producers. The exodus of young men and women from rural areas has given rise to a new series of megacities, most of which are in developing countries. By the year 2030, there could be as many as forty-one megacities (cities with populations of 10 million people or more).

When rural dwellers move to cities, they tend to change their dietary composition and food-consumption patterns. Qualitative changes in dietary consumption standards are positive, for the most part, and are a result of copying the diets of what is considered a more prestigious group or positive educational activities of modern nutritional scientists

working in developing countries. During the last four decades of the twentieth century, a tremendous shift took place in overall dietary habits as Western foods became increasingly available and popular throughout the world. While improved nutrition has contributed to a decrease in child mortality, an increase in longevity, and a greater resistance to disease, it is also true that conditions including morbid obesity, Type II diabetes, and hypertension are on the rise.

Strategies for Increasing Food Production

To meet the food demands and the food distribution needs of the world's people in the future, several strategies have been proposed. One such strategy calls for the intensification of agriculture—improving biological, mechanical, and chemical technology and applying proven agricultural innovations to regions of the world where the physical and cultural environments are most suitable for rapid food production increases.

The second step is to expand the areas where food is produced so that areas that are empty or underused will be made productive. Reclaiming areas damaged by human mismanagement, expanding irrigation in carefully selected areas, and introducing extensive agrotechniques to areas not under cultivation could increase the production of inexpensive grains and meats.

Finally, interregional, international, and global commerce should be expanded, in most instances, increasing regional specializations and production of high-quality, high-demand agricultural products for export and importing low-cost basic foods. A disequilibrium of supply and demand for certain commodities will persist, but food producers, regional and national agricultural planners, and those who strive for regional economic integration must take advantage of local conditions and location or create the new products needed by the food-consuming public in a one-world economy.

Perspectives

Humanity is entering a time of volatility in food production and distribution. The world will produce enough food to meet the demands of those who can afford to buy food. In many developing countries, however, food production is unlikely to keep pace with increases in the demand for food by growing populations.

Factors that could lead to larger fluctuations in food availability include weather variations such as those induced by El Niño and climate change, the growing scarcity of water, civil strife and political in-stability, and declining food aid. In developing countries, decision makers need to ensure that policies promote broad-based economic growth—and in particular agricultural growth—so that their countries can produce enough food to feed themselves or enough income to buy the necessary food on the world market.

William A. Dando

AGRICULTURE

The first agricultural revolution, which drastically changed the lifestyle of a significant proportion of the human population, occurred in Asia. This revolution involved the domestication of plants and animals. Carl O. Sauer, a cultural geographer who spent a lifetime studying the origin and diffusion of agriculture, believed that vegeculture first developed in Southeast Asia more than 14,000 years ago. In vegeculture, a part of a plant—other than the seed—is planted for reproduction. The first plants domesticated in Southeast Asia were taro, yam, banana, and palm. Early vegeculture may have developed independently in other parts of the world as well.

Seed agriculture, now the most common type of agriculture, uses seeds for plant reproduction. It originated in the Middle East about 1,400 years ago, in the basins of the two major rivers of present-day Iraq, the Tigris and the Euphrates. Wheat and barley were probably the first crops cultivated there. The first domestication of animals took place in the Middle East some 8,000 years ago—well after the origin of crop agriculture. The farmers of the Middle East probably first combined domesticated plants and animals into an integrated system. Although many plants and animals were likely domesticated simultaneously in different parts of the world, rice, oats, millet, sugarcane, cabbage, beans,

Canola fields in Luoping in Yunnan.

SELECTED AGRICULTURAL PRODUCTS OF ASIA

eggplant, onions, horses, pigs, buffalo, and chickens were all domesticated originally in Asia.

In modern times, agriculture is the dominant economic activity in most Asian countries. It contributes more than one-third of the national income, provides employment for about two-thirds of the workforce, and accounts for nearly half of export earnings in many countries of Asia. Exceptions to this can be found in several countries of the Middle East—for example, Jordan, Oman, Kuwait, and Saudi Arabia—and in East Asia—Singapore, Japan, and South Korea, specifically—where agriculture accounts for less than 5 percent of the gross domestic product, and often far less than 5 percent of the working population is engaged in agriculture. In the Middle East, agriculture is not the leading sector of the economy because of climate and topographic conditions and the presence of huge oil resources. Agriculture has diminished importance in

countries such as Japan and South Korea because of rapid industrialization and urbanization.

Asia supports about 60 percent of the global population on only about 23 percent of the world's arable land. As a result, Asian agriculture is far more intensive than on any other continent. Despite the population pressure on arable land, Asia has made remarkable progress in agricultural productivity. In Asia as a whole, food production has outpaced the growth of population. In most Asian countries, particularly in the low-income countries of South Asia, per capita food availability has risen.

Agrarian Structure

Most people in Asia are small farmers, owning an average of about 3.4 to 4.4 acres (1.4 to 1.8 hectares) of land per family. A farm this size can support a family of five to six persons, but average farm size varies among Asian countries, ranging from less than 0.6 acre (0.24 hectare) in Bangladesh to 5.6

acres (2.3 hectares) in Pakistan. Even in a highly developed country like Japan, most farmers own less than 3.7 acres (1.5 hectares) of land. Topographic and climate conditions, to a large extent, determine farm size in Asia. Agricultural potential is limited in Nepal, for example, because of the Himalayas, and in Saudi Arabia because of the Arabian Desert. In these countries, average farm size is larger relative to countries like Bangladesh, where a vast, fertile floodplain receives abundant rainfall.

Another feature of Asian agrarian structure is the inequitable distribution of farmland. For example, in India more than 30 percent of cultivated land is owned by fewer than 5 percent of farming families, and little land redistribution has occurred. Farm holdings in most Asian countries are highly fragmented, and tenancy is widespread. Most share-croppers do not own farmland, but some try to supplement what they produce on their own land. Sharecropping arrangements in Asia favor land-owners over sharecroppers: Produce generally is divided 50-50, and the tenant provides all labor and fertilizer. Fragmentation of farms inhibits agricultural mechanization, and land consolidation efforts have had limited success in most Asian countries.

Most Asian farmers are subsistence farmers, cultivating crops for family consumption. They practice traditional farming methods; almost all farm operations are done manually or with the help of draft animals. Exceptions are found in Japan, South Korea, and Taiwan, where small-scale equipment similar to garden tractors is widely used. Asian farmers have started to use chemical fertilizer; for water, they largely depend on rain. As a result,

LEADING AGRICULTURAL PRODUCTS OF ASIAN COUNTRIES WITH MORE THAN 15 PERCENT OF ARABLE LAND

Country	Products	Percent of Arable Land
Armenia	Fruit, vegetables, livestock	16
Azerbaijan	Cotton, grain, rice, grapes, fruit, vegetables, tea, tobacco, cattle, pigs, sheep, goats	24
Bangladesh	Rice, jute, tea, wheat, sugarcane, potatoes, beef, milk, poultry	60
Cambodia		22
India	Rice, wheat, oilseed, cotton, jute, tea, sugarcane, potatoes, cattle, water buffalo, sheep, goats, poultry, fish	53
Korea, South	Rice, root crops, barley, vegetables, fruit, cattle, pigs, chickens, milk, eggs, fish	15
Myanmar	Paddy rice, corn, oilseed, sugarcane, pulses, hardwood	16
Pakistan	Cotton, wheat, rice, sugarcane, fruits, vegetables, milk, beef, mutton, eggs	40
Philippines		19
Sri Lanka		21
Syria	Cotton, wheat, barley, lentils, chickpeas, beef, lamb, poultry, eggs, milk	25
Thailand	Rice, rubber, corn, tapioca, sugarcane, soybeans, coconuts	33
Turkey	Cotton, tobacco, cereals, sugar beets, fruits, olives, pulses, citrus, livestock	26
Vietnam	Rice, corn, potatoes, rubber, soybeans, coffee, tea, bananas, poultry, pigs, fish	23

Source: "The World Factbook—Central Intelligence Agency." cia.gov, 2018, www.cia.gov/library/publications/resources/the-world-factbook

Rice terraces in Yuanyang County, Yunnan, China.

yields are low, which compels farmers to cultivate the land intensively. Double-cropping is the norm; some farmers even grow three crops a year. Therefore, only a small fraction of the arable land in humid regions of Asia remains fallow, and the proportion of waste land is small. Farming is labor-intensive, and the extended family is the main source of labor. This helps to explain why family size is large in agrarian countries of Asia.

The area under agriculture differs strikingly from country to country in Asia. Countries with favorable topographic and climatic conditions usually have more land in farming. In 2016, more than 70 percent of the land is used for farming in Bangladesh; in 2019, only 12.5 percent in Japan. Mountains restrict the expansion of farm land in Japan, confining it primarily to the coastal areas. In 2019, only 11.4 percent of the land in Jordan was in agricultural use. Even within a given country, marked variation often is found in the proportion of land used for farming. Most of the farm land in China is concentrated in the eastern third of the country, while crop production is limited in the arid, mountainous western two-thirds.

Agricultural Regions
The distribution of agricultural systems corresponds closely to variations in the environmental setting and, particularly, to water availability. In the tropical rain forests of Southeast Asia, the mountainous and hilly parts of South Asia, and in southern China, a type of primitive agriculture known as shifting cultivation or slash-and-burn agriculture is practiced. This form of subsistence agriculture can support only a low population density. Shifting cultivators select a site for cultivation and then clear the area by cutting the trees, bushes, and other plants. These are then dried and burned, providing the soil with needed organic materials.

Shifting cultivators plant different crops such as rice, maize, millet, yams, sugarcane, oilseeds, potatoes, taro, vegetables, and cotton on one site. This practice is called intertillage, and crop yields are usually low. As a result, a large area of land is required to support a small population. Because shifting cultivators do not use fertilizer, soil nutrients rapidly decrease, and the land quickly becomes too infertile to nourish crops. Shifting cultivators abandon their fields and establish new ones every few years. The land devoted to shifting cultivation is declining at a rapid rate worldwide because of the demand for forest resources for other uses.

Rice is the staple food of more than half of the world's population, and 90 percent of it is grown in the coastal and deltaic plains, and in the river valleys of Asia's monsoonal areas. This region encompasses a broad geographic area characterized by a distinctive climate, stretching from Japan in the east, through Indonesia in the south, and west to Pakistan. Rice farming there is practiced mostly at the subsistence level, and farming practices are traditional. Wheat and other food crops are grown in some parts of Asia during the dry winter season. Many rice farmers also produce cash crops for market, such as jute and sugarcane in Bangladesh and India. They may also raise cattle, pigs, and poultry, cultivate vegetables, and raise fish in ponds and irrigated rice fields, particularly in Southeast Asia.

Farmers in the colder, drier parts of Asia (northeastern China, northern Japan, southeastern East Asia, northeastern Southeast Asia, and the western half of South Asia) and in the river valleys of the Middle East practice a system of intensive subsistence agriculture called peasant grain-and-livestock farming or dry agriculture. The dominant grain crops are wheat, barley, sorghum, millet, oats, and corn, while cotton, tobacco, and sugarcane are grown as cash crops. Farmers also raise cattle, pigs, sheep, and goats. In arid areas, such as the Middle Eastern river valleys, irrigation helps support dry farming. Traditional water-lifting devices, such as the shaduf (a counterweighted, lever-mounted bucket), and the naria (waterwheel), permit limited double-cropping in the dry season near the rivers of the Middle East.

Plantation crops such as tea, rubber, coconuts, and coffee are grown in Asia. Tea is indigenous to China, which in 2017 was the world's largest producer, with 40 percent of the global production, followed by India with 21 percent, and Sri Lanka and

Vietnam. Tea is Sri Lanka's largest export crop, accounting for about one-third of annual exports by value. Tea is grown in the central highlands of Sri Lanka and in the hilly regions of northeastern India and Bangladesh. Rubber is grown in Malaysia, Indonesia, Cambodia, India, and Sri Lanka. Beginning with rubber plants introduced from Brazil, today Thailand, Indonesia, and Vietnam account for much of the total world production.

Coconuts are grown in Malaysia, Indonesia, India, the Philippines, and Sri Lanka, usually in the coastal areas. This tree crop is grown primarily in small, family-owned plantation farms. Its fiber is processed for rope and housing material, and the fruit is pressed to obtain oil. Most coconut products are sold to the domestic market; the rest are exported. In 2010, Vietnam, Indonesia, the Philippines, and India were world-leading exporters of coconuts and coconut products.

Although Asia is not known for coffee production, coffee is grown in the southern states of India and in China. Sugarcane is cultivated in India, Thailand, the Philippines, and Bangladesh. In 2019, six of top ten cotton-producing countries were in Asia: India, China, Pakistan, Turkey, Uzbekistan, and Turkmenistan. In 2016, Malaysia and Indonesia were the world's leading producers of palm oil. In 2017, China was by far the world's top tobacco producer. India, Indonesia, and Pakistan were other important Asian producers.

In the arid and semiarid parts of South Asia and the Middle East, and in the dry and cold western two-thirds of East Asia, nomadic herders graze cattle, sheep, goats, and camels. Nomadic herding requires a great amount of land per person, and nomadic herders move from place to place with their livestock in search of forage. As in other places, nomadic herding is declining in Asia.

A distinctive type of subsistence agriculture, called Mediterranean agriculture, is practiced along the Mediterranean coast of the Middle East and in the northern part of Turkey that borders the Black Sea. Traditional Mediterranean agriculture is based on wheat and barley cultivation in the rainy winter season. Farmers of this region also cultivate vine and tree crops such as grapes, olives, and figs, and raise small livestock, particularly sheep, goats, and pigs.

The coastal areas and inland river valleys of East, Southeast, and South Asia are the agricultural cores of Asia. More than half of the crop area of these regions is used to cultivate food crops such as rice and wheat. Rice is the principal food crop of all Asian countries located east and north of India and for the people of southern and eastern India. Wheat is the primary food crop of northern and western India and all Asian countries located west of India.

People of the wheat-producing region consume rice as a secondary staple. The top-ten rice-producing countries of the world are all in Asia. India is first in total production, followed by China, Indonesia, Bangladesh, Thailand, Vietnam, Myanmar, the Philippines, Cambodia, and Pakistan. China and India are first and second, respectively, in both area and total world production of wheat. Pakistan also ranks high in total production of wheat.

Many Asian countries are not self-sufficient in food production. China, Saudi Arabia, the Philippines, Iran, Indonesia, and Iraq all imported more than 1 million metric tons of rice annually in 2017. In good years, Bangladesh produces a surplus of rice, but it often needs to import rice to maintain government food grain stocks. Both India and Bangladesh export high-quality rice to the Middle East and North America. In 2017, Asia's main rice-exporting countries were India, Thailand, Vietnam, Pakistan, Myanmar, and Cambodia.

In 2017, Indonesia, the Philippines, Japan, Turkey, and South Korea were the major wheat-importing countries of Asia. Pakistan will need to import wheat in the near future because of high population growth and a forecast slowdown in agricultural production.

With the exception of Japan, South Korea, China, and Taiwan, yields of all crops—particularly rice and wheat—are low in Asia compared to world standards. Although crop yields increased significantly in the past 50 years, typical yields in Asia remained low for several reasons. Fertilizer use and the areas under irrigation are among the lowest in the world. Also, most Asian farmers practice traditional farming methods, where high yields are atyp-ical. Another major obstacle to increasing crop yields is the preponderance of small farmers. Because farmers on small farms do not have access to assured irrigation, nor can they afford modern agricultural technology, their average yield is generally much lower than that of medium and large farms.

Green Revolution

A dramatic growth in food production in Asia began with the Green Revolution in the late 1960s, particularly for wheat and rice. The Green Revolution involved the integrated use of high-yielding varieties of seeds, irrigation water, and chemical fertilizer. Cultivation of the new varieties of rice and wheat caused an impressive increase in the use of fertilizer and the expansion of irrigation, particularly the exploitation of groundwater through tube wells. The use of fertilizer in Asia increased more than 200 percent between the mid-1970s and early 1990s. East and South Asia dominated all other developing regions in the intensity of fertilizer use. In 1994-1995, fertilizer use per hectare (2.5 acres) averaged more than 475 pounds (215 kilograms) in East Asia and 170 pounds (77 kilograms) in South Asia. These levels remained virtually unchanged in 2020 for both East Asia and South Asia.

With proper and timely application of required inputs, high-yield wheat yields can be tripled, and high-yield rice yields can be doubled. High-yield varieties led to the increase in the cost of cultivation per unit of land. Since the increase in yield has been much higher than the increase in cost, the cost per ton of output has declined. With growth in rice supplies outpacing demand, farmers have been able to sell rice and wheat at a lower price while maintaining profits, thus sharing the benefits of technological progress with consumers. Because Green Revolution technology is capital-intensive, it has provided more benefits to wealthy, large farmers than to poor, small farmers. Thus, expansion of this technology is exacerbating social-class polarization and income inequalities.

In Asia, the Middle East has been least affected by Green Revolution technology. In East Asia, Japan has achieved the greatest success from this technology. The Philippines and Sri Lanka have enjoyed similar success in Southeast Asia and South Asia, respectively. High-yield rice accounts for more than 70 percent of the total rice area in these countries. Green Revolution technology has had the least significant impact in Bhutan and Nepal.

The impact of the Green Revolution on different regions within individual countries has not been uniform. For example, the availability of irrigation water and lack of rural infrastructure have restricted the expansion of high-yield varieties in India. However, states that have been able to use the new technology have made an enormous contribution to their nations' food supply. Most of the gain in food production in India came from Punjab, Haryana, Tamil Nadu, and Andhra Pradesh. Indian states moderately affected by the Green Revolution are West Bengal, Orissa, Bihar, and Kerala for rice and Uttar Pradesh for wheat.

In South Asia, well-watered and wheat-growing areas have fared best. Wheat has benefited more than rice because the former is grown only in the dry season and the new agricultural technology requires significant irrigation to realize potential yields. Irrigation must be reliably controlled, which is more feasible in the dry season than in the wet season. Rice is grown in both the dry and rainy seasons, and high-yield rice is more successful in the dry season.

Criticisms of the Green Revolution have concentrated on the negative impacts of increased use of fertilizer and pesticides, which cause surface water pollution. With high-yield seeds, three crops a year can be cultivated. Adopting this practice has two consequences: It causes overuse of land, a major source of land degradation; and it leads to increasing monocultures of rice and wheat, reducing the genetic diversity of food crops.

Without the Green Revolution, feeding current Asian populations at prevailing nutritional standards would have been impossible. New agricultural practices have enabled Asia to avoid famines that had been widely predicted for the 1970s and onward. The new rice and wheat varieties also have stimulated agricultural employment, because more people are needed to cultivate, harvest, and handle the increased production. Use of tractors has displaced some laborers, but there are still relatively few farm machines and their presence in most Asian nations cannot be attributed solely to the Green Revolution.

Livestock and Poultry

About two-thirds of all farm families in Asia raise livestock. Livestock provide milk, meat, eggs, hides, and skins, and are used extensively for plowing land and for short-distance transport. Cattle dung is an important source of organic fertilizer for crop fields and is used as cooking fuel in South Asia. Livestock generally are raised by small and landless farmers who support their families by selling milk and other dairy and poultry products in local markets.

India, with about 300 million cattle and 1.36 million buffalo in 2018, ranks first in the world in numbers. Because Hindus believe in the sanctity of the cow, virtually none are slaughtered and in some states, slaughter is illegal. Not surprisingly, the cattle population of India includes a high proportion of aged animals, and cattle there are among the world's least productive. In 2008, China had more goats (149 million), pigs (474 million), sheep (146 million), and chickens (4.6 billion) than any other country, and more than 75 percent of the world's domesticated ducks.

The production of meat, milk, and eggs has increased in recent years in Asian countries as rising income and rapid urbanization have caused a demand for a diversified diet, particularly in the Middle East, East Asia, and Southeast Asia. As a result, per capita consumption of meat, milk, and eggs has increased rapidly. Turkey, China, and Japan are major importers of livestock and livestock products. No Asian country is a major exporter of these products. In 2018, China, Vietnam, and Indonesia exported leather and leather goods; Turkey, Iran, and India are major exporters of high-quality wool.

Fish

Fish is an important source of animal protein for the people of East and Southeast Asia. In South Asia, fish is widely consumed only in Bangladesh and the Maldives. Fish consumption is less widespread in the Middle East. The principal source of fish differs greatly among the fish-producing countries of Asia. More than 97 percent of fish caught in Japan is from marine water, but about 80 percent of the fish harvest in Bangladesh comes from inland freshwater sources—rivers, lakes, marshlands, and ponds. Fish farming in the irrigated rice fields of East and Southeast Asia has been common for millennia. Japan and China also practice aquaculture—raising fresh-water fish in artificial ponds, growing seaweed in aquariums, and harvesting oysters, prawns, and shrimp in shallow bays.

Fish provides an important supplementary source of agricultural income for poor villagers. Also, many people are employed in the fishing sector. For example, in Bangladesh nearly 1 million people are directly employed in fishing and another 10 million in fish marketing and processing.

The main fish-producing countries of Asia are China, India, Vietnam, Japan, the Philippines, Bangladesh, South Korea, and Myanmar. China leads the world in fish production, with an annual catch of about 81.5 million tons. In 2016, China accounted for more than 40 percent of world fish production. Japan, long the world leader, now accounts for only 2 percent. Japan's fishing industry is nevertheless larger than that of any of the long-time fishing nations of northwestern Europe, although somewhat smaller than that of the United States. Despite Japan's large open-ocean fishing fleet, most of its catch comes from waters near its own coast. In contrast, much of Taiwan's fish catch comes from deep-sea fishing.

The major fish-exporting countries of Asia are China, Vietnam, and India. Frozen prawn and shrimp are the major fish exports. Other significant export items are frozen frog legs, frozen lobster tails, dried fish, and shark fins, much of which is exported to Japan, the Middle East, and Western countries. Despite the potential for development, the fish industry is least developed in Middle Eastern countries.

In the Maldives, about 30 percent of the workforce is involved in fishing. In 2019, dried and fresh fish made up virtually 100 percent of that country's exports. The Maldives also has the highest consumption of fish and fishery products of any country in the world. In 2013, fish consumption there averaged more than 365 pounds (166 kilograms) per person per year, slightly less than nine times the average for the world as a whole.

Forestry

Forests of significant economic importance are found primarily in northeastern East Asia and Southeast Asia. These forests differ in the types of trees they contain and in the type of market or use they serve. The softwood forests of northeastern East Asia cover most of Japan and parts of North Korea. Trees grown there are used for construction lumber and to produce pulp for paper. Tropical hardwood forests cover all Southeast Asian countries and the south central part of China, several places in India, and the northern part of Iran. Trees grown in those forests are used primarily for fuel wood and charcoal, although an increasing quantity of special-quality woods are cut for export as lumber.

Forested areas differ widely in Asian countries. Forests are nearly nonexistent in most of the Middle East. In Saudi Arabia, forests covered 4,150 square miles (10,748 sq. km.) in 2011—only 0.5 percent of the country's land area. Only two Middle Eastern countries, Turkey and Lebanon, have forests covering 10 percent or more of their land area. In contrast, forests are the major land use in many countries of East and Southeast Asia. Forests account for nearly 68.5 percent of the total land area in Japan, 67.9 percent in Laos, 63.9 South Korea, 56.5 percent in Cambodia, 51.7 percent in Indonesia, and 25.9 percent in the Philippines. Except in Nepal and Bhutan, forests cover less than 20 percent of the total area of South Asian countries. Cambodia, Malaysia, Indonesia, the Philippines, Thailand, and Myanmar export large quantities of forestry products. Myanmar is known for its high-quality teak.

Overexploitation of hardwoods and conversion of forest lands for other uses have become serious concerns. Rates of forest conversion are most rapid in continental Southeast Asia, averaging about 1.3 percent a year. Deforestation has important local, regional, and global consequences, ranging from increased soil and land degradation to greater food insecurity, escalating carbon emissions, and loss of biodiversity. Small-scale, poor farmers clearing land for agriculture to meet food needs and the gathering of wood to be used for cooking account for the majority of the deforestation in Southeast Asia. Commercial logging and urban expansion account for most of the remaining deforestation.

Throughout Asia, agricultural growth and food production have increased somewhat in the twenty-first century. This trend has occurred despite the lack of new land to bring under cultivation. The increase in crop output has largely come from an increase in yields. The momentum of the Green Revolution has generally waned, although the production potential of the new technology has not been realized fully. Rice and wheat yields are still relatively low in many Asian countries, primarily because of low use of modern agricultural technology. For example, the use of chemical fertilizers in South Asian countries has not reached the levels of neighboring regions.

A study in Pakistan showed that the difference between potential yields under experimental conditions and the national average yield was 82 percent for cereals. Additionally, only slightly more than half of all farmers were now cultivating high-yield varieties of rice and wheat. Increasing that percentage could significantly increase food production. Applications of newer technologies also could increase food production. Genetically engineered crops can give even higher yields than current strains of rice and wheat. Although there is opposition to genetically engineered crops, further research may remove some of the negative perceptions associated with them. Agricultural scientists are also researching crops suitable for arid and semiarid areas and working to devise integrated pest-management systems that reduce reliance on chemical pesticides.

Demand for fruits, vegetables, meat, fish, milk, and eggs is growing with the increased urbanization and industrialization of Asia. This is reducing the demand for cereal crops. In Japan, South Korea, and Taiwan, consumption of rice has already begun to decline. Increased production of both food and nonfood crops is required to feed the growing populations of most Asian countries. While increasing agricultural production, Asian policymakers must also promote environmentally sound technologies and implement effective land reforms to address the problems of inequality and poverty caused by landlessness. Better crop management and better management of irrigation water are also needed to sustain agricultural growth in Asian countries.

Bimal K. Paul

INDUSTRIES

Asia comprises nations of varying physical dimensions and in varying stages of development. Japan and Singapore are considered to be developed nations. Indonesia, South Korea, Malaysia, Taiwan, and Thailand are newly industrialized nations. Saudi Arabia, Kuwait, Iran, and Iraq in the Middle East are resource-rich nations. Most other nations in Asia are developing nations. China and India, the world's two most populous nations, do not fall neatly into any category; these giants exhibit a conflicting pattern, with the coexistence of highly advanced sectors and fairly backward sectors.

China

Industrial development is a fairly recent phenomenon in China. In fact, agriculture still supported 21.7 percent of the Chinese people in 2017, although it represented only 7.9 percent of the country's GDP. But in the past 40 years, and especially since the turn of the twenty-first century, China's aggressive drive for industrial development has borne fruit in countless ways, and today more than 40 percent of the workforce is employed by industries. As a result, China has achieved the distinction of being the world's largest economy.

In 2020, along with being the world's largest economy, China is now the world's largest exporter of goods (since 2010) and the world's largest trading nation. In the early 2020s, China's main exports are computers and computer-related machinery, telecommunications equipment, clothing, furniture, textiles, rice, and tea. Its major imports include integrated circuits and other computer components, medical and optical equipment, metals, cars, and soybeans. China's most recent five-year economic plan seeks to accelerate growth by moving the economy from one that relies on exports to a more consumer-driven model.

Chinese industries show a geographical diffusion, from a few historically established areas to new growth centers. The most prominent industrial region with ties to the colonial past is the northeast plain. This strategic region, bordering Russia, Korea, and Mongolia, is rich in industrial raw materials, including iron ore, petroleum, coal, aluminum ore, lead, and zinc. The northeast region led China in manufacturing .in the post-World War II era. Its major industrial centers include Changchun, Harbin, and Shenyang, where petrochemical complexes, automobile assembly plants, iron and steel mills, and heavy machinery plants are located. In the early 1960s, a major oil field was discovered in Daqing, which has now built a giant petrochemical complex to process the local crude. By 2020, Daqing was the world's fourth most productive oil field. Nevertheless, the region as a whole is considered the Chinese "rust belt," and its aging, inefficient industries have made it an increasingly smaller player in Chinese industrial economy.

Headquarters of Alibaba Group in Hangzhou. Alibaba is the world's largest retailer and e-commerce company, and one of the largest Internet and AI companies.

Tianjin is one of China's leading industrial centers in the north plain, and includes Beijing, the nation's capital. In this core region, vast farmlands provide raw materials for local food processing plants. Tianjin has greatly diversified its industrial base by adding chemical, iron and steel, heavy machine, and textile plants as foreign investment flows in. Zhongguancun, in the northwestern part of Beijing, has emerged as China's "Silicon Valley."

The Yangtze River basin in the south is anchored by Sichuan province, the most populous and critical farming region in China. The region has rich deposits of iron ore, coal, copper, and lead. The Yangtze River provides inexpensive river transportation for the whole basin. Areas with a strong industrial base in the region include the city of Wuhan and the Sichuan Basin. Wuhan is a center for diverse heavy industries. Sichuan's Chengdu hosts food processing industries, textiles, and precision instruments, while Chongqing is home to iron and steelmaking and machine-building industries. The Three Gorges Dam on the Yangtze River supplies inexpensive hydroelectric power.

The West and Pearl River basin is China's southern coastal area, which includes the nation's most dynamic growth cores: Hong Kong and Shenzhen. Unlike other regions, this area has few natural resources. Nevertheless, Hong Kong, long a British colony but now a special administrative district of China since 1997, is the engine of the region's growing industrial economy. Hong Kong's versatile industrial economy churns out electrical equipment, appliances, apparel, and inexpensive consumer goods. These well-established sectors have been strongly supplemented by high-tech information-sector industries. Shenzhen, Hong Kong's next-door neighbor, is the nation's fastest-growing area and a center of diverse export-oriented industries such as electronics and labor-intensive manufacturing.

Notable industrial growth has taken place in all government-designated areas: six Special Economic Zones and dozens of Open Cities. In these areas, industries enjoy preferential treatment, including low tax rates and a variety of eased regulations.

Taiwan's export-oriented industries produce computer peripherals, medical equipment, tele-

Aircraft carrier Liaoning, the first aircraft carrier commissioned into the People's Liberation Army Navy Surface Force.

communications equipment, and petrochemicals. The United States, Hong Kong, and Japan are Taiwan's major trading partners. Industries located in Taipei and Kaohsiung produce home appliances, chemicals, automobiles, and synthetic textiles for foreign markets. With China now a major competitor in the labor-intensive products market, capital-rich Taiwan has increasingly turning to capital-intensive high-tech industries. Personal computers and telecommunications equipment are prominent growth sectors.

Japan

Asia's most advanced and productive industrial economy is Japan. During the Meiji Restoration in the middle of the nineteenth century, Japan not only transformed its feudal society into a Western-style modern society, but also built a foundation for a modern industrial economy. In 2015, 26.2 percent of the Japanese workforce was employed in the industrial sector. Japan exports various capital-intensive, high-technology goods, including automobiles, consumer electronics, optical products, ships, and telecommunications equipment. Japan's industry is so productive that it competes globally in most sectors that are not labor-intensive. Its export markets are global in nature, but the big markets of the United States, China, and South Korea are the most important.

Japan's most industrialized centers are in the gigantic Tokaido megalopolis, which extends from the middle Hōnshū region to northern Kyūshū. There, industrial plants are located along the coastal areas facing the Pacific Ocean, enabling industries to use cost-effective sea transportation to import raw materials from foreign sources and export final products to foreign markets.

Japan's foremost manufacturing belt is the Tokyo-Yokohama-Kawasaki region in the central Kanto Plain. Proximity to the nation's capital, Tokyo, gives the region easy access to various service institutions. The region's well-developed transportation infrastructure and education institutions are other advantages. Among the wide variety of industries located there are steel mills, auto manufacturers, optical goods producers, and shipyards.

The Osaka-Kobe-Kyoto region in the southern Kansai District is the second-largest industrial region in Japan and the country's second-most populated region. During the early part of the twentieth century, consumer-goods industries, shipbuilding, and heavy manufacturing industries were developed. After World War II, the region underwent a major transformation. The region's industrial economy now is highly diversified—so much so, in fact, that if the Osaka-Kobe-Kyoto region were a country, it would have the sixteenth-largest economy in the world. Appliances, petrochemical products, automobiles, and shipbuilding are major industries.

Nagoya in the Nobi Plain is Japan's third-largest industrial belt, and leads the nation in textile manufacturing. The region's other industries include oil refining, petrochemical products, and shipbuilding. The southernmost island of Kyūshū also has a fairly diverse industrial base, and has gained new prominence now that China is Japan's leading trade partner. Hōkkaidō, the northern island, is rich with coal deposits, but its adverse climate and sparse population discourage industrial development.

Korea

The arrival of a modern industrial economy in Korea was externally engineered. Imperial Japan introduced heavy, strategic industries in northern Korea during the 1930s. In southern Korea, Japan mainly established simple consumer goods industries. In a sense, Korea's industrial base functioned reasonably when it was one unified unit. However, the division of the nation after the Korean War of the early 1950s has been painful to both sides.

In South Korea, a systematic, government-led industrialization drive took place beginning in the late 1960s. Today, the nation is one of the major export-oriented industrial powers, with nearly 40 percent of the nation's workforce employed in the industrial sector. Computers, semiconductors, automobiles, consumer electronics, ships, and telecommunications equipment are the nation's key export items. China, the United States, Vietnam, Hong Kong, and Japan are its major trading partners.

South Korea's mineral resources are scant, except for coal. Its small territory requires industries to locate even in rural areas. The Seoul-Incheon region, the nation's industrial core, has the largest number of diverse industrial operations. The other inland industrial center is around the city of Daegu, a traditional hub of textile and high-tech industries. Busan, the nation's second-largest city and the sixth-largest port in the world, has long been known for its rubber and chemical, automobile, and electronics industries. Ulsan, on the southeast coast, is South Korea's industrial powerhouse, with robust chemical, automobile, and shipbuilding industries. Ulsan is home to the world's largest automobile-assembly plant, the world's largest shipyard, and the world's third-largest oil refinery. Yeosu, a city on the south coast, has a large, integrated petrochemical complex and its supporting industries.

In North Korea, the communist regime has emphasized military-oriented industries rather than consumer goods. Those industries are dispersed throughout the nation, with the heaviest concentration of key industries in Pyongyang and its immediate environs. Major machinery, plastics, and iron and steel mills are located there. In Kimchaek, a northeast coastal city, iron and steel mills, power-generating plants, and shipyards are located. Other important industrial cities include Huichon and Nampo.

Southeast Asia

The region's modern industrial economy has a relatively short history. During the colonial period, its industry was limited to processing domestic raw materials for export. In the 1960s, Southeast Asia embarked on a drive for import-substitution industrialization, which progressed to an aggressive export-oriented industrial expansion. To facilitate exports, the region's industrial leaders established free trade zones in their coastal regions. With strong foreign direct investment, the industrial growth rate has remained dynamically high in Indonesia, Malaysia, Singapore, Thailand, and Vietnam. The region's other nations have yet to launch systematic industrialization drives.

Indonesia

Indonesia has a large population base and rich natural resources. It produces petroleum, timber, rubber, palm oil, and tin. Foreign investors have been attracted to this labor-rich country's low wages and abundant natural resources. The results were promising until the nation encountered a major financial crisis and subsequent political upheaval in the late 1990s. Wide-ranging economic reforms were enacted in the early twenty-first century. Today, Indonesia has the largest economy in Southeast Asia, and about 41 percent of the workforce is employed by the industrial sector. Major trading partners are China, the U.S., Japan, India, and Singapore.

The overwhelming majority of foreign direct investment are in the Jakarta region of Java. Export-oriented industries in the region include electronics, textiles, medical instruments, and appliance manufacturing. Food processing and light manufacturing plants are also concentrated in the capital city region. Manufacturing is limited almost entirely to Java and Sumatra. The government's efforts to encourage local resource-based industries in the outer islands have been relatively ineffective.

Malaysia

Malaysia has been notably successful in industrialization. Its natural resource base is relatively strong, with a rich endowment of rubber, palm oil, tin, timber, and petroleum. In 2017, 37.6 percent of the workforce was employed in the industrial sector. Multinational corporations view the nation as a favorable investment location because of its relatively well-developed infrastructure, low-wage skilled workers, and liberal government policies. Foreign investment has been responsible for the rise of Malaysia's export-oriented electronics and textiles industries. The nation also has developed auto-assembly and other capital-intensive industries for domestic consumption, and has emerged as a world leader in the manufacture of solar-energy equipment. As in most countries of the region, Malaysia's industries have tended to locate in coastal cities, which have easy access to foreign markets. The states of Penang, Selangor, and Johor have been the main recipients of foreign investment and industrial growth. Malaysia's industrial economy, hit hard by the financial crisis of 1997, rapidly recovered. By 2020, Malaysia was the third-biggest economy in Southeast Asia China, the U.S., and Singapore are its major trade partners.

Singapore

This city-state has successfully created a highly competitive industrial economy. Its success results from productive workers; harmonious labor-management relations; an excellent infrastructure, including an airport and a seaport; transparent public services; and generous tax concessions. As a result, Singapore is the most heavily industrialized nation in the region, with more than 25 percent of the workforce employed by industries. Important exports include pharmaceuticals and high-end medical and scientific equipment. The recent emphasis on technology-intensive industries, such as electronics and aerospace, has been highly successful. China, Malaysia, and the U.S. are among its major trade partners.

Thailand

In 2017, about 36 percent of the Thai workforce was employed in industry. Thailand has some oil deposits and offers a qualified, low-wage workforce. Its export-oriented industrialization has been supported by the influx of investment from mostly Asian nations, including Japan, Hong Kong, and Taiwan. Its growing export industries, including electronics, footwear, and automobile assembly, are largely concentrated in Bangkok, the capital city, and its immediate environs. Today, Thailand produces more than 40 percent of the world's hard-disk drives. China, Japan, and the U.S. are Thailand's main trade partners.

Other Southeast Asian Countries

The Philippines is a newcomer to the government-led industrialization wave in the region. Its traditional industrial base includes food processing and the manufacture of light consumer goods. The Philippines remains largely dominated by the agricultural sector, and manufacturing is secondary in the national economy. Infrastructure shortcomings have proven to be a major impediment to growth. About 18 percent of the workforce is employed by industries. Export-processing zones have been established in an effort to attract such industries as electronics, textiles, and other light manufacturing.

Other members of the Association of Southeast Asian Nations (ASEAN)—Vietnam, Laos, Cambodia, and Myanmar—face even tougher economic challenges. The region as a whole has rich industrial potential, with quality, low-wage workers and fairly diversified mineral deposits. Only Vietnam has an extensive economic reform policy—*doi moi*—in place. Despite infrastructure challenges, Ho Chi Minh City remains poised to rise as Indochina's dominant industrial center. A special economic zone has been established nearby. Vietnam's latest reforms focus on free-market goals. Neither Laos nor Cambodia has formulated a comprehensive industrialization policy. Burma's internal politics has prevented it from welcoming foreign investment.

India

The largest nation in South Asia in terms of both population size and physical dimension is India. Its natural resource endowment is strong only in certain sectors, such as iron ore and coal deposits. In the early 2020s, India's lack of infrastructure presented a formidable barrier to competitive industrial development. In 2017, only 23 percent of more than 1.3 billion Indians were employed by industries. Major export items include engineering and IT products. Steel products, cement, and refined petroleum are promising growth sectors. Major export partners are the United States, the United Arab Emirates, Hong Kong, and China.

The Indian subcontinent received scant support for industrial development during its colonial period. It achieved its independence with little industrial foundation except for the light industries of textile and food processing for domestic consumption. Large-scale industries were developed after World War II and are located mainly in three industrial core areas: Kolkata, Mumbai, and Chennai.

India's industrial core areas are spread throughout the nation, but the most extensive are anchored by the nation's leading cities in populous coastal zones. Secondary or small industrial regions are mostly located inland. This indicates that the development of India's major industrial regions was influenced more by external imperatives than by access to raw materials.

The Bihar-Bengal district in the east, which includes Kolkata, is India's leading industrial area. This region is rich with cotton and jute industries. Its industrial economy has diversified with the addition of engineering and chemical industries. Major iron and steel works are located in the nearby Chota-Nagpur region, taking advantage of its rich coal and iron ore deposits. The populous Bihar and Bengal states provide inexpensive labor for this industrial region, while the port of Kolkata links it to world markets via sea routes. Industrial infrastructure in the region remains inadequate, however.

The Kolkata region's western counterpart is the Mumbai-Ahmadabad region, which represents Maharashtra and Gujarat states. Mumbai (Bombay), India's key commercial center and busiest port, has a diversified industrial base ranging from traditional textiles to pharmaceuticals. The textile industry utilizes locally produced cotton, abundant labor, and inexpensive hydroelectric power. The region has successfully pursued high-tech, export-oriented industries as the future of the labor-intensive, low-value-added textile sector becomes increasingly uncertain.

Like most of India's old industrial centers, Ahmadabad was a key textile center. Its drive for industrial diversification has proved to be effective, as a growing number of petrochemical and pharmaceutical industries have moved to the area. The coastal zone in Maharashtra and Gujarat states has received a disproportionate amount of foreign investment. The growth of export-oriented industries is expected to continue.

India's third major industrial region is Chennai (Madras) on the southeast coast. More than 40 percent of India's automobile industry and 45 percent of its auto-components industry are centered in and around Chennai, earning the city its "Detroit of In-

dia" nickname. Chennai is also the site of much textile manufacturing and the source of more than half of India's leather exports.

In the past two decades, Chennai has gained an international reputation for its robust computer-technology industries. The local abundance of quality engineers and scientists has prompted a rapid expansion of the electronics, telecommunications, aeronautics, and computer-software industries. About 80 miles (128 km.) to the west is Bangalore—the so-called Silicon Valley of India—where thousands of information-technology companies and the branches of major U.S. high-tech firms are located. The region's software industry has spread to nearby Mysore and southward to Madurai, fed by graduates from the region's excellent colleges and universities. The job market throughout the region has been further stimulated by innumerable tech jobs outsourced from the U.S. and Canada.

In addition to its major industrial regions, India has a number of relatively small, mostly inland, industrial regions. Anchored by Darjeeling, Delhi, Nagapur, and Vijayawada, these are based mainly on traditional craft industries and the manufacture of light consumer goods.

Other South Asian Countries

Pakistan has a semi-industrialized economy, its growth continually stymied by political instability. Conflicts with neighboring India have been only one of many hurdles the nation has had to overcome since independence. Pakistan's industries are limited largely to food processing and light consumer goods manufacturing, which employed about 19 percent of the workforce in 2016. Export-oriented textile industries have formed in recent years in the nation's two major population centers of Lahore and Karachi, the latter city home to about one-third of Pakistan's manufacturing.

Bangladesh is a developing market economy. The agricultural sector, which is productive but subject to unpredictable monsoons, nevertheless supports the country's large population. The nation is challenged by not only population pressure but also a nearly total lack of basic industrial infrastructure,

such as roads and bridges. Bangladesh's main industries are cotton and jute processing; garment
making is by far its largest export sector. Industries
employ about 20 percent (2016) of the workforce.
The nation's quality workers, hydroelectric power
potential, and relative domestic tranquility hold
great promise for future endeavors in information
technology industries

Sri Lanka possesses advantageous factors for industrial development. It has productive agriculture
and is well-located in the center of the Indian
Ocean. Instead of orderly industrial development,
however, it faced a disastrous three-decade civil war.
Since the war ended in 2009, the economy has improved somewhat. Its industrial economy is limited
to basic food processing and light manufacturing.

Afghanistan suffers from decades of war, terrorism, and violent turmoil. These have effectively
prevented Afghanistan from realizing normalcy in
economic life. Even under the best of circumstances, Afghanistan faces formidable challenges
for industrial development. It lacks skilled workers,
quality learning institutions, and basic infrastructure. Afghanistan's limited food processing and
light craft industries are headquartered in Kabul,
the capital city.

Southwest Asia

The region's one common characteristic is that each
nation has yet to enter the globally competitive industrial economy. There are many factors for this
lag, but the most significant one is the region's in-
ability to overcome chronic political and social
problems.

Lebanon, a small nation bordered by Israel and
Syria, has only recently emerged from a devastating
civil war. Beirut has more or less recovered, although a systematic drive for industrial development is still in the future. Only limited light manufacturing is found, mostly in urban areas.

The violent and destructive civil war in Syria has
caused the economy to decline by more than 70 percent from 2010 to 2017. Complicating matters is a
poor natural-resource base, except for some petroleum deposits and potential hydroelectric power
generation.

The Arabian Peninsula's main resource, oil, is not
evenly distributed. The Gulf States and Saudi Arabia are richly endowed, while both Yemen and
Oman are poor in oil reserves. Oil-rich Arab states
are eager to process their own raw materials, producing fertilizers, chemicals, and plastics. They
pursue expansion of their petrochemical sector, for
which they have overwhelming advantages. The
Gulf States also have food processing and traditional craft industries, which are centered in a few
urban areas. These countries are developing economic diversity schemes to implement before the
oil runs out.

Iraq has a significant potential for industrial development. In addition to large oil reserves, its fertile river basin provides a productive farming sector. What has been missing is political and social
tranquility and responsible government leadership. Iraq's decades of wars, sanctions, and turmoil
have devastated the national economy. Light manufacturing and food processing are centered around
Baghdad, the capital.

Iran has many attributes favorable for industrial
development: a large population base, rich petroleum deposits, and strategic location—all factors
attractive to potential investors. Petrochemical
plants and light manufacturing are located in urban
centers and the coastal areas. The country's industrial capacity has still not reached the levels it
achieved before 1979's Islamic revolution.

Turkey, a secular Islamic nation, is a member of
the North Atlantic Treaty Organization (NATO). In
spite of its proximity to the European core area, the
Turkish economy is still largely agricultural. Nevertheless, its industrial potential is strong. Its textile
industry grew significantly as domestic cotton production increased. The Ataturk Dam, the first of a
series of dams on the Euphrates River, is providing
more water resources for industry and irrigation.
Drawn-out negotiations to join the European Union were suspended in 2016, a blow to Turkey's
efforts to develop more diverse export-oriented
industries.

Israel is more European than any other nation in
Southwest Asia. It graduates quality workers from
its higher-education institutions. In addition, many

experienced scientists and engineers migrated to Israel from the former Soviet Union and Eastern Europe. This elite group lay the foundation for Israel's sophisticated industrial economy. Genetic engineering and computer software are among the promising industries in Israel. If and when it successfully concludes its peace negotiations with Arab neighbors, Israel can take full advantage of its other significant asset: a location surrounded by populous Arab states.

Central Asia

Five former Soviet republics that gained independence after 1990—Kazakhstan, Turkmenistan, Uzbekistan, Kyrgyzstan, and Tajikistan—occupy the vast but generally arid territory of Central Asia. The land is suitable only for limited farming or herding. The Soviet-era industries created massive environmental problems that each of these countries has been forced to confront.

Kazakhstan, has the largest and strongest economy in Central Asia. It also holds the distinction of being the first former Soviet republic to repay its entire debt to the International Monetary Fund. During the Soviet era, Moscow located its extensive space and missile-development facilities in Kazakhstan. These facilities and the surrounding area—about 2,317 square miles (6,000 sq. km.) in south-central Kazakhstan, are leased back to Russia. Kazakhstan's market-based industries depend greatly on the oil deposits found in the Tengiz basin and offshore in the Caspian Sea. Pipeline industries and oil refining are expanding rapidly. The Kazakh government, mindful of the country's dependence on its oil reserves, is taking steps to diversify its economy.

Turkmenistan's economy is based on agriculture, particularly irrigated cotton growing. In the industrial sector, oil and natural gas extraction and refinement are prominent, as are craft industries and food processing for domestic consumption.

Uzbekistan's economy is similar to Turkmenistan's. The high-tech industrial economy has not touched the nation. Its industry consists of a few Soviet-era heavy industries, food processing, and light manufacturing. Some foreign investment has been forthcoming in the manufacturing sector, including automobile assembly and electronics. Natural gas, gold, and cotton are the main exports.

Kyrgyzstan has implemented economic reforms, but has yet to reach a level of economic prosperity. Its economy has remained virtually unchanged since its independence. Industries are limited to food and cotton processing. In Tajikistan, mineral-extraction and small handcraft industries are the backbones of the industrial economy. Tajikistan remains the poorest of the former Soviet republics. In 2017, nearly 35 percent of the country's GNP came from Tajik working in Russia and sending money to their families back home.

Walter B. Jung

> ## DISCUSSION QUESTIONS:
> ## ECONOMIC GEOGRAPHY—INDUSTRIES
>
> Q1. What factors led to China's huge economic growth? How do China's economic dynamics differ from those in Japan and South Korea? How do China's energy resources figure in the country's economy going forward?
>
> Q2. Why is textile manufacturing the dominant industry of so many Asian countries? How did this industry evolve? How does it tie in with a given country's traditional agricultural products?
>
> Q3. Where is the "Silicon Valley" of India? How did that region gain that level of recognition? Compare and contrast India's Silicon Valley to the Silicon Valley of China? How do they differ? In what ways are they similar? What factors make it advantageous for Western corporations to outsource computer and IT work to companies in India's Silicon Valley?

ENGINEERING PROJECTS

Asia encompasses lands from eastern Russia and the southern Caucasus through the Indian subcontinent, most of the Middle East, and Southeast Asia to China, the Korean peninsula, and Japan. It includes diverse topographies, natural resources, and patterns of economic development. The variety of engineering projects that have been completed or are under development in Asia in the first decades of the twenty-first century reflects this diversity. The different nations' approaches to fundamental issues of economic growth, environmental preservation, and the use of modern technologies also vary greatly.

Canals and Terracing

One of the oldest types of engineering works, and the most widely used in Asia, is the development of water control and distribution systems necessary to support both dense populations and intensive agriculture. In East and Southeast Asia, where the lowlands and coastal plain zones are dominated by rice cultivation, canals and simple trench-type irrigation routes conserve water from the monsoon rains while simultaneously maintaining the required water levels in flooded fields. Farther inland, arable land is at a premium. In countries such as China, Vietnam, the Philippines, and Indonesia, arable land shortages have given rise to extensive terracing of otherwise unsuitably steep hillsides. Such terracing creates additional planting areas and prevents erosion. The 2,000-year-old Rice Terraces of the Philippine Cordilleras have been designated as a UNESCO World Heritage Site because of the

Wind turbines in Hunan, China.

227

Kobe Airport in Osaka Bay, Japan.

historical significance of this technique. In Western Asia, remnants of a first-century agricultural terrace system have been excavated in the desert around Jordan's famous archaeological site of Petra, which was an important trading center on Middle Eastern caravan routes. Petrans had a sophisticated rainwater management system that included water channels, ceramic pipelines, underground cisterns, and terrace walls made of waterproof cement. Today, terrace farming continues to be widely used in Southeast and East Asia for growing rice, barley, and wheat, and remains a key part of the agricultural systems of these regions.

Navigable canals in China have played a vital role in transportation and commerce since antiquity. The world's longest and oldest human-made waterway, China's Grand Canal, was first built between 400 and 300 BCE to transport grain from agricultural regions to cities and to China's large standing armies. Even today, the southern part of the ancient canal is in heavy use and continues to play an important role in the country's economy. The water-

way was dredged in the 2010s to increase its cargo capacity.

In the Middle East, leaders have been interested in linking the Mediterranean and Red seas ever since the days of the Egyptian pharaohs. In the modern era, French engineers began construction on what became the Suez Canal in 1859; with 1.5 million Egyptian laborers working on the canal, it opened for navigation in 1869. Since then, the Suez Canal has provided a vital link between Europe and Asia, greatly reducing transport times and therefore the costs of raw materials and manufactured goods.

The annual capacity of the Suez Canal is some 25,000 vessels, which carry around 13 percent of the volume of world trade. Its length is 120.11 miles (193.3 km.). In 2015, Egypt, which owns the canal, completed a major expansion. It included the deepening and widening of several sections of the canal to accommodate larger ships and the construction of a second shipping lane along part of the main waterway, which allows the canal to accommodate two-way traffic along much of its length. At the same time, Egypt also invested more than $1 billion to build seven tunnels under the Suez Canal to connect the Egyptian mainland with the Sinai peninsula. The first four of the tunnels were completed in 2019. This project is part of Egypt's strategy to turn the Sinai and the Suez Canal region into a major logistical and industrial hub.

Dams

The Three Gorges Project in the People's Republic of China, which was started in 1994 and reached its full capacity in 2012, is the largest hydroelectric dam engineering project in history. With its series of dams on the Yangtze River and a 400-mile-long (645 km.) reservoir, the project produces an unprecedented 22,500 megawatts of electricity where it is needed most—in China's relatively underdeveloped interior provinces. However, the project has also set a number of unwelcome records: it put 150,000 acres (60,700 hectares) of land permanently under water—land that supported 140 towns, 1,350 villages, 16 major archaeological sites, and 1.2 million people, who had to be relocated to other regions. Its environmental impact was also massive: the construction and the flooding of hundreds of factories and waste dumps caused massive pollution; the banks of the reservoir and the river have been eroding, resulting in landslides; and the dam critically diminished water supply to downstream residential centers and ecosystems. Nevertheless, Chinese companies have been replicating this model both domestically and internationally, in Laos and Cambodia. Within China, huge hydropower cascades are being constructed or planned upstream of the Three Gorges Dam on the Yangtze River and its tributaries as well as on Nu (Salween) River and the Upper Mekong.

Since the nineteenth century, civil engineers have longed to harness the power of the Mekong

Dujiangyan, an irrigation project completed in 256 BCE during the Warring States period of China by the State of Qin. Although a reinforced concrete weir has replaced Li Bing's original weighted bamboo baskets, the layout of the infrastructure remains the same and is still in use today to irrigate over 5,300 square kilometers of land in the region.

River, which rises in Tibet and flows through China, Thailand, Laos, Cambodia, and Vietnam. France in the 1950s and the United States in the 1960s each planned expansive damming projects on the Lower Mekong, both to protect fragile local agricultural industries and to provide electricity as a means to accelerate industrial and manufacturing growth. These plans were ultimately shelved by the Vietnam War, which eventually involved both Laos and Cambodia. After the war, military and political crises made resurrecting such large-scale construction projects impossible.

At the turn of the twenty-first century, as stability returned to the region, the development of the Mekong River again became a priority. On the Upper Mekong in China (where the river is known as the Lancang), a huge complex of more than twenty dams—most with hydroelectric generating plants—was begun in 1993; eight of them have been completed by 2020.

Thailand, Laos, and Cambodia have started building a series of dams and hydroelectric stations on the Lower Mekong with the help of China. Several of these projects, including the large Xayaburi Dam in northern Laos, have been completed in the 2010s. Dam building on the Mekong is controversial: there are serious ecological concerns and numerous complaints from downstream water users.

Desalination Plants

The process of desalination makes seawater and brackish water usable for human consumption, industrial applications, and irrigation. In the coastal regions of Western Asia, where freshwater is in very short supply but seawater is abundant, desalination plants provide the only viable solution to the vital problem of usable water scarcity.

Several different desalination technologies have been developed in the twentieth century. They are based on either thermal processes conducted in a series of closed tanks or on reverse osmosis carried out with the use of membrane filters. By 2020, about 20,000 desalination plants—producing over 3.4 billion cubic feet (95 million cubic meters) of potable water per day—were in operation in some 120 countries; about half of this desalted water was gen-

erated in the Middle East. The largest producers were Saudi Arabia, the United Arab Emirates, and Kuwait.

Kuwait was the first country in the world to adopt desalted water as its main water source. Its first water treatment plant was built in 1954. Saudi Arabia's Ras Al-Khair plant, when it was completed in 2015, became the world's largest desalination facility. Israel's Sorek plant is also one of the world's largest. In 2019, almost 60 percent of domestic water in Israel was generated through desalination. In other parts of Asia, multiple desalination facilities have been built in India, China, and Singapore, and their number has been growing. Singapore plans to meet a third of its water needs through desalted water by 2060.

Skyscrapers

Record-holding skyscrapers have been viewed in many countries as manifestations of national pride and symbols of progress. As of 2020, sixteen of the twenty tallest buildings on the planet were located in Asian countries, and Dubai's Burj Khalifa was the tallest of them all. Topped out at 2,716.5 feet (over 828 meters), it was designed by the same American architectural firm that designed the Willis Tower in Chicago (one of the previous world record holders) and One World Trade Center in New York City (the tallest building in the United States and the Western Hemisphere).

Burj Khalifa is a modular, Y-shaped structure with twenty-six levels arranged in a spiral pattern; they decrease in size as the tower spirals upward. The design is derived from the spiral minarets of Islamic architecture. The decision to build the Burj Khalifa, originally named Burj Dubai, was an essential part of the Dubai government's strategic plan to diversify the oil-based economy of the emirate and to achieve international recognition for Dubai as a tourist destination and service hub. The tower is a centerpiece of a large-scale development with hotels, condominiums, and shopping centers, including the world's second-largest mall. The strategy has largely paid off. The skyscraper is now one of the most recognizable buildings on the planet, and the Dubai Mall—with its high-end shops, numerous restaurants, high-tech cinemas, and unique amusement parks—gained the title of the most-visited building worldwide.

Before the Burj Khalifa was completed in 2009, the title of the world's tallest building in the twenty-first century was held successively by the Petronas Twin Towers in Malaysia's capital city of Kuala Lumpur and the Taipei 101 in Taiwan. The postmodernist Taipei 101 was designed to bring to mind the patterns of an Asian pagoda, a bamboo stalk, and a stack of ancient Chinese money boxes (a symbol of wealth). It incorporates innovative features that enable the structure to withstand large earthquakes and tropical storms. It also boasts the title of the largest "green building" in the world.

One of the unique features of Kuala Lumpur 's Twin Towers is the two-story skybridge that connects the towers. The bridge, weighing 750 tons and located at a height of 558 feet (170 meters), is the world's highest double-decker skybridge and the second-highest skybridge between separate buildings. (The highest one, called "Crystal," connects the top of four skyscrapers in the Raffles City complex in Chongqing, China.)

Bridges

With a few American exceptions, all of the world's longest bridges are located in China, Southeast Asia, and the Middle East. But unlike the supertall skyscrapers, these extremely long bridges are not products of a desire to establish a world record. They are parts of nationwide projects for building the modern highway and railway systems, and their length is determined by infrastructural demands and challenges of topography.

As of 2020, the world's longest bridge was China's Danyang-Kunshan Grand Bridge, at a length of 102.4 miles (164.8 km.). Completed in 2010, it is a viaduct on the Beijing-Shanghai High-Speed Railway that runs in the Yangtze River Delta over its numerous rivers, canals, lakes, and rice paddies. The railway, which connects two major economic zones in China, also boasts three other of the world's ten longest bridges. The design of these elevated viaducts was chosen to avoid building numerous individual structures for crossing bodies of water, roads, and railways.

The Bang Na Expressway in Thailand runs through the country's capital, Bangkok. It is a six-lane box girder bridge designed for heavy automobile traffic. Completed in 2000, it remains the world's longest car bridge, measuring over 34 miles (55 km.).

Kuwait's mega bridge project, the Sheikh Jaber Al-Ahmad Al-Sabah Causeway, is part of the emirate's national plan for developing the northern part of Kuwait. The world's biggest maritime causeway project, it is composed of two bridges with a combined length of 30 miles (over 48 km.). The route spans the Bay of Kuwait, passing through its fragile natural habitats. During the 2013-2019 construction of the causeway, extensive measures were taken to minimize damage to marine flora and fauna, especially the habitats of the bay's iconic green tiger shrimps. The effort, which included installing artificial reefs away from the construction zones, was highly successful.

Mining

The unearthing of valuable metals and stones is an old industry in Asia and the Middle East. Precious stones are mined in Myanmar (Burma) and in mainland Southeast Asia. In Cambodia, ruby mining has become one of the most lucrative industries. However, the simple dredge-digging techniques cause serious erosion of topsoil and the deposit of excessive amounts of soil into rivers. As a result, Cambodia's rivers and the floodplain of the Tonle Sap, Southeast Asia's largest lake, have experienced rapid sedimentation, causing rivers to change course and damaging aquatic life.

Mining for gold by drilling shafts deep underground is a widespread and profitable business in Asia, but the technology needed to cut through solid rock is expensive. Most successful mining companies are based in North America, China, or Australia, and they operate in Asia and the Middle East either in cooperation with local companies or on lands they have purchased. In the Philippines, which reportedly holds the world's second-largest gold deposits, the preferred practice of building big open-pit mines has been a subject of nationwide debate and fierce local resistance. Toxic tailings (residue) from these mines have been polluting farmlands and waters for decades, and in 1996 produced one of the world's worst mining and environmental disasters. At the Marcopper mine, a fracture in the drainage tunnel of a large pit containing toxic mine tailings caused the flooding of the area with contaminated wastewater, killing farm and aquatic animals and destroying crops. Thousands of people were displaced. Despite this disaster, large-scale open-pit mining in the Philippines has continued. New technologies, including ground-based radars, have been introduced there in recent years to monitor the integrity of mining sites and to identify hazards.

China derives more than 70 percent of its energy from coal, and coal mining has been steadily growing. However, the development of advanced and safe coal-mining technologies has been slow, and severe mining accidents still frequently occur, especially in underground mines. In recent years, the country has started introducing the so-called "intelligent mining" model in some of its coalfields. It involves using artificial intelligence and robotic machinery for the unmanned operation of mines, with monitoring from surface control centers. China plans to significantly expand the use of intelligent mining technology by 2030.

Since the turn of the twenty-first century, Saudi Arabia has been rapidly developing its abundant mineral resources, the largest in Western Asia. Large mines and processing plants have been built throughout the country for the extraction and processing of gold, bauxite, copper, phosphates, and other minerals. The largest of them is the Ras Al-Khair complex, built between 2009 and 2014 in cooperation with Alcoa, the U.S. aluminum giant. It includes an alumina refinery, a smelter, and one of the most advanced rolling mill plants in the world.

Petroleum and Natural Gas Projects

Until the mid-1970s, Asian industrial development was powered primarily by relatively small electricity plants that used either local coal, hydropower, or imported oil from the Middle East. As coal became more costly, rivers more controlled, and imported oil more expensive, Asian countries began seeking alternative means to produce electricity. Major oil and natural gas exploration efforts began in the late 1970s, financed chiefly by North American and European oil companies.

Among the Southeast Asian countries found to have large offshore petroleum and gas reserves are the Philippines, the small nation of Brunei, Malaysia, and Indonesia. These countries started earning huge amounts of revenue by selling their excess production to other Asian markets, especially to energy-hungry China, Japan, and South Korea. Offshore oil and gas are pumped to the processing terminals located on mainland shores, from where they are transported away via pipelines or shipped by tankers.

Kazakhstan, Turkmenistan, and Azerbaijan, formerly parts of the Soviet Union, started rapidly developing their large oil and gas resources after gaining independence in 1991; they soon became some of Asia's top exporters of these fuels. The Kashagan offshore oil field, discovered in 2000 in Kazakhstan's sector of the Caspian Sea, was the world's largest oil discovery in three decades. It also became one of the most challenging oil megaprojects in history due to extreme climate and weather conditions and very shallow waters in the northern part of the Caspian. Developed by an international consortium, the field started producing oil for export in 2016.

Almost all of Azerbaijan's natural gas is produced in its two offshore fields in the southern part of the Caspian. Largely completed by 2018, the Shah Deniz field complex is one of the largest natural gas development projects in the world. It is a founding link for the planned Southern Gas Corridor, an initiative of the European Union for building a major gas supply route from the Caspian and Middle Eastern gas fields to Europe. The project involves most countries of Central Asia, Transcaucasia, and the Middle East. It consists of several pipelines, the first of which, the Trans-Anatolian gas pipeline running through Turkey, was opened in 2018. The project is strategically important for reducing Europe's dependence on Russian oil and gas.

During the first two decades of the twenty-first century, some of the world's largest oil and natural gas pipeline projects were completed by Russia and China. Russia's enormous fossil fuel resources are located mostly in the Asian part of the country, in Siberia. The Yamal-Europe pipeline (built between 1992 and 2005) transports gas from Western Siberia to Germany and Austria, and the Eastern Siberia-Pacific Ocean pipeline (constructed in 2006-2012) brings Russian crude oil to China, Japan, and South Korea. The Kazakhstan-China pipeline, built between 2000 and 2009, is China's first direct oil import pipeline and Kazakhstan's first oil pipeline built after independence. China's West-East Gas Pipeline Project has been developed since 2002, and the first three of its four pipelines were completed by 2018. The huge network runs from the western gas fields through much of the country to the industrial east of China.

The Middle East—a transcontinental region centered on Western Asia—includes five of the world's top ten petroleum-producing countries. Oil and gas production in this region is cheaper than in most other parts of the world because both onshore and offshore resources are located near the surface and pooled in vast continuous fields. By the 2020s, the region was generating about 30 percent of the world oil production. In the first decades of the twenty-first century, Iran, Iraq, Kuwait, and the United Arab Emirates developed new oil and gas fields to expand their production. At the same time, Saudi Arabia has shifted its industry from the upstream sector (exploration and extraction) to the downstream sector (oil refining, gas purification, and production of petrochemicals). New refineries, chemical plants, and terminals have been built throughout the country, including the huge Petro Rabigh Integrated Refining and Petrochemical Complex, a joint venture with Japan.

Discovered in 2010 in the Mediterranean Sea off the coast of Israel, the Leviathan gas field is one of the world's largest offshore gas finds of the century. It holds enough gas to meet Israel's domestic needs for at least 40 years. By 2020, Israel had largely completed developing the Leviathan field and started a commercial gas flow.

Nuclear Energy

The 2011 disaster at the Fukushima nuclear power plant in Japan has affected but not stopped the growth of nuclear-power production in Asia. Japan, whose fossil fuel resources are very limited, became heavily reliant upon nuclear power. In 2010, nuclear power met over 30 percent of the country's electricity needs via thirty-three nuclear power plants with fifty-five reactors, and several more were under construction. But then, in 2011, disaster struck at the Fukushima Daiichi nuclear plant when a tsunami damaged the plant's reactors, which released radiation. After a series of explosions at containment facilities, irradiated water leaked into the surrounding landscape and the Pacific Ocean. In the aftermath of the accident, all of Japan's reactors were shut down. By 2020, only nine were back in operation, with seventeen more in the process of restart approval. A number of reactors were permanently retired. Japan revised its power-development strategy from heavy reliance on nuclear energy to a more diversified approach that included the revival of coal-fired power projects. It also started reprocessing used nuclear fuel to recover uranium and plutonium for re-use in electricity production.

In neighboring South Korea, where nuclear energy expansion had been a strategic priority for decades, the Fukushima disaster had even more dramatic consequences. Driven by public opinion, the South Korean government elected in 2017 to gradually decommission all of the country's twenty-four operable reactors and phase out nuclear power altogether over some forty-five years.

In spite of all this, in 2019 Asia still had more nuclear reactors than any other continent. Its 130 operable reactors were located in Japan, South Korea, mainland China, Taiwan, India, and Pakistan. Bangladesh, Turkey, and the United Arab Emirates have continued pursuing their plans to introduce nuclear power into their electricity mix.

Mainland China alone had forty-five reactors, and over fifty are under construction or at the planning stage. Nuclear power has been viewed in China as a principal "clean energy" alternative to the heavily polluting coal-fired plants which now dominate the country's energy sector. Pakistan had five commercial nuclear plants, with two under construction and over thirty planned to be built by 2050.

India has remained committed to building new nuclear power plants as part of the country's long-standing nuclear power program, launched in the 1950s. In 2020, India operated twenty-two nuclear reactors and more than twenty were being built or planned. Because its uranium resources are very limited, India has been developing a nuclear fuel cycle based on thorium, of which it has one of the largest reserves in the world.

In 2020, Iran had one nuclear reactor and had started construction on the first of two new, Russian-designed ones. The country has a major program for developing uranium enrichment, which was concealed from the rest of the world for many years. The program has long raised concerns that it might be intended for non-peaceful uses. Around 2000, Iran started building sophisticated enrichment facilities at Natanz, which now include both above-ground and underground plants. Operations at Natanz have been under international safeguards, though monitoring has been constrained by the Iranian authorities. The Iranian government has been insisting that the country needs to independently develop all of the elements of the nuclear fuel cycle, including uranium enrichment, in order for it to achieve energy independence.

Russia, all of whose large nuclear power plants are located in the European part of the country, has recently started building low-capacity, floating nuclear power stations to be used in the Far North—the large part of Russia (and the northernmost part of Asia) located mainly above the Arctic Circle. In 2019, the world's first floating nuclear power plant arrived at its permanent location in the Chukotka region. China was not far behind with its own modular floating reactor. These power stations can be used on islands and in remote coastal areas as well as for offshore oil and gas exploration.

Computer Hardware

Japan, a twenty-first century world leader in computer electronics, was hit hard by an economic stagnation of 1991-2001, known as "Japan's Lost Decade." It lost a large part of its market share to Chinese and South Korean companies. Computer hardware brands from mainland China and Taiwan—Lenovo, Acer, and Asus—have become major worldwide competitors of leading U.S. brands. In 2005, Lenovo acquired the personal-computer business from U.S.-based IBM. By 2018 the company had the largest share of the global personal-computer market, having surpassed U.S.-based HP, Dell, and Apple. Taiwanese Acer and Asus also acquired their first hardware designs from American companies, and within a few years they became top global computer vendors. These companies are best known for their desktop computers, laptops, and tablets, while South Korean companies LG and Samsung have gained worldwide recognition for their smartphones and smartwatches based on Google's Android operating system.

All leading computer brands, including the American ones, have been manufacturing and assembling most of their products in Asia—mainland China, Taiwan, India, Thailand, Malaysia, and the Philippines. In recent years, Taiwanese Foxconn Technology Group became the world's largest contract manufacturer of consumer electronics. Among its many products made for major world brands are Apple's iPhones and iPads, Amazon's Kindle e-readers, Microsoft's Xbox gaming consoles, and Japanese gaming consoles Nintendo and PlayStation. Foxconn maintains factories across the Asian continent and also has some in Europe and Latin America.

Software and Information Technology (IT)

Most of the world's largest software companies still operate out of the United States, and they had long regarded their Chinese counterparts as little more than copycats. But as a result of China's software revolution of the twenty-first century, that country has emerged as a major computer-technology competitor for the United States. The country produced three giant tech conglomerates—Baidu, Alibaba, and Tencent—collectively known as BAT. Tencent specializes in social media and online entertainment—music, mobile games, and multiplayer online games. It has become the world's largest gaming company. Alibaba is the world's largest online retailer and e-commerce company. All three companies have significant artificial intelligence (AI) components, with Baidu being one of the largest AI developers in the world. China's government considers the AI sector a national priority and has set a goal for the country to become the world's leading AI power by 2030.

India has become the world's biggest exporter of information-technology-related services. Educational institutions in India turn out large numbers of well-trained programmers and software engineers. The relatively low pay scale and the common use of the English language in India give the coun-

try a competitive advantage in IT-related services over other countries. Services supplied by India to European and North American markets include corporate technical support, Web hosting, cloud computing, systems administration, and business process outsourcing. The Indian city of Bangalore has become known as the Silicon Valley of India. Many of the country's high-tech companies, technology institutes, and research centers are located there. When it was inaugurated in 2000, TIDEL Park in the city of Chennai became Asia's single largest IT park. Hyderabad's HITEC City is also a global IT hub and a major bioinformatics center. By 2020, the IT-dominated services sector in India accounted for 40 percent of the country's GDP and employed 25 percent of its workforce.

THE SILK ROAD

In 700 CE, the great empires of the West and the East were linked by the Silk Road, an ancient trade network of caravan trails across the steppes and deserts of Central Asia. The Byzantine Empire, in what is modern Turkey, and the Tang Dynasty of China communicated with each other and traded in such valuable goods as silk, rare plants, herbal medicines, spices, and gold. Camel caravans left Chang'an, Tang China's largest city and capital, in early spring, and several months later, after numerous stops, arrived in Byzantium. Stops along the way included Dunhuang and the fertile Kashgar oasis in western China and Tashkent in the fertile Fergana Basin of Central Asia. After detouring south to avoid the Kara-Kum desert of modern Turkmenistan, the route wound its way to the thriving Persian city of Herat in modern Afghanistan. In Herat, traders connected with Indian traders.

The route followed the Caspian Sea coast on its way to Baghdad in the valley of the Tigris and Euphrates Rivers. After following the Euphrates upstream toward Damascus, the silk caravans moved to the ancient Mediterranean port of Tyre, and then overseas to Byzantium. In the thirteenth century Marco Polo traversed the Silk Road frequently and documented its activity. Eventually the Silk Road was made obsolete by inexpensive and relatively easy ocean shipping. By the end of the eighth century, the sea route from Canton (modern Guangzhou) to the Middle East was already well developed.

Aerospace Engineering: Asian Space Race

A number of Asian countries have developed advanced technologies in the fields of aerospace and avionics engineering. Japan, China, India, Pakistan, South and North Korea, and several Middle Eastern countries, including Israel, Iran, and Egypt, have advanced design and construction capabilities for building aircraft, rockets, satellites, and missiles. Some of these countries have been applying these capabilities for both military and peaceful purposes. Indonesia, for example, has used its large revenues from oil exports to create civil aviation and military aerospace industries. The Pakistan Aeronautical Complex, a major designer and builder of military aircraft, also services civilian aircraft and produces avionics systems for civilian usage.

In the twenty-first century, the aviation industry in Asia has been booming, especially across Southeast Asia, where it provides aircraft repair and maintenance to international commercial carriers and manufactures airplane parts for major aircraft makers such as Boeing and Airbus. Several Asian countries also manufacture their own aircraft—small passenger jets, military planes, helicopters, and unmanned aerial vehicles.

India, Japan, and China have built some of the largest satellite networks in the world. These countries have multiple satellite-building and launching facilities. Israel, Iran, Pakistan, and North Korea also have extensive experience in satellite operation, while nations such as South Korea, Singapore, Bangladesh, and Turkey only started to develop their own satellites in the first decades of the twenty-first century. The various indigenous satellites are used for telecommunications, surveying, remote sensing, meteorology, and other applications. Some are also employed for military communications, navigation, and intelligence gathering. India and China have made their launch vehicles available to foreign companies and organizations, with India emerging as a leading provider of rideshare services for small satellites. In 2017, India launched 104 of those on a single launch, setting a world record. Most of them were tiny imaging nanosatellites designed by a U.S.-based Earth observation company; the flock was launched with the aim of creating a composite image of the whole planet surface on a daily basis.

Virtually all aerospace technologies have dual-use capabilities, in that they can be applied to either peaceful or military purposes. On the extreme side, North Korea has been devoting a disproportionately large segment of its aerospace research and engineering to the development of conventional and nuclear-armed missiles. South Korea's government pays scientists and engineers to design and build rockets for research on ozone gases and solar radiation, but experts note that the same rockets could be used to deliver bombs. On the other hand, the development of high-tech capabilities generated in the aerospace sector also has spin-off effects, since they promote the rapid growth of associated industries.

An example of this phenomenon is China, which has invested in scientific research laboratories, world-class educational facilities, and aerospace technologies for more than sixty years. As a result, China has a highly developed aviation industry, computer and electronics sectors, telecommunications, and robotics.

In 2003, China became the third country in the world, after the Soviet Union (later, Russia) and the

Jiuquan Satellite Launch Center, Gansu Province, China.

United States, to conduct a human spaceflight mission. It has also developed its own space station, to be launched into space in the 2020s. India, too, has been planning to launch a crewed space mission and build its own space station. At the same time, Japan became one of the founding members of the International Space Station project; it has been sending its astronauts to space on Russian and American spacecraft. China, India, and Japan also conducted many advanced uncrewed space missions to carry out lunar, Mars, and asteroid exploration.

These and other Asian countries that have national space programs (Iran, Israel, North Korea, South Korea, Indonesia, and Turkey) have been actively competing for notable achievements in aerospace, a phenomenon sometimes characterized as Asia's space race.

Laura M. Calkins

TRADE

Asia stretches from Turkey and the eastern shores of the Mediterranean Sea and Red Sea to the Pacific Ocean, from the Indian Ocean to the Arctic Ocean through the Himalaya mountains and the Tibetan Plateau. This vast area is rich in natural resources, especially cultivatable land, energy sources such as oil and natural gas, valuable minerals, and vital sea lanes. For centuries, Asia has been the source of products highly valued by the rest of the world, including coffee, cotton, drugs, dyestuffs, perfumes, petroleum, precious stones, silks, spices, and tea.

Asian Marketplace

Asia provides a large and important marketplace. Containing approximately 60 percent of the world's population, it is both a huge source of labor and a market for consumer and industrial goods. Financiers in Singapore, Tokyo, and other money centers have long provided capital for enterprises throughout the world.

Technological revolutions have periodically altered the balance of power among trading regions within Asia. For example, the Chinese invention of the compass during the Qin Dynasty (221-206 BCE) enabled its sailors to venture beyond the China coast and to navigate the South China Sea and Indian Ocean. In the process of exploring these waters, Chinese mariners discovered the Spice Islands (in present-day Indonesia). The West did not discover the compass for another 400 years. The Arabs acquired the compass from the Chinese and then introduced it to Mediterranean sailors around 1250 CE. The device was crucial to the success of the Spanish and Portuguese maritime exploration of the fifteenth century.

Other technologies that changed trading patterns include the spread of underwater telegraph cables that allowed virtually instantaneous communications over oceans, and colonial railroad networks that connected port cities with interior regions where foodstuffs and raw materials are produced for export. Today's telecommunications systems have reconfigured the world's major financial markets so that New York, London, and Tokyo are the tripolar core of the world's economy. Trade patterns also change as shifts occur in the availability of and demand for agricultural products and natural resources such as gold and silver.

The trade patterns in Asia date to prehistory. Prehistoric human migration caused the spread of bananas, coconuts, and other plants from Southeast Asia to Africa in the west and the Pacific Islands in the east. Silk was widely traded even before the opening of the Silk Road; the fabric found in an Egyptian tomb has been dated to 1070 BCE.

Spices were among the first goods traded. The earliest caravans carrying luxury goods and other precious cargoes linking China, Europe, and the Middle East began during the Han Dynasty (206 BCE-220 CE). The most prudent way to travel was to join a caravan headed to where the merchant wanted to go. Traders often alerted each other to economic opportunities in distant lands. For example, the Han Emperor Wu'ti dispatched Zhang Qian (also known as Chang-ch'ien) to negotiate an alliance with the Xiongnu (also known as the Hsiung-nu) people. Zhang's mission revealed that numerous trade opportunities existed despite the difficult journey west through the Hexi (or Gansu) Corridor, which skirted the harsh Taklimakan Desert and into Central Asia. The Chinese were anxious to acquire horses from the nomads.

Long-Distance Trade

Some long-distance trade took place in the precapitalist world prior to 1400. For the most part, however, trade was limited to a series of overlapping regional circuits of trade. Middlemen took possession of the goods and shipped them by land or sea to their ultimate destination. The ancient Phoenicians distributed spices throughout the Mediterranean Sea until the Egyptian city of Alexandria became a commercial center. Starting in the tenth century, the Venetians became the major supplier of spices to Europe. Widespread direct trade did not become a reality until the fourteenth century, when Genghis Khan and his successors created the largest land-based empire in history, one that included Central Asia, China, Korea, much of the Middle East, and Russia. The unification of so many states under Mongol leadership created the conditions for long-distance trade to thrive.

Such ideal trading conditions existed during other periods. Arabs long acted as middlemen in trade between Asians and Africans. Asia began dominating the world economy as early as the seventh century, when Islam began to conquer Egypt, Iran, Iraq, and Syria. These conquests virtually guaranteed that traders would have security to conduct their business and that prosperity would spread along with Islam. The military, political, and spiritual head of Islam (the caliph) guaranteed—for the first time since the decline of the Roman Empire—that traders could conduct business safely between the Mediterranean Sea and the Indian Ocean.

Merchants doing business over long distances did so because of the lure of riches. Middlemen sometimes had a virtual stranglehold on the most lucrative trade routes, for example, the Turks in the fifteenth century. These merchant-adventurers risked not only enormous sums of money to buy goods and supplies for themselves and their camels and horses but also chanced their own lives to make journeys that might last a year and a half to two years. Bandits, disease, military and political turmoil, starvation, storms, thirst, and taxation were formidable obstacles to success. Increasing prosperity led merchants to search for new sources to meet the demand for luxury goods. Traders needed large sums of money to finance their journeys. Cash-starved merchants turned to markets in Asia, where they could exchange goods rather than silver.

Merchants transported their goods along trade routes such as the famed Great Silk Road, which reached from northwest China through the oases in Central Asia to commercial centers in Persia and ports on the Mediterranean Sea. Along the way, they stopped at caravansaries (way stations) in oases. These caravansaries provided travelers the opportunity to rest and restock their supplies. The oases often became prosperous trading centers.

Goods to be transported over land and sea had to be light in weight and not readily perishable. Travel by caravan was dangerous, expensive, and slow. The average caravan traveled about 20 miles (32 km.) per day across rivers, deserts, and mountains. Traders used camels, donkeys, horses, mules, and oxen to transport their goods. Merchant middlemen such as the Persians and Venetians then carried the goods from Asia either by caravan or ship to Europe. No evidence exists that direct trade occurred between Asia and Europe prior to the Mongol conquests of the thirteenth century.

Mongols

The Mongol Empire founded by Genghis Khan created conditions that promoted trade between Asia and Europe. During much of the Yuan Dynasty (1279-1368), the Mongols provided the military might (the Pax Mongolica) that linked Eurasia from China to the borders of Western Europe. The tales of Marco Polo (1254-1324), the Venetian adventurer and travel writer, alerted Europeans to a seemingly vast array of new products. Polo's tales of the wealth of western and central Asia inspired the voyages of Christopher Columbus, Ferdinand Magellan, Sir Francis Drake, and Vasco da Gama.

The arrival in 1498 of Vasco da Gama at Kozhikode (Calicut), a port city on India's Malabar Coast, inaugurated a period of Western maritime dominance that lasted until the end of World War II. For the first time, Asian goods reached Europe directly, where demand for luxury goods from the East skyrocketed. To meet the demand, European merchants began searching for a faster sea route to Asia. The Portuguese mariner Bartolomeu Dias passed the Cape of Good Hope on the southern tip of Africa in 1488 and sailed into the Indian Ocean.

The event launched the beginning of an international market that bypassed the caravan routes. Magellan's 1519-1522 maritime expedition—the first successful circumnavigation of the globe—was a Portuguese end-run around both the Muslim and Venetian strangleholds over Mediterranean trade routes. The direct sea route to India shifted the preponderance of trade from Mediterranean ports such as Venice and Genoa to Amsterdam, Antwerp, Lisbon, and London.

The Chinese were familiar with the lucrative business opportunities offered by the Pacific Ocean. Overland trade was expensive, because goods had to be repeatedly loaded and unloaded during caravan journeys. Merchants found that shipping their goods by sea was less expensive, relatively safer, and permitted the transport of bulkier products. Shipping by sea replaced the overland caravan routes in Central Asia as political turmoil increased in regions such as northwest China, the Middle East, and Persia. When political turmoil increased, the economies of these regions declined. Trade increased in peaceful regions. For example, Russian and Chinese merchants conducted their business over the stable routes that ran north through southern Siberia and north central Asia.

Governments and financiers quickly discovered the necessity of developing and guarding their trading partnerships and territories. In 1600 the English East India Company was established as a monopolistic trading company and agent of British imperialism in India. Two years later, the United Provinces of the Netherlands established the Dutch East India Company and granted it a monopoly on trade in the Indian and Pacific oceans.

The Treaty of Nerchinsk (1689) and the Treaty of Kyakhta (1727) permitted the Russians to trade with the Chinese. Russian traders swarmed into China to trade furs, leather, and woolens for cotton, rhubarb, silk, tea, and tobacco. The Chinese-Russian overland route increased chances that trading would be profitable. European traders who carried their goods to the southeast coast of China also hurt the Central Asian overland trading routes.

Traders crisscrossed Asia and Europe over a number of routes. Caravans went from China to Kashgar and from India to the Middle East. European merchants journeyed to Alexandria, Beirut, and ports on the Black Sea and the Mediterranean Sea to trade for goods from the East. Other European traders went from Russia via the Volga River and the Caspian Sea to Persia, and from there joined caravans going to China and India to the Strait of Hormuz to the Indian Ocean.

Trade between Asia and Europe was not balanced—a situation that existed for many centuries. Goods from Asia were worth relatively more than those from Europe. During the fifteenth century, China began substituting silver for depreciated paper and copper currency. British, Dutch, and Portuguese investors recognized the opportunities for trade in Asian markets where capital was scarce and the demand for imports and exports was great, such as in the Mediterranean and Indian Ocean regions. Entrepreneurs bought luxury goods such as porcelain and silk in China and Malaysia. Traded goods also included gold for coining money, iron, timber, and slaves bound for the Middle East.

The internationalization of Asian trade had a far-reaching influence on the economies of many regions. Rice growing, for example, spread from East Asia to India. Cotton farming began in India and made its way to the Middle East. The Europeans learned papermaking and other technologies from the Chinese.

Contemporary Trade Patterns

During the twentieth century, trade patterns shifted from colonial economies to entrepreneurial state-led capitalism in Southeast Asia. The economies of China, India, and Japan became increasingly entwined with the global economy. A majority of the world's population lives in Asia, and the continent represents a vast labor pool. Non-Asian investors have invested capital to build manufacturing plants in Asia, where labor is plentiful and relatively inexpensive. By 2020, "Factory Asia" accounted for about 50 percent of global manufacturing output and 30 percent of world exports, with China accounting for half of both.

The world's largest exporter, China has been sending most of its goods to the United States, its

The Port of Singapore is the busiest transshipment and container port in the world, and is an important transportation and shipping hub in Southeast Asia.

largest trading partner. In the twenty-first century, the United States has maintained a large and rapidly growing deficit in its trade with China. In 2018, the deficit stood at US$378.6 billion. That year, the United States shipped 8.4 percent of its overall exports to China, while China shipped 19.2 percent of its overall exports to the United States. The United States and other countries have long maintained that China sets protectionist tariffs against foreign goods, fails to enforce intellectual property rights, and undervalues its currency to gain unfair trade advantages. In 2018, the administration of U.S. President Donald J. Trump started by setting its own protectionist tariffs and other trade barriers on goods from China, Japan, and other countries with which the United States had trade deficits; the affected countries reciprocated in kind.

At the same time, China's rising labor costs and its increasing orientation toward the domestic market positively affected other Asian export-oriented economies. In recent years, the shares of other parts of Asia in the global trade in goods and services have been growing. India now rivals China in electronics exports and exceeds it (and the rest of the world) in the export of information-technology-related services. Bangladesh, Vietnam, and India have established highly successful garment industries, and they are now second only to China as the world's largest garment exporters.

Today, Japan and South Korea maintain their places among the world's most technologically advanced and globally connected societies. Their manufacturing sectors survived the 1991-2001 period of economic stagnation in Japan (known as "Japan's Lost Decade") and the 1997-1998 Asian financial crisis, which hit South Korea hard. In 2018, Japan and South Korea occupied the fourth and the sixth place, respectively, among the world's largest exporters. Both countries export a wide variety of manufactured goods, including electronics, machine parts, and automobiles. South Korean automobile producers now compete with Japanese and American carmakers worldwide.

Trade within Asia had long remained marginal. Economies of neighboring countries were too similar for these countries to offer much to each other. That situation began to change around the turn of the twenty-first century, as Asian countries started to diversify their economies and began trading with each other in intermediate goods such as computer components that are made in one country and assembled in another one. China and advanced Southeast Asian economies began building factories in other Asian countries; wealthy Middle Eastern countries also began investing in the Asian service sectors, including the real estate, finance, wholesale, and retail markets.

Middle Eastern producers long dominated the exports of oil and natural gas via the Organization of the Petroleum Exporting Countries (OPEC), of which they are leading members. In 2019, crude oil from OPEC made up 30 percent of the world oil supply. But as the United States and other major oil-consuming countries gained energy self-sufficiency, Saudi Arabia, the United Arab Emirates, Kuwait, and others started diversifying their economies beyond crude oil exports and embarked on long-term infrastructure projects in their own countries. These included building petrochemical complexes and related transportation networks as well as developing services industries and tourism. Such projects have largely relied on skill and labor from China, India, Indonesia, and other Asian countries.

The former Soviet republics of Central Asia still trade predominantly within the former Soviet bloc, for both historical and geographical reasons. However, they have been steadily increasing exports of oil, gas, and the exchange in other commodities

The Shanghai Stock Exchange building in Shanghai's Lujiazui financial district.

with culturally-related Turkey and Iran as well as with China and the European Union. Trade and investment flows between Africa and Asia and between Latin America and Asia have also increased significantly in recent years. These trends were accompanied by a relative decline of Asian exports to and imports from Europe, largely due to the declining trend in many European economies.

In general, both regional and international trade and exports have been growing in Asia at rates exceeding world averages. This trend will continue into the foreseeable future as Asian economies keep developing and diversifying.

Fred Buchstein

TRANSPORTATION

In 2019 the world's population reached 7.7 billion people. More than half those people—4.4 billion—lived in Asia. They were not spread evenly across the world's largest landmass but concentrated in different places, usually in places capable of growing an abundance of food. Substances needed for other items in daily life, such as iron ore for steel, oil for energy, or timber for construction, are not always located in the same places. They must be acquired from elsewhere, often somewhere far away or difficult to reach. Each of the places where people concentrate, and the places where the resources they need are located, represent either an origin or a destination for movements of people and goods. The number of routes that connect these places depends on how often they are used and how hard it is to get to each particular destination. The 4.4 billion people scattered throughout Asia in about six major centers must contend with the world's highest mountains, vast areas of high, rugged plateau, sprawling desert, and ice-choked tundra separating them from each other and the resources they need.

Trade and Travel Routes

Any area over which hundreds of millions of Asian people live, work, buy, and sell is a trade and travel route itself. Two regions in particular stand out: Eastern China, roughly one-third of China's landmass, and the Indian subcontinent (India, Pakistan, Bangladesh, Nepal, and Bhutan) each contained about 1.7 billion people in 2018. Each of those regions had more than five times the U.S. population in roughly the same space as the United States east of the Mississippi River. Most people live along the seacoasts, but densities are high inland, too, with people spaced evenly across the area. Only in Europe are there as many people packed into an area that size. There are numerous roads and railroad lines throughout these territories. In 2017, for example, India had the second-largest road network in the world (after the U.S.), connecting all major cities and state capitals. Also essential are the many miles of rivers capable of handling larger boats and barges, and thousands of miles of coastline for moving both people and cargo.

Population is concentrated to a lesser extent in a few smaller places. Extensive areas of dense settlement can be found in Korea and Japan; along the Turkish coast; in Transcaucasia; along and near the Tigris and Euphrates rivers in the Middle East; in and around the intermountain Fergana Basin of Central Asia; along navigable rivers in Myanmar, Thailand, and Vietnam; and on certain islands of the Philippines and Indonesia. More options for motorized travel exist in these places. Roads and rail lines are more plentiful, and cities are better connected to other cities within these regions. River and coastal shipping are also viable options for moving people and goods.

Overland routes connect each of these high-population areas with the others, unless, like Indonesia and the Philippines, they are islands. More traffic flows between these places over oceans, unless, as in Central Asia, there is no ocean access, or, as in Asian Russia, the ocean is frozen much of the year. The most important ocean route is that connecting the Persian Gulf to Japan. A variety of bulk raw materials, including coal, metal ore, and grain, moves along this route, although petroleum and petroleum products from the Middle East and Indonesia are the main commodities transported to Japan, with stops along the way.

The most extensive overland connection across Asia is the Trans-Siberian Railway and Road System, traversing Asian Russia from the Ural Mountains to Vladivostok on the Sea of Japan. This system connects the Russian heartland and the rest of

Europe to Russia's only seaport with year-round open-ocean access. It also connects eastern China and Korea to Russia and Europe by rail and road. Russia transports timber, metal ore, oil, coal, grain, and manufactured products that it either extracts from its own Siberian territory or imports from other places, such as Japan, China, and the United States.

The Trans-Siberian Railway system enters Asian Russia from the west, in a series of multiple-track lines and a pipeline, near Yekaterinburg and Chelyabinsk, on the eastern slope of the Urals. The system continues eastward approximately 3,700 miles (6,000 km.) through the Siberian cities of Omsk and Novosibirsk before bending southward to avoid Lake Baikal. The route next hugs the Mongolia-China border to Khabarovsk, then southward to Vladivostok. The full line was electrified in 2002. East of Ozero Baykal, the system links up with the southeast and northeast Chinese railroad network. An extension to Raijin, North Korea, was inaugurated in 2011.

Oil also moves to Europe and Russia from Asia by pipelines and tankers. The longest pipeline extends from the Siberian Ozero Baykal region to Chelyabinsk, then on westward to Nizhniy-Novgorod near Moscow. Several pipelines cross the Arabian Peninsula and Syrian Desert from the Persian Gulf to six different ports on the Red and Mediterranean seas. More than a dozen pipelines are operational in the Caspian Sea basin, moving oil throughout Transcaucasia and to China and beyond. Tankers move oil to Europe across the Mediterranean and Black seas, and from the Persian Gulf around Africa.

Infrastructure

No other railroad crosses Asia like the Russian Trans-Siberian system, nor is the Indian subcontinent connected to Russia or Europe by rail. A number of roads cut through some forbidding territory to connect Asian population clusters, but road-hauling of freight is too expensive for long-distance trade. Therefore, the road network that connects Asia is more important for serving re-mote areas and for short hauls of less than 620 miles (1,000 km.).

The actual amount of roadways and railroad track per square mile is greatest in the Indian subcontinent and Japan. China follows closely behind, but even with many roadway routes and railroads, large parts of its national territory are barely accessible. Numerous roads are available for local traffic throughout Southeast Asia, but countries in that region have few railroads, and even fewer that connect to each other or to other population centers. Consequently, those countries have no suitable long-distance overland freight-hauling connection to other population centers. Any bulk material that is exported from Southeast Asia usually moves to coastal ports down one of four main river systems, then is shipped out via ocean transport.

Coastal shipping of cargo is important in Southeast Asia and successfully competes for freight with overland carriers in countries such as India, China, and Russia, which have extensive coastlines. India has eight major ports along its national shore, and China has three. Thousands of smaller boats move cargo between hundreds of smaller ports. Coastal shipping is an important part of Russia's internal transportation system, when it is possible to move things that way. Russia's most useful coast is at or near the Arctic Circle and iced-in for much of the year.

Most Middle Eastern overland traffic bound for Europe moves through Turkey. Rail and numerous road connections extend from Abadan, Iran, at the mouth of the Tigris and Euphrates rivers, through Turkey's Bosporus, the narrow opening to the Black Sea. Southeast Asia, and the more densely settled parts of the Middle East, have about the same number of roads per square mile available for local traffic movement. The countries of Central Asia, having supplied cotton, textiles, and oil to the former Soviet Union before the 1990s, are well-connected to Russia by rail. The Central Asian countries are served internally by a much sparser road network and are connected to other Asian population centers only over long distances by road.

The Asian air corridor runs almost exclusively between Japan and Turkey along the southern rim of the Asian continent, from Istanbul to Riyadh to Mumbai (Bombay), then Bangkok, Hong Kong, Tokyo, and points in between. Traffic also flies to Europe from Japan and Beijing by way of Moscow, but it more frequently follows the southern corridor with more stops. In 2020, Asia had thirty national airlines and the second, fourth, and fifth busiest airports in the world: Beijing, Tokyo, and Dubai.

Navigable Rivers

There are approximately two dozen river systems throughout Asia large enough to handle ships, freighters, or barges. Cargo and people move along these river corridors rather easily, although rivers carry less traffic now than before railroads were built. Several Asian rivers can accommodate large ships for a considerable distance inland. The most heavily used are classified as inland waterways.

China and Asian Russia far outdistance other Asian countries in miles of inland waterway, and each has about 2.5 times as many miles of navigable river as does the United States. The five great river systems that drain the Siberian Far East—the Ob, Yenisei, Lena, Kolyma, and Argun—are each more than 1,000 miles (1,600 km.) long. The Ob is the sixth longest in the world at 3,360 miles (5,400 km.). Each flows through vast territories that contain immense stores of natural resources, but do not move nearly as much cargo as the Trans-Siberian Railway. These river basins lie almost entirely north of 50° north latitude, so they are frozen much of the year. Moreover, the rivers flow northward to the Arctic Ocean, which is also frozen much of the year. Ocean-going tankers are virtually useless, even with Russia's large fleet of ice-breakers.

China has the third- and fifth-longest rivers in the world. The Yangtze River is navigable for most of its 3,900 miles (6,275 km.)—a few hundred miles longer than the Mississippi-Missouri river system in the United States—and tracks down the middle of China's densely populated southern territory. The Yellow (Huang He) River is navigable for much of its 3,350 miles (5,400 km.), flowing through a vast area north of the Yangtze valley. The area between the Yangtze and Yellow rivers, near the eastern Chinese coast, is connected by a human-made naviga-

ble river system. China's Da Yunhe, or Grand Canal, connects Shanghai to Tianjin. It crosses both major river systems and even briefly follows an older abandoned course of the Yellow River.

In India, the Ganges River system forms a broad, heavily used, densely settled lowland. Millions of people use the Ganges system's numerous tributaries to move goods throughout the region. The Indian subcontinent has six other navigable river systems. The Brahmaputra flows around the Himalayas to the east and into northeastern India before connecting with the Ganges. The Indus River in Pakistan runs the length of the country from Islamabad to Hyderabad before emptying into the Arabian Sea just southeast of Karachi. Many of its larger tributaries start near where the Ganges starts, but flow in the other direction. The Mahanadi, Krishna, Godavari, and Narmada rivers drain different parts of central India.

In Southeast Asia, four south-flowing rivers are significant transportation routes. The Irrawaddy River is westernmost among these, running through the total length of central Myanmar and emptying into the Andaman Sea near Yangon (Rangoon). The Salween River is next, to the east, originating in the Tibetan Plateau near the headwaters of China's Yangtze. The Salween forms part of the Myanmar-Thailand border. Thailand's Chao Phraya is shorter than the others, but more people move along it. Bangkok, with 8.3 million people, is located at Chao Phraya's mouth and is the focus of settlement along its densely settled lower course. The easternmost of the four rivers, the Mekong, starts near the Salween and Yangtze rivers in the Tibetan Plateau, making its way through rugged territory before forming a broad lowland in Cambodia and Vietnam.

The Middle East contains vast deserts and rugged mountains and plateaus. Between these areas lies a twin river system that has served the transportation needs of the region throughout most of history. The Tigris and Euphrates rivers originate in the mountains of eastern Turkey and flow southeast to the Persian Gulf. These twin river systems parallel each other through most of Iraq, usually separated by less than 125 miles (200 km.) before reaching the Persian Gulf.

Natural Barriers

Several physical features make moving from place to place in Asia difficult. Extensive regions of tundra and permanently frozen ground, numerous mountain ranges—including the highest in the world—and vast expanses of desert minimize the number of routes crossing certain territories. They also redirect other crossing routes to places where slope is less severe, or more water is available along the route—in other words, where traveling is easier.

The tundra of the Russian Far East is the most impenetrable barrier to transportation. The west Siberian lowland, an expanse of spring and summer marshland bigger than Germany, and the high, rugged east Siberian plateau, starting just east of the Yenisei River, both present problems for the construction of routes. These formidable barriers are less difficult to overcome, however, than the permanently frozen ground (permafrost) that creates enormous engineering challenges, some still insurmountable.

Mountain barriers are found all over the Asian landmass. The most challenging of these are the Himalayas at the northern edge of the Indian subcontinent and the numerous chains that extend northward, with elevations up to a world record of 29,035 feet (8,850 meters) at Mount Everest. Elevations greater than 20,000 feet (6,100 meters) are common. The high, vast Tibetan Plateau, with average elevations generally exceeding 15,000 feet (4,570 meters), forms a similarly impenetrable extension of these lofty mountains. These ranges isolate the Indian subcontinent from western China and the Central Asian countries of Turkmenistan, Uzbekistan, Tajikistan, and Kyrgyzstan. The Himalayas are so forbidding to traffic that only two roads cross them, one through Nepal, the other through the Indian state of Sikkim. The few routes from the Indian subcontinent northward are mostly funneled through the slightly less dramatic Hindu Kush Mountains of Afghanistan.

The most populated areas of eastern China are also isolated from Central Asia, the most direct route to Russia. Between China and Central Asia lie not only the Altai and Tian Shan mountain ranges, but the Gobi and Taklamakan deserts of Mongolia and Xinjiang Uyghur provinces in China. Few roads cross the 1,500-mile (2,400-km.) stretch of rugged dryland territory that lies between eastern China and Kazakhstan; even fewer roads cross the additional 1,200 miles (1,930 km.) from Kazakhstan's eastern border to the Ural Mountains.

Throughout the rest of Asia, numerous lower mountain ranges make motorized travel difficult, but they do not reduce travel to the same extent as do the Himalayas. The Caucasus Range, between the Black and Caspian seas, makes travelers hug the coast of either inland sea to get from Armenia to Russia. Similarly, the Plateau of Iran offers travelers between the Tigris-Euphrates and Indus-Ganges River valleys few route options. In Southeast Asia and Southwest China, the high Yunnan Plateau slows travel between the Indian subcontinent and China. Southward-extending mountain ranges (Annamese, Arakan, and others) reduce traffic within the region in general.

Deserts are numerous throughout Asia and almost always seriously limit movement across them. The Gobi Desert, Asia's largest, severely restricts movement within Mongolia and northward out of China.

INFLUENCE OF A COUNTRY'S SHAPE ON ITS TRANSPORTATION SYSTEMS

The countries of Vietnam, Myanmar, Thailand, Korea, and Malaysia—like Chile in South America—must contend with their country's elongated shapes. Their national territories, or parts of them, are several times longer than they are wide. The elongated shape of these nations restricts route options available for traversing their lengths. Malaysia, Indonesia, the Philippines, and, to a lesser extent, Japan are all made up of many islands. This fragmented condition requires a combination of boats and trucks or trains to serve the transportation needs of all their inhabitants. A fragmented transportation system adds to the cost of moving things around, because they often must be loaded and unloaded several times, from truck to boat to train, before they reach their final destination.

The fastest train service measured by peak operational speed is the Shanghai Maglev Train which can reach 431 km/h (268 mph).

The Taklamakan of western China restricts transit into Central Asia from China. The Kyzylkum and Karakum deserts of Uzbekistan and Turkmenistan, an area about the size of France, are crossed by only two roads and three railroads. The Dasht-e Kavir Desert in northeast Iran and western Afghanistan, about the size of Texas, is crossed by only a sparse network of a few major roads, with no railroad penetration. The Arabian Peninsula, larger than Iran, is almost all desert. Saudi Arabia has invested in nearly 2,500 miles (4,022 km.) of paved expressways, specially built to resist the year-round high temperatures. Also up and running is a relatively complex train network connecting the centrally located capital of Riyahd with various port cities on the Persian Gulf and the Red Sea, and extending to points in neighboring Jordan and Bahrain. High-speed passenger service between Mecca and Medina was inaugurated in 2018. Train lines are proposed to improve connectivity with the other Gulf states. Infrastructure is lacking in the southern half of this peninsula, where in some places people and goods move only along caravan trails.

James Knotwell and Denise Knotwell

DISCUSSION QUESTIONS:
TRANSPORTATION

Q1. When was the Trans-Siberian Railway built? What distances does it cover? What impacts has it likely had on Siberia and other areas in far eastern Asia? What makes the Trans-Siberian Railway unique compared to the other rail lines in Asia?

Q2. How have the mountain ranges of Asia affected the building of transportation infrastructure in Asia? What impact have topographical features had on the ease or the difficulty with which connections among countries have been possible? How has aviation helped meet Asia's transportation challenges?

Q3. What role do the rivers of Asia play in the transportation of goods and people? Why does so much of the trade among Asian countries take place via rivers, lakes, or oceans? How are Asian goods transported to markets in Europe?

COMMUNICATIONS

For most of human history, communication and transportation have amounted to practically the same thing. The physical presence of a person was necessary to deliver a message. Famously, the first marathon runner was dispatched in 490 BCE to announce the Greek victory over the Persians at the Battle of Marathon and then died after running about 25 miles (40 km.) to Athens. People still run marathon races but not in order to communicate messages.

Early History

Speech is the most important form of human communication. Scientists disagree on how, why, and when our ancestors developed sophisticated vocal communication, but they agree that it was one of the main reasons why our human ancestors were able to progress from scattered tribes to builders of civilizations. After speech, writing was the next system of communication invented. The early history of writing, from approximately 3000 to 1000 BCE, unfolded in Asia.

The earliest known examples of writing were discovered by archaeologists in ancient Sumer, in the part of Southwest Asia that is now southern Iraq. Evidence suggests that it developed as a way for people to keep a record of trades in goods by making marks on clay containers or tablets. These marks

The Diamond Sutra, printed in 868, is the world's first widely printed book to include a specific date of printing.

later evolved into pictograms (pictures depicting things that were traded or counted) and still later, into logograms (symbols representing words). Scientists now believe that writing did not spread throughout the world from Sumer but that the hieroglyphic writing systems of ancient Egypt and China were developed independently around the same time.

The phonetic alphabet, was the next major development in the history of communication. It emerged in the ancient Near East and was developed by Semitic-speaking peoples who inhabited the Mediterranean coast of present-day Syria, Israel, and Lebanon between 1600 and 800 BCE. While the first systems of writing represented things, actions, or ideas by using a separate picture or symbol for each of them, the phonetic alphabet only represented the consonant sounds of human speech. Because it required far fewer symbols, more people could memorize the basics of this system. It vastly increased communication because more people could use it.

This writing system became known as the Phoenician alphabet. The Greeks adapted it to include vowel sounds, and all European alphabets are its direct descendants through the Greek alphabet, as are the many languages of Central and Southeast Asia that are based on the Latin and Cyrillic scripts. The Arabic and Hebrew writing systems are also descendants of the Phoenician alphabet through its other adaptation, the Aramaic. Modern Chinese and Japanese writing systems still use characters developed in ancient China.

Postal Service

The first advanced systems for delivering written messages over distance were also developed in Asia. Although some evidence exists of organized mail in Egypt as early as 2000 BCE, the system of relay stations, or staging posts, where horse-riding messengers would get fresh horses—similar to the Pony Express in the nineteenth century American West—was first developed to a high level under the Mongol emperors of the Chou Dynasty in China about 1000 BCE.

Relay delivery systems existed in Persia about 600 BCE and in Byzantium. They were later absorbed into the Arabian Empire based in Baghdad. A post station system was developed with a 2,000-mile (3,200 km.) network under the Mogul emperor Akbar in India in the sixteenth century CE. However, it collapsed in a subsequent period of political instability. The Chinese postal system was described by the Italian adventurer Marco Polo, after his visit in the thirteenth century. It flourished through the Middle Ages. Systems of this type also existed in. other Eastern and Western empires and were important governmental institutions. These systems, however, were used to deliver only official correspondence.

Postal communication services for the general population began to develop in the modern period when literacy became more widespread and new transportation means were introduced. Initially, such postal services were decentralized, limited, and unregulated. The next major improvement in postal service, making it available to everyone, did not occur until the middle of the nineteenth century. The transformation of mail delivery was made possible by a crucial report on the subject submitted to the British government by the colonial official Rowland Hill. Like most important innovations, the Hill reforms appear obvious in retrospect. He suggested three basic changes: a standard charge for each piece of mail within the particular jurisdiction involved, a system of prepayment by postage stamp, and introduction of official pre-printed envelopes.

The Hill reforms were instituted in Great Britain in 1837 and spread quickly to India and elsewhere in the British Empire. Inhospitable terrain and local social problems aside, mail delivery became available to the general public in Asia within decades. But every pair of countries that exchanged mail had to negotiate a postal treaty with each other. This changed when the General Postal Union was formed in 1875. Most Asian nations were already members when China joined the organization, by that time called the Universal Postal Union (UPU), in 1914. The UPU regulates international postal service worldwide. After the foundation of the United Nations, it became a specialized agency of the UN.

Japan was the leader in Asia in developing the physical infrastructure for mail delivery. Japan had been closed to all aspects of Western culture for more than 250 years, but the Meiji rulers began to pursue Western technology when they came to power in 1868. Postal service was instituted in 1871, and railroad, truck, and air freight delivery systems followed. Japan soon eclipsed the head start previously gained in those parts of Asia touched by the British Empire. Russia and China underwent major social revolutions in 1917 and 1949, respectively, before creating nationwide postal services for the general public.

At the beginning of the twenty-first century, physical communications sent by mail were eclipsed by email, instant messaging, texting, and social media. Almost all text-based correspondence is now conducted online. At the same time, Internet shopping opened new opportunities for postal services, as items bought online are delivered by post and courier delivery services.

The Telegraph

The first practical telecommunication system, invented in France in 1792 by Claude Chappe, was optical telegraph, or a semaphore system. It consisted of a series of towers arranged across the countryside, each in the line of sight of the next. Individuals in the towers relayed information in code by moving their arms and flags and blinking shutters. By 1840 there were about 1,000 of these towers in service. Most were in Europe, but a network was also operated by the British in India. Optical telegraph systems were used until the 1850s, when they were replaced by electrical telegraph systems.

This type of the telegraph became possible after engineers and inventors gained a better understanding of the nature of electricity. An early system based on a galvanometer was developed by a Russian diplomat of German origin, Baron Pavel Schilling, in the 1820s. After he demonstrated the system for Russia's Czar Nicholas I in 1836, an operational line was planned, but Schilling died

soon after, and no one carried his work forward. The eccentric American artist and inventor, Samuel Morse, and his assistant, Alfred Vail, solved the problem of maintaining signal strength through a metal wire over distance, which made long-distance communications commercially viable.

Crucial to communicating by telegraph was the character encoding system, also invented by Morse and Vail, rendering the letters of the Roman alphabet as dots and dashes. Known as Morse Code, it was adopted internationally, and the capacity to send and receive it was required on all oceangoing ships until 1999. However, it worked poorly for written languages that used diacritical marks or languages that are ideographic, like many in Asia. Therefore, different Morse-type codes have been developed for Arabic, Burmese, Chinese, Hebrew, Japanese, Korean, Russian, and Turkish.

The telegraph spread quickly throughout Asia, reaching Russia via the Black Sea in 1853 and the British colonies of India and Hong Kong between 1864 and 1870. The progressive Meiji government of Japan, determined to modernize the country, hired a British engineer in 1869 to design the country's first telegraph line. By 1891, Japan had thousands of miles of telegraph lines.

The Chinese government initially resisted telegraphy, which it considered a tool of Western interference. Only by the end of the 1870s did it relent to pressure from Russia, Britain, and France to grant them concessions for building telegraph lines. Telegraph would dominate long-distance communication in Asia, as in the rest of the world, until the rise of the telephone.

The Press

The printing press was another Western technological innovation that depended on the phonetic alphabet. After the invention of the printing press by Johannes Gutenberg in the 1440s in Germany, innumerable copies of any given text could be produced. Knowledge and ideas could be circulated to a much wider audience. Modern newspapers, made possible by the combination of the printing press and the telegraph, developed in Europe, from where newspaper publishing spread throughout the world.

Many of the first newspapers in Asian countries were published in English or other European languages. Asia's oldest newspaper was established in India, which was then a British colony. Founded in 1822, it was published in Gujarati and English.

Once established, newspapers and magazines in Asia flourished. Even in the first decades of the twenty-first century, when the print media in Western countries were in a steep decline under the onslaught of the electronic media, subscriptions to newspapers and magazines in Asia continued to grow. Some of the results of Asia's fast economic growth over the preceding decades were higher literacy rates and higher incomes, which created a whole new readership for the printed media and drove up paid subscriptions. In 2017, some seventy of the world's 100 largest daily newspapers were based in Asia.

Radio and Television Broadcasting

Asia's large and ethnically diverse populations differ in their cultural backgrounds and the languages through which they communicate. However, radio and television provide some common links. They are the main mass media in all countries of the region.

Besides their own programming, most Asian countries also relay radio programming in the border areas from neighboring countries and television programming from other Asian countries and international broadcasters. Satellite television is widely available; by 2020, Myanmar remained one of the last underdeveloped telecommunications markets in Asia with limited access to satellite television. In China, however, foreign-made TV programs must be approved prior to broadcast. Only in North Korea does the government completely prohibit listening to foreign radio broadcasts and jams them; television sets are pre-tuned to government stations.

The extent of government control over broadcast media varies throughout Asia. In North Korea, China, Vietnam, most Central Asian republics, Nepal, and Myanmar, all national broadcasting is owned by or affiliated with the government. In Singapore, four of the country's six terrestrial TV stations are government-owned, and the remaining two are owned by the military. On the other end of the spectrum are South Korea and Japan, where broadcasting is largely privately owned and independent of the government. In Pakistan, the state-owned broadcaster, Pakistan Television Corporation, enjoyed a monopoly on broadcasting until the early 2000s. The end of the state monopoly prompted a boom in private TV broadcasting and electronic media, which acquired substantial political influence in Pakistan through their own news-

The Ministry of Communications and Information in Singapore oversees the development of Infocomm, media and the arts.

casts and commentary. The influx of foreign TV programming, however, has raised widespread concerns over "cultural invasion" of the country, especially with respect to programs from India and the Middle East.

Public broadcasting in Asian countries is funded by their respective governments, either directly or—in Japan and South Korea—through license fees paid by receivers of broadcasts. The Chinese public broadcaster, China Central Television, is a government agency. It is part of the "Big Three" governmental media organizations, along with Xinhua news agency and the *Renmin Ribao* (People's Daily) newspaper, the official organ of the ruling Communist Party. India's public service broadcaster, Doordarshan, has a monopoly on terrestrial television broadcasting and operates about twenty national, regional, and local services. (Privately owned TV stations in India are distributed by cable or via satellite services.) All India Radio, the national public radio broadcaster, is the largest radio network in the world. In 2019, it operated some 420 radio stations and offered programming in twenty-three languages and some 180 dialects.

Censorship

Communications in Asia have been more tightly controlled than those in the West in several ways. First, traditional societies generally tolerate little opposition or nonconformity, and in Asia, many societies still bear a large imprint of traditional values and customs.

Second, although the Marxist ideology of the former Soviet Union and the People's Republic of China was theoretically based on scientific socialism, the reality has been different. The objective pursuit of truth has frequently taken a back seat to the dictates of a political elite. The Soviet Union collapsed around 1990 under the weight of this contradiction, but a new political elite soon emerged in post-Soviet Russia and reestablished government control over communication media through the selective implementation of defamation laws, media restrictions, and harassment of opposition activists. China has been thriving economically since the late 1970s by granting freedom of

expression in economic matters, but in the first decades of the third millennium, political communication there was still tightly controlled by the Communist Party. Communication media were also heavily censored in other Communist and post-communist countries—North Korea, Vietnam, and the former Soviet republics of Central Asia.

Finally, communication of information and ideas in much of the Middle East is subject to pressure from fundamentalist Islam. Separation of church and state is alien to fundamentalist Islamic ideology, as is debating moral or political issues. The truth is understood to be contained in the Qur'an (Koran), and any deviation from its fundamentalist interpretation is a crime. The holiest Muslim shrines are located in Saudi Arabia, which bases its entire civil and criminal code on the Qur'an. Foreign magazines, television programs, and other media are censored there, as are social media and foreign movies in Iran, where the government itself is controlled by the Islamic clergy.

Israel too must deal with the pressure of fundamentalism. Since its founding in 1948, Israel has been in a state of war with its Arab neighbors, interrupted by long periods of uneasy truce. This military situation creates intense pressure to repress freedom of expression for the sake of national security; the media in Israel are subject to military censorship. In addition, Jewish fundamentalism is similar to Islamic fundamentalism in its willingness to suppress deviant views. It constitutes a powerful lobby in Israel. Despite these two potential sources of repression, however, Israel maintains a pluralistic media environment and a climate of lively political debate, which distinguishes it from most of its neighbors.

Telephone Technology

The telephone was a result of successive improvements of the electrical telegraph, done by many people; thus, credit for the invention of the telephone is frequently disputed. Alexander Graham Bell of Scotland is most often cited as the inventor of the telephone because he was the first to successfully patent it in 1876. Within just one year, tele-

The Japanese Experiment Module (Kibō) at the International Space Station.

phone service was introduced in many countries, including Japan. In the 1980s, technologically advanced Southeast Asian nations were among the first to install fiber-optic cable lines in metropolitan areas. Since it requires extensive landline infrastructure, fixed-line telephone service was much slower to reach remote, mountainous regions of Asia. In Bhutan, for instance, the first telephone was installed only in 1963.

It was the necessity for landline infrastructure that caused the downfall of the fixed-line telephone in the twenty-first century. Mobile telephone technology did not descend from the traditional telephone but rather from the radio. It traces back to two-way radios that were permanently installed in taxis, police and military vehicles, and trains. Like the radios of the time, the earliest mobile phones received their signals from single powerful transmitters with a limited radius and a fixed number of channels. The development of low-powered signal transmission through multiple base stations, or cells, truly revolutionized long-distance communi-

cation. The principles of cellular technology and related microelectronic components were developed in the United States, but it was in Japan and the Scandinavian countries where the first cellular systems were built. For some time, these systems were mainly installed in cars. In 1973, ninety-seven years after Bell demonstrated his telephone, American engineer Martin Cooper unveiled his handheld cellular mobile phone.

While it took over ninety years for landline phones to reach 100 million consumers, it took cell phones just seventeen years to reach the same number. China imported its first mobile phone telecommunication facilities in 1987, and fourteen years later, in 2001, the world's most populous country also had the largest number of mobile phone subscribers in the world. In 2018, the cellular phone penetration rate, that is, the number of active mobile phones per 100 people, was the world's-highest in Chinese-administered Hong Kong. The United Arab Emirates, Thailand, Kuwait, Vietnam, Singapore, and Kazakhstan were also on the top of the list. The world's largest manufacturers of cellular phones were China, India, Japan, and South Korea.

The Internet and Social Media

Unlike traditional telephones, mobile phones are used not only—and even not predominantly—for verbal communication. Connected to the Internet, smartphones are used for getting news and information, online shopping and banking, entertainment, and communicating via social media. Most mobile phones—as well as components for desktops, laptops, and tablets—are manufactured in Asia. Ironically, in many Asian countries, the cost of these devices and that of Internet services place mobile phones out of reach for large segments of the population. High illiteracy levels in rural regions also prevent millions of people from using the Internet. In prosperous Asian countries such as Japan, South Korea, Singapore, Israel, Saudi Arabia, Qatar, Kuwait, and the United Arab Emirates, Internet-connected computers and smartphones are household items. In 2019, the Internet penetration in these countries was close to 100 percent of the population, while for the Asian region as a whole it was only around 50 percent—lower than the world average.

Moreover, the Internet and social media usage in Asia depends on the socio-political situation in each country. In North Korea, the population does not have access to the Internet or social media. Many Internet and social media sites and services have been blocked in China, both for reasons of political control and in order to grant advantages to Chinese Internet enterprises and social media providers. In the Central Asian countries of Tajikistan, Turkmenistan, and Uzbekistan, state control of Internet communications is strict as well, and Internet access, in general, is patchy due to under-developed infrastructure. In many other Asian countries, even the ones with the highest Internet usage, self-censorship on the Internet is a common practice. Wary of potential repercussions, both the journalists and social media users largely refrain from posting or endorsing dissenting views.

Conclusion

A number of countries in Asia are among the world leaders in the implementation of telecommunication technologies. Various communication systems and industries are well developed there, from radio and television broadcasting to electronics manufacturing. At the same time, communication media in much of the region are government-controlled. Many Asian countries score very low on the Press Freedom Index, a widely used ranking of countries compiled annually by the international organization Reporters Without Borders. It lists countries according to media independence, self-censorship, legislative framework, and the freedom of flow of information on the Internet. In recent decades, countries such as North Korea, China, Vietnam, Saudi Arabia, Iran, Iraq, Egypt, Turkey, Singapore, Bangladesh, and Central Asian republics have been on the very bottom of the list.

Steven Lehman

GAZETTEER

Places whose names are printed in SMALL CAPS *are subjects of their own entries in this gazetteer.*

Abkhazia. Republic in northwestern GEORGIA on the BLACK SEA coast. Total area of 3,343 square miles (8,660 sq. km.) with a 2015 population of 243,206. Capital is SOKHUMI. It became an autonomous republic within Georgia in 1930 but declared its independence in 1993. Agriculture, particularly tobacco, tea, silk, and fruit, is the predominant economic activity. Also known as the Abkhaz Republic.

Absheron Peninsula. Peninsula in AZERBAIJAN that extends 37 miles (60 km.) into the CASPIAN SEA and reaches a maximum width of 19 miles (30 km.). The national capital of Azerbaijan, BAKU, lies on the peninsula's southern coast.

Aden, Gulf of. Arm of the ARABIAN SEA situated between the countries of YEMEN and Somalia, commanding the southern entrance to the RED SEA in the MIDDLE EAST. Yemen's main seaport, Aden stands near the gulf's connection with the Red Sea.

Aegean Sea. Arm of the MEDITERRANEAN SEA situated between western TURKEY and the peninsular part of Greece. Covers 80,000 square miles (207,200 sq. km.). Contains more than 2,000 mostly uninhabited islands.

Afghanistan. Landlocked country in SOUTH ASIA. Total area of 253,000 square miles (655,270 sq. km.) with a 2018 population of nearly 35 million, predominantly Muslim. Capital is KABUL. Made up of dry mountains, plateaus, and basins. Has been a crossroad for eastern, southern, southwestern, and central Asia and attracted many invaders over the centuries. Nearly 80 percent of Afghans engage in agriculture, but most production is of a subsistence nature and Afghanistan must import food. Major export items include natural gas, hides, dried fruits, and cotton. The U.S. and its allies invaded Afghanistan following the September 11, 2001, terrorist attacks, overthrowing the fundamentalist Taliban regime. (The Taliban had granted sanctuary to the attack's mastermind, Osama bin Laden.) U.S. and allied troops in varying numbers have been stationed in Afghanistan ever since, and the Taliban remains active.

Agra. City in SOUTH ASIA widely known for the Taj Mahal, the royal mausoleum built by the fifth Mogul emperor Shah Jahan in memory of his wife, Mumtaz Mahal. Situated 125 miles (200 km.) southeast of New DELHI on the right bank of the Yamuna River in the Indian state of UTTAR PRADESH. Established in 1566 by Akbar; was a Mogul capital until 1658. Population was 1,585,704 in 2011.

Ajmer. Most sacred Muslim pilgrimage site in INDIA. Population was 542,321 in 2011. Founded in the twelfth century in the state of Rajasthan, it contains the tomb of the Muslim saint Moinuddin Chishti, a palace of Akbar, and Jain temple. Formerly a Mogul military base.

Al-Akhdar Mountains. Crescent-shaped range of mountains in the MIDDLE EAST. Located in OMAN, the southeastern corner of the ARABIAN PENINSULA. Maximum elevation is 9,997 feet (3,047 meters). The northern end of the range overlooks the Strait of HORMUZ.

Alania. Republic in southwestern Russia, on the northern flank of the GREATER CAUCASUS range. Also known as North Ossetia, Alania is bordered on the south by GEORGIA and on the north by the Sunzha and Terek ranges. Total area of 3,100 square miles (8,000 sq. km.) with a 2009 population of about 700,000. Capital is Vladikavkaz. The majority of its inhabitants are Ossetians who speak an Indo-Iranian language.

Aligarh. City in UTTAR PRADESH, INDIA, which is an important religious center to Muslims. Population was 911,223 in 2011. The Aligarh Muslim University, founded there in 1875, is the leading educational institution for Indian Muslims.

Amnok-kang River. See YALU RIVER.

Amritsar. City in the Punjab state of INDIA. Had a population of 1,183,549 in 2011. Site of the Golden Temple and the most sacred religious center of the Sikhs. Modern Sikh nationalism was founded there. Was center of a Sikh empire in the early nineteenth century.

Amu Darya River. River formed by the joining of the Vakhsh and Panj (Pyandzh) rivers. It flows

879 miles (1,415 km.) west-northwest to the southern shore of the ARAL SEA. In its upper course, it forms part of the border between AFGHANISTAN and TAJIKISTAN, UZBEKISTAN, and TURKMENISTAN; in its lower course, between Uzbekistan and Turkmenistan. Ancient name was Oxus.

Anatolian Plateau. Interior of west central TURKEY in the MIDDLE EAST. Surrounded by slightly higher mountains that catch most of the moisture in the prevailing winds, causing the plateau to be dry and have a hot steppe climate. ANKARA is located there.

Andaman and Nicobar Islands. Union territory of INDIA, located in the Bay of BENGAL. Total area of the territory is 3,185 square miles (8,250 sq. km.); had a population of 380,581 in 2011. Tamil and Bengali settlers from the Indian mainland account for most of the inhabitants. The islands have become a tourist center, with regular air and sea transportation connecting Port Blair, the capital, with KOLKATA and CHENNAI. A tsunami in 2004 killed some 7,000 people.

Andaman Sea. Body of water in SOUTHEAST ASIA; joined to the SOUTH CHINA SEA by the Strait of Malacca. Covers 308,000 square miles (798,000 sq. km.). Fed by the IRRAWADDY, Sittang, and SALWEEN rivers.

Angkor. Capital of the ancient Khmer kingdom of CAMBODIA. Noted for its ancient temples, particularly Angkor Wat, which was built in the early twelfth century and abandoned in 1443. Today, unchecked tourism is stressing Angkor's delicate archaeological remains.

Anhui. Province in southern CHINA. Total area of 54,000 square miles (139,900 sq. km.), with a 2010 population of 559,500,468. The capital city, HEFEI, had a 2000 population of 3 million. Northern Anhui is in the Huai River basin and has a temperate monsoon climate, causing flooding, drought, and salinity problems. The south is hilly and mountainous, and lies in the subtropical zone along the YANGTZE (Chang Jiang) River. Central Anhui is low-lying with numerous lakes.

Ankara. Capital of TURKEY in the MIDDLE EAST. Population in 2018 was 5.4 million. Established by former president Kemal Ataturk, who placed it eastward in the interior of the country to signify his government's focus away from the populated ISTANBUL and coastal areas.

Annam Highlands. Part of a region in VIETNAM called "Annam" ("pacified South" from ancient local history). Located south of the RED RIVER Delta, they form the backbone of the country. Also known as Truong Son.

An'p'ing. Town, seaport, and district of T'AI-NAN in southwestern TAIWAN. The oldest Chinese settlement in southern Taiwan.

Anyang. City in KYONGGI province in the northwestern area of South KOREA. Population was 595,644 in 2016. About 19 miles (31 km.) southwest of SEOUL, and its largest industrial satellite. Industries include the manufacture of textiles, pottery, paper, and bricks; also has KOREA's largest motion picture studios. Two ancient temples, Yombul-am and Jungcho-sa, built in the ninth century, are found there.

Aqabah, Gulf of. Narrow and deep extension of water at the northern end of the RED SEA in the MIDDLE EAST. Part of the JORDAN RIFT VALLEY, with depths greater than 5,900 feet (1,800 meters). Located between the SINAI PENINSULA and the northwestern corner of SAUDI ARABIA. At its extreme northern end are the neighboring port and resort towns of Elat in ISRAEL and Aqabah in JORDAN.

Arabian Desert. Extension of the Sahara Desert that crosses the ARABIAN PENINSULA. Comprises three desert regions: the RUB' AL KHALI in its southeast corner, the NAJD in its center, and the SYRIAN DESERT in the north.

Arabian Gulf. See PERSIAN GULF.

Arabian Peninsula. Large projection of the southwest Asian landmass that separates SOUTH ASIA from North Africa, of whose exceptionally arid Sahara Desert it is an extension. The approximately 1-million-square-mile (2.6-million-sq.-km.) peninsula is bordered by the RED SEA on the west, the PERSIAN GULF on the east, and the Gulf of ADEN and ARABIAN SEA on the south. The peninsula and its offshore islands contain SAUDI ARABIA, YEMEN, OMAN, the UNITED ARAB EMIRATES, QATAR, BAHRAIN, and KUWAIT.

Arabian Sea. Northwestern part of the INDIAN OCEAN, located between the ARABIAN PENINSULA and the subcontinent of INDIA.

Arafura Sea. Shallow arm of the Pacific Ocean in eastern INDONESIA. When its floor was above sea level, it formed a land bridge between Australia and Southeast Asia. Torres Strait on the east connects it to the Coral Sea.

Arakan Mountains. Major range in MYANMAR that separates it from INDIA.

Arakan Yoma. Mountain chain in west MYANMAR. Divides the coast from the interior (Upper Myanmar). Most of its mountains are elevated between 3,000 and 5,000 feet (915 and 1,525 meters); Mount Victoria has the highest peak at 10,016 feet (3,053 meters). Crystalline rock core surrounded by sedimentary rock. Evergreen and bamboo forests are found on the coastal slope; teak forests are found in the east. Annual rainfall is higher along the coast (200 inches/5,100 millimeters) than in the east. Wildlife includes bears, leopards, tigers, and elephants.

Aral Sea. Saltwater lake straddling the boundary between KAZAKHSTAN to the north and UZBEKISTAN to the south. Once the world's fourth-largest body of inland water. Because of the diversion of the AMU DARYA and SYR DARYA rivers as part of Soviet-era irrigation projects, the Aral shrank to a fraction of its original size. Restoration programs by UNESCO and the World Bank have had only modest success.

Ararat, Mount. Ancient volcanic mountain in western TURKEY. Reaches a height of 16,948 feet (5,166 meters). In the Bible, the landing place of Noah's Ark.

Aravalli Range. Oldest mountain range in SOUTH ASIA. Resulted from mountain-building activity during the Precambrian period. Runs 497 miles (800 km.) northeast to southwest from DELHI to GUJARAT, India. Highest point is Mount Abu (3,799 feet/1,158 meters), where a sacred Jain temple is located.

Armenia. Country of the Transcaucasia region, formerly a republic of the Soviet Union. Total area of 11,500 square miles (29,800 sq. km.) with a 2018 estimated population of 3 million. Capital is YEREVAN. Old Armenia extended over what is now the northeastern part of TURKEY and the Republic of Armenia.

Asir Mountains. Extension of the HEJAZ MOUNTAINS in the MIDDLE EAST. In YEMEN, they rise to 10,000 to 12,000 feet (3,000 to 3,600 meters). At those altitudes, the climate changes from desert and steppe to dry winter subtropical.

Aso, Mount. Volcano in central KYŪSHŪ, Japan; the central feature of Aso-Kuju National Park. Its elevation is 5,223 feet (1,592 meters). It has the largest active crater in the world, measuring 71 miles (114 km.) in circumference.

Assam. State in northeastern INDIA. Covers 30,285 square miles (78,438 sq. km.), with a population of 31.2 million in 2011. Part of the ASSAM-BURMA RANGES, an eastern subdivision of the HIMALAYAS. Contains low hills and the floodplain of the BRAHMAPUTRA RIVER. Rice is the main crop of the plain; tea is grown on the surrounding hill slopes.

Assam-Burma Ranges. Mountain ranges in SOUTH ASIA. Oriented in a north-south direction along the INDIA-MYANMAR border, they consist of several hill systems, such as the Naga and Lushai, which occasionally exceed 6,890 feet (2,100 meters). Covered with thick forests; mostly settled by indigenous peoples.

Mount Ararat, as seen from Nakhchivan.

Ayeyarwady River. See IRRAWADDY RIVER.

Azad Kashmir. See JAMMU AND KASHMIR.

Azerbaijan. Country of the eastern Transcaucasia region, formerly a republic of the Soviet Union. Total area of 33,436 square miles (86,600 sq. km.) with a 2018 estimated population of 10 million. Capital is BAKU. Its area includes the disputed enclave of NAGORNO-KARABAKH and the geographically detached region of NAKHICHEVAN.

Baghdad. Capital and largest city of IRAQ, in the MIDDLE EAST. Population in 2011 estimated at 6.2 million. Rose to prominence during the Umyyad period in response to a focus of Muslim conquest eastward into INDIA. Baghdad's infrastructure and economy have incurred much damage from the wars, sanctions, and violence that have beset the city since 1991.

Bago. See PEGU.

Baguio. Summer capital and mountain resort in the PHILIPPINES. Located on LUZON ISLAND. Population was 318,676 in 2010. Known as the City of Pines; noted for gold mining and handicrafts. Designed in 1905 by a U.S. architect and home to the Philippine Military Academy and two universities.

Bahrain. Country in the MIDDLE EAST. Located on several islands in the PERSIAN GULF, the largest of which is the island of Bahrain, about 15 miles (24 km.) from the coast of the ARABIAN PENINSULA. Total area of 286 square miles (741 sq. km.) with an estimated population in 2018 of more than 1.4 million people, who are ethnically Arabs. Capital is Manama, located on the northern end of the largest island. About 75 percent of the population are Shia Muslims and 25 percent Sunni Muslims. Economy in ancient times was based on pearl diving and more recently on petroleum and natural gas. With most of its petroleum and natural gas deposits depleted, Bahrain successfully developed the first "post petroleum" economy in the Middle East.

Baku. Capital of AZERBAIJAN. It lies on the western shore of the CASPIAN SEA and the southern side of the ABSHERON PENINSULA. Population was 1.2 million in 2015. The basis of Baku's economy is petroleum. It is also a major cultural and educational center. Azerbaijanis are the largest ethnic group.

Bali. One of the Lesser SUNDA ISLANDS in southern INDONESIA. Total area of 2,147 square miles (5,561 sq. km.) with a 2015 population of 4.1 million. Capital is Denpasar. Located between JAVA and the island of LOMBOK, its mountain ranges include Mount Agung, a volcano that erupted most recently in 2017. Famous for its beaches and culture. Produces rice, sugarcane, coffee, copra, tobacco, fruits, and vegetables. In 2017, large oil and gas reserves discovered.

Baliem River. One of thirty major rivers in INDONESIA. Runs 250 miles (400 km.). Rises in the Jayawijaya Mountains; drains into the ARAFURA SEA.

Banda Sea. Body of water in INDONESIA. Is about 600 miles (960 km.) long and 300 miles (480 km.) wide. One of several seas connecting the Indian and Pacific oceans. Parts of it are very deep, but submerged reefs and strong currents are hazards for ships.

Bandar Seri Begawan. Capital of the Sultanate of BRUNEI. Located 298 miles (480 km.) north of the equator, on the northwest coast of the island of BORNEO, in SOUTHEAST ASIA. Covers 2,226 square miles (5,765 sq. km.), 80 percent of which is mainly rain forest. Population was 27,285 in 2002. Its climate is marked by heavy rain, with a six-month season, and extreme humidity.

Bangalore. City in south INDIA, with a population of about 8.4 million in 2011. An army cantonment during the British rule, it has become known as the Silicon Valley of India because of the presence of domestic and foreign electronic and high-technology industries.

Bangkok. Capital, administrative, economic, and cultural center, and chief seaport of THAILAND. Located on the CHAO PHRAYA River. Once called the Venice of the East because of its many canals; these have been turned into paved roads, but the floating market remains. Produces rice, textiles, electronics, and computer components. Also known as Krung Thep (City of Angels).

Chao Phraya River as it passes through Bang Kho Laem and Khlong San districts.

Bangladesh. The most densely populated large country in the world; located in SOUTH ASIA. Total area of 55,598 square miles (143,998 sq. km.) with a 2018 population of 159 million. Capital is DHAKA. About 90 percent of the population are Muslims, 9 percent Hindus, and the rest Buddhists and Christians. Has fertile alluvial soils; agriculture contributes more than one-third of the national income and provides employment for about two-thirds of the workforce. Jute, tea, fish products, garments, and hides are major export items. Was known as East Pakistan from 1947 until it separated from the rest of Pakistan in 1971.

Bataan Peninsula. Mountainous area in northern PHILIPPINES. Bounded by MANILA BAY and the SOUTH CHINA SEA. The scene of heavy fighting between the Allies and JAPAN in World War II.

Beijing. National capital of CHINA. Located on the NORTH CHINA PLAIN. Population was 16,446,857 in 2010. Was capital of numerous dynasties and is a major classic Chinese city because of its antiquity and cultural value. Houses many imperial buildings and historic sites, including the Forbidden City, the Temple of Heaven, and TIANANMEN SQUARE. Name means "northern capital." Its many industries include iron, steel, textiles, machinery, chemicals, silk, porcelain, glass, and lacquer. The terminus of the Grand Canal, it has railway links to NANJING and TIANJIN, and a major airport.Site of the 2008 Olympic Games. Formerly known as Peking in the West.

Beirut. Capital of LEBANON. Located on the coast of the MEDITERRANEAN SEA; population estimated at 1.2 million in 2004. The city was devastated by a long civil war, but in the past few years has worked feverishly to reclaim its status as the resort and banking center of the MIDDLE EAST.

Bekaa Valley. in LEBANON between the LEBANON MOUNTAINS and the Anti-Lebanon Mountains, in the MIDDLE EAST. Geologically, the northern part of the JORDAN RIFT VALLEY.

Benares. Most sacred city to Hindus; stands on the banks of the GANGES RIVER in SOUTH ASIA between DELHI and KOLKATA, India. Population was more than 1.4 million in 2011; attracts about 1 million pilgrims annually. For Hindus, the most important part of a pilgrimage to Benares is bathing in the holy waters of the Ganges River. Sacred to Buddhists as well, the city is also known as Varanasi.

Bengal, Bay of. Arm of the Indian Ocean. Fed by many large rivers, mainly from INDIA. The western part has poor harbors, while the eastern has better ones.

Bhopal. City in north-central INDIA (population 1.9 million) that experienced a devastating toxic gas explosion at the U.S.-owned Union Carbide pesticide plant in 1984. The explosion killed at least 4,000 people; many others became blind or suffered respiratory disease or cancer. Increased birth deformities have been another consequence of this unprecedented disaster. Today, perhaps ironically, the city is rated as one of the greenest and cleanest cities in India.

Bhutan. Small, landlocked Himalayan state located between INDIA and TIBET (CHINA). Total area of 14,824 square miles (38,394 sq. km.) with a 2018 population of 766,397. Capital and largest city is Thimphu, with fewer than 80,000 residents. Most Bhutanese are Buddhists who live in the foothills and river valleys. Isolation and inaccessibility preserve traditional ways of life in this mountainous buffer state. Government is a constitutional monarchy. Diplomatic relations between Bhutan and the U.S., the United Kingdom, and most other countries are handled via New Delhi.

Bilauktaung Range. One of the mountain chains in southeast MYANMAR that form a border with THAILAND. Highest elevation is 6,798 feet (2,072 meters). Its center contains granite and granodiorite dating back millions of years. Tin, tungsten, and antimony are found here. Also known as the Tenasserim or Tavois Mountains.

Biwa, Lake. Largest freshwater lake in JAPAN. Located in Shiga *ken* (prefecture), west central Hōnshū. Approximately 50 miles (80 km.) long; covers 261 square miles (676 sq. km.). Its only outlet is the Yodo River, which flows from its southern end, then past the Seta River southwest to Osaka Bay. It is deepest (about 338 feet/103 meters) in the northwest corner. The lake's surface rises in the springtime when snow melts and rain falls, and in the autumn because of typhoons. It is a breeding ground for freshwater fish, including trout, and a reservoir for KYOTO and Otsu. Well known for its natural beauty; has been the subject of much Japanese poetry. Name refers to a Japanese musical instrument that the lake resembles in shape.

Black Sea. Large inland sea situated at the south eastern extremity of Europe. Russia and GEORGIA border it on the northeast and east, TURKEY on the south. Connected to the MEDITERRANEAN SEA through Turkey's BOSPORUS. The sea covers an area of about 178,000 square miles (461,000 sq. km.), with a maximum depth of more than 7,250 feet (2,210 meters).

Bo Hai. Northwestern inlet of the YELLOW SEA. Also known as Po Hai.

Bombay. See MUMBAI.

Borneo. One of the largest islands in the world, with an area of about 292,000 square miles (756,000 sq. km.); located southeast of the MALAY PENINSULA in the Greater SUNDA group of the MALAY ARCHIPELAGO. The island is politically divided among several countries; its largest segment is the Indonesian territory of KALIMANTAN. Along the northwest coast and northern tip lie SARAWAK and SABAH (formerly North Borneo), which in 1963 joined the Malaysian Federation. The Islamic Sultanate of BRUNEI is situated be-

tween them. Total population of the island was about 21 million in 2014.

Bosporus. Twenty-mile-long (32 km.) strait in TURKEY, linking the Sea of MARMARA with the BLACK SEA. With its sister strait of the DARDANELLES, it controls the entrance and exit to the Black Sea and separates Europe, on the west, from Asia, on the east.

Brahmaputra River. One of the three major rivers that drain the INDO-GANGETIC PLAIN of SOUTH ASIA. Runs 1,800 miles (2,900 km.). Beginning in the KAILAS RANGE of the HIMALAYAS of southwest TIBET, it flows east, then turns south and enters northeastern INDIA where it becomes known as the Brahmaputra ("Son of Brahma"). Where the Brahmaputra meets the Tista River in northern BANGLADESH, the combined rivers become known as the Jamuna River. This joins the GANGES RIVER at Gowalnanda and then is known as the Padma River. Later, the Padma joins the MEGHNA RIVER and finally empties into the Bay of BENGAL.

Brunei. Independent Islamic sultanate, officially known as the State of Brunei, that occupies a 2,226-square-mile (5,765-sq.-km.) enclave on the northwestern coast of the island of BORNEO. Under British protection from 1888 to 1983, it became fully independent in 1984. The climate is very tropical with uniform temperatures and high humidity throughout the year. Brunei's natural resources include major oil and natural gas fields. Population was 450,565 in 2018. Capital is BANDAR SERI BEGAWAN.

Bukhara. City and administrative center of Bukhoro *oblast* (province) in UZBEKISTAN. Population was 274,721 in 2016. Founded not later than the first century CE, it has historically been an important trade and crafts center and a center of Islamic culture.

Burma. See MYANMAR.

Busan. See PUSAN.

Byzantium. See Istanbul.

Calcutta. See KOLKATA.

Cambodia. SOUTHEAST ASIA country lying in the southwestern portion of the Indochinese Peninsula. It was part of French Indochina from the late nineteenth century until 1953. THAILAND borders it on the west and VIETNAM on the southeast. Its area is 69,898 square miles (181,035 sq. km.). Its population was 16,449,519 in 2018. Capital is PHNOM PENH.

Canton. See GUANGZHOU.

Caspian Sea. Largest inland water body in the world; a landlocked body of salt water in the MIDDLE EAST. Covers about 123,700 square miles (320,400 sq. km.) with a maximum depth of 3,200 feet (975 meters). Located about 92 feet (28 meters) below sea level. Forms the northern coastline of IRAN and sits atop sizable petroleum reserves exploited by the countries of AZERBAIJAN, GEORGIA, and RUSSIA. The Caspian Sea Convention of 2018 recognized it as a body of water with special legal status with regard to its five surrounding states.

Cathay. Medieval European name for CHINA.

Caucasus. Also known as Caucasia, the broad neck of land that separates the Black and Caspian seas. The region covers about 170,000 square miles (440,000 sq. km.), encompassing GEORGIA, AZERBAIJAN, ARMENIA, (the three Transcaucasian Republics) and part of southern RUSSIA. The CAUCASUS MOUNTAINS divide the region into Ciscaucasia in the north and Transcaucasia in the south.

Caucasus Mountains. Mountain region located between the CASPIAN and BLACK Seas, spread over the countries of GEORGIA, ARMENIA, and AZERBAIJAN. The mountains reach heights in excess of 18,500 feet (5,600 meters).

Cebu. Island province in central PHILIPPINES. Part of the CENTRAL VISAYAS region; includes small islands. A mountainous area with fertile soil, it produces coal, tobacco, sugar, cotton, coffee, hemp, and maize. Population of nearly 3 million in 2015.

Celebes. See SULAWESI.

Central Asia. Vast inland region between RUSSIA on the north and SOUTHWEST and SOUTH ASIA on the south. The term is generally understood to encompass the former Central Asian republics of the old SOVIET UNION: KAZAKHSTAN, KYRGYZSTAN, TAJIKISTAN, TURKMENISTAN, and UZBEKISTAN.

Ceylon. See SRI LANKA.

Chang Jiang Hills and Basins. Region in southern CHINA. Located between the middle and lower reaches of the YANGTZE (Chang Jiang) River in the north and the NANLING MOUNTAINS in the south. Consists of southwest-northeast-flowing rivers and their basins and surrounding hills.

Chang Jiang Plain. Series of alluvial plains and numerous lakes in CHINA. Covers 62,000 square miles (160,000 sq. km.) along the YANGTZE (Chang Jiang) River basin between Yichang AND HUBEI, in the west, and SHANGHAI in the east. With ample heat and precipitation, it is a major center for growing rice and many other commercial crops.

Chang Jiang River. See YANGTZE River.

Changbai Mountains. Series of northeast-southwest-trending ranges and broad river valleys in eastern northeastern CHINA. Includes numerous volcanic crater lakes. Elevations vary between 6,000 and 9,000 feet (1,830 and 2,740 meters).

Changchun. Capital city of northeastern CHINA's JILIN province. It had a 2010 population of 3,411,209.

Chang-hua. Administrative center of Chang-hua county (*hsien*) in west central TAIWAN. Located southwest of T'AI-CHUNG, in the middle of the western coastal plain. Population was 235,022 in 2014. Founded at the beginning of the seventeenth century and fortified in 1734. In the nineteenth century, it became the chief market and commercial center for Taiwan's central region.

Changsha. Capital city of southern CHINA's HUNAN province. It had a 2010 population of 3,193,354.

Chao Phraya. Chief river in THAILAND. Rises in the north and flows south for 227 miles (365 km.), emptying into the Gulf of THAILAND. A major rice-producing area.

Chardzhou. See TURKMENABAD.

Cheju Island. Smallest province (*do*) of South KOREA. Located in the EAST CHINA SEA, 75 miles (120 km.) southwest of Cholla-nam province, of which it was once a part. Total area is 712 square miles (1,845 sq. km.) including the main islands and several associated islands; population was 513,260 in 2005. Cheju Island is 40 miles (64 km.) across, east to west, and 16 miles (26 km.) from north to south. Off its coasts, skilled women divers gather seaweed and shellfish. Agricultural products include sweet potatoes, most of the barley used for beer brewing in South Korea, oranges in the south, and mushrooms in the upland. Also called Jeju Island.

Cheju Island seashore.

Chengdu. Capital city of southwestern CHINA's SICHUAN province. It had a 2010 population of 6,316,922.

Chennai. Major port city in southern INDIA, with a population of 8,653,521 in 2011. A major industrial center specializing in the production of textiles, chemicals, and an array of light manufacturing products. Developed by the British as an outpost, but became the seat of the British East India Company until 1773. Formerly named Madras. In 2015, heavy rains flooded much of the city. By 2019, drought was the new crisis facing Chennai.

Cherrapunji. City at the base of the HIMALAYAS in the Meghalaya state (*see* SHILLONG PLATEAU) of INDIA. Has the world's heaviest rainfall—the yearly average is 450 inches (1,143 centimeters), of which more than nine-tenths falls in the summer.

Chi-lung. Municipality in northern TAIWAN. The principal port of TAIPEI, from which it is 16 miles (26 km.) southwest. Population was 390,764 in 2002. In 1626 was occupied by the Spanish, who built a fort at the harbor entrance. In 1638 incorporated into the Chinese province of Fukien, which lies across the TAIWAN STRAIT from the city. Then taken over by the Dutch; later returned to CHINA. In 1860 opened to foreign trade as a treaty port. During the Japanese occupation of Taiwan it became a modern, industrialized city and was chiefly an export port. An important fishing port and center of import for Taiwan. Also known as Keelung.

China. Largest of all Asian countries in area with about 3.7 million square miles (9.58 million sq. km.), and the largest country in the world in population with 1,384,688,986 people in 2018. China occupies nearly the entire EAST ASIAN landmass, stretching about 3,100 miles (5,000 km.) from east to west and 3,400 miles (5,470 km.) from north to south. Only Russia and Canada cover larger areas. A complex country in every way imaginable, China has almost every variety of climate and topography; it has the greatest contrast in temperatures between its northern and southern borders of any nation in the world. Although the great majority of the populace is Han Chinese, few countries have as wide a variety of indigenous peoples as does China. Even among the Han there can be great differences among regions; the only linguistic commonality may be the written Chinese language. China is culturally complex and has more than 4,000 years of recorded history. China is unique among nations in its longevity and resilience as a discrete politico-cultural unit. Invaders have always been absorbed into the fabric of its culture. China's relative isolation from the outside world over the centuries made possible the flowering and refinement of its culture. That same isolation, however, left China ill-prepared to cope with the technologically superior nations, which confronted it starting in the mid-nineteenth century. In more recent times that situation has changed substantially. In fact, by 2020, along with having the world's largest economy, China is now the world's largest exporter of goods (since 2010) and the world's largest trading nation. Capital is BEIJING.

China, Republic of. See TAIWAN.

China Sea. Part of the western Pacific Ocean, bordering the Asian mainland. Comprises two parts—the SOUTH CHINA SEA (Nan Hai) and THE EAST CHINA SEA (Tung Hai). Bounded on the west by Asia, on the south by a rise in the sea floor between SUMATRA and BORNEO, and on the east by Borneo, the PHILIPPINES, and TAIWAN. Its northern boundary reaches from the northern edge of Taiwan to the coast of Fukien Province, CHINA.

Chittagong. Major seaport and second-largest city of BANGLADESH. Located about 125 miles (200 km.) southeast of the capital of DHAKA. Had a population of about 2.6 million in 2011. The port was used by Arakan, Arab, Persian, and Portuguese mariners.

Chongqing. City in southwestern CHINA. Population was 6.2 million in 2010. Originally part of SICHUAN province; in 1997, designated as a separate city, directly administered by the central government. Site of the national government of the Republic of China (TAIWAN) during the Sino-Japanese War (1937-1945).

Choshui River. Longest river in TAIWAN. Runs 116 miles (186 km.). Heavily exploited for hydroelectricity.

Chota Nagpur Plateau. Upland region in central INDIA, covering parts of three states: southern Bihar, northeastern Madhya Pradesh, and northern Orissa. Populated largely by indigenous groups. Elevation ranges between 2,000 and 3,000 feet (610 and 915 meters). Composed of extensive broken hills and valleys and drained by several rivers. One-third of the plateau is covered with forests; rich in mineral resources, especially iron ore and coal.

Chung-yang Range (Central Range). Mountain group in eastern TAIWAN. Runs north to south, comprising the Chung-yang Range, the Yu Mountains, and the A-li Mountains. Extends about 170 miles (275 km.), the length of the island of Taiwan, and 50 miles (80 km.) in width. On the eastern shore, the mountains drop sharply to the ocean; their western edges slope gently into rolling hills and level land. Twenty-seven peaks are more than 9,850 feet (3,000 meters) high. The highest peak, Yu Shan (also known as Mount Hsin-kao; formerly Mount Morrison), is 13,114 feet (3,997 meters) high. Much of the A-li range is faulted, giving rise to structured depressions. Rainfall usually exceeds 150 inches (3,800 millimeters) annually. Range gives rise to the Cho-shui, Ta-chia, Kao-p'ing, and Hsin-Wu-lu rivers. At lower levels are broad-leaved trees; at higher levels, mixed forests; at the highest levels, coniferous trees. Animal life includes deer, wild boar, bears, monkeys, goats, wildcats, panthers, and birds such as pheasants, geese, flycatchers, and larks. Groups of aborigines live in sparsely settled areas.

Colombo. Port on the Indian Ocean near the mouth of the Kelani River; capital, primate city, and major industrial and financial center of SRI LANKA. Population of the greater Colombo area was about 2.4 million in 2006.

Constantinople. See ISTANBUL.

Coromandel Coast. Region stretching about 450 miles (720 km.) along the east coast of Tamil Nadu and Andhra Pradesh states in south INDIA. Lacks good harbors.

Corregidor Island. Rocky island at the entrance to MANILA BAY in the PHILIPPINES. Located in northern Philippines, it divides the bay into the Boca Chica and Boca Grande Channels. Once a Spanish stronghold, it was captured by the United States from JAPAN in World War II.

Dacca. See DHAKA.

Damascus. Capital of SYRIA, in the MIDDLE EAST, and one of the oldest cities in the world. Located in southwestern Syria, approximately 50 miles (80 km.) from the 1990s border with ISRAEL. Estimated population was 1.8 million in 2011. The city has sustained much damage during Syria's prolonged civil war.

Dardanelles. Forty-mile-long (64 km.) strait in TURKEY, 1 to 4 miles (1.6 to 6.4 km.) wide, linking the Sea of MARMARA with the AEGEAN SEA.

With its sister strait of the BOSPORUS, it controls the entrance and exit to the BLACK SEA.

Dasht-e Kavir. Salt desert basin in north central IRAN, in the MIDDLE EAST. Has a large erosional surface with salt- and mud-filled depressions.

Dasht-e Lut. Desert basin in southeastern IRAN, in the MIDDLE EAST. Somewhat smaller, but with rougher terrain, than the DASHT-E KAVIR to the north.

Dead Sea. Lowest water body on Earth. Located in the MIDDLE EAST, at more than 1,310 feet (400 meters) below mean sea level. It is about ten times as salty as the oceans, because it is the drainage terminus of the JORDAN RIVER basin, with no exit to the oceans, and has a high evaporation rate. Its salts are being extracted by both ISRAEL and JORDAN. Environmentalists warn that continued loss of water could cause the sea to dry up by 2050.

Deccan Plateau. Triangular-shaped area of modest uplands covering most of peninsular INDIA. Composed of hard rocks and formed by layers of volcanic lava. Tilts to the east, so that its highest areas are in the west and the major rivers (Godavari, KRISHNA, and Cauvery) flow eastward into the Bay of BENGAL. Bounded on the east and west by two moderately high mountain ranges, the EASTERN and WESTERN GHATS, respectively. "Deccan" means "south."

Delhi. Urban center containing the capital of INDIA. The original city of Delhi, sometimes referred to as Old Delhi, was the site of seventeen earlier historical imperial capitals, but it did not become important until Shah Jahan made it capital of the Mogul empire in 1638. In 1911 the British colonial government left KOLKATA and built a new capital adjacent to Old Delhi that it called New Delhi. After independence in 1947, the Indian government retained New Delhi as its capital. Delhi and New Delhi together form the third-largest metropolis of India, with more than 16 million people (2011).

Demilitarized Zone (DMZ). Buffer zone separating North KOREA from South KOREA, at roughly 38° north latitude. It runs east to west for about 160 miles (250 km.), roughly from the mouth of the Han River on the west coast of the Korean peninsula to a little south of the North Korean town of Kosong on the east coast.

Dhaka. Capital and largest city of BANGLADESH, located in the geographic center of the country. Covers 160 square miles (414 sq. km.), with a population of 8.9 million in 2011. It has Bangladesh's greatest industrial concentration. Dhaka's history dates back to 1000 CE. It was the Mogul capital of Bengal during the seventeenth century.

Djakarta. See JAKARTA.

DMZ. See DEMILITARIZED ZONE.

East Asia. Loosely applied term that is usually understood to include CHINA, JAPAN, KOREA, MONGOLIA, and TAIWAN.

East China Sea. Arm of the Pacific Ocean; part of the CHINA SEA. Covers about 290,000 square miles (752,000 sq. km.). Boundaries are the islands of CHEJU on the north, KYŪSHŪ on the northeast, the RYUKYU Island chain on the east, TAIWAN on the south, and the Chinese coastline on the west. Connected to the SOUTH CHINA SEA by the TAIWAN STRAIT.

East Korea Warm Current. Surface ocean current in the northward-moving branch of the TSUSHIMA CURRENT in the Sea of JAPAN. After flowing along the coast of KOREA, turns eastward and divides into the Tsugaru Warm Current and Soya Warm Current. The former enters the Pacific Ocean through the Tsugaru Strait. The latter enters the Sea of Okhotsk through the La Perouse Strait, bending to the southeast along the coasts of HOKKAIDO and HONSHŪ.

East Pakistan. See BANGLADESH.

Eastern Ghats. Mountain range in south INDIA. Paralleling the coast of the Bay of BENGAL, it forms the eastern border of the DECCAN PLATEAU. With a total length of 900 miles (1,450 km.), the range extends from the Mahanadi River to the Nilgiri Hills. Numerous rivers cutting across it are used for hydroelectric power generation and irrigation. *See also* WESTERN GHATS.

Edo. See TOKYO.

Elbrus, Mount. Highest peak of the CAUCASUS MOUNTAINS in southwestern Russia. It is an extinct volcano with twin cones reaching 18,510 and 18,356 feet (5,642 and 5,595 meters). Elbrus has a total area of 53 square miles (138 sq. km.). It is covered by twenty-two glaciers that feed the KUBAN RIVER and some of the headwaters of the TEREK.

Elburz Mountains. Mountains in northern IRAN, in the MIDDLE EAST, forming a crescent pattern just south of the CASPIAN SEA. Rising to about 18,600 feet (5,669 meters), they provide a green contrast to the desert basin to the south in central Iran.

English Central Range. See CHUNG-YANG RANGE.

Euphrates River. River in the MIDDLE EAST. Begins in the eastern highlands of TURKEY before entering the Mesopotamian plain, travels more than 1,500 miles (2,400 km.) before joining the TIGRIS RIVER. Its waters are an important source of irrigation for agriculture in SYRIA and IRAQ. Construction of several dams in the Turkish headwaters has reduced the amount of water flowing southward, causing great concern for the agricultural economies of the two countries.

Everest, Mount. World's tallest mountain, with a summit that reaches an elevation of 29,035 feet (8,850 meters). Located in NEPAL, on its border with CHINA in the HIMALAYAS, at 27°59′ north latitude, longitude 86°56′ east. Named after a British surveyor of the Himalayas. Edmund Hillary of New Zealand and Tenzing Norkay of Nepal became the first persons to reach its summit on May 29, 1953.

Farakka Barrage. Structure built across the GANGES RIVER in West Bengal, near the India-Bangladesh border. Stretches 7,351 feet (2,240 meters). Designed to divert water into the Bhagirathi-Hooghly River during the dry season in order to flush out KOLKATA harbor, which had experienced siltation problems. Came into operation in 1975, following a temporary accord between BANGLADESH and INDIA. In 1996 India and Bangladesh signed a landmark thirty-year agreement to share the waters of the Ganges.

Fergana Valley. Enormous depression between the Tien Shan and Gissar and Alay mountain systems, mainly in eastern UZBEKISTAN and partly in TAJIKISTAN and KYRGYZSTAN. The roughly triangular valley has an area of 8,500 square miles (22,000 sq. km.). In the twenty-first century, the area has been the site of deadly interethnic clashes.

Fertile Crescent. Agriculturally productive region of the MIDDLE EAST that gets its name from the crescent-like shape of the countries that make up MESOPOTAMIA and the LEVANT in what are now IRAQ and SYRIA.

Flores. Island in southeastern INDONESIA. Is 224 miles (360 km.) long and 35 miles (56 km.) wide. Extremely mountainous, with active volcanoes.

Formosa. Early Portuguese name for TAIWAN.

Fuji, Mount. Highest mountain in JAPAN, at 12,388 feet (3,776 meters). Near the Pacific Coast on central HŌNSHŪ, about 70 miles (115 km.) west of TOKYO. A volcano, it last erupted in 1707. The most important feature of Fuji-Hakone-Izu National Park. On its northern slope are the five lakes (Fujigoko), which were formed by lava flow. Considered sacred by the Japanese people and surrounded by many temples and shrines. Name, of Ainu origin, means "everlasting life." Also known as Fujiyama.

Fujian. Province in southern CHINA. Located across the TAIWAN STRAIT from TAIWAN. Total area of 46,000 square miles (120,000 sq. km.), with a 2017 population of 38 million. The province's capital city, FUZHOU, had a 2017 population of 7.6 million. Mostly hilly and mountainous, with narrow river plains, irregular coastlines, and many offshore islands. Has the nation's highest forest coverage rate and is known for various subtropical products. Most major population centers are near the coast.

Fujiyama. See FUJI, MOUNT.

Fuzhou. Capital city of southern CHINA's FUJIAN province. Its 2017 population was 7.6 million.

Galilee, Sea of. Freshwater lake in northeastern ISRAEL in the MIDDLE EAST. Covers about 64 square miles (166 sq. km.); lies about 690 feet (210 meters) below sea level. Control of the lake is extremely important because it supplies a significant amount of fresh water to Israel and Jordan. Also known as Lake Kinneret or Lake Tiberias.

Ganga River. See Ganges River.

Ganges River. Asian river that is the most sacred river to the Hindus of South Asia. Rising in the Gangotri glacier in the Himalayas, it follows a generally southeast path through a vast plain in northern India and Bangladesh and falls to the Bay of Bengal. Runs about 1,560 miles (2,510 km.), passing through holy bathing sites at Haridwar and Varanasi (Benares), both located in India. According to mythology, bathing in the Ganges washes away all sins. Also known as the Ganga River.

Gansu. Province in China. Borders Mongolia in the northwest. Total area of 173,100 square miles (450,000 sq. km.), with a 2010 population of 25.6 million. The capital city, Lanzhou, had a 2010 population of 2,438,595. East central Gansu is on the Loess Plateau; western Gansu is mostly mountainous and desert. Farming and population centers are mainly in the river valleys and oases. Erosion and drought are constant problems.

Gaza Strip. Narrow strip of coastal land in the Middle East, 26 miles (42 km.) long and about 6 miles (10 km.) wide. Population in 2017 was 1,899,291. Controlled by Israel after being conquered in the Arab-Israeli War of 1967. In 2000, was administered by the Palestinian National Authority, which pressed Israel to permit creation of an independent Palestinian state. Although a desert area, its good soils and freshwater wells support a productive agricultural economy, primarily orchards of citrus, olives, and nuts. Gaza, its major city, is extremely overcrowded, with more than half a million Palestinian Arabs. Violence in the area had caused devastating economic consequences.

Georgia. Country of Caucasus, formerly a constituent republic of the Soviet Union. Total area of 26,911 square miles (69,700 sq. km.) with a 2018 population of 4.9 million. Capital is Tbilisi. The country is mostly mountainous. Two-thirds

of the people are Georgians who may have always lived in that region of the Caucasus.

Ghats. See Eastern Ghats, Western Ghats.

Goa. Former Portuguese colony in South Asia. Integrated into India as a union territory in 1961; became a state in 1987. Total area of 1,430 square miles (3,704 sq. km.), with a 2011 population of 1.45 million. Capital is Panaji. About 40 percent of the state's people are Roman Catholic; Portuguese is one of the dominant languages. Its beaches have made tourism its main industry.

Gobi Desert. Arid region in East Asia covering about 500,000 square miles (1.3 million sq. km.), mostly in Mongolia and China's Inner Mongolia. Most of the area is covered by stony and gravel terrain and shifting sandy deserts. The highest sand dune reaches 1,380 feet (420 meters). Generally a plateau between higher mountains, its surface is mostly rolling gravel plains, with intervals of low, flat-topped ridges and isolated hills. Maximum elevation is 5,000 feet (1,524 meters); minimum elevation is 3,000 feet (914 meters). Remains of ancient civilizations (including the Upper Paleolithic, Neolithic, and Bronze Age) have been found there.

The sand dunes of Khongoryn Els, Gurvansaikhan NP, Mongolia.

Godwin-Austen. See K2.

Golan. Basaltic plateau to the east of the Huleh Valley and the Sea of Galilee in the Middle East. Captured by Israel during the Six-Day War of 1967, it has remained a subject of dispute between Israel and SYRIA. Both a fertile farming area and a strategic physical feature for both countries, as it overlooks the Huleh Valley and forms a transportation corridor to DAMASCUS. Several tributary streams flow off the Golan into the Sea of Galilee, supplying much-needed water to the Jordan Valley. Also called the Golan Heights.

Great Indian Desert. See THAR.

Greater Caucasus. Major range of the CAUCASUS MOUNTAINS. The range extends west-east for about 750 miles (1,200 km.) from the Taman Peninsula on the BLACK SEA to the ABSHERON PENINSULA on the CASPIAN SEA.

Greater Hinggan Mountains. See HINGGAN MOUNTAINS, GREATER.

Guangdong. Province in southernmost mainland CHINA. Total area of 76,000 square miles (197,000 sq. km.), with a 2010 population of 104 million. Capital is GUANGZHOU. Climate is subtropical-tropical. Mostly hilly and mountainous, with plains along the coasts, where most population centers are located. China's major foreign investment and export-producing region.

Guangxi. Autonomous region in southern CHINA. Located in the mountainous Guangxi Basin and borders VIETNAM in southwest. Total area of 85,100 square miles (220,400 sq. km.), with a 2010 population of 46 million. Capital, Nanning, had a 2010 population of 2,660,833. Populated by various non-Chinese groups, notably the Zhuang. Known for its karst landscape.

Karst landscape around Yangshuo in Guangxi.

Guangzhou. Capital city of southern CHINA's GUANGDONG province. Located in southern Guangdong and at the mouth of the PEARL RIVER. Population was 10.6 million in 2010. China's major trading post with the West in the nineteenth century; the center of the foreign investment that fueled the spectacular growth of Guangdong province in the late twentieth century. Guangzhou was at the center of a series of political maneuvers that led to the Opium War in 1840, which opened a new era in Chinese history, characterized by foreign control and exploitation.

Guiyang. Capital city of southwestern CHINA's GUIZHOU province. It had a 2010 population of 2,520,061.

Guizhou. Province in southwestern CHINA. Situated on the mountainous northern YUNNAN-GUIZHOU PLATEAU. Total area of 67,200 square miles (174,000 sq. km.), with a 2010 population of 34.7 million. Capital, GUIYANG, had a 2010 population of 2,520,061. Most people live in the central area.

Gujarat. Western state of INDIA; formed in 1960 from the north and west (predominantly Gujarati-speaking) portion of former Bombay State as a result of the Bombay Reorganization Act. Covers 75,685 square miles (196,024 sq. km.); population at the 2011 census was 60.38 million. One of the most industrialized states of India and the center of the Indian cotton-textile industry. Contains large crude oil and gas reserves. Building one of Asia's largest solar-energy parks.

Gunanag Kinabalu. See KINABALU, MOUNT .

Hai River. River rising in Taihang Shan Mountains of eastern CHINA. Flows northeast for 826 miles (1,329 km.) before emptying in the BO HAI inlet of the YELLOW SEA east of TIANJIN.

Haikou. Capital city of CHINA's HAINAN province. It had a 2010 population of 2 million.

Hainan. Smallest province in CHINA. Located on Hainan Island, off the coast of GUANGDONG. Total area of 13,240 square miles (34,300 sq. km.), with a 2010 population of 8,671,485. Its capital city, HAIKOU, had a 2010 population of 2 million. Mostly in the tropical zone, Hainan Island has a mountainous core flanked by coastal plains. Main population centers are along the northern and southern coasts.

Haiphong. Seaport and third-largest city in VIETNAM; located on the Cam River, a part of the RED RIVER Delta. Population was 839,000 in 2019. Developed by the colonial French, it was badly damaged by air raids during colonial wars and bombed again during the Vietnam War. Chief industries are coal and zinc mining.

Hangzhou. Capital city of southern CHINA's ZHEJIANG province. Located in northern Zhejiang, on the southern terminus of the Grand Canal. Had a 2010 population of 5,578,288. The capital of numerous ancient dynasties; a major classic Chinese city because of its antiquity and rich cultural value. Known for its ancient buildings, scenic gardens and lakes, and traditional handicrafts.

Hanoi. Capital and second-largest city in VIETNAM; located in the northern part of Vietnam, about 47 miles (75 km.) from the Gulf of TONKIN on the west bank of the RED RIVER. Population was 3,472,000 in 2019. Produces tools, chemicals, and textiles. Once the capital of French Indochina.

Life on the streets of central Hanoi, Vietnam.

Harbin. Capital city of CHINA's HEILONGJIANG province. Population was 4,596,313 in 2010.

Hebei. Province in northern CHINA. Located on the northern NORTH CHINA PLAIN. Total area of 78,200 square miles (202,700 sq. km.), with a 2010 population of 71.9 million. The capital city, SHIJIAZHUANG, had a 2010 population of 2,770,344. Western and northern Hebei are mountainous; southern Hebei is a plain. Most of Hebei has cold, dry winters and hot summers with concentrated rainfall. Spring droughts are common. Has rich reserves of coal, petroleum, and iron ore. Major population centers are widespread except in the north and northwest.

Hefei. Capital city of southern CHINA's ANHUI province. It had a 2010 population of 3 million.

Heilongjiang. Province in northeastern CHINA. Borders Russia to the north and east. Total area of 179,000 square miles (463,600 sq. km.), with a 2010 population of 38.3 million. The capital city, HARBIN, had a 2010 population of 4,596,313. Mountainous in the central section and east, its vast fertile plains to the west and northeast give it the nickname "breadbasket of the north." Farming suffers from low temperatures and a short growing season. Main population centers are in the central-south segment.

Hejaz Mountains. Mountains extending along the RED SEA coastline of SAUDI ARABIA in the MIDDLE EAST. Elevations of about 8,000 feet (2,440 meters) in both their northern and southern extremities.

Henan. Province in north central CHINA. Total area of 64,480 square miles (167,000 sq. km.), with a 2010 population of 94 million. The capital city, ZHENGZHOU, had a 2010 population of 3,677,032. Western Henan is mountainous. Central eastern Henan is plains with a monsoon climate, which contributes to flooding, drought, salinity problems, and river channel swings. Henan is at the core of traditional Chinese culture. Major population centers are widespread except in the southwest.

Hermon, Mount. Small mountain at the southern end of the Anti-Lebanon Mountains in the MIDDLE EAST. Reaches to 9,232 feet (2,814 meters)—high enough to retain a small amount of snow and ice at its summit throughout the year. Summit is the location of an Israeli observation and listening military post. Also known as Jabal ash-Shaykh.

Hexi Corridor. Piedmont plain located north of the QILIAN MOUNTAINS in northwestern CHINA. A segment of the ancient Great SILK ROAD, it stretches 620 miles (1,000 km.).

Hida Range. Mountain range on central Hōnshū, the largest island in the Japanese archipelago. Consists of granite with crystalline rocks containing feldspar. The highest peaks are more than 10,000 feet (3,000 meters) high.

Himalayas. Highest mountain range in the world; separates the INDO-GANGETIC PLAIN from the TIBETAN PLATEAU. Extending along the northern frontiers of PAKISTAN, INDIA, NEPAL, BHUTAN, and MYANMAR for about 1,550 miles (2,500 km.), the Himalayas comprise three parallel ranges: the Greater Himalayas in the north, the Lesser Himalayas in the middle, and the Outer Himalayas in the south. The Greater Himalayas have the highest elevations among the three ranges and feature the tallest mountain on Earth, Mount EVEREST. The Himalayan mountain system is geologically young and subject to severe earthquakes. Extensive tourism is common throughout the Himalayas. These ranges are associated with many legends in Asian mythology. Other ranges intersecting the Himalayas include the HINDU KUSH and the KARAKORAM.

Hindu Kush. High mountain system extending 450 miles (725 km.) in a generally east-west direction from northern PAKISTAN into northeast AFGHANISTAN. Receives heavy snowfall, and most of its peaks are covered with snow through-

Kalash girls in the Kalasha Valleys of the Hindu Kush.

out the year. Crossed by many high-elevation passes, through which Alexander the Great, Genghis Khan, Timur, and Babar came to INDIA. The Hindu Kush, meaning "mountains of India," were called the Caucasus Indicus by the ancient Greeks.

Hinggan Mountains, Greater. Mountain range in eastern INNER MONGOLIA, China. The mountains trend northeast-southwest and have an average elevation of 3,280 feet (1,000 meters). Eastern slope is steep with erosion surfaces; gentle western slope merges with the INNER MONGOLIAN PLATEAU.

Hinggan Mountains, Lesser. Mountain range in northernmost northeastern CHINA. Mostly lower hills, 1,310 to 1,970 feet (400 to 600 meters) in elevation, the mountains trend northwest-southeast. Separated from the Greater HINGGAN MOUNTAINS in the northwest by the Songhau Jiang Plain.

Hiroshima. Capital city of Hiroshima prefecture in JAPAN. Located on Hiroshima Bay of the Inland Sea. Population was 1.2 million in 2015. Founded as a castle town in the sixteenth century; was a military center from the mid-nineteenth century onward. Because of its value as a military center, U.S. forces bombed it on August 6, 1945, killing as many as 70,000 people. An annual service to honor people who died from the atomic bomb blast is held at Peace Memorial Park, located at the center of the bomb blast. In 2016, Barack Obama became the first sitting U.S. president to visit the city since the 1945 bombing.

His-Sha Ch'un Tao. See PARACEL ISLANDS.

Ho Chi Minh City. Largest city and chief industrial center of VIETNAM; located just northeast of the Mekong Delta, with a population of about 7 million in 2019. Handles almost all the trade of the southern part of Vietnam. Produces textiles, ships, machinery, pharmaceuticals, and consumer goods. Known as Saigon until 1975, when it was renamed after Vietnam's modern founder.

Hohhot. Capital city of CHINA's autonomous INNER MONGOLIA region. It had a 2010 population of 1,497,110.

Hōkkaidō. Northernmost and second-largest of the four main islands of the Japanese archipelago. Bordered by the Sea of JAPAN on the west, the Sea of Okhotsk on the north, and the Pacific Ocean on the east and south. Along with a few small adjacent islands, it forms a Japanese *do* (province). The island covers 30,314 square miles (78,513 sq. km.); the province covers 32,247 square miles (83,520 sq. km.). Hōkkaidō makes up about 22 percent of the total area of JAPAN. Its 2010 population was about 5.5 million. SAPPORO, the administrative headquarters, is an important industrial, commercial, and tourism center. The economy of the island depends on the iron, steel, wood-pulp, dairy, and fishing industries, and on agriculture. The island also has Japan's largest coal deposits.

Sapporo City, part of Hōkkaidō.

Holy Land. Widely used name for the land of ancient PALESTINE in the MIDDLE EAST, with sites sacred to Jews, Christians, and Muslims. Although primarily associated with the modern country of ISRAEL by most people, for others it includes JORDAN, LEBANON, SYRIA, and sometimes even parts of TURKEY.

Hong Kong. Administrative region, and a tourist, trade, and banking center of CHINA. Passed back to the People's Republic in 1997 after more than 150 years of British rule. Has a mainland, including Kowloon and part of the New Territories, and more than 200 islands, the largest being Lantau. Bordered on the north by GUANGDONG Province and on the east, west, and south by the SOUTH CHINA SEA. Has little fresh water of its own; imports 80 percent of its water needs from Guangdong. A major financial and commercial center for southern China and Asia. Population in 2009 was about 7 million.

Hōnshū. Largest island in the Japanese archipelago. Located between the Pacific Ocean on the east and the Sea of JAPAN on the west. Extends about 800 miles (1,287 km.) in a northeast-southwest curve; total area is 87,804 square miles (227,413 sq. km.). Coastline covers 6,226 miles (10,018 km.). Includes about three-fourths of the prefectures of JAPAN and the major industrial areas of TOKYO-YOKOHAMA and OSAKA-KOBE. Mount FUJI and Lake BIWA are also on HŌNSHŪ.

Hormuz, Strait of. Strategic choke point commanding the entrance to the oil-rich PERSIAN GULF. Located between the Persian Gulf and the Gulf of OMAN; controlled from either side by OMAN and IRAN.

Hsin-chu. Administrative center for Hsin-chu county in TAIWAN. Located southwest of TAIPEI, about 6 miles (10 km.) from the country's west coast. Population was 448,207 in 2019. First settled and walled in the eighteenth century. Was an important military base during the Japanese occupation of Taiwan (1895-1945). An important marketing and distribution center for the rice, tea, and citrus farms in the area. A large petroleum field lies nearby.

Hua-lien. Largest, least densely populated of TAIWAN's counties. Located in east central Taiwan; population was 333,392 in 2014, including about 9,000 aborigines. The CHUNG-YANG RANGE and the Han-an Range run north to south over the western and eastern parts of the county, respectively. Between these two ranges is the T'ai-tung River valley. The Hua-lien and Wu rivers provide power for hydroelectricity used throughout the area. Iron ore, asbestos, sulfur, and copper are extracted there, and marble is quarried.

Huang He River. See YELLOW RIVER.
Huang He Sea. See YELLOW SEA.

Hubei. Province in east central CHINA. Situated along the middle YANGTZE (Chang Jiang) River. Total area of 72,400 square miles (187,500 sq. km.), with a 2010 population of 57.2 million. Its capital city is WUHAN. Mostly hilly and mountainous; has river plains in the east, with numerous lakes and many population centers. Hubei is subtropical and is known for its productive farming.

Huleh Valley. Area north of the Sea of GALILEE in ISRAEL in the MIDDLE EAST. Formerly contained a lake, which was drained by Israel to expand its farmland.

Hunan. Province in south central CHINA. Located along the middle YANGTZE (Chang Jiang) River valley. Total area of 81,300 square miles (210,500 sq. km.), with a 2010 population of 65.7 million. Its capital city, CHANGSHA, had a 2010 population of 3,193,354. In the north is the vast Lake Dongting and surrounding plains. Southern Hunan is hilly and mountainous. Hunan is subtropical and has frequent summer droughts. The main population centers are in the north-central area.

India. Second-most populous country in the world (after China); located in SOUTH ASIA. A former British colony (1757-1947), India dominates the geography of South Asia in area, population, and centrality. Total area of 1.27 million square miles (3.29 million sq. km.), about one-third the size of the United States—with a 2018 population of nearly 1.3 billion. Capital is New DELHI. Although Hinduism is practiced by 80 percent of its people, India, the world's largest democracy, strictly follows secular ideology. Other religious groups represented in the country are Muslims (14 percent), Christians (2 percent), Sikhs (1.7 percent), Buddhists (0.7 percent), and Jains (0.4 percent). India has more than 850 languages and dialects in use. This predominantly agricultural country is rapidly becoming a major services and industrial nation and had become self-sufficient in food production at the end of the twentieth century. The country consists of twenty-five states and several union territories.

Indochina. French colonial-era name for the portion of SOUTHEAST ASIA that includes CAMBODIA, LAOS, and VIETNAM.

Indo-Gangetic Plain. One of the three major physiographic regions of SOUTH ASIA. Located between the HIMALAYAS in the north and the peninsular plateaus in the south, covering parts of PAKISTAN, INDIA, and BANGLADESH. This vast plain of 325,000 square miles (840,000 sq. km.) is composed of fertile alluvial soils deposited by three of the world's major river systems—the INDUS, the GANGES, and the BRAHMAPUTRA, all originating in the Himalayas. It is densely populated and intensively cultivated.

Indonesia. Island country located off the coast of the SOUTHEAST ASIA mainland in the Indian and Pacific oceans. It is a vast archipelago whose 17,508 islands—more than half of which are uninhabited—spread across the equator over a distance equivalent to one-eighth of Earth's circumference. With a total land area of about 735,358 square miles (1.9 million sq. km.), it is the largest country in Southeast Asia, and its population of 262,787,403 (2018) places it among the most populous nations in the world. Almost 75 percent of Indonesia's land area is included in the three largest islands of BORNEO, SUMATRA, and the IRIAN JAYA portion of NEW GUINEA. The climate is tropical and uniform year round. The people are mostly ethnic Malays and who practice Islam. The country is resource-rich including oil.

Mount Semeru and Mount Bromo in East Java. Indonesia's seismic and volcanic activity is among the world's highest.

Indus River. Longest of the three rivers that pass through the INDO-GANGETIC PLAIN of SOUTH

Asia; runs about 1,800 miles (2,900 km.). Originates in the southwestern Tibetan HIMALAYAS and flows west across the Ladakh region of northern INDIA, then southwest through PAKISTAN to the ARABIAN SEA southeast of KARACHI. The upper Indus flows through deep gorges, while the lower Indus flows through fertile plains of the lower Punjab, extending south to the coast. All of Pakistan's major rivers flow into it. In the southern Punjab, the British attempted to harness the Indus water in the nineteenth century, when they established what came to be known as the canal colonies. Use of the water of the upper Indus and its tributaries was a source of conflict between Pakistan and India until 1960, when the two nations signed a treaty.

Ingushetia. Republic within southwestern Russia. Total area of 1,242 square miles (3,217 sq. km.). Population (2010) was 412,529. Capital is Nazran. The crestline of the GREATER CAUCASUS range forms its southern boundary with GEORGIA. Declared its independence from the former Soviet Union in 1991.

Typical Ingush medieval castle. A majority of towers and walls were destroyed by the Russian army in 19th and 20th centuries.

Inner Mongolia. Autonomous region of CHINA. Borders MONGOLIA and Russia in the north. Total area of 454,650 square miles (1.18 million sq. km.) with a 2010 population of 24.7 million. Its capital city, HOHHOT, had a 2010 population of 1,497,110. The eastern section is mountainous; the central section is a plateau with rich pastoral land; in the west are deserts. In the southwest,

the fertile plains along the YELLOW (Huang He) RIVER are sandwiched between the Yinshan Mountains to the north and the ORDOS PLATEAU to the south.

Inner Mongolian Plateau. Located in CHINA's central Inner Mongolian Autonomous Region. Mostly a rolling upland and steppe grassland. In the south are the Yinshan Mountains, a divide separating the crop-growing region to the south from the pastoral land to the north.

Iran. Country in the MIDDLE EAST, situated between the CASPIAN SEA in the north and the PERSIAN GULF and ARABIAN SEA to the south. Total area of 636,371 square miles (1,648,195 sq. km.) with an estimated population in 2018 of 83 million, most of whom are Shia Muslims. Capital is TEHRAN. The population is not Arab but mostly Persian, speaking the Farsi language. Minority groups include Azerbaijanis and Kurds who live in the northern ZAGROS MOUNTAINS. Its government since the late 1970s has been controlled by Shia religious leaders. Its oil-based economy has been repeatedly stressed by sanctions imposed by Western countries over its nuclear program and its support of terrorist groups elsewhere in the Middle East. Formally known as the Islamic Republic of Iran.

Iraq. Country in the MIDDLE EAST, on the lower TIGRIS and EUPHRATES RIVER basin of MESOPOTAMIA. Total area of 167,235 square miles (438,317 sq. km.), with an estimated population in 2018 of 40.2 million. Capital is BAGHDAD. Since the early 1980s, Iraq has been involved in a war with neighboring IRAN (1980-1988), the Gulf War (1990-1991), the Iraq War (2003-2011), civil war (2014-2017), and various violent insurgencies. The U.S. maintains a military presence on Iraqi soil.

Irian Jaya. Province in eastern INDONESIA; located just south of the equator; divided by the MAOKE MOUNTAINS. Climate is tropical, with heavy monsoon rains, but the highlands are cooler. Formerly known as West New Guinea or West Irian.

Irrawaddy River. Major river of MYANMAR (Burma) in SOUTH ASIA. Flows 1,300 miles (2,095 km.)

from the eastern Himalayan ranges of southern CHINA, running southward until it empties into the ANDAMAN SEA. It created the Irrawaddy delta, which is one of the world's great rice-producing regions. The river is the economic lifeline of Myanmar. Also spelled Ayeyarwady.

Islamabad. Capital of PAKISTAN since 1959. Population has increased rapidly, with just over 1 million people in 2017. It is a well-planned city with well-defined residential areas and industrial-commercial zones. City suffered major damage in a 2005 earthquake.

Israel. Country in the MIDDLE EAST, established in 1948 as a Jewish state after the withdrawal of British forces from the British Palestine Mandate. Bounded by LEBANON on the north, SYRIA and JORDAN on the east, EGYPT on the southwest, and the MEDITERRANEAN SEA on the west. Total area of 8,552 square miles (22,072 sq. km.); population of more than 8.4 million, of whom about half are Palestinian Arabs in the occupied territories. Capital is JERUSALEM. Since independence, Israel has expanded its territory beyond that originally allocated under a United Nations partition plan through several wars with neighboring Arab states. This expansion has resulted in the uprising of Arab Palestinians in GAZA and the occupied areas of the WEST BANK.

Istanbul. City, port, and major transportation center in the MIDDLE EAST; was the capital of successive Greek, Byzantine, and Turkish empires. It commands the BOSPORUS straits, which permit entry and exit to the BLACK SEA to the north and the MEDITERRANEAN SEA by way of the DARDANELLES to the south. Also controls a major land access between Europe and Asia. Although no longer the political capital of TURKEY, its estimated population of 13.8 million people in 2012 makes it an economic and cultural center for the Turkish-speaking world. Formerly Byzantium; later called Constantinople.

Jabal ash-Shaykh. See HERMON, MOUNT.

Jaffna. Small port city on Jaffna Peninsula at the northern tip of SRI LANKA. With a 2007 population of 78,781, the peninsula and the city are densely inhabited. Most residents are Tamil-speaking people who have been fighting since 1984 to establish an independent Tamil state with headquarters in Jaffna. War was officially declared to have ended in 2009, but disputes over land persist.

Jaipur. Capital and largest city of the state of Rahasthan, INDIA; was capital of the former Indian state of Jaipur. Had a population of 3 million in 2011. A popular tourist center; the palace of Maharaja Sawaii Jai Singh II is the main attraction. Founded in 1727, it is internationally known for its ivory and enamel work and glass and marble carvings.

Jakarta. Capital and largest city of INDONESIA; former capital of the Dutch East Indies. Located on the northwest coast of JAVA Island, on a flat, alluvial plain. Population was 10.1 million in 2015. Has a hot, humid climate, with average daily temperatures of 69° to 92° Fahrenheit (21° to 33° Celsius) and an average annual rainfall of 71 inches (1,800 millimeters). Also known as Djakarta.

Banjir Kanal Barat (west flood-control canal) in Jakarta.

Jammu and Kashmir. Region in SOUTH ASIA. In 1947 British India was divided into two independent countries: The Muslim-dominated parts became PAKISTAN and Hindu-dominated parts became INDIA. The rulers of 562 princely states of British India were asked whether they would join India or Pakistan. The Hindu king of Muslim-dominated Kashmir wanted to remain an autonomous state, which led to a Muslim uprising in Kashmir. The king asked for assistance from India, and Pakistan's forces came to the aid of the Muslims in Kashmir. After more than a year's fighting and the intervention of the United Nations, a cease-fire line was accepted by both countries in 1949. Pakistan effectively controls the western part of this line, known as Azad (free) Kashmir. India holds the rest, which is called Jammu and Kashmir.

Japan. Country off the east coast of Asia. Comprises a string of islands curving northeast-southwest for about 1,500 miles (2,400 km.). Bounded on the east by the Sea of JAPAN, separating it from the Korean Peninsula; on the north by the Sea of Okhotsk and the La Perouse Strait, separating it from the Sakhalin Islands; on the northeast by the southern Kuril Islands; on the east and south by the Pacific Ocean; and on the southwest by the EAST CHINA SEA, separating it from CHINA. The capital is TOKYO. Total area of the country is 145,914 square miles (377,915 sq. km.). Population was about 126 million in 2018. The four main islands are Hōkkaidō, Hōnshū, Shikōkū, and Kyūshū. Among the smaller islands groups are the RYUKYU (Nansei) Islands south and west of Kyūshū, which include OKINAWA and the Bonin (Ogasawara) and Volcano (Kazan) groups south and east of central Hōnshū. More than 80 percent of Japan is mountainous. Also known as Nihhon and Nippon.

Japan, Sea of. Marginal sea of the western Pacific Ocean that is bounded by JAPAN on the east and the Russian mainland on the west. Its surface area is approximately 386,100 square miles (1,000,000 sq. km.). It has an average depth of 5,750 feet (1,750 meters) and a maximum depth of 12,300 feet (3,750 meters). The sea contributes greatly to the mild climate of Japan because of the moderating effect exerted by its relatively warm waters. Called the East Sea by South Koreans.

Japanese Alps. Mountains in the central area of the island of Hōnshū, the largest island in the Japanese archipelago. The name was first applied to the HIDA RANGE in the late nineteenth century; it includes the Kiso and Akaishi ranges to the south. These mountains, some of which have peaks higher than 10,000 feet (3,000 meters),

are popular for skiing and mountain climbing. Also known as Nippon Arupusa.

Java. Island in INDONESIA. Has a volcanic mountain chain that has been the scene of disastrous eruptions. Rivers in the low coastal plain are swift, narrow, and shallow. The southern alluvial plain is dry one season and overflows in the rainy months. Wide variations in temperature and rainfall. Produces teak, bamboo, oak, and rubber. Population in 2005 was 128 million.

Java Sea. Portion of the western Pacific Ocean between the islands of JAVA and BORNEO. The sea has a total surface area of 167,180 square miles (433,000 sq. km.) and a comparatively shallow average depth of 151 feet (46 meters). The oil fields of northern Java extend under the sea.

Jeddah. City in the MIDDLE EAST. Located on the eastern coast of the RED SEA; the port of entry for millions of Muslims on pilgrimage to MECCA about 45 miles (72 km.) inland. In 2010, it had a population of 3.4 million people.

Jeju Island. See CHEJU ISLAND.

Jerusalem. MIDDLE EAST city that is a major holy center to Judaism, Christianity, and Islam. The proclaimed capital of ISRAEL and also the designated capital of a future Palestinian state. Located about 25 miles (40 km.) from the MEDITERRANEAN SEA near the crest of the Judean highlands. In 2016 it had an estimated population of 882,652.

Jiangsu. Province in southern CHINA. Straddles the NORTH CHINA PLAIN and the lower CHANG JIANG PLAIN. Total area of 39,614 square miles (102,600 sq. km.), with a 2010 population of 78.6 million. Its capital city is NANJING. Major population centers are widespread, except in the northeast. Mostly plains, with numerous lakes and dense irrigation canals. Ample heat and precipitation contribute to its productive agriculture, which has led to its sophisticated light industries and traditional handicrafts. Cities in southern Jiangsu are known for their scenic gardens and traditional buildings traversed by ponds and canals.

Jiangxi. Province in southern CHINA. Located in the south of the middle and lower CHANG JIANG

Plain. Total area of 63,630 square miles (164,800 sq. km.), with a 2010 population of 44.5 million. The capital city of NANCHANG had a 2001 population of 1,386,500. Southern Jiangxi is mountainous; northern Jiangxi consists of the vast Lake Poyang and surrounding plains. The province is subtropical. The Jinggangshan area in western Jiangxi was the site of the communist base during the civil war in 1928-1937.

Jilin. Province in central northeastern CHINA. Borders North KOREA and Russia in the southeast. Total area of 72,200 square miles (187,000 sq. km.), with a 2010 population of 27.4 million. Its capital city, CHANGCHUN, had a 2010 population of 3,411,209. Southeastern Jilin is mountainous; the central-western portion is fertile plains; the west is swampy grassland. Jilin has a long winter and a short summer with a brief growing season. Has ample land resources and produces a major food surplus for the nation.

Jinan. Capital city of northeastern CHINA's SHANDONG province. It had a 2010 population of 3,527,566.

Jordan. Country in the MIDDLE EAST formed when the British divided its original Palestinian Mandate into two separate mandates. Total area of 37,738 square miles (97,740 sq. km.) with a population estimated at 9.7 million in 2017. Capital is Amman. Consisting mostly of desert. In 1946 Britain gave the Trans-Jordan Mandate its independence and permitted the establishment of a kingdom under the rule of the Hashemite family. Jordan gave up all claims to the Israeli-occupied WEST BANK and signed a peace treaty with ISRAEL in 1994. Hundreds of thousands of Syrians fled to Jordan in the second decade of the twenty-first century, putting an enormous strain on Jordan's resources. Jordan has the fifth-largest oil-shale reserves in the world.

Jordan Rift Valley. Down-faulted valley in the MIDDLE EAST, between ISRAEL and JORDAN, through which the JORDAN RIVER flows. Geologically, it is a graben that extends from the Gulf of AQABA through the DEAD SEA and the Sea of GALILEE

into the BEKAA VALLEY in LEBANON. Also called the Levant Rift System.

Jordan River. MIDDLE EAST river. Flowing 200 miles (320 km.) between the Sea of GALILEE and the DEAD SEA, it forms the border between ISRAEL's occupied WEST BANK and the country of JORDAN. Vital to the agricultural industry in the Jordan Valley.

Junggar Basin. Basin in northern XINJIANG Autonomous Region, China. Covers 69,000 square miles (179,000 sq. km.). Flanked by the Altay Mountains in the north and the TIAN SHAN MOUNTAINS in the south. The sandy Gurbantunggut Desert is in the basin.

Kabul. Largest city, capital, and economic and cultural center of AFGHANISTAN. Had an estimated population of 4 million in 2017. Strategically located in a high narrow valley and surrounded by rugged, treeless mountains, it is the nodal point of all major roads in the country. Has been the site of devastating internal and external conflicts throughout its 3,000-year history.

Kaema Plateau. Region in the northeastern part of North KOREA. Covers about 15,500 square miles (40,000 sq. km.), with an average elevation of 3,300 feet (1,000 meters). Bounded on the north by Mount PAEKTU; on the west by the NANGNIM MOUNTAIN Range; on the east by the Sea of JAPAN; on the south by the northern tip of the Taebaek Mountains. Abundant forests are located there. Beans, oats, potatoes, millet, and other crops are produced in the highlands.

Kaifeng. City in central-eastern portion of CHINA's HENAN province. Population was 725,573 in 2010. Capital city of numerous ancient dynasties; a major classic Chinese city because of its antiquity and rich cultural value. Northwest of the city is the sacred Buddhist peak Songshan.

Kailas Range. Mountain range of the Trans-Himalayas in SOUTH ASIA. Attracts Hindu pilgrims, who consider it to be the dwelling place of the god Shiva. Mount Kailas (22,028 feet/6,714 meters), the highest peak of the range, is located in southwest TIBET (CHINA). Buddhists believe that Mount Kailas is the sacred center of the universe, from which flow the four great rivers: the

Sutlej, the INDUS, the GANGES, and the BRAHMAPUTRA. In reality, the sources of these rivers lie within a 60-mile (97-km.) radius of Mount Kailas.

Kalimantan. Region on the south part of the island of BORNEO that is part of INDONESIA. Bounded by the SOUTH CHINA SEA, Celebes Sea, MAKASSAR STRAIT, and Java Strait. Population was 13.8 million in 2010. Contains four provinces, with physical features ranging from lush, swampy lowlands and eroded mountain ranges to navigable rivers. Its tropical climate allows for variations, but rainfall averages 150 inches (3,810 millimeters) annually. Rich in resources such as oil, liquefied natural gas, timber, coal, prawns, white pepper, rattan, rubber, and gemstones.

Kamchatka. Peninsula in eastern RUSSIA that divides the Sea of Okhotsk on its west from the Pacific Ocean and Bering Sea on its east. It is about 750 miles (1,200 km.) long, north to south, and about 300 miles (480 km.) across at its widest point, with an area of approximately 140,000 square miles (370,000 sq. km.). It has 1,127 known volcanoes, of which twenty-eight are still active. Most of its inhabitants are ethnic Russians.

Kampuchea. See CAMBODIA.

Kanchenjunga. Third-highest mountain peak in the world. Located in SOUTH ASIA on the India-Nepal border, it has an elevation of 28,208 feet (8,598 meters).

Kandahar. Second-largest city in AFGHANISTAN, in SOUTH ASIA. Located 285 miles (459 km.) southwest of the capital of KABUL, with a 2013 estimated population of 409,700. An important distribution center for Afghanistan's imports and exports, especially wool and fruits.

Kao-hsiung. Second-largest city and major international port in southwestern TAIWAN. Covers 59 square miles (153 sq. km.); population was 1.5 million in 2006. Settled during the latter part of the Ming Dynasty. Opened as a treaty port in 1863. Harbor was developed during Japanese occupation (1898-1945). Rice, sugar, bananas, citrus fruits, and pineapples from the Taiwanese southern agricultural areas are shipped from the port, along with industrial products such as steel, aluminum, petrochemicals, cement, bricks, and tiles.

Kaolan. See LANZHOU.

Karachi. Chief port, largest city, and economic hub of PAKISTAN. Population was more than 21 million people in 2011. Developed in the early eighteenth century; in 1843 it passed to the British, who made it the seat of the Sind government. Was the capital of Pakistan from 1947 until 1959, when ISLAMABAD assumed that role During that period Karachi absorbed tens of thousands of *Muhajirs* (Muslim refugees from INDIA and their descendants) following independence in 1947. The city has grown rapidly and suffers serious problems, including the worst slums in the country and interethnic conflict between Muhajirs and non-Muhajirs.

Karakalpakstan. Autonomous republic in UZBEKISTAN, situated southeast and southwest of the ARAL SEA. Total area of 64,000 square miles (165,600 sq. km.) with a 2017 population of 1.8 million. Capital is Nukus. The Karakalpaks are a Turkic people who are closely related to the Kazaks. During the eighteenth century they settled in the region of the AMU DARYA RIVER. Also known as Karakalpak Autonomous Republic.

Karakoram. Southeastern extension of the HINDU KUSH mountains, between the INDUS and Yarkand rivers in northern PAKISTAN and south central Asia. Runs 300 miles (483 km.), has an average elevation of 20,000 feet (6,096 meters), and contains some of the world's highest peaks—sixty-one peaks are more than 22,966 feet (7,000 meters) high. The Karakoram spreads over disputed territory held by CHINA, INDIA, and Pakistan.

Karakum Desert. Great sandy region in CENTRAL ASIA covering about 115,830 square miles (300,000 sq. km.) of the area of TURKMENISTAN. Karakum means "black sand." The name is also applied to another, smaller desert situated to the northeast of the ARAL SEA in Kazakhstan and known as the Aral Karakum.

Karun River. Largest river in IRAN, in the MIDDLE EAST. Flows north through the ZAGROS MOUNTAINS, then turns west to connect with the SHATT AL-ARAB and the PERSIAN GULF.

Kashmir. See JAMMU AND KASHMIR.

Kathmandu. Capital and largest city of NEPAL in SOUTH ASIA. Population was about 1 million in 2011. Located in the fertile KATHMANDU VALLEY of central Nepal. Landmarks include palaces of the politically dominant Rana family, numerous pagoda-shaped temples, and many other small temples.

New Road is the shopping district of Kathmandu.

Kathmandu Valley. Region in SOUTH ASIA containing NEPAL's three largest cities—KATHMANDU, Patan, and Bhaktapur. Situated in a lush alluvial plain 16 miles (26 km.) long and 12 miles (19 km.) wide, surrounded by some of the world's highest mountains, Nepal's large cities threaten to absorb most of the valley as they continue to expand. Formerly called Nepal Valley.

Kavkazsky Nature Reserve. Natural area set aside at the western end of the CAUCASUS MOUNTAINS in southwestern Russia. Established in 1924, with an area of 1,017 square miles (2,633 sq. km.). The reserve's fauna includes brown bear, lynx, red deer, chamois, wolf, and the European bison.

Kazakhstan. Largest country of CENTRAL ASIA. Total area of 1,052,089 square miles (2,724,900 sq. km.) with a 2018 population of 18,744,548. Capital is Nursaltan. Obtained its independence from the former Soviet Union in 1991. The Kazaks, a Turkic-speaking people, have inhabited the region for centuries but now make up less than half the population.

Kedah. Oldest state in Malaysia; bordered by THAILAND, PENANG, Perlis, and PERAK. Covers 3,660 square miles (9,479 sq. km.), including the Langkawi Islands in the Strait of Malacca. Population was 2 million in 2015.

Keelung. See CHI-LUNG.

Khiva. City in south central UZBEKISTAN. Population was 56,600 in 2004. Dates from the sixth century CE.. Was a center of CENTRAL ASIA's Islamic culture and of a notorious slave market from the seventeenth to the nineteenth centuries.

Khyber Pass. Mountain pass in SOUTH ASIA, 33 miles (53 km.) long, between AFGHANISTAN and PAKISTAN. Indo-Aryans entered the Indian subcontinent from Central Asia through this pass around 2000 BCE.. Turks also came to the upper Indus Valley through this pass, and later the Moguls streamed through it on their way to INDIA.

Kinabalu, Mount. Highest peak (13,455 feet/4,101 meters) in the MALAY ARCHIPELAGO of Southeast Asia. Part of Kinabalu National Park in SABAH. Has a granite plateau about 2,600 feet (800 meters) long that drops into cliffs, precipices, and gullies. Name means "sacred place of the dead." Also called Gunanag Kinabalu.

Kinneret, Lake. See GALILEE SEA OF.

Kobe. City in the OSAKA-Kobe industrial complex, the second-largest industrial area in JAPAN. Population was 1.5 million in 2010. Began as a fishing village; settled during the Paleolithic period. Became an important international port in the eighteenth century. In 1995 an earthquake destroyed large areas of Kobe and its suburbs, killing about 5,500 people and seriously damaging its port and transportation system.

Kolkata. Formerly called Calcutta. State capital and port city of West Bengal, India. Covers 229 square miles (539 sq. km.); population in 2011 was more than 14 million, including 170,000 homeless. Founded by the British on a swampy riverbank of the Hooghly River in 1690 and the first capital of British India. Kolkata was the largest city in India until 1991, when it was surpassed by Mumbai (formerly Bombay). Jute processing has been the traditional industry of Kolkata since the British period.

Korea. Peninsular region in East Asia covering approximately 85,000 square miles (220,150 sq. km.) that comprises North and South KOREA.

Korea, North. Officially the Democratic People's Republic of Korea. Socialist republic that borders the Sea of JAPAN, the YELLOW SEA, and the Republic of Korea (South KOREA) at roughly 38° north latitude. Covers 46,540 square miles (120,538 sq. km.). Population was 25 million in 2018. Capital city is Pyongyang. North Korea's totalitarian government is notorious for its human-rights violations. The country's enormous military, its nuclear-weapons program, and its provocative behavior continue to be of great concern to its East Asia neighbors and to the U.S.

Korea, South. Officially the Republic of Korea; a country on the Korean Peninsula. Separated from the Japanese island of Tsushima by the KOREA STRAIT; borders the Sea of JAPAN, (East Sea) the Korea Strait, the YELLOW SEA, and North KOREA at roughly 38° north latitude. Area of 38,502 square miles (99,720 sq. km.). Population was 51.4 million in 2018. In recent decades, South Korea has emerged as one of the world's leading sources of innovation and new technology. In 2020, among other superlatives, the country is recognized as having the world's 12th-largest economy and a superb health-care system; South Korea even holds the distinction of having the world's fastest Internet-connection speeds.

Korea Strait. Section of the northwest Pacific, extending northeast from the EAST CHINA SEA to the Sea to the Sea of JAPAN, between the south coast of the Korean Peninsula and the Japanese islands of Kyūshū and Honshū. It is about 300 feet (90 meters) deep and split into two parts by the Tsushima Islands.

Krakatoa. Small volcanic island in southwestern INDONESIA in the Sunda Strait, between JAVA and SUMATRA. Its infamous 1883 eruption was one of the most violent volcanic disasters in history;

more than 37,000 people perished. Also called Krakatau.

Krasnodar. Region that extends northward from the crestline of the CAUCASUS MOUNTAINS across the plains east of the BLACK SEA and the Sea of Azov as far as the Gulf of Taganrog. The plains form about two-thirds of the region. It is a major agricultural region. The western Caucasus range that reaches 12,434 feet (3,790 meters) in Mount Psysh occupies the southern third of the region.

Krishna River. River in SOUTH ASIA, flowing southeast through the states of Karnataka and Andhra Pradesh, emptying into the Bay of BENGAL. Runs 800 miles (1,287 km.). Rises in the state of Maharashtra in western INDIA. The Krishna source is sacred to Hindus; the river is named for the god Krishna.

Krung Thep. See BANGKOK.

K2. Second-highest mountain peak in the world; (after Everest) located in the KARAKORAM Range within the Pakistani-controlled part of JAMMU AND KASHMIR. Elevation is 28,250 feet (8,611 meters). The "K" is derived from the first letter of KARAKORAM, and "2" indicates that it was the second peak in the range to be measured. Also called Godwin-Austen.

Kuala Lumpur. Capital and principal economic, administrative, and cultural center of Malaysia; located in the southern MALAY PENINSULA, at the confluence of two muddy rivers. Population was 1.3 million in 2000. Has a hot, humid climate, with high annual rainfall. Produces tin, rubber, and electrical products. Founded by Chinese settlers as a tin-mining camp in 1857.

Kuban River. River in southwestern Russia, 563 miles (907 km.) long and draining 23,600 square miles (61,000 sq. km.). It rises from glaciers on Mount ELBRUS in the GREATER CAUCASUS and flows to the Sea of Azov. The river gave its name to a Cossack group that settled along its northern bank in the late eighteenth and early nineteenth centuries.

Kunming. Capital city of southwestern CHINA's YUNNAN province. It had a 2010 population of 3,278,777.

Kura River. Largest river in the CAUCASUS MOUN-TAINS. Flows 940 miles (1,510 km.) through TURKEY, GEORGIA, and AZERBAIJAN, and drains an area of 72,500 square miles (188,000 sq. km.). Just above TBILISI, a dam has been built along the Mtskheta narrows. Farther down-stream, the narrows near Mingäçevir in Azerbaijan are the site of another dam and hy-droelectric station.

Kurdistan. Mountainous mass of ancient volcanoes and landlocked salt lakes in northeastern TUR-KEY, northwestern IRAN, and northern IRAQ. In-cludes Mount ARARAT of biblical fame. Home to the Kurdish people, who are scattered among the three countries and have aspirations for their own independent state.

Kuroshio. Northeasterly flowing continuation of the Pacific North Equatorial Current between LUZON in the PHILIPPINES and the east coast of JAPAN. Flows past TAIWAN and the RYUKYU Is-LANDS, touches the east coast of Kyūshū, then flows northeast through the KOREA STRAIT to parallel the west coast of Hōnshū. In the Sea of JAPAN, it becomes the TSUSHIMA CURRENT; later becomes the North Pacific Current. It loses much of its force west of the Hawaiian Islands as the south-flowing Kuroshio countercurrent joins the North Equatorial Current, sending much of the warm water back to the Philippine Sea. Warms the southern coastal regions of Japan as far north as TOKYO. Sometimes referred to as the Japanese Current.

Kuwait. Country in the MIDDLE EAST, on the north-west coast of the PERSIAN GULF. Total area of 6,880 square miles (17,818 sq. km.) with an esti-mated population in 2018 of 2,916,467. Capital is Kuwait. Had its origins in the 1899 treaty with the British recognizing Kuwait as a protectorate. In 1961 the protectorate status was terminated and Kuwait became an independent state under the rule of the al-Sabah family. With its large re-sources of petroleum, the country quickly be-came a modern state. Invaded by IRAQ in 1990. Combined U.S., European, and Arab coalition forces defeated the Iraqi army and liberated Ku-wait in the joint operation known as Desert Storm.

Kwai River. River in SOUTH ASIA, rising from the BILAUKTAUNG RANGE (Tenasserim) on the MYANMAR border. It flows 150 miles (241 km.) in a southeasterly direction and drains west central THAILAND. During World War II, the river was the site of a notorious Japanese prisoner-of-war camp, whose British prisoners were forced to build a bridge across the river. Their story be-came the subject of the Academy Award-winning film *The Bridge on the River Kwai* (1957). Also called Khwae Noi River.

Kyonggi. Province in the northwestern section of South KOREA. Bounded by the DEMILITARIZED ZONE to the north, the provinces of Kangwon to the east, and Kyongsang-puk and Ch'ungch'ong-nam to the south, and the YEL-LOW SEA to the west. Total area is 4,196 square miles (10,867 sq. km.); population was 12.3 mil-lion in 2014. Suwon is the provincial capital. SE-OUL is in its geographic middle but was sepa-rated from it administratively in 1946. ANYANG, Bucheon, SONGNAM, and Uijeongbu have devel-oped as Seoul's satellites, and Incheon as its seaport.

Kyoto. City in JAPAN; administrative center of Kyoto prefecture (*fu*) on west central Hōnshū. Located 30 miles (48 km.) northeast of OSAKA, a major Japanese industrial city, and about the same dis-tance from NARA, an ancient city of Japanese culture. Population was 2.6 million in 2010. Built on a gradual north-to-south slope, averag-ing 180 feet (55 meters) above sea level. The capital of Japan from 794 to 1868. Kyoto is a center of Japanese culture and Buddhism. Most Japanese people try to visit Kyoto at least once in their lifetime; about one-third of them visit there every year.

Kyrgyzstan. Mountainous country of CENTRAL ASIA that gained independence from the former So-viet Union in 1991. Total area of 77,201 square miles (199,951 sq. km.) with a 2018 population of 5,849,296. Capital is Bishkek. The Kyrgyz, who were one of the great nomadic groups of

Central Asia, make up half of the population and speak a Turkic language.

Nomads in Kyrgyzstan.

Kyūshū. Southernmost and third-largest of the islands of JAPAN. Bordered by the EAST CHINA SEA on the west and the Pacific Ocean on the east. Population was 13.2 million in 2010. Separated from Hōnshū by the Shimonoseki Strait and from KOREA by the Tsushima Strait, or Eastern Channel. Comprises a series of complex volcanic ranges; the site of Mount Aso, one of the world's largest active volcanoes. Rice, tobacco, sweet potatoes, and citrus fruit are grown there. Name means "nine provinces" and refers to the nine provinces into which the island was divided in ancient times.

Kyzylkum. Red sand desert in KAZAKHSTAN and UZBEKISTAN. Located between the SYR DARYA and the AMU DARYA rivers southeast of the ARAL SEA, with an area of about 115,000 square miles (298,000 sq. km.).

Laguna de Bay. Largest lake in the PHILIPPINES; located on LUZON ISLAND, southeast of MANILA. Crescent-shaped, it is 32 miles (51 km.) long and 25 miles (40 km.) wide, covering 344 square miles (891 sq. km.). Dotted by small islands, its main outlet is the Pasig River. Once an important center for early Chinese traders.

Lahore. Second-largest city and cultural nucleus of PAKISTAN, in SOUTH ASIA. Located close to the border with INDIA, with a population of more than 11 million in 2017. Received hundreds of thousands of refugees and grew rapidly after the partition in 1947. Founded about 2,000 years ago, Lahore reached its peak in the sixteenth century when it became the most famous Mogul city and the cultural focus of Islam. As a seat of royalty, Lahore was adorned with numerous magnificent buildings, including the Badshahi Mosque and the tomb of the Mogul emperor Jahangir.

Lakes. See under individual names.

Lakshadweep. Thirty-six small islands off the coast of the state of Kerala, INDIA. Covers 12 square miles (32 sq. km.) and had a population of 64,473 in 2011, most of whom are Muslims. A union territory of India, formerly known as Laccadive.

Lanzhou. Capital of CHINA's GANSU province; an old walled city. It had a 2010 population of 2,438,595. Provides an overland connection between Central Asia and the heartland. A major transportation and industrial center, famous for its petrochemical and nuclear plants and products such as machinery, railroad equipment, fertilizer, aluminum, and rubber. Founded before 6 BCE.. Also known as Kaolan, Lan-chou, and Lanchow.

Laos. Landlocked SOUTHEAST ASIA country located in the center of the Indochinese peninsula. It was part of French Indochina from the late nineteenth century into the 1950s. It is mountainous, with more than 90 percent of its land more than 600 feet (180 meters) above sea level. The climate is subequatorial and monsoonal. Only about 4 percent of its total area of 91,429 square miles (236,800 sq. km.) is suitable for agriculture. Rice is the principal crop. Its population was 7,234,171 in 2018. Capital is VIENTIANE.

Laptev Sea. Marginal sea of the Arctic Ocean off the coast of northern SIBERIA. The Taimyr Peninsula bounds it on the west and the New Siberian Islands on the east. Its area is about 251,000 square miles (650,000 sq. km.). Its average depth is 1,896 feet (578 meters), and the greatest depth is 9,774 feet (2,980 meters).

Lebanon. Mountainous coastal country in the MIDDLE EAST at the eastern end of the MEDITERRANEAN SEA. Total area of 4,015 square miles (10,400 sq. km.) with an estimated population of

6,100,075 in 2018. Capital is BEIRUT. Part of the French Mandate after World War I, Lebanon achieved independence in 1943. Its fragile coalition government of Christian Maronites and Sunni Muslims fell apart with the intervention of outside forces, including the Palestinian Liberation Organization in 1975 and the Israeli invasion in 1978. Long torn by a multitude of fractious political-militia groups, Lebanon has finally achieved a level of stability in recent years, although refugees from neighboring war-torn Syria—more than 1 million by 2016—are taxing its resources.

Lebanon Mountains. MIDDLE EAST mountain range paralleling the Mediterranean coast of LEBANON and SYRIA. Its highest point is 10,131 feet (3,088 meters) at Qurnet es Sauda. Rough terrain is well-watered on its western slopes by precipitation-laden winds coming off the MEDITERRANEAN SEA. Home to various minority groups that have found refuge from powerful enemies. A parallel range to the east—the Anti-Lebanon Mountains—form part of the border with Syria.

Lesser Hinggan Mountains. See HINGGAN MOUNTAINS, LESSER.

Levant. Geographic term for the coastal countries of the eastern MEDITERRANEAN SEA that form the western end of the FERTILE CRESCENT—SYRIA, LEBANON, ISRAEL, and JORDAN.

Levant Rift System. See JORDAN RIFT VALLEY.

Lhasa. Capital city of TIBET. It had a 2001 population of 130,000. City dominated by the Potala,

The Potala Palace in Lhasa.

former residence of the Dalai Lama. Economic, cultural, and transportation hub of Tibet.

Liaodong Peninsula. Peninsula in northeastern CHINA projecting between the BO HAI inlet and Korea Bay in the YELLOW SEA. Consists largely of low hills whose elevations decrease from north to south.

Liaoning. Province in southern northeastern CHINA, bordering North KOREA to the east. Total area of 58,300 square miles (151,000 sq. km.), with a 2010 population of 43,746,323. Its capital city is SHENYANG. Low hills dominate the east and west. Central Liaoning consists of the Liao River Plain. Has cold winters and hot summers, during which most rainfall occurs. Known for its rich coal, petroleum, and iron ore reserves, which contribute to its massive heavy-industrial capacity. Major population centers are in its central south portion.

Litani River. Small but important river flowing from SYRIA through LEBANON'S BEKAA VALLEY before turning west and emptying into the MEDITERRANEAN SEA. The longest river in Lebanon, and a vital source for the country's water supply, irrigation, and hydroelectricity.

Loess Plateau. Region in northern CHINA with an area of 116,000 square miles (300,440 sq. km.). The plateau is covered with loess deposits varying from 230 feet (70 meters) thick in the southeast to 1,070 feet (326 meters) thick in the west. Severe water erosion has created a highly fragmented surface in some areas.

Lombok. One of the Lesser SUNDA ISLANDS in southern INDONESIA. Located east of BALI across Lombok Strait. Covers 1,826 square miles (4,729 sq. km.), with a population of more than 3 million in 2010. Contains volcanic mountains. Colonized by the Dutch.

Louangphrabang. Former royal capital of LAOS. A port on the MEKONG RIVER, and a trade center for rice, rubber, teak, and handicrafts. Population of 55,250 in 1995. Also known as Luang Prabang.

Lower Burma. Coastal region of MYANMAR. Comprises the Arakan, Irrawaddy, PEGU, and

Tenasserim (*see* BILAUKTAUNG RANGE) divisions. Has numerous islands with good harbors.

Luang Prabang. See LOUANGPHRABANG.

Lucknow. Educational, commercial, and cultural center of the middle Ganges Valley of INDIA, and the capital of UTTAR PRADESH state. Located 260 miles (420 km.) southeast of New DELHI on the Gumati River, with a population of 2.9 million in 2011. Known for its zoological gardens, parks, and National Botanical Gardens.

Luoyang. Historic city in the western part of CHINA's HENAN province. Its population was nearly 1.6 million in 2010. Capital city of numerous ancient dynasties; a major classic Chinese city because of its antiquity and rich cultural value. Known for its historical buildings, Chinese roses, and ceramic ware originating in the T'ang Dynasty (618-907).

Luzon Island. Largest of the Philippine Islands. Site of LAGUNA DE BAY, Lake TAAL, and MANILA BAY, which shelters an excellent harbor. Covers 40,420 square miles (104,688 sq. km.), with a 2010 population of 48.5 million. Coastal region is mountainous, but the central interior has a broad, fertile plain that supports the cultivation of rice, corn, sugarcane, and fruit. MANILA and QUEZON CITY are on the island.

Luzon Strait. Part of a main shipping route between TAIWAN and the PHILIPPINES, connecting the SOUTH CHINA SEA to the west with the PHILIPPINE SEA to the east. Extends more than 200 miles (320 km.). In its southern part are a series of channels dotted with islands.

Maale. See MALDIVES.

Macao. Special administrative region of CHINA, adjacent to southern GUANGDONG. Population was 650,834 in 2016. Became a Portuguese colony in the sixteenth century and was returned to China in 1999. World's largest gambling center and an important tourism destination for southern China and Southeast Asia.

Madras. See CHENNAI.

Madura. Island in southwest INDONESIA in East Java province. Separated from the island of JAVA by the Madura Strait. Population was more than 3.6 million in 2010. Noted for salt panning and cattle breeding; exports salt, teak, copra, and coconut oil.

Makassar Strait. Southeast Asian channel separating BORNEO from the Indonesian island of SULAWESI. Average width is 9.3 miles (15 km.). Site of sea and air battles between the Allies and JAPAN during World War II. Also known as Selat Makasar.

Malabar Coast. Densely populated coastal plain of INDIA. Stretches in a north-south direction for about 525 miles (845 km.) from GOA to Kanniyakumari (Cape Comorin), the southernmost point on the DECCAN PLATEAU. Because it is located immediately west of the WESTERN GHATS, the Malabar Coast receives abundant rainfall from the southwest monsoon. Exceptionally fertile soil, abundant rainfall, and warm climate make this one of the major rice-growing areas of India.

Malaweli Ganga River. Longest river in the island nation SRI LANKA, at only 208 miles (335 km.). Begins in the Central Province on the Hatton Plateau, flows northeast through extensive tea and rubber plantations, and discharges into the Bay of BENGAL. A multipurpose river development plan was launched in the late 1970s to irrigate new land with water from the Malaweli Ganga and produce hydropower. Hydroelectricity from six dams on the river supply Sri Lanka with more than 40 percent of its electricity.

Malay Archipelago. Largest island group in the world, located southeast of Asia and northwest of Australia, between the Indian and Pacific oceans. Includes NEW GUINEA, SULAWESI, BORNEO, the Philippine Islands, THE SUNDA ISLANDS, the MOLUCCAS, and numerous smaller islands. Many volcanoes, active and extinct, dot this mountainous area, which is extremely fertile and luxuriant with tropical vegetation.

Malay Peninsula. Extension of the mainland of SOUTHEAST ASIA; includes parts of MYANMAR, THAILAND, and Malaysia. About 700 miles (1,120 km.) long, with a maximum width of 200 miles (320 km.). Bounded by bodies of water (gulfs, seas, and straits); has rich vegetation and mineral deposits.

Maldives. Island nation and the smallest country of SOUTH ASIA, both in land area and in population. Covers only 115 square miles (298 sq. km.), with a population of 392,473 in 2018. Located about 300 miles (480 km.) southwest of the southern tip of INDIA. It comprises 2,000 coral atolls, of which 200 are inhabited. Almost all the people are Muslims; about 25 percent of them live on the capital island, Maale. Fish, coconuts, and tourism are the main industries of the country.

Mali River. River in northern MYANMAR that flows 200 miles (322 km.) south from the northern hill slopes to unite with the NMAI RIVER and form the IRRAWADDY RIVER.

Maluku. See MOLUCCAS.

Mandalay. One of the largest cities in MYANMAR; capital of Mandalay Division, and a major trading and communications center. The cultural and religious center of Buddhism, boasting many monasteries and over 700 pagodas. Population was 1,225,133 in 2014. Noted for silk weaving, jade cutting, brewing, distilling, and silver work. Built in 1850, it was the capital of the independent kingdom until captured by the British in 1885.

Mangla Dam. See TARBELA DAM.

Manila. Capital, second-largest city, (after Quezon City) and chief seaport of the PHILIPPINES, and the second-largest (after Jakarta, Indonesia) metropolis in SOUTHEAST ASIA. Population of 1.65 million in 2010. Located on the eastern shore of MANILA BAY, in the violent typhoon belt. Produces textiles, clothing, and electronics equipment.

Manila Bay. Large inlet of the SOUTH CHINA SEA in northern PHILIPPINES. Is 35 miles (56 km.) long and can be entered by a channel that is 11 miles (18 km.) wide. The Spanish fleet was destroyed here in 1898 during the Spanish-American War.

Maoke Mountains. Ancient mountain chain of sandstone rocks and slate in INDONESIA. Part of a system that crosses NEW GUINEA from east to west, they drop sharply to a marshy plain along the ARAFURA SEA. The lower slopes contain coniferous forests, alpine vegetation, and an ever-green rain forest. Also called Pegunungan Maoke.

Marmara, Sea of. Body of water in western TURKEY between the AEGEAN SEA and the BLACK SEA. Linked to these by the straits of the DARDANELLES and the BOSPORUS. Covers about 4,300 square miles (111,140 sq. km.).

Mary. City and administrative center of Mary *oblast* (province) in TURKMENISTAN. Population was 123,000 in 2009. A center for the huge Shatlyk gas field and a transport junction.

Matsu Island. Main island in the nineteen-island Matsu group, which makes up the Taiwanese county (*hsien*) of Lien-kiang. Located in the EAST CHINA SEA off the Min River Estuary, about 130 miles (210 km.) northwest of CHI-LUNG, TAIWAN. Population was 9,359 in 2004. Originally a port of Fukien province, CHINA; in 1949, occupied by the Nationalist Chinese as they fled from the communists. Remained under Taiwanese military rule until 1972, when it became a regular county of Taiwan. Fishing is the main industry; the hilly land supports some agriculture, including vegetable, grain, hog, and chicken production.

Mecca. Islam's holiest city. Located in southwestern SAUDI ARABIA near the RED SEA, it has more than 1.5 million inhabitants. (2010) When an attempt was made in 622 to kill the Prophet Muhammad in Mecca, he and his followers fled to Yathrib, which they renamed MEDINA. In 630 he and his followers returned to Mecca and destroyed its idols, turning its traditional temple into a mosque. All Muslims are expected to make a *hajj* (pilgrimage) there at least once in their lifetimes, usually during the month of Ramadan.

Medan. Capital of INDONESIA's North Sumatra Province on the island of SUMATRA. Population was 2 million in 2010. Has rail and road connections to Lake TOBA resort area. Tobacco and tea-processing factories and machinery plants have made their mark on the city, which also produces fiber products, ceramics, tile, soap, rubber, tobacco, tea, oil palms, and coffee. Once the capital of the East Coast Dutch residency.

Medina. Second-holiest of Islamic cities, (after MECCA) located in western SAUDI ARABIA, north of MECCA. Its population in 2010 was 1,100,093. Formerly called Yathrib; renamed by Muhammad and his followers who fled there from Mecca in 622 after an attempt on Muhammad's life. Many Muslims make pilgrimages there, often as part of their pilgrimages to Mecca. The Arabic word for "city."

Mediterranean Sea. Body of water extending from the Strait of GIBRALTAR eastward to the coasts of TURKEY and the LEVANT area of the MIDDLE EAST. Covers about 1,158,000 square miles (3 million sq. km.) with an average depth of about 5,000 feet (1,525 meters).

Meghalaya Plateau. See SHILLONG PLATEAU.

Meghna River. One of the three major rivers of BANGLADESH, SOUTH ASIA. Begins in the SHILLONG PLATEAU in the Indian state of ASSAM and enters Bangladesh from the northeast in the form of two tributaries, the Surma and Kushiyara. Those reunite to form the main channel, which flows southwest to meet the Padma at Chandpur and finally empties into the Bay of BENGAL.

Mekong River. Major river in CAMBODIA. Flows for 2,600 miles (4,180 km.), starting in the glacial meltwaters of eastern TIBET, crossing CHINA's YUNNAN province, continuing through Cambodia, LAOS, and VIETNAM, before ending in the SOUTH CHINA SEA. Forms a border between MYANMAR and Laos, and most of the border between Laos and THAILAND. Marked by swift rapids and precipitous gorges, it is navigable for only 1,000 miles (1,600 km.) downstream from its source. Thick silt deposits expand its delta annually, fostering rice production in the Lower Mekong Valley.

Mesopotamia. Geographic term for the fluvial plain in IRAQ and SYRIA encompassing the EUPHRATES and TIGRIS rivers and their tributaries. Means "land between the rivers."

Middle and Lower Chang Jiang Plain. See CHANG JIANG PLAIN.

Middle East. Term loosely applied to the predominantly Muslim countries of Southwest Asia and North Africa. Usually understood to include IRAN, ISRAEL, and all Arab countries of the region; sometimes used to include TURKEY. Replaced earlier term, Near East. Both terms reflected European perspectives on the region's location relative to Western Europe.

Moluccas. Island group and province in eastern INDONESIA. Separated from SULAWESI by the Molucca Sea and from IRIAN JAYA by the Halmahera Sea. Was called the Spice Islands because of its abundant nutmeg, mace, and cloves. Also produces cassava, taro, yams, sweet potatoes, copra, and coffee. Its 1,000 islands, ranging from tiny atolls to large, mountainous islands, contain many active volcanoes. Has a tropical climate and many Austral-Asian species of wildlife. Also known as Maluku.

Mongolia. Mountainous landlocked republic located in East Asia between northen CHINA and Russia, covering 603,908 square miles (1,564,116 sq. km.). Estimated population in 2018 was 3,103,428. Also known as Outer Mongolia. See also INNER MONGOLIA.

Mumbai. Formerly known as Bombay, INDIA's largest and most prosperous city. Had a population of 18,394,912 in 2011. Capital of Maharashtra state of India, India's busiest international seaport, and the country's primary air transport gateway. The headquarters of nearly all commercial banks in India are located there. Also the center of an active film industry (often called Bollywood), which produces more feature films than Hollywood.

Myanmar. Asian country formerly known as Burma; located along the eastern coasts of the Bay of BENGAL and the ANDAMAN SEA; variously considered part of both SOUTH ASIA and SOUTHEAST ASIA. With an area of 261,228 square miles (676,578 sq. km.)—about the size of the U.S. state of Texas—the country is exceptionally mountainous. Only about one-sixth of its land is suitable for farming, and about a tenth of that land is irrigated. Half of the country is forested. Myanmar has rich deposits of such minerals as silver, rubies, and tungsten. Total population was 55,622,506 in 2018, about 70 percent of whom are ethnic Burmans. The western and northern hills have diverse tribal groups such as the Kachin in the north. Buddhism is the religion of 90 percent of the people. Government moved the capital to the new city of Naypyidaw in 2005.

Nagasaki. Largest city and capital of Nagasaki prefecture in JAPAN. Located on the western side of Kyūshū, at the mouth of the Urakami-gawa, the Urakami River. Population was 443,766 in 2010. Second of Japan's ports to be opened to international trade and the only Japanese port permitted to trade with the outside world during the Tokugawa Shogunate (1603-1867). In the early twentieth century, it became a major shipbuilding city; for this reason, U.S. forces dropped an atomic bomb on the city in 1945, killing more than 70,000 people. Although much of the city was destroyed, many old buildings remain, including a Chinese temple built in 1629 that is an excellent example of Ming Dynasty architecture.

Nagorno-Karabakh. Region of southwestern AZERBAIJAN. Covers 1,700 square miles (4,400 sq. km.). The population is about 80 percent Armenian. There have been various levels of armed conflict between ARMENIA and Azerbaijan over the region since the collapse of the Soviet Union. Armenia has control over Nagorno-Karabakh, and in 2017 renamed it Artsakh.

Najd. Central region of the ARABIAN DESERT and the ARABIAN PENINSULA in the MIDDLE EAST. Located between the SYRIAN DESERT to the north and the RUB' AL KHALI to the south. Essentially uninhabited, it once was important for widely scattered albeit small mineral deposits of gold, silver, copper, lead, and zinc.

Nakhichevan. Exclave and autonomous republic of AZERBAIJAN. Bounded by ARMENIA to the north and east, IRAN to the south and west, and TURKEY to the west. Total area of 2,124 square miles (5,501 sq. km.) with a 2006 population of 376,400. The republic, especially the capital city of Nakhichevan, has a history dating back to about 1500 BCE.

Namangan. City and administrative center of Namangan *oblast* (province) in UZBEKISTAN. Population was 475,700 in 2014. Known for its food and other light industries, especially the processing of cotton.

Nanchang. Capital city of southern CHINA's JIANGXI province. It had a 2001 population of 1,386,500.

Nangnim Mountains. Mountains stretching north to south, west of the KAEMA HIGHLANDS, in the central area of North KOREA. They form the watershed between Kwanbuk, the northeastern part of the Korean Peninsula, and Kwanso, the northwestern part.

Nanjing. Capital city of CHINA's JIANGSU province. Located on the lower reaches of the YANGTZE (Chang Jiang) River with a 2010 population of 5,827,888. Was capital of numerous ancient dynasties and the Republic of China before it moved to TAIWAN. The site of the "Rape of Nanjing" between late 1937 and early 1938, when about 300,000 Chinese soldiers and civilians were massacred by invading Japanese.

Nankan. See MATSU ISLAND.

Nanking. See NANJING.

Nanling Mountains. Major geographic divide in southern CHINA separating the YANGTZE (Chang Jiang) River from the ZHU JIANG drainage systems.

Nansei Islands. See RYUKYU ISLANDS.

Nansei Shoto Trench. See RYUKYU TRENCH.

Nara. Capital of Nara-ken (prefecture) on southern Hōnshū. Located in the hilly northeastern edge of the Nara Basin, 20 miles (30 km.) east of OSAKA. Population was 366,591 in 2010. JAPAN's capital from 710 to 784, when it was called Heijo-kyo. Its many ancient Japanese Buddhist temples and artifacts, including the Seven Great Temples of Nara, are the basis of its tourist industry.

Narmada River. River in SOUTH ASIA that rises in Madhya Pradesh of central INDIA and flows west through the state of GUJARAT to the Gulf of Khambhat. Runs 800 miles (1,300 km.). Is sacred to Hindus, and many holy bathing sites line its banks. During the 1990s the river was the proposed site for India's largest dam. However, the plan called for relocating thousands of people and posed many ecological problems, so the World Bank—its major funding organization—withdrew from the project A long legal battle ensued, but in 2000, India's Supreme Court ruled that work on the dam could go forward. The Sardar Sarovar dam began operating in 2007.

Naryn. City and administrative center of Naryn *oblast* (province) in KYRGYZSTAN. Population was 40,000 in 2019. Located along the Naryn River at an elevation of 6,725 feet (2,050 meters). Founded as a fortified point on the trade route from Kashgar in Sinkiang in CHINA to the Chu River Valley, Naryn became a city in 1927. Has a number of small industries and a music and drama theater.

Near East. See MIDDLE EAST.

Negev. Desert region in southern ISRAEL inhabited by nomadic herders. Its northwestern corner has been more agriculturally productive through irrigation made possible with Israel's National Water Carrier, a canal system bringing fresh water from as far away as the Sea of GALILEE.

Nepal. Himalayan state of SOUTH ASIA. Total area of 56,827 square miles (147,181 sq. km.) with a 2018 population of 29,717,587. Capital is KATHMANDU. The Himalayas cover 90 percent of the land area of Nepal, and its terrain consists mostly of mountain slopes. Nepal contains the world's highest mountain, Mount Everest. Agriculture is practiced only in the southern low-lying Terai Plain, where more than half of the population lives. About 90 percent of Nepalese are Hindus and a similar percentage of people speak Nepali, a language related to Indian Hindi. With the Himalayan peaks as the main attraction, Nepal has a substantial tourist industry. The Gurkhas, who hail from Nepal, have distinguished themselves in the British and Indian armies and in UN peacekeeping operations. The monarchy, which existed for 240 years, was abolished in 2008. Nepal's constitution (adopted in 2015) made the country a parliamentary republic.

Nepal Valley. See KATHMANDU VALLEY.

New Delhi. See DELHI.

New Guinea. Second-largest island in the world. Located in Southeast Asia, east of the MALAY ARCHIPELAGO and north of Australia. Divided into Papua New Guinea in the east and the Indonesian province of IRIAN JAYA in the west. Covers 305,000 square miles (790,000 sq. km.). Population in 2011 was 7 million.

Nicobar Islands. See ANDAMAN AND NICOBAR ISLANDS.

Ningxia. Autonomous region in northwestern CHINA. Total area of 25,640 square miles (66,400 sq. km.), with a 2010 population of 6,301,350. Its capital city, YINCHUANG, had a 2010 population of 1,290,170. Comprising uplands in the north and the LOESS PLATEAU in the south, it has mostly harsh arid lands with the exception of the well-watered YELLOW RIVER valley. Has a large Muslim population.

Nippon. See JAPAN.

Nippon Arupusa. See JAPANESE ALPS.

Nmai River. Southeast Asian river in Upper MYANMAR Flowing south from the southeast corner of TIBET, it runs 300 miles (483 km.) and joins the MALI RIVER.

North China Plain. Floodplain in northern CHINA. Formed by the lower reaches of the YELLOW RIVER (Huang He) and its tributaries, it covers 127,000 square miles (329,000 sq. km.). The plain is mostly flat land below 164 feet (50 meters). The massive mud and silt load of the Yellow River causes rapid deposition, leading to an elevated river bed, unstable course, frequent dam breaks, and floods.

North Korea. See KOREA, NORTH.

North Korea Cold Current. Surface ocean current flowing southward east of KOREA near Vladivostok, RUSSIA, which forms a small counterclockwise swirl in the Sea of JAPAN.

North Ossetia. See ALANIA.

North Vietnam. See VIETNAM.

Northeast China Plain. Floodplain in northeastern CHINA. Comprises the fertile Songhua Jiang Plain in the north and the fossil-energy-rich Lower Liao He Plain in the south. A series of low uplands forms the divide between the two river systems.

Nursultan. Selected as Kazakhstan's new capital (replacing Almaty) in 1994. From 1998 to 2019 it was called Astana. Renamed Nursultan in 2019. Population in 2018 was 1,032,475.

Okinawa. Prefecture in JAPAN comprising the RYUKYU ISLANDS, of which Okinawa Island is the largest. Area is 454 square miles (1,176 sq. km.); population was 1.3 million in 2010. Once a semi-independent kingdom influenced by the Chinese and Japanese. Okinawa Island is 70 miles (112 km.) long and about 7 miles (11 km.) wide. During World War II, a bloody battle was fought on Okinawa Island between the United States and Japan. The United States held Okinawa from 1945 to 1972. Many Japanese people have objected to U.S. bases remaining on Okinawa. Sweet potatoes, rice, and soybeans are grown there; sake (rice wine), textiles, and lacquerware are manufactured; petroleum is drilled offshore.

Oman. Country in the MIDDLE EAST, located at the southeastern corner of the ARABIAN PENINSULA. Includes the tip of the Musandam peninsula, which juts into the Strait of HORMUZ, the strategic choke point for entering and exiting the PERSIAN GULF. Total area of 119,499 square miles (309,500 sq. km.), with an estimated population of 4,613,241 in 2017. Capital is Muscat. Began modernizing in 1970, when Crown Prince Qaboos, the Western-educated son of the sultan, overthrew his father. Has allowed the United States to use an island off its coast as a military base.

Oman, Gulf of. Arm of the ARABIAN SEA in the MIDDLE EAST. Located at the entrance to the Strait of HORMUZ between OMAN and IRAN.

Ordos Plateau. Situated in northern CHINA between the northern bend of the YELLOW RIVER (Huang He) and the Great Wall in the south. A rolling arid upland covering 15,800 square miles (41,000 sq. km.).

Orontes River. MIDDLE EAST river that begins in the Anti-Lebanon Mountains of SYRIA. Flows northward past the cities of Homs (known for its ancient waterwheels) and Hama before turning westward into the Ghab Valley, a northern extension of the BEKAA VALLEY in LEBANON. Passes into TURKEY, turns west across the northern part of the LEBANON MOUNTAINS (Jabal an Nusayriyah), and empties into the MEDITERRANEAN SEA near Antioch (Antakya). Also known in Arabic as the Asi.

Osaka. City in the third-largest urban, industrial conglomerate in JAPAN, and capital of the Osaka prefecture. Located in the west central area of Hōnshū nshū. Population was 2,691,185 in 2015. Osaka-Kobe industrial area is rivaled only by Tokyo-Yokohama in production and growth. Settled during the Paleolithic period. By 300 CE, it was an important political center. Many ancient burial mounds, including the tomb of Japanese emperor Nintoku, are there.

Outer Mongolia. See MONGOLIA.

Pacific Rim. Modern term for the nations of Asia, North and South America, Oceania, and Australia that border, or are in, the Pacific Ocean. Used mostly in discussions of economic growth.

Paektu, Mount. Highest mountain on the Korean Peninsula. Located at the edge of the KAEMA PLATEAU in North KOREA. An extinct volcano topped by a large crater; 9,022 feet (2,744 meters) at its highest point.

Pakistan. Second-most populous Muslim nation in the world, located in SOUTH ASIA. Total area of 307,374 square miles (796,095 sq. km.) with a 2018 population of 207,862,578. Capital is ISLAMABAD. Consists of arid lowlands and high mountains. Ninety-six percent of its people belong to the Islamic faith, and Islam is the state religion. Despite religious uniformity, the country suffers ethno-linguistic fragmentation. Urdu is the official language but is spoken by only 10 percent of the population. Its agricultural economy depends completely on the INDUS RIVER and its tributaries. Textiles are its largest industry. BANGLADESH, one of the two provinces of Pakistan from 1947 to 1971, broke away from Pakistan in December 1971. Today, Pakistan is striving to shore up its democratic institutions amid many internal and external challenges.

Pakxe. Railroad and river port in southwestern LAOS. Population was 119,848 in 2010. Main industry is the manufacture of building materials.

Palawan. Island province of western PHILIPPINES, between SULU and SOUTH CHINA seas, in Southeast Asia. Covers 4,550 square miles (11,785 sq. km.), with a population of 849,469 in 2015. An isolated and little-developed region with beautiful beaches and unique flora and fauna.

Palestine. Historic region of the MIDDLE EAST. Great Britain applied the biblical name of Palestine to one of the territories mandated to it by the League of Nations after World War I. Part of this territory is now the modern state of ISRAEL. Mandate Palestine was about 150 miles (240 km.) long and 80 miles (128 km.) wide, stretching from the MEDITERRANEAN to the ARABIAN DESERT, and from the Litani River on the north to Egypt in the south. After declaration of the state of Israel in 1948, the status of Palestine Arabs remained unsettled, and creation of a separate Palestinian state became a volatile issue that has kept the entire Middle East unsettled ever since. In 2020, Palestine officially consisted of only the territory surrounding JERUSALEM on the WEST BANK of the JORDAN RIVER and the GAZA STRIP.

Paracel Islands. Group of 130 small, uninhabited coral islands and reefs in the SOUTH CHINA SEA. Located 250 miles (400 km.) east of central VIETNAM and about 220 miles (350 km.) southeast of CHINA. Each covers less than one square mile. Claimed by TAIWAN, Vietnam, and China, because of the discovery of oil in the South China Sea. Also called His-sha and Xisha.

Pearl River. See ZHU JIANG RIVER.

Pegu. Capital city of Pegu Division in MYANMAR. Population was 491,434 in 2014. A rail junction located in a rice-producing region, it has teak forests in the nearby Pegu Yoma Mountains. Formerly a major port, it is now a fishing area and Buddhist pilgrimage center. Also known as Bago.

Pegunungan Maoke. See MAOKE MOUNTAINS.

Peking. See BEIJING.

Penang. Northwestern Malaysian state made up of two parts: Penang Island in the Strait of Malacca, and Seberang Perai, on the peninsular mainland. Its two sectors are joined by the longest bridge in SOUTHEAST ASIA. The third-smallest of thirteen Malaysian states, with an area of 399 square miles (1,033 sq. km.). Main crops are rice and rubber. Also called Pinang.

Peng-hu Islands. Group of islands that are a county of TAIWAN. Located about 30 miles (50 km.) off Taiwan's west coast. Entire group occupies 49 square miles (127 sq. km.). Population was 101,758 in 2014. Largest island is P'eng-hu, on which half the country's population lives. Volcanic in origin, comprises basalt surrounded by coral reefs. On average, 100 to 130 feet (30 to 40 meters) above sea level; highest peak is 157 feet (48 meters) tall. Climate is warm; from June until September, rainfall averages 35 inches (890 millimeters). The islands have no rivers, so water shortages occur during the dry season. Islanders' chief occupation is fishing. Also known as the Pescadores, the Portuguese word for "fishermen."

Perak. State in Malaysia bordering THAILAND. Covers 8,099 square miles (20,976 sq. km.). Hosts a naval base; produces silver, rice, palm oil, rubber, timber, and pineapples.

Persia. See IRAN.

Persian Gulf. Shallow marginal sea of the Indian Ocean located between the ARABIAN PENINSULA and southwestern IRAN; covers about 92,500 square miles (240,000 sq. km.). It is about 615 miles (990 km.) long and 125 to 185 miles (201 to 300 km.) wide, and reaches depths of more than 330 feet (100 meters). Its seafloors contain vast quantities of petroleum and natural gas, which has made the states bordering its shores wealthy. Known as the Arabian Gulf to most people in Arab nations.

Pescadores Islands. See PENG-HU ISLANDS.

Philippines. SOUTHEAST ASIA archipelago nation comprising some 7,100 islands and islets that lie about 500 miles (800 km.) off the mainland of Southeast Asia and cover about 115,800 square miles (300,000 sq. km.). Population was

105,893,381 in 2018. The country takes its name from King Philip II of Spain, which began colonizing the islands in the sixteenth century. The islands remained under Spanish rule until the Spanish-American War of 1898 and then were under U.S. control for an additional forty years. The Philippines is the fourth most populous country in which English is an official language; the Philippines and Timor-Leste are the only predominantly Roman Catholic countries in Southeast Asia. The Filipino people, however, are Asian in consciousness and aspiration. It has great extremes of wealth and poverty. Educationally, it is among the most advanced of Asian countries, having a high literacy rate. MANILA is the capital and second-largest city (after Quezon City) in the country.

Phnom Penh. Capital of CAMBODIA. A port at the junction of the MEKONG, Bassac, and Tonle Sap rivers; has an outlet to the SOUTH CHINA SEA through the Mekong Delta in VIETNAM. Founded by the Khmers in the fourteenth century, it succeeded ANGKOR as capital in the mid-fifteenth century. Brutally occupied and badly damaged during the mid-1970s by the Khmer Rouge, the city nevertheless retains some classic Buddhist and French colonial architecture. Also called Nam-Vang.

Central Market in Phnom Penh.

Pinang. See PENANG.

P'ing-tung. Southernmost county of TAIWAN, bordered by KAO-HSIUNG county (*hsien*) on the northwest, by T'ai-tung county on the northeast, and by the LUZON STRAIT on the southwest. Total area is 1,072 square miles (2,776 sq. km.);

population was 839,001 in 2016. The Central Range mountains, with an elevation of 2,300 feet to 10,000 feet (700 to 3,050 meters) cover its southeastern part. The center of sugar refining in Taiwan. Other crops raised there include paddy rice, tobacco, sweet potatoes, bananas, and pineapple.

Po Hai. See BO HAI.

Pokhara. Second-largest city of NEPAL in SOUTH ASIA, with an estimated population of 402,995 in 2011. It is a popular tourist center and has an airport with regular air service to KATHMANDU.

Pontic Mountains. Belt of mountains in the MIDDLE EAST, along TURKEY'S BLACK SEA coast. Altitudes vary from 5,000 to 13,000 feet (1,500 to 3,960 meters). The northern slopes receive abundant rainfall and are forested.

Pusan. Largest port and second-largest city of South KOREA; capital of Kyongsang-nam *do* (province). Located at the southeast tip of the Korean Peninsula. Population was 3,414,950 in 2010. Located on a bay at the mouth of the Naktong River, across the Korea Strait from the Japanese island of Tsushima. Under Japanese occupation (1910-1945), Pusan developed into a modern port. Temporary capital of South Korea during the Korean War. During the Koryo Dynasty (the tenth to fourteenth centuries), it was named Pusanpo (Korean *pu*, meaning "kettle," *san*, meaning "mountain," and *po* meaning "harbor") for the kettle-shaped mountain behind the city. Also known as Busan.

Pyongyang. Capital of the Democratic People's Republic of Korea (North KOREA). Located on the TAEDONG RIVER, 30 miles (48 km.) inland from the Korea Bay of the YELLOW SEA. Population was 2,581,076 in 2008. A major textile and food processing center; many museums are located there. Thought to be the oldest city on the Korean Peninsula. Founded in 1122 BCE, but its recorded history did not begin until 108 BCE when a Chinese trading company was begun in the area. The city was completely rebuilt following the Korean War, and the skyline is noted for its many tall buildings.

Pyongyang, North Korea.

Qatar. Sheikhdom occupying a peninsula in the PERSIAN GULF. Total area of 4,473 square miles (11,586 sq. km.) with a population of 2,363,569 in 2018. Capital is Doha. Became independent in 1971 after being a British protectorate. Its economy is based on revenues from oil; when that declines, it will be replaced by revenue from its vast reserves of natural gas.

Qilian Mountains. Mountain range in the northeast of the TIBETAN PLATEAU, between the HEXI CORRIDOR and the Qaidam Basin, CHINA. Mountains stretch 620 miles (1,000 km.) and trend northwest-southeast. The highest peak is the Tuanjie Peak at 21,800 feet (6,644 meters).

Qinghai. Province in CHINA. Situated on northeastern TIBETAN PLATEAU. Total area of 278,400 square miles (721,000 sq. km.), with a 2010 population of 5,626,723. Its capital city, XINING, had a 2010 population of 1,153,417. Mainly a pastoral animal-herding region with some mineral extraction. Lake Qinghai, the largest salt lake of China, is in the northeast. Its population is clustered mainly in the northeast.

Qinling-Daba Mountains. Mountain range extending between the eastern TIBETAN PLATEAU and the southern NORTH CHINA PLAIN. Forms part of the physical barrier dividing northern and southern CHINA.

Quelpart Island. See CHEJU ISLAND.

Quemoy Islands. Group of twelve islands under the jurisdiction of TAIWAN. Located in the Amoy Bay, about 184 miles (296 km.) northwest of KAO-HSIUNG, Taiwan. Population was 97,364 in 2010. Principal island is Quemoy. When the Nationalist Chinese were driven from the mainland of CHINA by the communists in 1949, they occupied the Quemoy Islands. In 1958 the Chinese bombarded Quemoy, and it was placed under Taiwanese military rule until 1992. Farming is the main occupation.

Quezon City. Largest city, and government and tourist center in the PHILIPPINES. Located in Rizal Province on LUZON ISLAND, adjacent to MANILA. Primarily a residential city, with a population of 2,761,720 in 2010. Named after Manuel Louis Quezon y Molina, first president of the Philippine Commonwealth.

Araneta City is one of the major commercial centers in Quezon City.

Rangoon. See YANGON.

Rann of Kutch. Region of salt flats that is submerged for half of the year, located in SOUTH ASIA south of the THAR Desert. Covers 9,000 square miles (23,310 sq. km.) over northwestern GUJARAT in INDIA and southern Sind province in PAKISTAN. The scene of Indo-Pakistan wars in 1965 and 1971. An international tribunal settled the dispute over the region in 1968. Also spelled Rann of Kachchh.

Red Basin. See SICHUAN BASIN.

Red River. Principal river in VIETNAM that rises in the mountains of CHINA's YUNNAN Province, flows southeast, passes through HANOI, and then empties into the Gulf of TONKIN. Runs 750 miles (1,200 km.) and has a rich delta.

Red Sea. Body of water separating the ARABIAN PENINSULA from the continent of Africa. Covers about 169,000 square miles (437,700 sq. km.); part of a long rift valley extending from Mozambique in Africa to TURKEY, the eastern side of which is moving in a turning eastern motion away from the rest of Africa and the Mediterranean coast.

Republic of China. See TAIWAN.

Republic of Korea. See KOREA, SOUTH.

Riau. Archipelago in western INDONESIA, SOUTHEAST ASIA. Located southeast of the MALAY PENINSULA and east of SUMATRA. Its islands vary in size and shape from small coral reefs to large mountainous terrain. Bauxite, rubber, rice, and pepper are produced there.

Ring of Fire. Nearly continuous ring of volcanic and tectonic activity around the margins of the Pacific Ocean.

Riyadh. Capital and largest city of SAUDI ARABIA, in the MIDDLE EAST. Located about 240 miles (390 km.) inland from the PERSIAN GULF. In 2010, the city had a population of 5,188,286.

Rub' al Khali. Largest expanse of sandy desert in the world. Located in the southeastern corner of the ARABIAN PENINSULA in the MIDDLE EAST. Covers about 250,000 square miles (650,000 sq. km.). Few people inhabit this truly hostile area. The name means "Empty Quarter."

Ryukyu Islands. Western Pacific island chain belonging to JAPAN, from whose KYŪUSHŪ Island it stretches in a shallow arc about 650 miles (1,050 km.) toward TAIWAN. Total population in 2010 was 1,392,818. Largest island is OKINAWA. Also called the Nansei Islands.

Ryukyu Trench. Ocean trench that runs north along the eastern edge of the RYUKYU ISLANDS in the Philippine Sea between TAIWAN and the Japanese archipelago. Located about 60 miles (100 km.) south of OKINAWA. Covers 52,000 square miles (135,000 sq. km.); 1,398 miles (2,250 km.) long; averages 37 miles (60 km.) wide; maximum depth is 24,629 feet (7,507 meters). Mostly covered by red clay. Also called the Nansei-Shoto Trench.

Sabah. State of Malaysia, on the island of BORNEO in Southeast Asia, adjacent to SARAWAK. Covers 28,425 square miles (73,620 sq. km.), with a 2010 population of 3,206,742. Capital is Kota Kinabalu. Has an indented coastline 900 miles (1,450 km.) long. Heavily forested mountains and generally swift streams, broken by rapids, characterize its geography, which allows for large bays and natural harbors.

Sahul Shelf. Indonesian extension of the northern coastal shelf of Australia, which includes NEW GUINEA and Aru Islands. Has shallow seas. Because it is more closely linked to Australia than to Asia, its animals are similar to those found in Australia.

Saigon. See HO CHI MINH CITY.

Salween River. River in MYANMAR; part of the border between Myanmar and THAILAND. Runs 1,750 miles (2,816 km.). Rises in CHINA and enters northeastern Myanmar, flowing south before emptying into the Gulf of Martaban. Because of its numerous rapids, is navigable only 74 miles (120 km.) above its mouth. In 2013 the Chinese government moved forward on plans to build a number of hydroelectric dams on the river, despite environmental concerns.

Samarkand. City in east-central UZBEKISTAN; one of the oldest cities of CENTRAL ASIA. Population was 530,400 in 2018. Economy is primarily agricultural. In the fourth century BCE, it was the capital of Sogdiana, and was captured in 329 BCE by Alexander the Great. In 1365 CE, it became

Shirdar Madrasa in Samarkand.

the capital of the empire of Timur (Tamerlane) and became the most important economic and cultural center in Central Asia.

Sapporo. Administrative and commercial center of Hōkkaidō. Located on the Ishikari-gawa (Ishikari River) in Hōkkaidō Territory. Population was 1.9 million in 2010. Otaru on the Sea of JAPAN is its outport. Foodstuffs, sawmilling, printing, and publishing are major commercial activities. A popular center for skiing and winter sports: The 1972 Winter Olympics were held there and it has an annual snow festival noted for its snow and ice sculptures. Also famous for its beautiful botanical gardens.

Sarawak. State of Malaysia, on the island of BORNEO in SOUTHEAST ASIA, adjacent to SABAH. Covers 48,050 square miles (124,450 sq. km.) with a population of 2,471,140 in 2010. Capital is Kuching. Principal products are rubber, petroleum, timber, pepper, sago, rice, gold, and bauxite. In 1963 joined with SINGAPORE, Sabah, and the Federation of Malaya to form the independent Federation of Malaysia.

Satpura Range. Mountain range in SOUTH ASIA, south of the Vindha Range and nearly parallel to it, separating the NARMADA and Tapti River valleys of south INDIA. Runs 600 miles (965 km.); several of its peaks rise above 3,281 feet (1,000 meters).

Saudi Arabia. Kingdom in the MIDDLE EAST. Located on the ARABIAN PENINSULA between the RED SEA and the PERSIAN GULF. Total area of 830,000 square miles (2,149,690 sq. km.) with an estimated population of 33,091,113 in 2018. Capital is RIYADH. Created by the daring conquest of central Arabia by Abd al-Aziz Ibn Saud. Discovery of oil in 1938 eventually propelled the country into the twentieth century. Its petroleum reserves are estimated to be more than 18 percent of the world's total supply. The Saud family has modernized the country, but it is still conservative with respect to its religious adherence to the Wahhabi movement within Islam. Future problems of the kingdom include its finite reserves of oil, the growing demand and competition in the oil industry, its growing population, and its dependence on foreign workers.

Seas. See under individual names.

Seikan Tunnel. Deepest undersea tunnel in the world. Runs for 33.4 miles (53.8 km.) under the Tsugaru Strait, linking Hōnshū with JAPAN's northernmost island of Hōkkaidō; contains a railway line. Construction began in 1964 and was completed in 1988; thirty-four people were killed by cave-ins and other disasters. Originally proposed because of the dangers to ferries traveling between Hōnshū and Hōkkaidō. For example, in 1954, 1,400 people were killed when a ferry capsized during a typhoon. By the time the tunnel was completed, air travel had become cheap and efficient, so the tunnel was no longer a necessity.

Selat Makasar. See MAKASSAR STRAIT.

Seoul. Capital of the Republic of Korea (South KOREA). Located in the northwestern part of the country on the Han-gang River, 37 miles (60 km.) from the YELLOW SEA. Population was 9,794,304 in 2010. Except for 1399-1405, was the capital of KOREA until 1948, when the country was formally divided into two countries. During the Yi Dynasty (1392-1910), city's official name was Hansong. During the Japanese occupation (1910-1945), it was known as Kyongsong. Its popular name has always been Seoul, but it received that name officially only when Korea was divided.

Sevan, Lake. Lake in ARMENIA and the largest lake in the CAUCASUS region. Covers 525 square miles (1,360 sq. km.). Located 6,285 feet (1,916 meters) above sea level in a mountain-enclosed basin, it is drained by the Hrazdan River into the Aras River and to the CASPIAN SEA.

Shaanxi. Province in northwestern CHINA. Total area of 73,600 square miles (195,800 sq. km.), with a 2010 population of 37,327,379. Its capital city is XI'AN. Northern Shaanxi is on the LOESS PLATEAU; southern Shaanxi is mountainous. Agriculture in the north suffers from erosion, drought, and flood. The fertile Wei River valley in central Shaanxi is a major farming region and the main population center.

Shandong. Province in northeastern CHINA. Total area of 59,189 square miles (153,300 sq. km.), with a 2010 population of 95,792,719. Its capital city, JINAN, had a 2010 population of 3,527,566. Western Shandong is plains; central Shandong is mountainous. In the east is the SHANDONG PENINSULA. Shandong lies in the temperate monsoon region, which contributes to spring droughts, summer and fall floods, and salinity problems.

Shandong Peninsula. Peninsula in the eastern part of northeastern CHINA's SHANDONG province, projecting between the BO HAI and the YELLOW SEA. Consists mainly of low hills about 980 feet (300 meters) in elevation. Has undulating coastlines and many deep-water harbors.

Shanghai. Largest city, port, and financial center in CHINA. Located at the mouth of the YANGTZE (Chang Jiang) RIVER. Its 2010 population of 20.2 million made it one of the largest cities in the world. It is also the major industrial, commercial, and cultural center of central and southern China.

Shanxi. Province in northern CHINA. Straddles the western NORTH CHINA PLAIN and eastern LOESS PLATEAU. Total area of 60,640 square miles (157,100 sq. km.), with a 2010 population of 35,712,101. Its capital city, TAIYUAN, had a 2010 population of 3,154,157. Most of Shanxi is uplands, and major population centers are found in river valleys. With cold, dry winters and hot summers with moderate rainfall, many areas have insufficient surface water. Renowned for its extremely rich coal reserves.

Shatt al-Arab. Short river in IRAQ in the MIDDLE EAST. The EUPHRATES, TIGRIS, and KARUN rivers flow into it before their waters reach the PERSIAN GULF. Runs 120 miles (193 km.). Control of the Shatt al-Arab is vital to Iraq for exporting its oil through the Persian Gulf.

Shenyang. Capital city of CHINA's northeastern LIAONING province. Situated in central-eastern Liaoning, with a 2010 population of 5,718,232. Was the capital of the early Jin Dynasty and is famous for imperial buildings and tombs. A major center of heavy industry.

Shenzhen. City in southern GUANGDONG, China. Population was 10,358,381 in 2010. Separated from HONG KONG by the Shenzhen River, it was a trademark city for experimenting with capitalist approaches to promoting economic growth by communist CHINA from 1980 to the mid-1990s.

Shijiazhuang. Capital city of CHINA's HEBEI province. Its 2018 population was 10,951,600.

Shikōkū. Smallest of JAPAN's four main islands. Separated from Hōnshū by the Inland Sea and from Kyūshū by the Bungo Strait. Area is 7,251 square miles (18,780 sq. km.). Population was 3,977,282 in 2010, with most concentrated in urban areas along the coast. Japanese prefectures there are Ehime, Kagawa, Kochi, and Tokushima. Much of the island is mountainous. Rice, barley, wheat, and mandarin oranges are grown in the northern part; salt is produced from seawater. Fishing is an important part of the economy. Industries include petroleum, nonferrous metals, textiles, wood pulp, and paper.

Shillong Plateau. Disconnected portion of the DECCAN PLATEAU in SOUTH ASIA. Located in Meghalaya state of northeast INDIA; bounded by the Jaintia, Khasi, and Garo Hills, which run east to west. The highest peak is Shillong at 5,000 feet (1,500 meters). Also called Meghalaya Plateau.

Sichuan. Province in southwestern CHINA. Total area of 188,032 square miles (487,000 sq. km.), with a 2010 population of 80,417,528. Its capital city, CHENGDU, had a 2010 population of 6,316,922. Western Sichuan is an upland; eastern Sichuan is a basin with the fertile Chengdu Plain as the main agriculture and population center.

Sichuan Basin. Region in SICHUAN, China. Crisscrossed by numerous rivers; extends northeast-southwest over 100,360 square miles (260,000 sq. km.). In the west is the fertile Chengdu Plain; in the central area, low hills; in the east, a series of ranges and valleys. Also called the Red Basin because of its reddish sandstone.

Sikkim. Former kingdom in SOUTH ASIA, wedged between NEPAL and BHUTAN in the HIMALAYAS. Total area of 2,740 square miles (7,096 sq. km.) with a 2011 population of 610,577. Capital is Gangtok. Became an Indian protectorate in 1950 after its king requested help to end political and social unrest. The people voted to end the monarchy in 1974; the following year, Sikkim united with INDIA and became a state. The overwhelming majority of the population are Buddhist. Most people practice subsistence agriculture; corn, rice, and wheat are the major crops.

The red panda is the state animal of Sikkim.

Silk Road. Historic trading route made up of an intricate network of roads stretching from CHINA to the MEDITERRANEAN SEA. Eastern merchants carried their goods by camel caravan to AFGHANISTAN, IRAN, and SYRIA. Route was developed in second century BCE, reached its peak around 750 CE, and was larged eclipsed by seagoing trade by 1200. Also historically known as the Jade or Emperor's Road. The term "Silk Road" was coined by the German geographer Ferdinand von Richthofen in the late nineteenth century.

Sinai Peninsula. Triangular peninsula that links North Africa with SOUTHWEST ASIA. The Gulf of SUEZ and the Suez Canal bound it on the west, and the Gulf of Aqaba and the NEGEV Desert on the east. It is mountainous and very arid. It has great religious significance due to the exodus of the Hebrews and Moses. It has also been the focus of Israeli-Egyptian combat in every military

confrontation between the two nations since 1949. Its area is 23,444 square miles (60,714 sq. km.). Population was 554,000 in 2012.

Singapore. Independent city-state officially known as the Republic of Singapore. Founded as a British trading colony in 1819, Singapore joined Malaysia in 1963 but withdrew two years later and became independent. It is the largest port in SOUTHEAST ASIA and one of the world's greatest commercial centers. The city's domination of Singapore Island led the Republic of Singapore to be referred to as a city-state. Its land area is 269 square miles (697 sq. km.). Its populace is 77 percent Chinese and 14 percent Malay. Its population was estimated to be 5,995,991 in 2018.

Singapore Strait. Important shipping channel in SOUTHEAST ASIA. Links the Indian Ocean with the SOUTH CHINA SEA, separating SINGAPORE Island from the RIAU archipelago.

Siwalik Hills. Southernmost range of the Outer HIMALAYAS in SOUTH ASIA. Averaging 2,953 to 3,937 feet (900 to 1,200 meters) in elevation, these hills are located in Himachal Pradesh and UTTAR PRADESH of INDIA.

Sokhumi. Capital of ABKHAZIA, GEORGIA. Population was 62,914 in 2011. Popular resort, located on the site of the ancient Greek colony of Dioscurias on the BLACK SEA coast. Formerly Sukhumi.

Songnam. City in KYONGGI province, South KOREA. Located 12 miles (19 km.) southeast of SEOUL. Population was nearly 1 million in 2012. From 1970 to 1975 its population increased by 50 percent, the highest rate of growth in the country. The South Korean government has encouraged the movement of industry from Seoul to Songnam.

South Asia. Term generally understood to include AFGHANISTAN, BANGLADESH, INDIA, NEPAL, PAKISTAN, SRI LANKA, and the MALDIVES.

South China Sea. Body of water in Southeast Asia. Covers an area of about 1.4 million square miles (3.6 million sq. km.) with average depths ranging from less than 1,000 feet (305 meters) to 16,460 feet (5,017 meters). Subject to heavy monsoons and typhoons. The MEKONG and Xi

Jiang rivers drain into it. The ports of Manila, Singapore, Bangkok, Ho Chi Minh City, Hong Kong, and Macao are located within it.

South Ossetia. Region in north-central Georgia on the southern slopes of the Greater Caucasus Mountains. About 90 percent of the region lies more than 3,300 feet (1,000 meters) above sea level. The major city is Tskhinvali. Population in 2011 was 55,000.

South Vietnam. See Vietnam.

Southeast Asia. In its narrower sense, this term encompasses the countries of mainland Asia east of South Asia and south of East Asia: Cambodia, Laos, Malaysia, Myanmar, Singapore, Thailand, and Vietnam. In its broader sense, it also includes the nations of Indonesia, Brunei, the Philippines, and Timor-Leste

Southeast Coast Hills and Mountains. Region in southern China and east of the South Chang Jiang Hills and Basins. Consists of a series of northeast-southwest-trending mountains, with numerous short rivers, narrow rivers and coastal plains, rugged surfaces, and irregular coastlines.

Southwest Asia. Term for the predominantly Muslim nations west of South Asia and south of Central Asia. Usually understood to include all the nations between Turkey, the Arabian Peninsula, and Iran. The term is often used interchangeably with Middle East; however, the latter term is also often understood to include the Arab nations of North Africa.

Spice Islands. See Moluccas.

Spratly Islands. Group of more than 100 islets, coral reefs, sandbars, and atolls in the South China Sea off Vietnam. Ownership is disputed by China, Taiwan, Brunei, Malaysia, the Philippines, and Vietnam. The largest island is Itu Aba. None of the islets, which lie atop oil and gas reserves, is permanently inhabited, except by seabirds and turtles. Located amid an important shipping lane.

Sri Lanka. Most religiously diverse country of South Asia. Total area of 25,332 square miles (65,610 sq. km.), with a 2018 population of 22,576,592. Capital is Colombo. Separated from southern India by 22-mile-wide (35 km.) Palk Strait. Some 70 percent of the people are Sinhalese-speaking Buddhists, and most of the rest are Tamil-speaking Hindus of Dravidian descent; both came originally from India, but at different times. Many Tamils were brought by the British during the second half of the nineteenth century to work on rubber and tea plantations. A civil war between the majority Sinhalese and the minority Tamils ended in 2009. Most of the people live in the wet, hilly southwest region. Sri Lanka is the world's second-largest producer of tea and the leading producer of high-quality graphite. In 2004, a tsunami killed 31,000 people. Known as Ceylon prior to 1972.

Suez, Gulf of. Narrow extension of water at the northern end of the Red Sea in the Middle East. Located between the Sinai Peninsula and the Egyptian Red Hills. Geologically younger, longer, and shallower than its neighboring Gulf of Aqabah.

Sukhumi. See Sokhumi.

Sulawesi. Indonesian island of active volcanoes east of Borneo and west of the Moluccas. Covers an area of 72,986 square miles (189,034 sq. km.), with an estimated population in 2019 of more than 19.5 million. Its four peninsulas are separated by deep gulfs. Its tropical climate enhances the cultivation of spices, fruit, corn, rice, tobacco, and sugar. Produces gold, copper, tin, sulfur, salt, diamonds, and other gems; exports coffee, copra, coconuts, and trepang (edible sea slugs). Also called Celebes.

Sulu Sea. Arm of the Pacific Ocean extending from the Philippine Islands to Borneo and Palawan Island.

Sumatra. One of the Sunda Islands in Indonesia. Located in the Indian Ocean; separated by the Strait of Malacca from the Malay Peninsula and by the Sunda Strait from Java. Covers 182,860 square miles (473,605 sq. km.), with an estimated population in 2019 of more than 58 million. Distinguished by a mountain chain and many lakes, earthquakes, and storms. Annual rainfall is between 90 and 185 inches (2,286 and 4,700 millimeters). Has fertile soil and deposits of bauxite and petroleum; produces rice, rubber, tea, coffee, coconuts, and spices. Also known as Sumatera. A 2004 earthquake off the coast resulted in a tsunami that devastated the island.

Sumbawa. One of the SUNDA ISLANDS in southern INDONESIA, near FLORES and Sumba Island. Covers 5,900 square miles (15,280 sq. km.). Has mountain ranges and volcanoes in the north. Its fertile soil supports farming and livestock breeding. Rice, corn, teak, and sappanwood are principal products. Population in 2014 was 1,391,340.

Sunda Islands. Major island group in the MALAY ARCHIPELAGO between the SOUTH CHINA SEA and the Indian Ocean. Divided into the Greater Sunda Islands—SULAWESI (also known as Celebes), SUMATRA, JAVA, BORNEO, and smaller islands—and the Lesser Sunda Islands—BALI, Sumba, FLORES, and TIMOR.

Sunda Shelf. Extension of SOUTHEAST ASIA that includes many islands of western INDONESIA, including JAVA and SUMATRA.

Sundarbans. Extensive tidal mangrove forest (more than 3,860 square miles/10,000 sq. km.) along the Bay of BENGAL in the south GANGES Delta; covers southwestern BANGLADESH and a southeastern portion of West Bengal, INDIA. This forest is the source of timber used for many purposes, including pulp for the domestic paper industry and poles for electric power distribution. Sundari is the main timber tree of this forest and gives the region its name. Also the home of the endangered Bengal tiger. Rising sea levels threaten the delicate ecosystem.

Syr Darya River. Longest river in CENTRAL ASIA. Flows 1,328 miles (2,237 km.) through UZBEKISTAN, TAJIKISTAN, and KAZAKHSTAN. Formed by the Naryn and Qaradaryo rivers joining in the eastern FERGANA VALLEY. From there, it flows northwest into the ARAL SEA.

Syria. Country in the MIDDLE EAST, located at the eastern end of the MEDITERRANEAN SEA. Total area of 71,037 square miles (185,985 sq. km.) with an estimated population of 19,454,263 in 2018. Capital is DAMASCUS. More than half of the country has a steppe climate. In 1946 achieved its independence from French Mandate rule; spent the next several decades in political chaos, experiencing many political coups. In 1970 Hafez al-Assad seized power; he was president of Syria until his death in 2000; succeeded by his son, Bashar al-Assad. Depends heavily on agricultural production (predominantly cotton) and petroleum, of which it has small reserves. The country has been the scene of a violent multisided civil war since 2011. Also known as the Syrian Arab Republic.

Syrian Desert. MIDDLE EAST desert. Located in northeastern corner of the ARABIAN PENINSULA, within the borders of SAUDI ARABIA, JORDAN, SYRIA, and IRAQ. Covers about 100,000 square miles (260,000 sq. km.). Still sustains a few nomadic herders, but they are dwindling in numbers.

Taal, Lake. Third-largest lake in the PHILIPPINES. Covers 94 square miles (243 sq. km.). Occupies the crater of an extinct volcano on LUZON ISLAND, south of MANILA; the site of Volcano Island and an active volcano that has erupted more than thirty-three times since 1572.

Tadzhikistan. See TAJIKISTAN.

Taedong River. River in North KOREA that begins in the NANGNIM MOUNTAINS in Hamgyong-nam province; flows 273 miles (439 km.) southward and empties into the Korea Bay at Namp'o. Forms a drainage basin of approximately 7,855 square miles (20,344 sq. km.). There are many river ports, such as Pyongyang along its banks. The upstream area is used for irrigation.

T'ai-Chung. Administrative seat for TAIWAN province in the country of Taiwan. Located in west central Taiwan. Area is 63 square miles (163 sq. km.); population was 990,041 in 2002. In the 1970s, a harbor and fishing port were developed on the coast to its west. From 1948 until 1977, the population more than tripled as refugees fled there to escape mainland CHINA. Today, it is a major market center for the bananas, rice, and sugar grown in the area.

Taihang Mountains. Mountain range in northern CHINA. Extends northeast-southwest, forming the boundary between the NORTH CHINA PLAIN and the LOESS PLATEAU.

T'ai-nan. One of TAIWAN's oldest settlements. Located in southwestern Taiwan. Han Chinese settled the town as early as 1590. The Dutch settled there in 1623 but were driven out by Cheng Ch'eng-kung (Koxinga) in 1662. When the Ch'ing Dynasty established control over Taiwan, T'ai-nan was the island's first capital. The main market for produce of the southern plain. Originally known as T'ai-yuan, also called Ta-yuan, and T'ai-wan, its name was eventually given to the entire country. Population in 2002 was 742,574.

Taipei. Largest city and capital of TAIWAN, the Republic of China. Located at the northern tip of Taiwan, about 16 miles (26 km.) from CHI-LUNG (Keelung), which is its port, and about 121 miles (195 km.) across the TAIWAN STRAIT from the People's Republic of China. Covers about 105 square miles (272 sq. km.). Population was 2.7 million in 2012. Has a humid, subtropical climate with warm summers and mild winters. Rain falls year-round, but the heaviest period is from October through March. Typhoons strike from June to October. Has a manufacturing economy, producing textiles, machinery, metals, electronics, and chemicals. Name means "northern terrace." In 1886, when Taiwan became a province of China, Taipei became the capital of Taiwan. When JAPAN acquired Taiwan after the Sino-Japanese War, the Japanese retained Taipei as Taiwan's capital. When World War II ended in a Japanese defeat, Taiwan reverted to Chinese ownership. In 1949, when the Chinese Nationalist government fled from mainland China, it set up a government on Taiwan, and Taipei became the capital of the Republic of China.

Taiwan. Country comprising a group of islands located about 100 miles (160 km.) off the southeast coast of the CHINA mainland. Total area of Taiwan, including the PENG-HU ISLANDS, is 13,892 square miles (35,980 sq. km.); population was 23,545,963 in 2018. About 75 percent of the people live in urban areas; the rest, in

small villages or on farms. The main island of Taiwan is about 245 miles (400 km.) long, north to south, and about 90 miles (145 km.) at its widest point. Government of Taiwan controls the main island of Taiwan and a group of islands known as the Peng-hu in the Pescadores Channel; claims the QUEMOY and MATSU groups in the TAIWAN STRAIT. In 1949, when the Communist party took control of the government of China, nationalists fled to Taiwan, set up the government of the Republic of China, and made TAIPEI its capital. The major cities in Taiwan are Taipei, KAO-HSIUNG, T'AI-NAN, and CHI-LUNG (Keelung). Named "Formosa" ("beautiful") by the Portuguese in the sixteenth century.

Taiwan Strait. Shallow channel in SOUTHEAST ASIA separating CHINA from TAIWAN, and connecting the SOUTH CHINA SEA with the EAST CHINA SEA. Is 115 miles (185 km.) wide and 230 feet (70 meters) deep.

Taiyuan. Capital city of northern CHINA's SHANXI province. It had a 2010 population of 3,154,157.

Tajikistan. Country of CENTRAL ASIA that gained independence from the former Soviet Union in 1991. Total area of 55,251 square miles (143,100 sq. km.) with a 2018 population of 8,604,882. Capital is Dushanbe. Its principal inhabitants are the Tajiks, who speak an Indo-Iranian language that is closely related to Persian. However, the Tajiks are culturally close to the Kazaks. Also known as Tadzhikistan.

Takla Makan Desert. See TAKLIMAKAN DESERT.

Taklimakan Desert. Harsh, barren desert of western CHINA that occupies most of the TARIM BASIN. Historically restricted communication and travel between China and CENTRAL ASIA. Also known as the Takla Makan Desert.

T'ao-yuan. Special municipality in TAIWAN. The Hsueh-shan Shan-mo Range extends over much of it. Population was 2.1 million in 2015, including Atayal aborigines in the mountains. The TAIPEI oil and gas fields are in the northeast; iron ore, coal, and nickel are extracted there. Was a county until 2014, when it was administratively combined with the municipality to form a special municipality.

Tarbela Dam. World's largest earth-filled dam. Constructed on the INDUS RIVER in the North-West Frontier Province of PAKISTAN in the 1960s under the Indus Basin Development Fund. This dam and the giant Mangla Dam, on the Jhelum River, provided the first significant water storage for the Indus irrigation system. The two dams also provide flood control and irrigation, regulate water flow for some of the link canals, and supply a large percentage of Pakistan's energy.

A U.S. Marine Corps CH-46 Sea Knight helicopter flies near the Tarbela Dam in Pakistan's Khyber-Pakhtunkhwa province, Aug. 27, 2010.

Tarim Basin. Vast basin in southern XINJIANG Autonomous Region, CHINA. Covers 168,340 square miles (436,000 sq. km.). Flanked by the TIAN SHAN MOUNTAINS to the north and the Kunlun and Altun Mountains to the south. Most of the basin consists of the extremely arid TAKLIMAKAN DESERT.

Tashkent. Capital of UZBEKISTAN; also the largest city and main economic and cultural center in CENTRAL ASIA. Population was 2,485,900 in 2018. Located in the northeastern part of Uzbekistan at an elevation of 1,475 to 1,575 feet (450 to 480 meters). Cotton is the chief crop of the region.

Taurus Mountains. Range of mountains in the MIDDLE EAST, bordering the MEDITERRANEAN SEA in southern TURKEY. Elevations vary from about 7,000 to 9,000 feet (2,133 to 2,743 meters). Rugged and scenic in places. Throughout history, commerce has crossed and armies have marched through several famous passes, such as the Cilician Gates. Also known as the Toros Mountains.

Tavois Mountains. See BILAUKTAUNG RANGE.

Tbilisi. Capital of the Republic of GEORGIA and a major cultural, educational, and industrial center. Population was 1,082,400 in 2016. Industries include the production of electric locomotives, machine tools, agricultural machinery, and electric equipment. Founded in 458 CE.

Tehran. Capital of IRAN in the MIDDLE EAST, and largest city in Southwest Asia. Situated in northern Iran just south of the ELBURZ MOUNTAINS. Had an estimated population in 2016 of 8,693,706.

Tel Aviv. First major urban settlement of Jews returning to their ancient homeland in the MIDDLE EAST. Located on the coast of the MEDITERRANEAN SEA near the ancient Arab port city of Jaffa. The political capital of ISRAEL before JERUSALEM. Population in 2013 was 2,436,800.

Tempe, Lake. Remnant of an inland sea in the center of South SULAWESI province, INDONESIA. Now less than 6 feet (2 meters) deep, and still shrinking. Fed by the Wallace River, it is an important source of fish.

Tenasserim Mountains. See BILAUKTAUNG RANGE.

Terek River. River that rises in northern GEORGIA and flows north and then east through Russia to empty into the CASPIAN SEA. Runs 370 miles (600 km.) and drains a basin of 16,900 square miles (43,700 sq. km.). It was the southern frontier of Russian settlement in the CAUCASUS for much of the nineteenth century.

Terrai. Southernmost physiographic region of NEPAL, SOUTH ASIA, running in an east-west direction parallel to the HIMALAYAS. This flat, fertile, narrow strip of land is Nepal's major crop-growing region. It covers only 17 percent of the total land area of Nepal but supports about 30 percent of its population and produces more than 60 percent of the country's gross national product. Also spelled Terai and Tarai.

Thailand. SOUTHEAST ASIA country in the western portion of the Indochinese Peninsula, with an area of 198,117 square miles (513,120 sq. km.).

It is bordered on the west by MYANMAR and on the east by CAMBODIA. Thailand is unusual among countries of the region in that it was never under European control or domination. It has a tropical monsoonal climate. Forests cover about one-fourth of the total land area, and the country has about 10 percent of the world's known reserves of tin. People speaking Tai languages, including Thai and Lao, account for nearly four-fifths of the total population. Its population was 68,615,858 in 2018. Capital is BANGKOK.

Thailand (Siam), Gulf of. Inlet of the SOUTH CHINA SEA. Located between the MALAY ARCHIPELAGO and the Southeast Asian mainland. Bounded by THAILAND, CAMBODIA, and VIETNAM, it receives the CHAO PHRAYA River.

Thar. Uplifted desert situated between the Indus and the Ganges plains of SOUTH ASIA. Has an area of 97,000 square miles (250,000 sq. km.) and covers the western part of the Indian State of Rajasthan and the northeastern part of Sindh, a province of PAKISTAN. Through the extension of canals, irrigation has reclaimed some land for agriculture along the north and western edges of the region. Also called the Great Indian Desert.

Thimphu. See BHUTAN.

Thirty-eighth Parallel. Location (north latitude) that separates North KOREA from South KOREA.

Three Gorges. Stretch of deep scenic gorges with looming rock walls, narrows, and bends on the YANGTZE (Chang Jiang) RIVER, between its confluences with the Fengjie, CHONGQING, Yichang, and Han rivers. The gorges are the site of the Three Gorges Dam, the world's largest dam and the world's largest hydroelectric power provider.

Three Gorges Dam. In China, the world's largest hydroelectric dam-engineering project in history. Begun in 1994, the immense dam complex reached its full capacity in 2012. With its series of dams on the Yangtze River and a 400-mile-long (645 km.) reservoir, the project produces an unprecedented 22,500 megawatts of electricity for China's relatively underdeveloped interior provinces. The construction phase displaced some 2 million people. The dam complex continues to have a disastrous environmental impact.

Tiananmen Square. Large open area in central BEIJING. On June 4, 1989, Chinese troops killed hundreds of people gathered in the square to demand democratic reforms, thereby setting back the country's reform movement and damaging its government's international reputation.

Tianjin. Third-largest city in CHINA; major industrial and commercial center and port on the HAI RIVER. Located on northeastern NORTH CHINA PLAIN. Population was 9,290,263 in 2010. Produces iron, steel, chemicals, motor vehicles, carpets, machinery, bicycles, and sewing machines. Formerly known as Tientsin.

Tian Shan Mountains. Mountain range in CENTRAL ASIA. Extends from central XINJIANG Autonomous Region, CHINA, to KYRGYZSTAN, KAZAKHSTAN, and UZBEKISTAN. Spans 1,553 miles (2,500 km.). The eastern and central portions, which extend 1,060 miles (1,700 km.), are located in Xinjiang. The mountains separate the JUNGGAR BASIN in northern Xinjiang from the TARIM BASIN in southern Xinjiang.

Tiberias, Lake. See GALILEE, Sea of.

Tibet. Autonomous region of CHINA. Situated on the TIBETAN PLATEAU and bordering SOUTH ASIA to the south and MYANMAR to the southeast. Total area of 471,662 square miles (1,221,600 million sq. km.), with a 2010 population of 3,002,165. Northern Tibet has wide basins with numerous lakes and pastoral land. Population is mainly confined to river valleys in the south. The Tibetan secessionist movement has been a factor contributing to political instability. Capital city is LHASA.

Tibetan Plateau. Region in southwestern CHINA. Covers 965,000 square miles (2.5 million sq. km.); averages 13,120 to 16,400 feet (4,000 to 5,000 meters) in elevation. On the surface are a series of near-east-west-trending mountains, including the Kunlun, Altun, QILIAN, Bayan Har, Tanggula, Nyainqentanglha, Gandise, and HIMALAYAS. Between these mountains are broad basins, plateaus, and lakes. In the southeast, the mountains take a sharp southern turn and rivers cut deeply between them, forming narrow gorges. The plateau is the source of numerous major Asian rivers, including the YANGTZE (Chang Jiang), YELLOW (Huang He), INDUS, GANGES, BRAHMAPUTRA, MEKONG, and SALWEEN.

Tientsin. See TIANJIN.

Tigris River. MIDDLE EAST river that rises in the mountains of eastern TURKEY and flows about 1,180 miles (1,900 km.) southward, through the Mesopotamian plain in IRAQ. There it merges with the EUPHRATES RIVER, then enters the PERSIAN GULF. Although its flow of water is not as dependable as that of the Nile River, it provides a necessary source of fresh water for field irrigation.

Timor. Largest and easternmost of the Lesser SUNDA ISLANDS. The eastern part of the island is the country of TIMOR-LESTE (East Timor), while the western section is part of the Indonesian province of East Nusa Tenggara. It is 280 miles (450 km.) long and 65 miles (105 km.) wide. Its two main cities are Dili, the capital of Timor-Leste, and Kupang, .in the western, Indonesian section of the island Averages 50 inches (1,270 millimeters) of rainfall annually; daily maximum temperatures can be 86° to 93° Fahrenheit (30° to 34° Celsius). Produces eucalyptus, sandalwood, teak, bamboo, rosewood, maize, corn, rice, coffee, copra, and fruit.

Timor-Leste. Asia's newest country. Occupies the eastern half of the island of Timor in the Lesser Sunda Islands. At 5,743 square miles (14,874 sq. km.), Timor-Leste is slightly larger than the U.S. state of Connecticut. It was formerly part of a Portuguese colony, and the Portuguese language is still widely spoken. Timor-Leste was annexed by Indonesia in 1976, which ruled it for more than two decades. In 2002, after a long struggle, Timor-Leste gained full independence. In 2018, Timor-Leste's population was 1,321,929; the major religion is Roman Catholicism. Timor-Leste derives more than 90 percent of its revenue from the exploitation of offshore oil and natural-gas deposits. Also called East Timor.

Toba, Lake. Largest lake in INDONESIA, in SUMATRA's highlands, about 110 miles (180 km.) south of MEDAN. Covers 442 square miles (1,145 sq. km.). A tourist resort surrounded by steep cliffs and sandy beaches.

Tokyo. Capital of JAPAN and one of the largest cities in the world. Located on Hōnshū, the main island of the Japanese archipelago. The metropolis covers 856 square miles (2,217 sq. km.). Population of the city was 13,159,388 in 2010; population of the entire metropolitan area was 34.5 million in 2012. Bordered by Saitama prefecture to the north, Chiba prefecture to the east, Yamanashi prefecture to the west, and Kanagawa prefecture to the southwest. The city is an administrative subdivision equal to a prefecture. Located at the head of Tokyo Bay, the site of the city was developed in ancient times. The small fishing village of Edo, located there for centuries, became the capital of the Tokugawa Shogunate (1603-1867). During the Meiji Restoration in 1868 the shogunate was abandoned and the imperial family moved to Edo, renaming it Tokyo, which means "eastern capital." The Imperial Palace, home of the Japanese emperor and the imperial family, is in Tokyo. The metropolitan area includes the capital, industrial sites, residential suburbs, and a large mountainous rural area to the west.

Tonkin, Gulf of. Arm of the SOUTH CHINA SEA in Southeast Asia. Shallow near the coast of VIETNAM but 650 feet (200 meters) deep in places. A major trade route and the site of two major ports, HAIPHONG in Vietnam and Beihai in CHINA. Contains thousands of islands, many uninhabited and one that is noted for pearl fishing. Receives several rivers, including the RED RIVER. Its seabed is mainly silt.

Toros Mountains. See TAURUS MOUNTAINS.

Trans-Jordan. See JORDAN.

Truong Son. See ANNAM HIGHLANDS.

Tsushima Current. Branch of the KUROSHIO Current.

Turkestan. City in southern KAZAKHSTAN. Population was 160,746 in 2017. An ancient center of the caravan trade, it became a religious center because of the twelfth century Sufi, Ahmed Yesevi, whose fourteenth century mausoleum is the city's chief monument.

Turkey. Secular state in the MIDDLE EAST, at the eastern end of the MEDITERRANEAN SEA, situated between it and the BLACK SEA and the mountainous area of the ZAGROS and Trans-Caucasus Mountains. Total area of 302,535 square miles (783,562 sq. km.) with an estimated population of 81,257,239 in 2018. Capital is ANKARA. Once the core area of the powerful Ottoman Empire, Turkey became a republic in 1922 under the leadership of Mustafa Kemal Ataturk. With a modest variety of mineral resources, it is one of the most powerful countries in the Middle East, although remaining nominally neutral in conflicts between Arab and non-Arab states. Turkey transitioned from a parliamentary government to a presidential system in 2018. The country straddles the arbitrary line separating Europe and Asia.

Turkmenabad. Formerly known as Chardzhou. Second-largest city in Turkmenistan and capital of Lebap region. Population was 253,000 in 2009. It is a rail junction and the largest port on the Amu Darya River. Superphosphates and astrakhan furs are local products.

Turkmenistan. Country of west CENTRAL ASIA that gained independence from the former Soviet Union in 1991. Total area of 188,456 square miles (488,100 sq. km.) with a 2018 population of 5,411,012. Capital is Ashgabat (Askhabad). About 90 percent of the country is desert. The Turkmen, a Turkic-speaking people, make up more than 70 percent of the population.

United Arab Emirates. Federation of seven independent emirates situated on the southern coast of the PERSIAN GULF in the MIDDLE EAST. Covers 32,278 square miles (83,600 sq. km.) with a population of 9,701,315 in 2018. Originally a British protectorate, the emirates formed a union in 1971 when given independence. From a modest economy based on camel breeding, farming, and pearl fishing, the country was rapidly propelled into the twentieth century by the discovery of vast reserves of petroleum and natural gas. Recognizing the illusionary aspects of overdependence on oil revenues, the country has begun to diversify its economy. The international conflicts that characterize the region, and its dependence on expatriate workers who make up about 80 percent of its workforce, remain major concerns. The Burj Khalifa, a skyscraper in Dubai, is the world's tallest building.

Upper Burma. Northern or inland part of MYANMAR. Comprises Magwe, MANDALAY, and Sagaing divisions. Largely agricultural.

Urmia, Lake. Large, shallow, salty landlocked lake in the northwestern corner of IRAN. Covers about 2,300 square miles (6,000 sq. km.). By the second decade of the twenty-first century, the lake was shrinking dramatically due to climate change and unwise irrigation practices.

Urumqi. Capital city of northwestern CHINA's XINJIANG autonomous region. It had a 2010 population of 2,853,398.

Ustyurt Plateau. Plateau in UZBEKISTAN and KAZAKHSTAN. Located between the ARAL SEA and the AMU DARYA RIVER in the east and the Mangyshlak Plateau and the Kara-Bogaz-Gol in the west. Covers about 77,000 square miles (200,000 sq. km.) and has an average elevation of about 500 feet (150 meters)

Uttar Pradesh. Most populous state of INDIA. Covers 75,062 square miles (194,411 sq. km.); the northern part of the state falls within the Himalaya zone. Population was 199,581,477 in 2011. Most of the state is in a low-lying alluvial plain, and agriculture dominates the economy. India's largest producer of rice, wheat, and pulses; several minerals, including coal, copper, and bauxite, are found there.

Uzbekistan. Country in CENTRAL ASIA that gained independence from the former Soviet Union in 1991. Total area of 172,742 square miles (447,742 sq. km.) with a 2018 population of 30,032,709. Capital is TASHKENT. It includes the historic cities of BUKHARA and SAMARKAND. The Uzbeks, who speak a Turkic language, make up more than 70 percent of the population.

Van, Lake. Large, salty, landlocked lake in eastern TURKEY. Covers 1,470 square miles (3,800 sq. km.).

Varanasi. See BENARES.

Vientiane. Port, capital, and the largest city of central LAOS. Located near the MEKONG RIVER, with a population of 510,000 in 2005. Trades in teak, gum, and textiles, and manufactures processed food, footwear, textiles, and building materials.

Vietnam. Southeast Asian country occupying the east coast of the Indochinese Peninsula. CAMBODIA and LAOS border it on the west and the Gulf of TONKIN and the SOUTH CHINA SEA on the east. It was part of French INDOCHINA from the late nineteenth century into the 1950s. After the French left in 1954, Vietnam split into North Vietnam and South Vietnam, and the United States became involved in what was initially a civil war between the two countries. In the mid-1960s, American troops became actively engaged in combat. South Vietnam fell in 1975 and was annexed and absorbed by North Vietnam. Vietnam has a tropical monsoonal climate. Almost one-third of the total land area is under tropical evergreen and subtropical deciduous forests. Vietnam has experienced strong growth as it transitions to a market-based economic model. Its area is 127,818 square miles (331,210 sq. km.). The population is almost entirely Vietnamese. Its population was 97,040,334 in 2018. Capital is HANOI.

Vindhya Range. Mountains forming the northern boundary of the DECCAN PLATEAU in SOUTH ASIA. Composed of massive sandstone and limestone beds, it runs in an east-west direction for 652 miles (1,050 km.) with an average elevation of 984 feet (300 meters). The range has been the historic dividing line between north and south INDIA, separating the Aryans from the Dravidians of the Deccan.

Visayas. Group of islands, beaches, and resorts at the geographical center of the PHILIPPINES. Bounded on the north by LUZON ISLAND and on the south by Mindanao.

West Bank. Israeli-occupied Arab lands of PALESTINE consisting of the ancient highlands of Judah and Samaria in the MIDDLE EAST. Encompasses about 2,260 square miles (5,860 sq. km.) between the coastal plain of ISRAEL and the Jordan Valley. Administered by the Kingdom of JORDAN until conquered by Israel in the Arab-Israeli War of 1967, it is an area of Palestinian aspirations and struggle for an independent state of Palestine.

Western Ghats. Mountain range in SOUTH ASIA, extending 994 miles (1,600 km.) along the western border of the DECCAN PLATEAU from the mouth of the Tapti River in the north to Cape Comorin in the south. They have an average elevation of 3,000 to 5,000 feet (915 to 1,525 meters). The highest peak is Doda Betta (8,652 feet/2,637 meters). See also EASTERN GHATS.

Wuhan. Capital city of CHINA's HUBEI province. Located in central eastern Hubei at the intersection between the YANGTZE (Chang Jiang) RIVER and the Beijing-Guangzhou Railway, it is a major transportation hub and industrial center. Population was 7,541,527 in 2010. The revolution against the Ch'ing Dynasty (1644-1911) began in Wuhan in 1911 with an armed rebellion, eventually overthrowing the Qing and ushering in a modern era in China. A deadly outbreak of the coronavirus, Covid-19, focused on, virtually shut down Wuhan in early 2020.

Xi'an. Capital city of northwestern CHINA's SHAANXI province. Located in central-southern Shaanxi. Population was 5,206,253 in 2010. A major classic Chinese city because of its antiquity and rich cultural value. Called Chang'an in historical times, Xi'an was the capital city of numerous ancient dynasties and has historical sites and architectural remains dating from as early as the Han Dynasty (206 BCE-CE 220) and well-preserved city walls from the Ming Dynasty (1368-1644). The famous terra-cotta armies that lie near the burial chamber of the Qinshi Huangdi (258-210 BCE) are located in a northeastern suburb.

Xining. Capital city of CHINA's QINGHAI province. It had a 2015 population of 1,153,417.

Xinjiang. Autonomous region in northwestern CHINA. Borders MONGOLIA in the northeast,

Russia in the north, KAZAKHSTAN and KYRGYZSTAN in the northwest, and TAJIKISTAN, AFGHANISTAN, and PAKISTAN in the southwest. Total area of 635,870 square miles (1,646,900 sq. km.), with a 2010 population of 21,815,815. Its capital city, URUMQI, had a 2010 population of 2,853,398. Exhibits basin and range topography. Xinjiang's mostly Turk Muslim population concentrates in the oases flanking the basins and in the river valleys. Secessionist movements have been a source of instability since historic times.

Yalu River. River forming the northwestern boundary between North KOREA and MANCHURIA. Runs 490 miles (790 km.), beginning in the Chang-pai Mountains and flowing south to the YELLOW SEA; drains 24,250 square miles (62,780 sq. km.). The Chinese provinces of Kirin and LIAONING are bordered by the river. An important source of hydroelectric power. Important tributaries are the Herchun, Changjin, and Tokro. Ya-lu is its Chinese name; Amnok-kang is its Korean.

Yangon. Cultural, commercial, industrial center, and largest city of MYANMAR. Seaport on the Rangoon River, one of the mouths of the IRRAWADDY. Formerly called Rangoon; in 1989, reverted to its original name of Yangon, which it acquired in 1755. Was replaced by Naypyidaw as capital in 2005. Had a population of 4,775,000 in 2014. Exports rice, teak, oil, and rubber. Major landmark is Shwedagon Pagoda, built over 2,000 years ago to hold eight sacred hairs of Buddha. Yangon means "the end of strife."

Yangtze River. Longest river in Asia and third-largest in the world. Originating in the Tanggula Mountains on the TIBETAN PLATEAU, it flows 3,720 miles (5,990 km.) to the EAST CHINA SEA. Its upper reaches flow through rugged terrain, carving deep gorges and rapids. Many tributaries irrigate the fertile Chengdu Plain. Between Fengjie, CHONGQING, Yichang, and HUBEI, in the famous THREE GORGES DAM complex. From Yichang onward are the fertile middle and lower plains. The river enters the delta at Zhenjiang, JIANGSU, and empties into the East China Sea at SHANGHAI, after draining an area of about 697,700 square miles (1.8 million sq. km.). Also known as the Yangzi River and as the Chang Jiang River.

Yellow River. Second-longest river in CHINA (after the Yangtze). Flows 3,000 miles (4,830 km.), with a drainage area of 290,350 square miles (752,000 sq. km.). Originates in the Bayan Har Mountains on the TIBETAN PLATEAU. In QINGHAI and GANSU, it flows in rugged terrain, leaving deep gorges; in NINGXIA and INNER MONGOLIA, it flows in uplands, forming fertile narrow river plains. Between the SHAANXI and SHANXI provinces, it cuts into the LOESS PLATEAU, causing severe erosion. Enters the lower reaches at Mengjin, HENAN. Alluvial deposits build most of the NORTH CHINA PLAIN but also cause dam breaks, floods, and changes of the river course. Empties into BO HAI Bay in SHANDONG province. Also called the Huang He River.

Yellow Sea. Extension of the EAST CHINA SEA that separates KOREA from northeastern CHINA; covers about 180,000 square miles (466,200 sq. km.); average depth about 140 feet (42 meters). Also known as the Huang He Sea.

Yemen. Country in the MIDDLE EAST, located in the southwestern corner of the ARABIAN PENINSULA. Controls the Bab al-Mandeb choke point that commands the entrance and exit to the southern end of the RED SEA. Total area of 203,850 square miles (527,968 sq. km.), with an estimated population of 28,667,230 in 2018. Capital is Sana. Formerly divided into two countries with opposing political ideologies; for two decades, they contended with one another, with the Yemen Arab Republic supported by SAUDI ARABIA and the People's Democratic Republic of Yemen supported by Egypt. In 1990 the two states formally merged into one country. The bloody civil war that has beset Yemen since 2015 shows little sign of abating.

Yerevan. Capital of ARMENIA. Population was 1,077,600 in 2018. Industries include acetylene, cars, plastics, synthetic rubber, tires, and turbines. The city's site has evidence of continuous human settlement since at least 5000 BCE.. The Romans, Parthians, Arabs, Mongols, Turks, Persians, Georgians, and Russians have ruled it in turn.

Yinchuang. Capital city of northwestern CHINA's autonomous NINGXIA region. It had a 2010 population of 1,290,170.

Yokohama. Port city and capital of Kanagawa-ken prefecture in JAPAN; the second-largest city in Japan. Located 20 miles (32 km.) southwest of TOKYO. Area is 167 square miles (433 sq. km.); population was 3,688,773 in 2010. The Tokyo-Yokohama industrial complex is one of the world's largest. Its business district, containing many banks and other businesses, is concentrated around the port.

Yunnan. Province in southwestern CHINA. Situated on the mountainous southern YUNNAN-GUIZHOU PLATEAU, bordering VIETNAM and LAOS in the southeast and MYANMAR in the southwest. Total area of 168,400 square miles (436,200 sq. km.), with a 2010 population of 45,966,766. Its capital city, KUNMING, had a 2010 population of 3,278,777. Dotted with scenic lakes, Yunnan falls mostly in the subtropical zone, but the southernmost area is tropical, with rich wildlife and vegetation resources.

Yunnan-Guizhou Plateau. Plateau in southwestern CHINA. Covers 270,270 square miles (700,000 sq. km.); consists mostly of rugged hills, dissected uplands, deep canyons, and basins.

Zagros Mountains. MIDDLE EAST mountain range extending along IRAN's western border from just east of the Mesopotamian plain to near the Strait of HORMUZ at the entrance to the PERSIAN GULF. The higher peaks range from around 10,000 to 14,100 feet (3,000 to 4,300 meters) and accumulate snow and ice during winter, effectively closing most of its passes to travel.

Zambales Mountains. Mountain range in northwestern PHILIPPINES. Home to Mount Pinatubo, which erupted in 1991 and 1992. Located on western LUZON ISLAND. Extends 100 miles (160 km.) into the BATAAN PENINSULA. Place of volcanic rock, minerals, short but rapid rivers, tall hardwoods, and pines.

Zhejiang. Province in southern CHINA. Located on the Lower YANGTZE (Chang Jiang) River Total area of 39,300 square miles (101,800 sq. km.), with a 2010 population of 54,426,891. Its capital city is HANGZHOU. Mostly hilly and mountainous, with limited plains in the north and more than 2,000 islands offshore. Has a subtropical climate. Known for silk, tea, bamboo shoot, and jute production. Sophisticated light industries developed over centuries include silk textiles, traditional crafts, and food processing. Most population centers are in the north and along the coast.

Zhengzhou. Capital city of north central CHINA's HENAN province. Its 2010 population was 3,677,032.

Zhu Jiang Delta. Delta on the southernmost coast of CHINA. Formed by the Xi Jiang, Bei Jiang, ZHU JIANG, and other tributaries; covers 4,250 square miles (11,000 sq. km.). With ample heat and rainfall and fertile farmland, it is a major agriculture area in subtropical and tropical China. Also called the Pearl River Delta.

Zhu Jiang River. Also called the Pearl River, the second-largest river system in CHINA in terms of annual runoff. Main rivers in the system include Xi Jiang in the west, Bei Jiang in the north, and Dong Jiang in the east. Zhu Jiang refers to the segment formed by the Dong Jiang and several other small tributaries. The entire Zhu Jiang system has a drainage area of 169,880 square miles (440,000 sq. km.) and empties into the SOUTH CHINA SEA.

*Thomas F. Baucom; Keith Garebian;
Dana P. McDermott; Bimal K. Paul;
Bin Zhou*

APPENDIX

The Earth in Space

The Solar System

Earth's solar system comprises the Sun and its planets, as well as all the natural satellites, asteroids, meteors, and comets that are captive around it. The solar system formed from an interstellar cloud of dust and gas, or nebula, about 4.6 billion years ago. Gravity drew most of the dust and gas together to make the Sun, a medium-size star with an estimated life span of 10 billion years. Its system is located in the Orion arm of the Milky Way galaxy, about two-thirds of the way out from the center.

During the Sun's first 100 million years, the remaining rock and ice smashed together into increasingly larger chunks, or planetesimals, until the planets, moons, asteroids, and comets reached their present state. The resulting disk-shaped solar system can be divided into four regions—terrestrial planets, giant planets, the Kuiper Belt, and the Oort Cloud—each containing its own types of bodies.

Terrestrial Planets
In the first region are the terrestrial (Earth-like) planets Mercury, Venus, Earth, and Mars. Mercury, the nearest to the Sun, orbits at an average distance of 36 million miles (58 million km.) and Mars, the farthest, at 142 million miles (228 million km.). Astronomers call the distance from the Sun to Earth (93 million miles/150 million km.) an astronomical unit (AU) and use it to measure planetary distances.

Terrestrial planets are rocky and warm and have cores of dense metal. All four planets have volcanoes, which long ago spewed out gases that created atmospheres on all but Mercury, which is too close to the Sun to hold onto an atmosphere. Mercury is heavily cratered, like the earth's moon. Venus has a permanent thick cloud cover and a surface temperature hot enough to melt lead. The air on Mars is very thin and usually cold, made mostly of carbon dioxide. Its dry, rock-strewn surface has many craters. It also has the largest known volcano in the solar system, Olympus Mons, which is 16 miles (25 km.) high.

Average temperatures and air pressures on Earth allow liquid water to collect on the surface, a unique feature among planets within the solar system. Meanwhile, Earth's atmosphere—mostly nitrogen and oxygen—and a strong magnetic field protect the surface from harmful solar radiation. These are the conditions that nurture life, according to scientists. It is widely accepted that Mars had abundant water very early in its history, but all large areas of liquid water have since disappeared. A fraction of this water is retained on modern Mars as both ice and locked into the structure of abundant water-rich materials, including clay minerals (phyllosilicates) and sulfates. Studies of hydrogen isotopic ratios indicate that asteroids and comets from beyond 2.5 AU provide the source of Mars' water. Like Earth, Mars has polar ice caps, although those on Mars are made up mostly of carbon dioxide ice (dry ice), while those on Earth are made up of water ice.

A single natural satellite, the Moon, orbits Earth, probably created by a collision with a huge planetesimal more than 4 billion years ago. Mars has two tiny moons that may have drifted to it from the asteroid belt. A broad ring from 2 to 3.3 AU from the Sun, this belt is composed of space rocks as small as dust grains and as large as 600 miles (1,000 km.) in diameter. Asteroids are made of mineral compounds, especially those containing iron, carbon, and silicon. Although the asteroid belt contains

FORMATION OF THE SOLAR SYSTEM

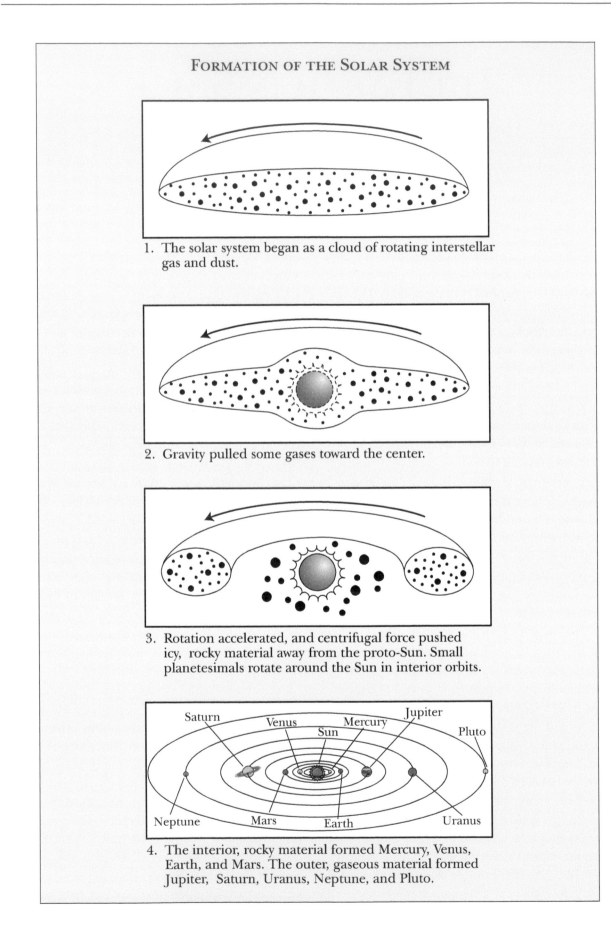

1. The solar system began as a cloud of rotating interstellar gas and dust.

2. Gravity pulled some gases toward the center.

3. Rotation accelerated, and centrifugal force pushed icy, rocky material away from the proto-Sun. Small planetesimals rotate around the Sun in interior orbits.

4. The interior, rocky material formed Mercury, Venus, Earth, and Mars. The outer, gaseous material formed Jupiter, Saturn, Uranus, Neptune, and Pluto.

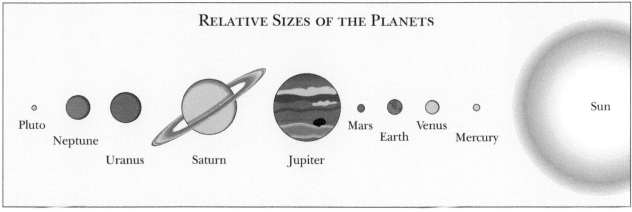

RELATIVE SIZES OF THE PLANETS

Pluto — Neptune — Uranus — Saturn — Jupiter — Mars — Earth — Venus — Mercury — Sun

Note: The size of the Sun and distances between the planets are not to scale.; Source: Data are from Jet Proulsion Laboratory, California Institute of Technology. The Deep Space Network. Pasadena, Calif.:JPL, 1988, p. 17.

enough material for a planet, one did not form there because Jupiter's gravity prevented the asteroids from crashing together. The belt separates the first region of the solar system from the second.

The Giant Planets

The second region belongs to the gas giants Jupiter, Saturn, Uranus, and Neptune. The closest, Jupiter, is 5.2 AU from the Sun, and the most distant, Neptune, is 30.11 AU. Jupiter is the largest planet in the solar system, its diameter 109 times larger than Earth's. The giant planets have solid cores, but most of their immense size is taken up by hydrogen, helium, and methane gases that grow thicker and thicker until they are like sludge near the core. On Jupiter, Saturn, and Uranus, the gases form wide bands over the surface. The bands sometimes have immense circular storms like hurricanes, but hundreds of times larger. The Great Red Spot of Jupiter is an example. It has winds of up to 250 miles (400 km.) per hour, and is at least a century old.

These planets have such strong gravity that each has attracted many moons to orbit it. In fact, they are like miniature solar systems. Jupiter has the most moons—eighteen—and Neptune has the fewest—eight—but Neptune's moon Triton is the largest of all. Most moons are balls of ice and rock, but Jupiter's Europa and Saturn's Titan may have liquid water below ice-bound surfaces. Several moons appear to have volcanoes, and a wispy atmosphere covers Titan. Additionally, the giant planets have rings of broken rock and ice around them, no more than 330 feet (100 meters) thick. Saturn's hundreds of rings are the brightest and most famous.

The Kuiper Belt

The third region of the solar system, the Kuiper Belt, contains the dwarf planet, Pluto. Pluto has a single moon, Charon. It does not orbit on the same plane, called the ecliptic, as the rest of the planets do. Instead, its orbit diverges more than seventeen degrees above and below the ecliptic. Its orbit's oval shape brings Pluto within the orbit of Neptune for a large percentage of its long year, which is equal to 248 Earth years. Two-thirds the size of the earth's moon, Pluto has a thin, frigid methane atmosphere. Charon is half Pluto's size and orbits less than 32,000 miles (20,000 km.) from Pluto's surface. Some astronomers consider Pluto and Charon to be a double planet.

The Kuiper Belt holds asteroids and the "short-period" comets that pass by Earth in orbits of 20 to 200 years. These bodies are the remains of

---

OTHER EARTHS

By the year 2000 astronomers had detected twenty-eight planets circling stars in the Sun's neighborhood of the galaxy. Planets, they think, are common. Those found were all gas giants the size of Saturn or larger. Earth-size planets are much too small to spot at such great distances. Where there are gas giants, there also may be terrestrial dwarfs, as in Earth's solar system. Where there are terrestrial planets, there may be liquid water and, possibly, life.

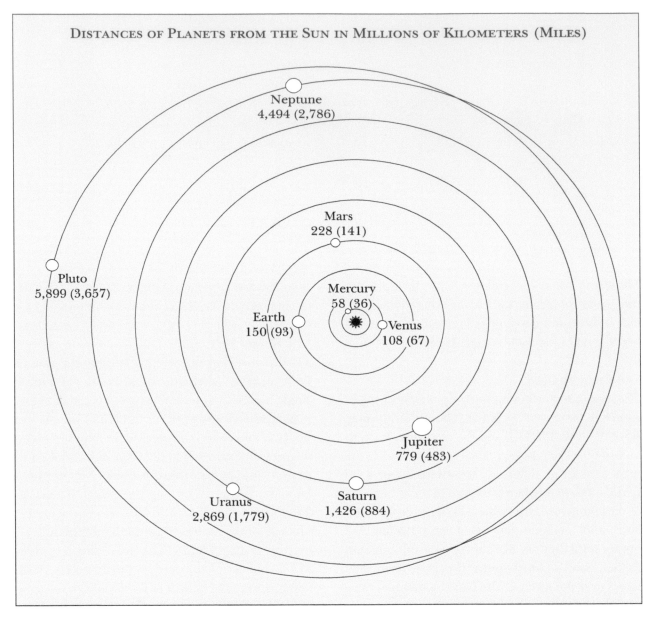

DISTANCES OF PLANETS FROM THE SUN IN MILLIONS OF KILOMETERS (MILES)

Neptune
4,494 (2,786)

Mars
228 (141)

Pluto
5,899 (3,657)

Mercury
58 (36)

Earth
150 (93)

Venus
108 (67)

Jupiter
779 (483)

Uranus
2,869 (1,779)

Saturn
1,426 (884)

planet formation and did not collect into planets because distances between them are too great for many collisions to occur. Most of them are loosely compacted bodies of ice and mineral—"dirty snowballs," as they were termed by the famous astronomer Fred Lawrence Whipple (November 5, 1906–August 30, 2004). An estimated 200 million Kuiper Belt objects orbit within a band of space from 30 to 50 AU from the Sun.

The Oort Cloud

In contrast to the other regions of the solar system, the Oort Cloud is a spherical shell surrounding the entire solar system. It is also a collection of com-

ets—as many as two trillion, scientists calculate. The inner edge of the cloud forms at a distance of about 20,000 AU from the Sun and extends as far out as 100,000 AU. The Oort Cloud thus gives the solar system a theoretical diameter of 200,000 AU—a distance so vast that light needs more than three years to cross it. No astronomer has yet detected an Oort Cloud object, because the cloud is so far away. Occasionally, however, gravity from a nearby star dislodges an object in the cloud, causing it to fall toward the Sun. When observers on Earth see such an object sweep by in a long, cigar-shaped orbit, they call it a long-period comet.

The outer edge of the Oort Cloud marks the farthest reach of the Sun's gravitational power to bind bodies to it. In one respect, the Oort Cloud is part of interstellar space.

In addition to light, the Sun sends out a constant stream of charged particles—atoms and subatomic particles—called the solar wind. The solar wind shields the solar system from the interstellar medium, but it only does so out to about 100 AU, a boundary called the heliopause. That is a small fraction of the distance to the Oort Cloud.

Roger Smith

EARTH'S MOON

The fourth-largest natural satellite in the solar system, Earth's moon has a diameter of 2,159.2 miles (3,475 km)—less than one-third the diameter of Earth. The Moon's mass is less than one-eightieth that of Earth.

The Moon orbits Earth in an elliptical path. When it is at perigee (when it is closest to Earth), it is 221,473 miles (356,410 km.) distant. When it is at apogee (farthest from Earth), it is 252,722 miles (406,697 km.) distant.

The Moon completes one orbit around Earth every 27.3 Earth days. Because it rotates at about the same rate that it orbits the earth, observers on Earth only see one side of the Moon. The changing angles between Earth, the Sun, and the Moon determine how much of the Moon's illuminated surface can be seen from Earth and cause the Moon's changing phases.

Volcanism

Naked-eye observations of the Moon from Earth reveal dark areas called *maria*, the plural form of the Latin word *mare* for sea. The maria are the remains of ancient lava flows from inside gigantic impact craters; the last eruptions were more than 3 billion years ago. The lava consists of basalt, similar in composition to Earth's oceanic crust and many volcanoes. The maria have names such as Mare Serenitatis (15° to 40°N, 5° to 20°E) and Mare Tranquillitatis (0° to 20°N, 15° to 45°E). Some of the smaller dark areas on the Moon also have names that are water-related: lacus (lake), sinus (bay), and palus (marsh).

Impact Craters

Observing the Moon with an optical aid, such as a telescope or a pair of binoculars, provides a closer view of impact craters. Impact craters of various sizes cover 83 percent of the Moon's surface. More than 33,000 craters have been counted on the Moon.

One of the easiest craters to observe from the Earth is Tycho. Located at 43.3°S, longitude 11.2 degrees west, it is about 50 miles (85 km.) wide. Surrounding Tycho are rays of dusty material, known as

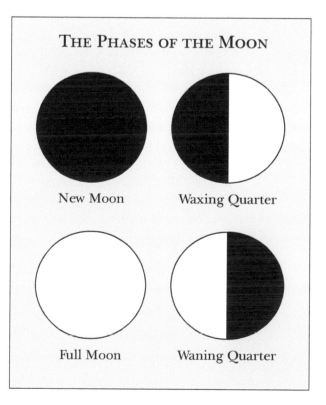

THE PHASES OF THE MOON

New Moon

Waxing Quarter

Full Moon

Waning Quarter

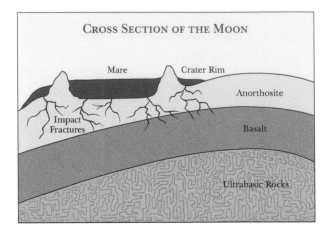

CROSS SECTION OF THE MOON

Mare   Crater Rim
Anorthosite
Impact Fractures
Basalt
Ultrabasic Rocks

ejecta, that appear to radiate from the crater. When an object from space, such as a meteoroid, slams into the Moon's surface, it is vaporized upon impact. The dust and debris from the interior of the crater fall back onto the lunar surface in a pattern of rays. Because the ejecta is disrupted by subsequent impacts, only the youngest craters still have rays. Sometimes, pieces of the ejecta fall back and create smaller craters called secondary craters. The ejecta rays of Tycho extend to almost 1,865 miles (3,000 km.) beyond the crater's edge.

Other Lunar Features

Near the crater called Archimedes is the Apennines mountain range, which has peaks nearly 20,000 feet (60,000 meters) high—altitudes comparable to South America's Andes.

The Moon also has valleys. Two of the most well known are the Alpine Valley, which is about 115 miles (185 km.) long; and the Rheita Valley, located about 155 miles (250 km.) from the Stevinus crater, which is 238 miles (383 km.) long, 15.5 miles (25 km.) wide, and 2,000 feet (609 meters) deep.

Smaller than valleys and resembling cracks in the lunar surface are features called rilles, which are thought to be places of ancient lava flow. Many rilles can be seen near the Aristarchus crater. Rilles are often up to 3 miles (5 km.) wide and can stretch for more than 104 miles (167 km.).

A wrinkle in the lunar surface is called a ridge. Many ridges are found around the boundaries of the maria. The Serpentine Ridge cuts through Mare Serenitatis.

Exploration of the Moon

Robotic spacecraft were the first visitors to explore the Moon. The Russian spacecraft Luna 1 made the first flyby of the Moon in January, 1959. Eight months later, Luna 2 made the first impact on the Moon's surface. In October, 1959, Luna 3 was the first spacecraft to photograph the side of the Moon not visible from Earth. In 1994 the United States' *Clementine* spacecraft was the first probe to map the Moon's composition and topography globally.

The first humans to land on the Moon were the U.S. astronauts Neil Armstrong and Edwin "Buzz" Aldrin. On July 20, 1969, they landed in the *Eagle* lunar module, during the Apollo 11 mission. Armstrong's famous statement, "That's one small step for man, one giant leap for mankind," was heard around the world by millions of people who watched the first humans set foot on the lunar surface, at the Sea of Tranquillity. The last twentieth century human mission to reach the lunar surface, Apollo 17, landed there in December, 1972. Astronauts Gene Cernan and geologist Jack Schmitt landed in the Taurus-Littrow Valley (20°N, 31°E).

Noreen A. Grice

THE SUN AND THE EARTH

Of all the astronomical phenomena that one can consider, few are more important to the survival of life on Earth than the relationship between Earth and the Sun. With the exception of small amounts of residual (endogenic) energy that have remained inside the earth from the time of its formation some 4.5 billion years ago and which sustain some specialized forms of life along some oceanic rift systems, almost all other forms of life, including human, depend on the exogenic light and energy that the earth receives directly from the Sun.

The enormous variety of ecosystems on Earth are highly dependent on the angles at which the Sun's rays strike Earth's spherical surface. These angles, which vary greatly with latitude and time of year, determine many commonly observed phenomena, such as the height of the Sun above the horizon, the changing lengths of day and night throughout the year, and the rhythm of the seasons. Daily and seasonal changes have profound effects on the many climatic regions and life cycles found on earth.

The Sun

The center of Earth's solar system, the Sun is but one ordinary star among some 100 billion stars in an ordinary cluster of stars called the Milky Way galaxy. There are at least 10 billion galaxies in the universe, each with billions of stars. Statistically, the chances are good that many of these stars have their own solar systems. Late twentieth century astronomical observations discovered the presence of what appear to be planets, large ones similar in size to Jupiter, orbiting other stars.

Earth's Sun is an average star in terms of its physical characteristics. It is a large sphere of incandescent gas that has a diameter more than 100 times that of Earth, a mass more than 300,000 times that of Earth, and a volume 1.3 million times that of Earth. The Sun's surface gravity is thirty-four times that of Earth.

The conversion of hydrogen into helium in the Sun's interior, a process known as nuclear fusion, is the source of the Sun's energy. The amount of mass that is lost in the fusion process is miniscule, as evidenced by the fact that it will take perhaps 15 million years for the Sun to lose one-millionth of its total mass. The Sun is expected to continue shining through another several billion years.

Earth Revolution

The earth moves about the Sun in a slightly elliptical orbit called a revolution. It takes one year for the earth

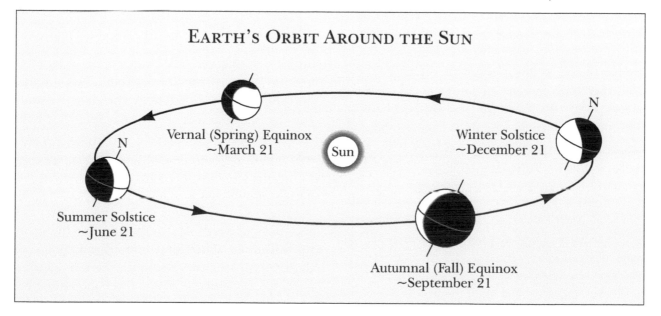

EARTH'S ORBIT AROUND THE SUN

Vernal (Spring) Equinox ~March 21

Sun

Winter Solstice ~December 21

Summer Solstice ~June 21

Autumnal (Fall) Equinox ~September 21

to make one revolution at an average orbital velocity of about 29.6 kilometers per second (18.5 miles per second). Earth-sun relationships are described by a tropical year, which is defined as the period of time (365.25 average solar days) from one vernal equinox to another. To balance the tropical year with the calendar year, a whole day (February 29) is added every fourth year (leap year). Other minor adjustments are necessary so as to balance the system.

Perihelion and Aphelion

The average distance between Earth and the Sun is approximately 93 million miles (150 million km.). At that distance, sunlight, which travels at the speed of light (186,000 miles/300,000 kilometers per second), takes about 8.3 minutes to reach the earth. Since the earth's orbit is an ellipse rather than a circle, the earth is closest to the Sun on about January 3—a distance of 91.5 million miles (147 million km.). This position in space is called perihelion, which comes from the Greek *peri*, meaning "around" or "near," and *helios*, meaning the Sun. Earth is farthest from the Sun on about July 4 at aphelion (Greek *ap*, "away from," and *helios*), with a distance of 152 million kilometers (94.5 million miles).

Axial Inclination

Astronomers call the imaginary surface on which Earth orbits around the Sun the plane of the ecliptic. The earth's axis is inclined 66.5 degrees to the plane of the ecliptic (or 23.5 degrees from the perpendicular to the plane of the ecliptic), and it maintains this orientation with respect to the stars. Thus, the North Pole points in the same direction to Polaris, the North Star, as it revolves about the Sun. Consequently, the Northern Hemisphere tilts away from the Sun during one-half of Earth's orbit and toward the Sun through the other half.

Winter solstice occurs on December 21 or 22, when the tilt of the Northern Hemisphere away from the Sun is at its maximum. The opposite condition occurs during summer solstice on June 21 or 22, when the Northern Hemisphere reaches its maximum tilt toward the Sun. The equinoxes occur midway between the solstices when neither the Southern nor the Northern Hemisphere is tilted toward the Sun. The

---

ECLIPSES

The Sun's diameter is 400 times larger than the moon's; however, the moon is 400 times closer to Earth than the Sun, making the two objects appear nearly the same size in the sky to observers on Earth. As the moon orbits Earth, it crosses the plane of the Earth-Sun orbit twice each month. If one of the orbit-crossing points (called nodes) occurs during a new or full moon phase, a solar or lunar eclipse can occur.

A solar eclipse occurs when the moon and the Sun appear to be in the exact same place in the sky during a new moon phase. When that happens, the moon blocks the light of the Sun for up to seven minutes. Because solar eclipses can be seen only from certain places on Earth, some people travel around the world—sometimes to remote places—to view them.

A lunar eclipse occurs when Earth is positioned between the Sun and the moon and casts its shadow on the moon. In contrast to solar eclipses, lunar eclipses are visible from every place on Earth from which the moon can be seen.

---

vernal and autumnal equinoxes occur on March 20 or 21 and September 22 or 23, respectively.

The axial inclination of 66.6 degrees (or 23.5 degrees from the perpendicular) explains the significance of certain parallels on the earth. The noon sun shines directly overhead on the earth at varying latitudes on different days—between 23.5°S and 23.5°N. The parallels at 23.5°S and 23.5°N are called the Tropics of Capricorn and Cancer, respectively.

During the winter and summer solstices, the area on the earth between the Arctic Circle (at 66.5°N) and the North Pole has twenty-four hours of darkness and daylight, respectively. The same phenomena occurs for the area between the Antarctic Circle (at 66.5°S) and the South Pole, except that the seasons are reversed in the Southern Hemisphere. At the poles, the Sun is below the horizon for six months of the year.

For those living outside the tropics (poleward of 23.5 degrees north and south latitude), the noon sun will never shine directly overhead. Hours of daylight will also vary greatly during the year. For example, daylight will range from approximately nine hours during the winter solstice to fifteen hours during the summer solstice for persons living near 40°N, such as in Philadelphia, Denver, Madrid, and Beijing.

Solar Radiation

Given the size of the earth and its distance from the Sun, it is estimated that this planet receives only about one two-billionth part of the total energy released by the Sun. However, this seemingly small amount is enough to drive the massive oceanic and atmospheric circulation systems and to support all life processes on Earth.

Solar energy is not evenly distributed on Earth. The higher the angle of the Sun in the sky, the greater the duration and intensity of the insolation.

To illustrate this, note how easy it is look at the Sun when it is very low on the horizon—near dawn and sunset. At those times, the Sun's rays have to penetrate much more of the atmosphere, so more of the sunlight is absorbed. When the Sun's rays are coming in at a low angle, the same solar energy is spread over a larger area, thereby leading to less insolation per unit of area. Thus, the equatorial region receives much more solar energy than the polar region. This radiation imbalance would make the earth decidedly less habitable were it not for the atmospheric and oceanic circulation systems (such as the warm Gulf Stream) that move the excess heat from the Tropics to the middle and high latitudes.

Robert M. Hordon

THE SEASONS

Earth's 365-day year is divided into seasons. In most parts of the world, there are four seasons—winter, spring, summer, and fall (also called autumn). In some tropical regions—those close to the equator—there are only two seasons. In areas close to the equator, temperatures change little throughout the year; however, amounts of rainfall vary greatly, resulting in distinct wet and dry seasons. The polar regions of the Arctic and Antarctic also have little variation in temperature, remaining cold throughout the year. Their seasons are light and dark, because the Sun shines almost constantly in the summer and hardly at all in the winter.

The four seasons that occur throughout the northern and southern temperate zones—between the tropics and the polar regions—are climatic seasons, based on temperature and weather changes. Winter is the coldest season; it is the time when days are short and few crops can be grown. It is followed by spring, when the days lengthen and the earth warms; this is the time when planting typically begins, and animals that hibernate (from the French word for winter) during the winter leave their dens.

Summer is the hottest time of the year. In many areas, summer is marked by drought, but other regions experience frequent thunderstorms and humid air. In the fall, the days again become shorter and cooler. This is the time when many crops are harvested. In ancient cultures, the turning of the seasons was marked by festivals, acknowledging the importance of seasonal changes to the community's survival.

Each season is defined as lasting three months. Winter begins at the winter solstice, which is the time when the Sun is farthest from the equator. In the Northern Hemisphere, this occurs on December 21 or 22, when the Sun is directly over the tropic of Capricorn. Summer begins at the other solstice, June 20 or 21 in the Northern Hemisphere, when the Sun is directly over the tropic of Cancer. The winter solstice is the shortest day of the year; the summer solstice is the longest.

Spring and fall begin on the two equinoxes. At an equinox, the Sun is directly above the earth's equator and the lengths of day and night are approximately equal everywhere on Earth. In the Northern Hemisphere, the vernal (spring) equinox occurs on March 21 or 22; in the Southern Hemisphere, it is the autumnal (fall) equinox. The Northern Hemisphere's autumnal equinox (and the Southern Hemisphere's vernal equinox) occurs September 22 or 23.

Seasons and the Hemispheres

The relationship of the seasons to the calendar is opposite in the Northern and Southern Hemispheres. On the day that a summer solstice occurs in the Northern Hemisphere, the winter solstice occurs in the Southern Hemisphere. Thus, when it is summer in the Southern Hemisphere, it is winter in the Northern Hemisphere, and vice versa.

The Sun and the Seasons

The reason why summers and winters differ in the temperate zones is often misunderstood. Many people think that winter happens when the Sun is more distant from Earth than it is in summer. What causes Earth's seasons is not the changing distances between the earth and the Sun, but the tilt of the earth's axis. A line drawn from the North Pole to the South Pole through the center of the earth (the earth's axis) is not perpendicular to the plane of the earth's orbit (the ecliptic). The earth's axis and the perpendicular to the ecliptic make an angle of 23.5 degrees. This tilts the Northern Hemisphere toward the Sun when the earth is on one side of its orbit around the Sun, and tilts the Southern Hemisphere toward the Sun when the earth moves around to the Sun's opposite side. When the Sun appears to be at its highest in the sky, and its rays are most direct, summer occurs. When the Sun appears to be at its lowest, and its rays are indirect, there is winter.

Local Phenomena

Local conditions can have important effects on seasonal weather. At locations near oceans, sea breezes develop during the day, and evenings are characterized by land breezes. Sea breezes bring cooler ocean air in toward land. This results in temperatures at the shore often being 5 to 11 degrees Fahrenheit (3 to 6 degrees Celsius) lower than temperatures a few miles inland.

At night, when land temperatures are lower than ocean temperatures, land breezes move air from the land toward the water. As a result, coastal regions have less seasonal temperature variations than inland areas do. For example, coastal areas seldom become cold enough to have snow in the winter, even though inland areas at the same latitude do.

Hailstorms

Hail usually occurs during the summer, and is associated with towering thunderstorm clouds, called cumulonimbus. Hail is occasionally confused with sleet. Sleet is a wintertime event, and occurs when warmer layers of air sit above freezing layers near the ground. Rain that forms in the warmer, upper layer solidifies into tiny ice pellets in the lower, subfreezing layer before hitting the ground.

Hail is an entirely different phenomenon. When cold air plows into warmer, moist air—called a cold front boundary—powerful updrafts of rising air can be created. The warm, moist air propelled upward by the heavier cold air can reach velocities approaching 100 miles (160 kilometers) per hour. Ice crystals form above the freezing level in the cumulonimbus clouds and fall into lower, warmer parts of the clouds, where they become coated with water. Picked up by an updraft, the coated ice crystals are carried back to a higher, colder levels where their water coatings freeze. This cycle can repeat many times, producing hailstones that have multiple, concentric layers of ice.

Hailstorms can be very damaging. Hail can ruin crops, dent car bodies, crack windshields, and injure people. The Midwest of the United States is particularly susceptible to hailstorms. There, warm, moist air from the Gulf of Mexico often meets much colder, drier air originating in Canada. This combination produces the extreme atmospheric instability necessary for that kind of weather.

Alvin S. Konigsberg

EARTH'S INTERIOR

EARTH'S INTERNAL STRUCTURE

Earth is one of the nine known planets in the Sun's solar system that formed from a giant cloud of cosmic dust called a nebula. This event is thought to have happened between 4.44 billion years ago (based on the age of the oldest-known Moon rock) and 4.56 billion years ago (the age of meteorite bombardment). After Earth's formation, heat released by colliding particles combined with the heat energy released by the decay of radioactive elements to cause some or all of Earth's interior to melt. This melting began the process of differentiation, which allowed the heavier elements, mainly iron and nickel, to sink toward Earth's center while the lighter, rocky components moved upward, as a result of the contrast in density of the earth's forming elements.

This process of differentiation was probably the most important event of Earth's early history. It changed the planet from a homogeneous mixture with neither continents nor oceans to a planet with three layers: a dense core beginning at 1,800 miles (2,900 km.) deep and ending at Earth's center, 3,977 miles (6,400 km.) below the surface; a mantle beginning between 3 and 44 miles (5-70 km.) deep and ending at Earth's core; and a crust going from Earth's surface to about 3-6 miles (5-10 km.) deep for oceanic crust and 22-44 miles (35-70 km.) deep for continental crust.

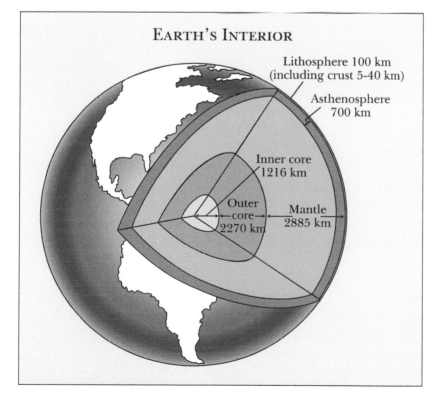

EARTH'S INTERIOR

Lithosphere 100 km (including crust 5-40 km)

Asthenosphere 700 km

Inner core 1216 km

Outer core 2270 km

Mantle 2885 km

Layering of the Earth

Earth's layers can be classified either by their composition (the traditional method) or by their mechanical behavior (strength). Compositional classification identifies several distinct concentric layers, each with its own properties. The outermost layer of Earth is the crust or skin. This is divided into continental and oceanic crusts. The continental crust varies in thickness between 22 and 25 miles (35 and 40 km.) under flat continental regions and up to 44 miles (70 km.) under high mountains. The oceanic crust is made up of igneous rocks rich in iron and magnesium, such as basalt and peridotite. The upper continental crust is composed mainly of alumino-silicates. The old-

PROPERTIES OF SEISMIC WAVES

Seismologists use two types of body waves—primary (P-waves) and secondary (S-waves) waves—to estimate seismic velocities of the different layers within the earth. In most rock types P-waves travel between 1.7 and 1.8 times more quickly than S-waves; therefore, P-waves always arrive first at seismographic stations. P-waves travel by a series of compressions and expansions of the material through which they travel. P-waves can travel through solids, liquids, or gases. When P-waves travel in air, they are called sound waves.

The slower S-waves, also called shear waves, move like a wave in a rope. This movement makes the S-wave more destructive to structures like buildings and highway overpasses during earthquakes. Because S-waves can travel only through solids and cannot travel through Earth's outer core, seismologists concluded that Earth's outer core must be liquid or at least must have the properties of a fluid.

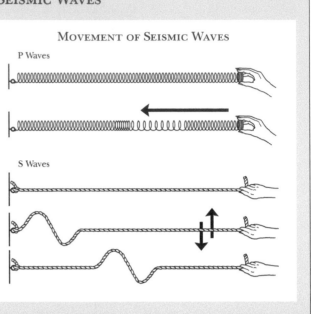

MOVEMENT OF SEISMIC WAVES

P Waves

S Waves

est continental crustal rock exceeds 3.8 billion years, while oceanic crustal rocks are not older than 180 million years. The oceanic crust is heavier than the continental crust.

Earth's next layer is the mantle, which is made up mostly of ferro-magnesium silicates. It is about 1,800 miles (2,900 km.) thick and is separated into the upper and lower mantle. Most of Earth's internal heat is contained within the mantle. Large convective cells in the mantle circulate heat and may drive plate-tectonic processes.

The last layer is the core, which is separated into the liquid outer core and the solid inner core. The outer core is 1,429 miles (2,300 km.) thick, twice as thick as the inner core. The outer core is mainly composed of a nickel-iron alloy, while the inner core is almost entirely composed of iron. Earth's magnetic field is believed to be controlled by the liquid outer core.

In the mechanical layering classification of the earth's interior, the layers are separated based on mechanical properties or strength (resistance to flowing or deformation) in addition to composition. The uppermost layer is the lithosphere (sphere of rock), which comprises the crust and a solid portion of the upper mantle. The lithosphere

is divided into many plates that move in relation to each other due to tectonic forces. The solid lithosphere floats atop a semiliquid layer known as the asthenosphere (weak sphere), which enables the lithosphere to move around.

Exploring Earth's Interior

Volcanic activity provides natural samples of the outer 124 miles (200 km.) of Earth's interior. Meteorites—samples of the solar system that have collided with Earth—also provide clues about Earth's composition and early history. The most ambitious human effort to penetrate Earth's interior was made by the former Soviet Union, which drilled a super-deep research well, named the Kola Well, near Murmansk, Russia. This was an attempt to penetrate the crust and reach the upper mantle. The reported depth of the Kola Well is a little more than 7.5 miles (12 km.). Although impressive, the drilled depth represents less than 0.2 percent of the distance from the earth's surface to its center.

A great deal of knowledge about Earth's composition and structure has been obtained through computer modeling, high-pressure laboratory experiments, and meteorites, but most of what is known about Earth's interior has been acquired by studying seismic waves generated by earthquakes and nuclear explosions. As seismic waves are transmitted, reflected, and refracted through the earth, they carry information to the surface about the materials through which they have traveled. Seismic waves are recorded at receiver stations (seismographic stations) and processed to provide a picturelike image of Earth's interior.

Changes in P- and S-wave velocities within Earth reveal the sequence of layers that make up Earth's interior. P-wave velocity depends on the elasticity, rigidity, and density of the material. By contrast, S-wave velocity depends only on the rigidity and density of the material. There are sharp variations in velocity at different depths, which correspond to boundaries between the different layers of Earth. P-wave velocity within crustal rocks ranges from 3.6-4.2 miles (6-7 km.) per second.

The boundary between the crust and the mantle is called the Mohorovičić discontinuity or Moho. At Moho, P-wave velocity increases from 4.2-4.8 miles (7-8 km.) per second. Beyond the crust-mantle boundary, P-wave velocity increases gradually up to about 8.1 miles (13.5 km.) per second at the core-mantle boundary. At this depth, S-waves are not transmitted and P-wave velocity, decreases from 8.1 to 4.8 miles (13.5 to 8 km.) per second, which strongly supports the concept that the outer core is liquid, since S-waves cannot travel through liquids. As P-waves enter the inner core, their velocity again increases, to about 6.8 miles (11.3 km.) per second.

Earth's interior seems to be characterized by a gradual increase with depth in temperature, pressure, and density. Extensive experimental and modeling work indicates that the temperature at 62 miles (100 km.) is between 1,200 and 1,400 degrees Celsius (2,192 to 2,552 degrees Fahrenheit). The temperature at the core-mantle boundary—about 1,802 miles (2,900 km.) deep—is calculated to be about 8,130 degrees Fahrenheit (4,500 degrees Celsius). At Earth's center the temperature may exceed 12,092 degrees Fahrenheit (6,700 degrees Celsius). Although at Earth's surface, heat energy is slowly but continuously lost as a result of outgassing, such as from volcanic eruptions, its interior remains hot.

Seismic Tomography and Future Exploration

Seismic tomography is one of the newest tools that earth scientists are using to develop three-dimensional velocity images of Earth's interior. In seismic tomography, several crossing seismic waves from different sources (earthquakes and nuclear explosions) are analyzed in much the same way that computerized axial tomography (CAT) scanners are used in medicine to obtain images of human organs. Seismic tomography is providing two- and three-dimensional images from the crust to the core-mantle boundary. Fast P-wave velocities have been correlated to cool material—for example, a piece of sinking lithosphere (cool rigid layer) such as in regions underneath the Andes Mountains (subduction zone); slow P-wave velocities have been correlated with hot materials—for example, rising mantle plumes of hot spots such as the one responsible for volcanic activity in the Hawaiian Islands.

Rubén A. Mazariegos-Alfaro

PLATE TECTONICS

The theory of plate tectonics provides an explanation for the present-day structure of the large landforms that constitute the outer part of the earth. The theory accounts for the global distribution of continents, mountains, hills, valleys, plains, earthquake activity, and volcanism, as well as various associations of igneous, metamorphic, and sedimentary rocks, the formation and location of min-

MAJOR TECTONIC PLATES AND MID-OCEAN RIDGES

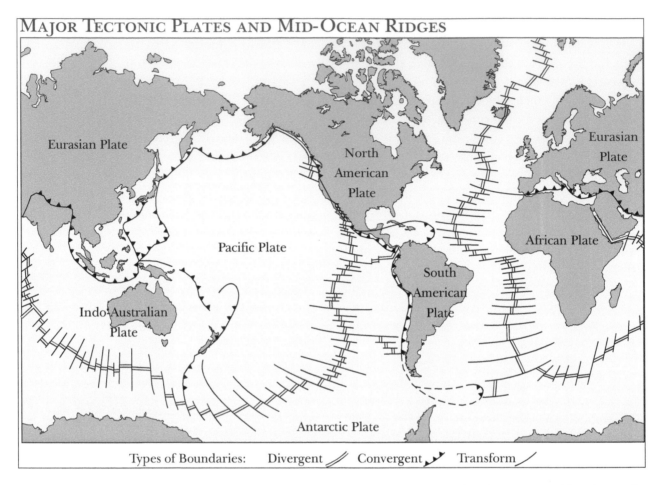

Types of Boundaries: Divergent   Convergent   Transform

eral resources, and the geology of ocean basins. Everything about the earth is related either directly or indirectly to plate tectonics.

Basic Theory

Plate-tectonic theory is based on an Earth model in which a rigid, outer shell—the lithosphere—lies above a hotter, weaker, partially molten part of the mantle called the asthenosphere. The lithosphere varies in thickness between 6 and 90 miles (10 and 150 km.), and comprises the crust and the underlying, upper mantle. The asthenosphere extends from the base of the lithosphere to a depth of about 420 miles (700 km.). The brittle lithosphere is broken into a pattern of internally rigid plates that move horizontally relative to each other across the earth's surface.

More than a dozen plates have been distinguished, some extending more than 2,500 miles (4,000 km.) across. Exhibiting independent motion, the plates grind and scrape against each other,

similar to chunks of ice in water, or like giant rafts cruising slowly on the asthenosphere. Most of the earth's dynamic activity, including earthquakes and volcanism, occurs along plate boundaries. The global distribution of these tectonic phenomena delineates the boundaries of the plates.

Geological observations, geophysical data, and theoretical models support the existence of three types of plate boundaries. Divergent boundaries occur where adjacent plates move away from each other. Convergent boundaries occur where adjacent plates move toward each other. Transform boundaries occur where plates slip past one another in directions parallel to their common boundaries.

The continents were formed by the movement at plate boundaries, and continental landforms were generated by volcanic eruptions and continental plates colliding with each other. The velocity of plate movement varies from plate to plate and even within portions of the same plate, ranging from 0.8 to 8 inches (2 to 20 centimeters) per year. The rates

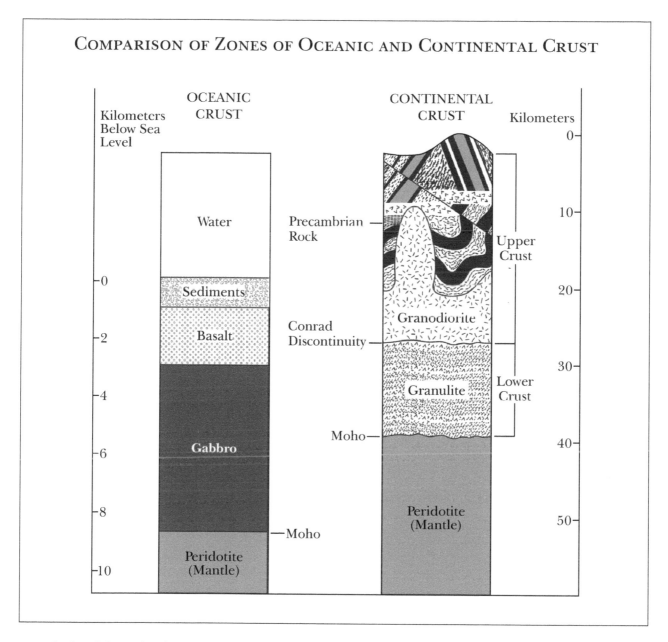

COMPARISON OF ZONES OF OCEANIC AND CONTINENTAL CRUST

are calculated from the distance to the midoceanic ridge crests, along with the age of the sea floor as determined by radioactive dating methods.

Convection currents that are driven by heat from radioactive decay in the mantle are important mechanisms involved in moving the huge plates. Convection currents in the earth's mantle carry magma (molten rock) up from the asthenosphere. Some of this magma escapes to form new lithosphere, but the rest spreads out sideways beneath the lithosphere, slowly cooling in the process. Assisted by gravity, the magma flows outward, dragging the overlying lithosphere with it, thus continu-

ing to open the ridges. When the flowing hot rock cools, it becomes dense enough to sink back into the mantle at convergent boundaries.

A second plate-driving mechanism is the pull of dense, cold, down-flowing lithosphere in a subduction zone on the rest of the trailing plate, further opening up the spreading centers so magma can move upward.

Divergent Plate Boundaries

During the 1950s and 1960s, oceanographic studies revealed that Earth's seafloors were marked by a nearly continuous system of submarine ridges,

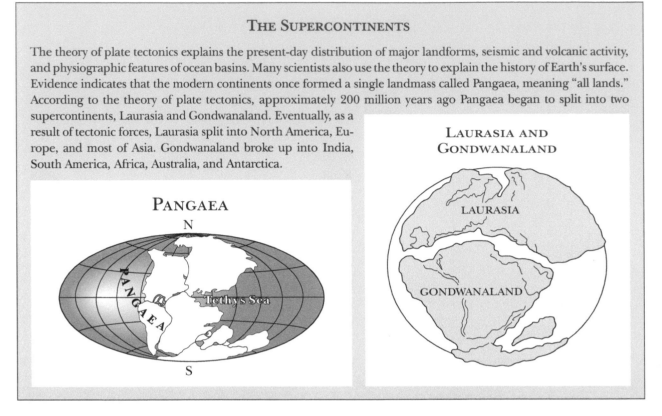

THE SUPERCONTINENTS

The theory of plate tectonics explains the present-day distribution of major landforms, seismic and volcanic activity, and physiographic features of ocean basins. Many scientists also use the theory to explain the history of Earth's surface. Evidence indicates that the modern continents once formed a single landmass called Pangaea, meaning "all lands." According to the theory of plate tectonics, approximately 200 million years ago Pangaea began to split into two supercontinents, Laurasia and Gondwanaland. Eventually, as a result of tectonic forces, Laurasia split into North America, Europe, and most of Asia. Gondwanaland broke up into India, South America, Africa, Australia, and Antarctica.

PANGAEA

LAURASIA AND GONDWANALAND

more than 40,000 miles (64,000 km.) in length. Detailed investigations revealed that the midoceanic ridge system has a central rift valley that runs along its length and that the ridge system is associated with volcanic and earthquake activity. The earthquakes are frequent, shallow, and mild.

Magnetic studies of the seafloor indicate that the oceanic lithosphere has been segmented into a series of long magnetic strips that run parallel to the axis of the midoceanic ridges. On either side of the ridge, the ocean floor consists of alternating bands of rock, magnetized either parallel to or exactly opposite of the present-day direction of the earth's magnetic field.

Midoceanic ridges, or divergent plate boundaries, are tensional features representing zones of weakness within the earth's crust, where new seafloor is created by the welling up of mantle material from the asthenosphere into cracks along the ridges. As rifting proceeds, magma ascends to fill in the fissures, creating new oceanic crust. Iron minerals within the magma become aligned to the existing Earth polarity as the rock cools and crystallizes. The oceanic floor slowly moves away from the oce-

anic ridge toward deep ocean trenches, where it descends into the mantle to be melted and recycled to the earth's surface to generate new rocks and landforms.

As the seafloor spreads outward from the rift center, about half of the material is carried to either side of the rift, which is later filled by another influx of molten basalt. When the polarity of the earth changes, the subsequent molten basalt is magnetized in the opposite polarity. The continuation of this process over geologic time leads to the young geologic age of the seafloor and the magnetic symmetry around the midoceanic ridges.

Not all spreading centers are underneath the oceans. An example of continental rifting in its embryonic stage can be observed in the Red Sea, where the Arabian plate has separated from the African plate, creating a new oceanic ridge. Another modern-day example of continental divergent activity is East Africa's Great Rift Valley system. If this rifting continues, it will eventually fragment Africa, producing an ocean that will separate the resulting pieces. Through divergence, large plates are made into smaller ones.

Convergent Plate Boundaries

Because Earth's volume is not changing, the increase in lithosphere created along divergent boundaries must be compensated for by the destruction of lithosphere elsewhere. Otherwise, the radius of Earth would change. The compensation occurs at convergent plate boundaries, where plates are moving together. Three scenarios are possible along convergent boundaries, depending on whether the crust involved is oceanic or continental.

If both converging plates are made of oceanic crust, one will inevitably be older, cooler, and denser than the other. The denser plate eventually subducts beneath the less-dense plate and descends into the asthenosphere. The boundary along the two interacting plates, called a subduction zone, forms a trench. Some trenches are more than 620 miles (1,000 km.) long, 62 miles (100 km.) wide, and 6.8 miles (11 km.) deep. Heated by the hot asthenosphere beneath, the subducted plate becomes hot enough to melt.

Because of buoyancy, some of the melted material rises through fissures and cracks to generate volcanoes along the overlying plate. Over time, other parts of the melted material eventually migrate to a divergent boundary and rise again in cyclic fashion to generate new seafloor. The volcanoes generated along the overriding plate often form a string of islands called island arcs. Japan, the Philippines, the Aleutians, and the Mariannas are good examples of island arcs resulting from subduction of two plates consisting of oceanic lithosphere. Intense earthquakes often occur along subduction zones.

If the leading edge of one of the two convergent plates is oceanic crust and the other is continental crust, the oceanic plate is always the one subducted, because it is always denser. A classic example of this case is the western boundary of South America. On the oceanic side of the boundary, a trench was formed where the oceanic plate plunged underneath the continental plate. On the continental side, a fold mountain belt—the Andes—was formed as the oceanic lithosphere pushed against the continental lithosphere.

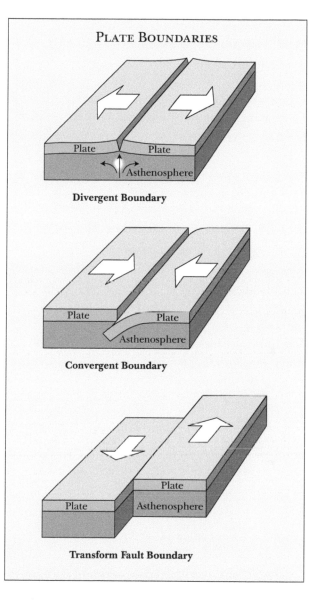

PLATE BOUNDARIES

Divergent Boundary

Convergent Boundary

Transform Fault Boundary

When the oceanic plate descends into the mantle, some of the material melts and works its way up through the mountain belt to produce rather violent volcanoes. The boundary between the plates is a region of earthquake activity. The earthquakes range from shallow to relatively deep, and some are quite severe.

The last type of convergent plate boundary involves the collision of two continental masses of lithosphere, which can result in folding, faulting, metamorphism, and volcanic activity. When the plates collide, neither is dense enough to be forced into the asthenosphere. The collision compresses and thickens the continental edges, twisting and deforming the rocks and uplifting the land to form unusually high fold mountain belts. The prototype example is the collision of India with Asia, resulting in the formation of the Himalayas. In this case, the earthquakes are typically shallow, but frequent and severe.

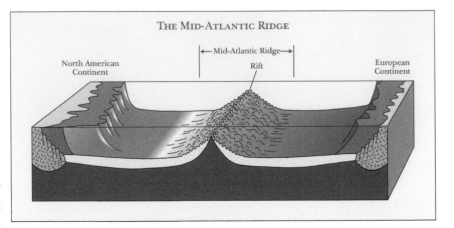

THE MID-ATLANTIC RIDGE

Transform Plate Boundaries

The actual structure of a seafloor spreading ridge is more complex than a single, straight crack. Instead, ridges comprise many short segments slightly offset from one another. The offsets are a special kind of fault, or break in the lithosphere, known as a transform fault, and their function is to connect segments of a spreading ridge. The opposite sides of a transform fault belong to two different plates that are grinding against each other in opposite directions.

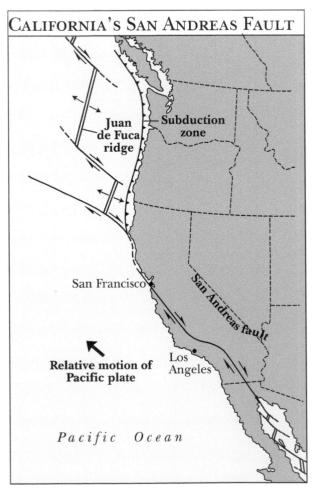

CALIFORNIA'S SAN ANDREAS FAULT

Transform faults form the boundaries that allow the plates to move relative to each another. The classic case of a transform boundary is the San Andreas Fault. It slices off a small piece of western California, which rides on the Pacific plate, from the rest of the state, which resides on the North American plate. As the two plates scrape past each other, stress builds up, eventually being released in earthquakes that can be quite violent.

Mantle Plumes and Hot Spots

Most plate tectonic features are near plate boundaries, but the Hawaiian Islands are not. In the late twentieth century, the only active volcanoes in the Hawaiian Islands were on the island of Hawaii, at the southeast end of the chain. Radiometric dating and examination of states of erosion show that, when proceeding along the chain to the northwest, successive islands are progressively older.

Evidently, the same heat source produced all the volcanoes in the Hawaiian chain. Known as a mantle plume, it has remained stationary while the Pacific plate rides over it, producing a volcanic trail from which absolute motion of the plate can be determined. Since mantle plumes do not move with the plates, the plumes must originate beneath the lithosphere, probably far below it. Resulting volcanoes are called hot spots to distinguish them from subduction-zone volcanoes. Iceland is a good example of a hot spot, as is Yellowstone. At least 100 hot spots are distributed around Earth.

Alvin K. Benson

VOLCANOES

Volcanoes form mountains both on land and in the sea and either do it on a grand scale or merely create minute bumps on the seafloor. Volcanoes do not occur in a random pattern, but are found in distinct zones that are related to plate dynamics. Each of the three types of volcanism on Earth is characterized by specific types of eruptions and magma compositions. Molten magma is the rock material below the earth's crust that forms igneous rock as it cools.

Types of Volcanoes

Geologists generally group volcanoes into four main kinds—cinder cones, composite volcanoes, shield volcanoes, and lava domes.

Cinder cones are built from congealed lava ejected from a single vent. As the gas-charged lava is blown into the air, it breaks into small fragments that solidify and fall as *cinders* around the vent to form a circular or oval cone. Most cinder cones have a bowl-shaped *crater* at the summit and rarely rise more than a thousand feet or so above their surroundings. Cinder cones are numerous in western North America and in other volcanic terrains of the world.

Composite volcanoes —sometimes called stratovolcanoes—include some of the Earth's grandest mountains, including Mount Fuji in Japan, Mount Cotopaxi in Ecuador, Mount Shasta in California, Mount Hood in Oregon, and Mount St. Helens and Mount Rainier in Washington. The essential feature of a composite volcano is a conduit system through which magma deep in the Earth's crust rises to the surface. They are typically steep-sided, symmetrical cones of large dimension built of alternating layers of lava flows, volcanic ash, cinders, blocks, and bombs. They may rise as much as 8,000 feet above their bases. Most have a crater at the summit that contains a central vent or a clustered group of vents. Lavas either flow through breaks in the crater wall or fissures on the flanks of the cone.

Shield volcanoes, the third type of volcano, are built almost entirely of fluid lava flows that pour out in all directions from a central vent, or group of vents, building a broad, gently sloping cone. They are built up slowly as thousands of highly fluid lava flows—basalt lava—spread over great distances, and then cool into thin sheets. Some of the largest volcanoes in the world are shield volcanoes. The Hawaiian Islands are composed of linear chains of these volcanoes including Kilauea and Mauna Loa on the island of Hawaii—two of the world's most active volcanoes. The floor of the ocean is more than 15,000 feet deep at the bases of the islands. As Mauna Loa, the largest of the shield volcanoes (and also the world's largest active volcano), projects 13,679 feet above sea level, its top is over 28,000 feet above the deep ocean floor.

Volcanic Composition

Volcanoes in the midocean ridges and plume environments draw most of their magmas from the earth's mantle and produce mainly dark, magnesium-rich basaltic magmas. When basaltic magmas accumulate in the continental crust (for example, at Yellowstone), the large-scale crustal melting leads to rhyolitic volcanism, the volcanic equivalent of granites. Arc magmas cover a wider range of magmatic compositions, ranging from arc basalt to light-colored, silica-rich rhyolites; the latter are commonly erupted in the form of the silica-rich volcanic rock known as pumice, or the black volcanic glass known as obsidian. Andesites, named after the Andes Mountains, are a common volcanic rock in stratovolcanoes, intermediate in composition between basalt and rhyolite.

Magmas form from several processes that lead to partial melting of a solid rock. The simplest is adding heat—for example, plumes carrying heat from deep levels in the mantle to shallower levels, where melting occurs. Decompressional (lowering the pressure) melting of the mantle occurs where the

SOME VOLCANIC HOT SPOTS AROUND THE WORLD

ocean floor is thinned or carried away by seafloor spreading in midocean ridge environments.

Genesis of Magma

Adding a "flux" to a solid mineral mixture may lower the substance's melting point. The most common theory about arc magma genesis invokes the addition of a low-melting-point substance to the arc mantle, a layer of mantle material at about 60 to 90 miles (100 to 150 km.) below the volcanic arc. The relatively dry arc mantle would usually start to melt at about 2,100 to 2,300 degrees Fahrenheit (1,200 to 1,300 degrees Celsius). However, the addition of water and other gases can lower the melting point of the mixture. The water and its dissolved chemicals are supposedly derived from the subducted slab, the former ocean floor that is pushed back into the earth.

The sequence of events is as follows: New basaltic ocean floor forms at midocean ridge volcanoes. The new hot magma interacts with seawater, leading to vents at the seafloor with their mineralized

deposits. The seafloor becomes hydrated, and sulfur and chlorine from seawater are locked up in newly formed minerals. During subduction, this altered seafloor with slivers of sediment, including limestone, is gradually warmed up and starts to decompose, adding a flux to the surrounding mantle rocks. The mantle rocks then start to melt, and these magmas with minor inherited oceanic materials start to rise and pond at the bottom of the crust. There the magmas sit and wait for an opportunity to erupt, while cooling and crystallizing. Thus, arc magmas bear a chemical signature of subducted oceanic components while their chemical compositions range from basalt to rhyolite.

Volcanic Eruptions

Volcanic eruptions occur as a result of the rise of magma into the volcano (from depths as great as several miles) and then into the throat of the volcano. In basaltic volcanoes, the magmas have relatively little gas, and the magma simply overflows and forms large lava flows, sometimes associated

VOLCANIC ERUPTION AND CALDERA FORMATION

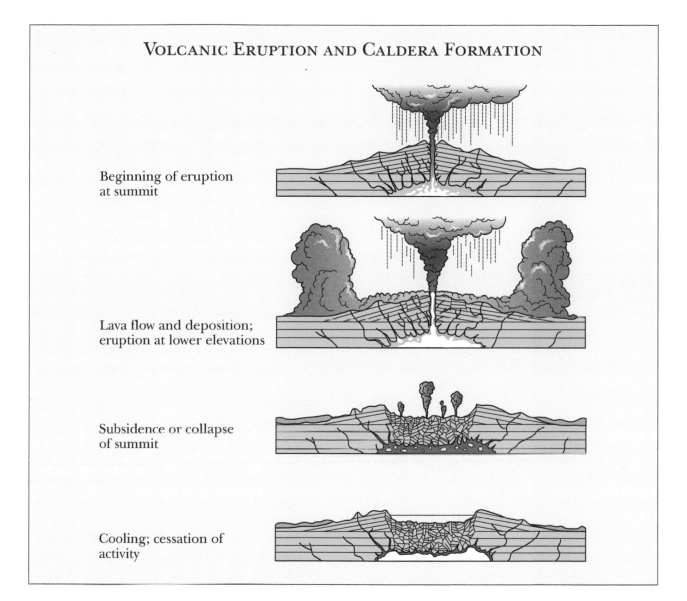

Beginning of eruption
at summit

Lava flow and deposition;
eruption at lower elevations

Subsidence or collapse
of summit

Cooling; cessation of
activity

with fire fountains. Stratovolcanoes can erupt regularly with small explosions or catastrophically after long periods of dormancy. Mount Stromboli, a volcano in Italy, erupts every twenty minutes, with an explosion that creates a column 650 to 980 feet (200 to 300 meters) high. Mount St. Helens in the U.S. state of Washington had a catastrophic eruption in 1980 after about 200 years of dormancy. It emitted an ash plume that reached more than 12 miles (20 km.) into the atmosphere.

After long magma storage periods in the crust, crystallization and melting of crustal material can lead to silica-rich magmas. These are viscous and can have high dissolved water contents—up to 4 to 6 percent by weight. When these magmas break out,

the eruption can be violent and form an eruption column 12 to 35 miles (20 to 55 km.) high. Many cubic miles of magma can be ejected. This leads to so-called plinian ash falls, with showers of pumice and ash over thousands of square miles, with the ash commonly carried around the globe by the high-level winds known as jet streams.

If the volume of ejected magma is large, the volcano empties itself and collapses into the hole, leading to a caldera—a volcanic collapse structure. The caldera at Crater Lake in Oregon is related to a large pumice eruption about 76,000 years ago. Basaltic volcanoes can also form collapse calderas when large volumes of lava have been extruded in a short time. Examples of famous basaltic calderas can be found in Hawaii's Mount Kilauea and the Galapagos Islands.

Volcanic Plumes

The dynamics of volcanic plumes has been studied from eruption photographs, experiments, and theoretical work. The rapidly expanding hot gases force the viscous magma out of the throat of the volcano, where it freezes into pumice. The kinetic energy of the ejected mass carries it 2 to 2.5 miles (3-4 km.) above the volcano. During this phase, air is entrained in the column, diluting the concentration of ash and pumice particles. The hot particles heat the entrained air, the mixture of hot air and solids becomes less dense than the surrounding atmosphere, and a buoyant column rises high into the sky.

The height of an eruption column is not directly proportional to the force of the eruption but is strongly dependent on the rate of heat release of the volcano. If little of the entrained air is heated up, the column will collapse back to the ground and an ash flow forms, which may deposit ash around the volcano. These types of eruptions are among the most devastating, creating glowing ash clouds traveling at speeds up to 60 miles (100 km.) per hour, burning everything in their path. The 1902 eruption of Mount Pelée on Martinique in the Caribbean was such an eruption and killed nearly 30,000 people in a few minutes.

Many volcanoes that are high in elevation are glaciated, and their eruptions lead to large-scale ice melting and possibly mixing of water, magma, and volcanic debris. Massive hot mudflows can race down from the volcano, following river valleys and filling up low areas. The 1980 Mount St. Helens eruption created many mudflows, some of which reached the Pacific Ocean, ninety miles to the west. A catastrophic mudflow event occurred in 1984 at Nevado del Ruiz, a volcano in Colombia, where 20,000 people were buried in mud and perished. When magma intrudes under the ice, meltwater can accumulate and then escape catastrophically, but such meltwater bursts are rare outside Iceland.

Minerals and Gases in Eruptions

The gas-rich character of arc magmas leads to fluid escape at various levels in the volcanoes, and these fluids tend to be rich in chlorine. They can transport metals such as copper, lead, zinc, and gold at high concentrations, and lead to the enrichment of these metals in the fractured volcanic rocks. Many of the world's largest copper ore deposits are associated with older arc volcanism, where erosion has removed most of the volcanic structure and laid the volcano innards bare. Many active volcanoes have modern hydrothermal (hot-water) systems, leading to acid hot springs and crater lakes and the potential to harness geothermal energy. Some areas in Japan, New Zealand, and Central America have an abundance of geothermal energy resources, which are gradually being developed.

Apart from the dangers of eruptions, continuous emissions of large amounts of sulfur dioxide, hydrochloric acid, and hydrofluoric acid present a danger of air pollution and acid rain. Incidences of emphysema and other irritations of the respiratory system are common in people living on the slopes of active volcanoes. The large lava emissions in Iceland in the eighteenth century led to acid fogs all over Europe. Many cattle died in Iceland during this period from the hydrofluoric acid vapors. High levels of fluorine in drinking water can lead to fluorosis, a disease that attacks the bone structure. The discharge of highly acidic fluids from hot springs and crater lakes can cause widespread environmental contamination, which can present a danger for crops gathered from fields irrigated with these waters and for local ecosystems in general.

Johan C. Varekamp

GEOLOGIC TIME SCALE

A major difference between the geosciences (earth sciences) and other sciences is the great enormity of their time scale. One might compare the magnitude of geologic time for geoscientists to the vastness of space for astronomers. Every geological process, such as the movement of crustal plates (plate tectonics), the formation of mountains, and the advance and retreat of glaciers, must be considered within the context of time.

Although certain geologic events, such as floods and earthquakes, seem to occur over short periods of time, the vast majority of observed geological features formed over a great span of time. Consequently, modern geoscientists consider Earth to be exceedingly old. Using radiometric age-dating techniques, they calculate the age of Earth as 4.6 billion years old.

Early miners were probably the first to recognize the need for a scale by which rock and mineral units could be compared over large geographic areas. However, before a time scale—and even geology as a science—could develop, certain principles had to be established. This did not occur until the late eighteenth century when James Hutton, a Scottish naturalist, began his extensive examinations of rock relationships and natural processes at work on the earth. His work was amplified by Charles Lyell in his textbook *Principles of Geology* (1830-1833). After careful observation, Hutton concluded that the natural processes and functions he observed had operated in the same basic manner in the past, and that, in general, natural laws were invari-

able. That idea became known as the principle of uniformitarianism.

The Birth of Stratigraphy

In 1669 Nicholas Steno, a Danish physician working in Italy, recognized that horizontal rock layers contained a chronological record of Earth history and formulated three important principles for interpreting that history. The principle of superposition states that in a succession of undeformed strata, the oldest stratum lies at the bottom, with successively younger ones above. The principle of original horizontality states that because sedimentary particles settle from fluids under gravitational influence, sedimentary rock layers must be horizon-

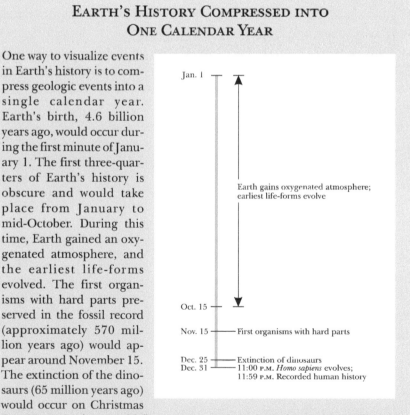

EARTH'S HISTORY COMPRESSED INTO ONE CALENDAR YEAR

One way to visualize events in Earth's history is to compress geologic events into a single calendar year. Earth's birth, 4.6 billion years ago, would occur during the first minute of January 1. The first three-quarters of Earth's history is obscure and would take place from January to mid-October. During this time, Earth gained an oxygenated atmosphere, and the earliest life-forms evolved. The first organisms with hard parts preserved in the fossil record (approximately 570 million years ago) would appear around November 15. The extinction of the dinosaurs (65 million years ago) would occur on Christmas Day. Homo sapiens would first appear at approximately 11 P.M. on December 31, and all of recorded human history would occur in the last few seconds of New Year's Eve.

tal; if not, they have suffered from subsequent disturbance. The principle of original lateral continuity states that strata originally extended in all directions until they thinned to zero or terminated against the edges of the original area of deposition.

In the late eighteenth century, the English surveyor William Smith recognized the wide geographic uniformity of rock layers and discovered the utility of fossils in correlating these layers. By 1815, Smith had completed a geologic map of England and was able to correlate English rock layers with layers exposed across the English Channel in France.

From the need to classify and organize rock layers into an orderly form arose a subdiscipline of modern geology—stratigraphy, the study of rock layers and their age relationships. In 1835 two British geologists, Adam Sedgwick and Roderick Murchison, began organizing rock units into a formal stratigraphic classification. Large divisions, called eras, were based upon well known and characteristic fossils, and included a number of smaller subdivisions, called periods.

The periods are often subdivided into smaller units called epochs. Each period is defined by a representative sequence of rock strata and fossils. For instance, the Devonian period is named for exposures of rock in Devonshire in southern England, while the Jurassic period is defined by strata exposed in the Jura Mountains in northern Switzerland.

Approximately 80 percent of Earth's history is included in the Crypotozoic era (meaning obscure life). Fossils from the Crypotozoic era are rare, and the rock record is very incomplete. After the Crypotozoic era came the Paleozoic (ancient life), Mesozoic (middle life), and Cenozoic (recent life) eras. Most of the life forms that evolved during the Paleozoic and Mesozoic eras are now extinct, whereas 90 percent of the life-

forms that evolved up to the middle Cenozoic era still exist.

The Geologic Time Scale

The geologic time scale is continually in revision as new rock formations are discovered and dated. The ages shown in the table below are in millions of years ago (MYA) before the present and represent the beginning of that particular period. It would be impossible to list all the significant events in Earth's history, but one or two are provided for each period. Note that in the United States, the Carboniferous period has been subdivided into the Mississippian period (older) and the Pennsylvanian period (younger).

The Fossil Record

The word "fossil" comes from the Latin *fossilium*, meaning "dug from beneath the surface of the ground." Fossils are defined as any physical evidence of past life. Fossils can include not only shells, bones, and teeth, but also tracks, trails, and bur-

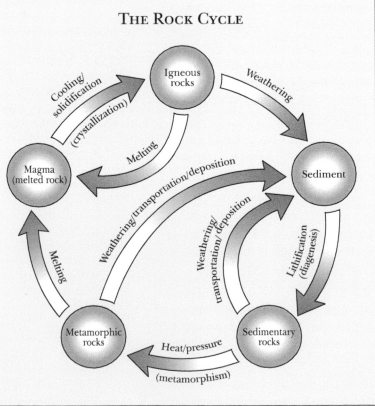

rows. The latter group are referred to as trace fossils. Fossils demonstrate two important truths about life on Earth: First, thousands of species of plants and animals have existed and later became extinct. Second, plants and animals have evolved through time, and the communities of life that have existed on Earth have changed.

Some organisms are slow to evolve and may exist in several geologic time periods, while others evolve quickly and are restricted to small intervals of time within a particular period. The latter, referred to as index fossils, are the most useful to geoscientists for correlating rock layers over wide geographic areas and for recognizing geologic time.

The fossil record is incomplete, because the process of preservation favors organisms with hard parts that are rapidly buried by sediments soon after death. For this reason, the vast majority of fossils are represented by marine invertebrates with exoskeletons, such as clams and snails. Under special circumstances, soft-bodied organism can be preserved, for instance the preservation of insects in amber, made famous by the feature film *Jurassic Park* (1993).

The Rock Cycle

A rock is a naturally formed aggregate of one or more minerals. Three types of rocks exist in the earth's crust, each reflecting a different origin. Igneous rocks have cooled and solidified from molten material either at or beneath Earth's surface. Sedimentary rocks form when preexisting rocks are weathered and broken down into fragments that accumulate and become compacted or cemented together. Fossils are most commonly found in sedimentary rocks. Metamorphic rocks form when heat, pressure, or chemical reactions in Earth's interior change the mineral or chemical composition and structure of any type of preexisting rock.

Over the huge span of geologic time, rocks of any one of these basic types can change into either of the other types or into a different form of the same type. For this reason, older rocks become increasingly more rare. The processes by which the various rock types change over time are illustrated in the rock cycle.

Larry E. Davis

Earth's Surface

Internal Geological Processes

The earth is layered into a core, a mantle, and a crust. The topmost mantle and the crust make up the lithosphere. Beneath this is a layer called the asthenosphere, which is composed of moldable and partly liquid materials. Heat transference within the asthenosphere sets up convection cells that diverge from hot regions and converge to cold regions. Consequently, the overlying lithosphere is segmented into ridged plates that are moved by the convection process. The hot asthenosphere does not rise along a line. This causes the development of a structure called a transform plate boundary, which is perpendicular to and offsetting the divergent boundary.

The topographic features at Earth's surface, such as mountains, rift valleys, oceans, islands, and ocean trenches, are produced by extension or compression forces that act along divergent, convergent, or transform plate boundaries. The extension and compression forces at Earth's surface are powered by convection within the asthenosphere.

Mountains and Depressions in Zones of Compression

Compression along convergent plate boundaries yields three types of mountain: island arcs that are partly under water; mountains along a continental edge, such as the Andes; and mountains at continental interiors, such as the Alps. At convergent plate boundaries, the denser of the two colliding plates slides down into the asthenosphere and causes volcanic activity to form on the leading edge of the upper plate. Island arcs such as the Aleutians and the Caribbean are formed when an oceanic plate descends beneath another oceanic plate.

Volcanic mountain chains such as the Andes of South America are formed when an oceanic plate descends beneath a continental plate. In both the island arc type and Andean type collisions, a deep depression in the oceans, called a trench, marks the place where neighboring plates are colliding and where the denser plates are pulled downward into the asthensophere. If the colliding plates are of similar density, neither plate will go into the asthenosphere. Instead, the edges of the neighboring plates will be folded and faulted and excess material will be pushed upward to form a block mountain, such as the mountain chain that stretches from the Alps through to the Himalayas. This type of mountain chain is not associated with a trench.

The Appalachians of the eastern United States are an example of the alpine type of mountain belt. When the Appalachians were forming 300 million years ago, rock layers were deformed. The deformation included folding to form ridges and valleys; fracturing along joint sets, with one joint set being parallel to ridges, while the other set is perpendicular; and thrust faulting, in which rock blocks were detached and shoved upward and northwestward.

Millions of years of erosion have reduced the height of the mountains and have produced topographic inversion in the foothills. Topographic inversion occurs because joints create wider fractures at upfolded ridges and narrower fractures at downfolded valleys. Erosion is then accelerated at upfolded ridges, converting ancient ridges into valleys, while ancient valleys stand as ridges. The Valley and Ridge Province of the Appalachians is noted for such topographic inversion.

West of the Valley and Ridge Province of the Appalachians is the Allegheny Plateau, which is bounded by a cliff on its eastern side. In general, plateaus are flat topped because the rock layer that covers the surface is resistant to weathering. The cliff side is formed by erosion along joint or fault surfaces.

The Sierra Nevada range, which formed 70 million years ago, is an example of an Andean type of mountain belt. Millions of years of erosion there has exposed igneous rocks that formed at depth. Over the years, the force of compression that formed the Sierras has evolved to form a zone of extension between the Sierras and the Colorado Plateau.

Mountains and Depressions in Zones of Extension

Extension is a strain that involves an increase in length and causes crustal thinning and faulting. Extension is associated with convergent boundaries, divergent boundaries, and transform boundaries.

Extension Associated with a Convergent Boundary

During the formation of the Sierra Nevada, an oceanic plate that was subducted beneath California declined at a shallow angle eastward toward the Colorado Plateau. Later, the subducted plate peeled off and molten asthenosphere took its place. From the asthenosphere, lava ascended through fractures to form volcanic mountains in Arizona and Utah, and lava flowed and volcanic ash fell as far west as California. The lithosphere has been heated up and has become buoyant, so the Colorado Plateau rises to higher elevations, and rock layers slide westward from it in a zone of extension that characterizes the Basin and Range Province.

In the extension zone, the top rock layers move westward on curved displacement planes that are steep at the surface and nearly horizontal at depth. When rock layers move westward over a curved detachment surface, the trailing edge of the rock layers roll over and are tilted toward the east so they do not leave space in buried rocks. On the other hand, a west-facing slope is left behind on a mountain from which the rock layers were detached. There-

fore, movement along one curved detachment surface creates a valley, and movement along several such detachment surfaces forms a series of valleys separated by ridges, as in the Basin and Range Province. The amount of the displacement along the curved surfaces is not uniform. For example, more displacement has created wide zones of valleys such as the Las Vegas valley in Nevada, and Death Valley in California.

Extension Associated with a Divergent Boundary

The longest mountain chain on Earth lies under the Pacific Ocean. It is about 37,500 miles (60,000 km.) long, 31.3 miles (50 km.) wide, and 2 miles (3 km.) high. The central part of this midoceanic ridge is marked by a depression, about 3,000 feet (1,000 meters) deep, and is called a rift valley. A part of the submarine ridge, called the East Pacific Rise, forms the seafloor sector in the Gulf of California and reappears off the coast of northern California, Oregon, and Washington as the Juan de Fuca Ridge. Another part forms the seafloor sector in the Gulf of Aden and Red Sea seafloor, part of which is exposed in the Afar of Ethiopia. From the Afar southward to the southern part of Mozambique is the longest exposed rift valley on land, the East African Great Rift Valley.

A rift valley is the place where old rocks are pushed aside and new rocks are created. Blocks of rock that are detached from the rift walls slide down by a series of normal fault displacements. The ridge adjacent to the central rift is present because hot rocks are less dense and buoyant. If the process of divergences continues from the rifting stage to a drifting stage, as the rocks move farther away from the central rift, the rocks become older, colder, and denser, and push on the underlying asthenosphere to create basins. These basins will be flooded by oceanic water as neighboring continents drift away. However, not all processes of divergence advance from the rifting to the drifting stage.

Extension Associated with Transform Boundary

The best-known example of a transform boundary is the San Andreas Fault that offsets the East Pacific

Rise from the Juan de Fuca Ridge, and is exposed on land from the Gulf of California to San Francisco. Along transform boundaries, there are pull-apart basins that may be filled to form lakes, such as the Salton Sea in Southern California. Another example is the Aqaba transform of the Middle East, along which the Sea of Galilee and the Dead Sea are located.

H. G. Churnet

EXTERNAL PROCESSES

Continuous processes are at work shaping the earth's surface. These include breaking down rocks, moving the pieces, and depositing the pieces in new locations. Weathering breaks down rocks through atmospheric agents. The process of moving weathered pieces of rock by wind, water, ice, or gravity is called erosion. The materials that are deposited by erosion are called sediment.

Mechanical weathering occurs when a rock is broken into smaller pieces but its chemical makeup is not changed. If the rock is broken down by a change in its chemical composition, the process is called chemical weathering.

Mechanical Weathering

Different types of mechanical weathering occur, depending on climatic conditions. In areas with moist climates and fluctuating temperatures, rocks can be broken apart by frost wedging. Water fills in cracks in rocks, then freezes during cold nights. As the ice expands and pushes out on the crack walls, the crack enlarges. During the warm days, the water thaws and flows deeper into the enlarged crack. Over time, the crack grows until the rock is broken apart. This process is active in mountains, producing a pile of rock pieces at the mountain base called talus.

Salt weathering occurs in areas where much salt is available or there is a high evaporation rate, such as along the seashore. Salt crystals form when salty moisture enters rock cracks. Growing crystals settle in the bottom of the crack and apply pressure on the crack walls, enlarging the crack.

Thermal expansion and contraction occur in climates with fluctuating temperatures, such as deserts. All minerals expand during hot days and contract during cold nights, and some minerals expand and contract more than others. This process continues until the rock loosens up and breaks into pieces.

Mechanical exfoliation can happen to a rock body overlain by a thick rock or sediment layer. If the heavy overlying layer over a portion of the rock body is removed, pressure is relieved and the exposed rock surface will expand in response. This expanding surface will break off into sheets parallel to the surface, but the remaining rock body remains under pressure and unchanged.

When plant roots grow into cracks in rocks, they enlarge the cracks and break up the rocks. Finally, abrasion can occur to rock fragments during transport. Either the fragments collide, breaking apart, or fragments are scraped against rocks, breaking off pieces.

Chemical Weathering

Water and oxygen create two common causes of chemical weathering. For example, dissolution occurs when water or another solution dissolves minerals within a rock and carries them away. Hydrolysis can occur when water flows through earth materials. The hydrogen ions or the hydroxide ions of the water may react with minerals in the

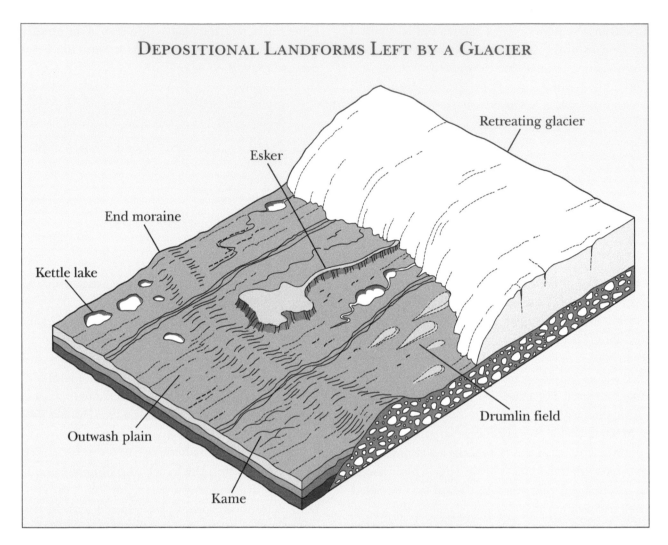

DEPOSITIONAL LANDFORMS LEFT BY A GLACIER

Retreating glacier

Esker

End moraine

Kettle lake

Outwash plain

Kame

Drumlin field

rocks. When this occurs, the chemical composition of the mineral is changed, and a new mineral is formed. Hydrolysis often produces clay minerals.

Some elements in minerals combine with oxygen from the atmosphere, creating a new mineral. This process is called oxidation. Some of these oxidation minerals are commonly referred to as rust.

Mass Movement

Weathered rock pieces (sediments) are transported (eroded) by one or more of four transport processes: water (streams and oceans), wind, ice (glaciers), or gravity. Mass movement transports earth materials down slopes by the pull of gravity. Gravity, constantly working to pull surface materials down, parallel to the slope, is the most important factor affecting mass movement. There is also a force in-

volved perpendicular to the slope that contributes to the effects of friction.

Friction, the second factor, is determined by the earth material type involved. For example, weathering may create cracks in rocks, which form planes of weakness on which the mass movement can occur. Loose sediments always tend to roll downhill.

The third factor is the slope angle. Each earth material has its own angle of repose, which is the steepest slope angle on which the materials remain stable. Beyond this slope angle, earth materials will move downslope.

Water, the fourth factor, affects the stability of the earth material in the slope. Friction is weakened by water between the mineral grains in the rock. For example, water can make clay quite slippery, causing the mass movement.

How Hydrology Shapes Geography

Water and ice sculpt the landscape over time. Fast-flowing rivers erode the soil and rock through which they flow. When rivers slow down in flatter areas, they deposit eroded sediments, creating areas of rich soils and deltas at the mouths of the rivers. Over time this process wears down mountain ranges. The Appalachian Mountain range on the eastern side of the North American continent is hundreds of millions of years older than the Rocky Mountain range on the continent's western side. Although the Appalachians once rivaled the Rockies in size, they have been made smaller by time and erosion.

Canyons are carved by rivers, as the Grand Canyon was carved by the Colorado River, which exposed rocks billions of years old. Ice also changes the landscape. Large ice sheets from past ice ages could have been well over 1 mile (1,600 meters) thick, and they scoured enormous amounts of soil and rock as they slowly moved over the land surface. Terminal moraines are the enormous mounds of soil pushed directly in front of the ice sheets. Long Island, New York, and Cape Cod, Massachusetts, are two examples of enormous terminal moraines that were left behind when the ice sheets retreated.

The rooting system of vegetation, the fifth factor, helps make the surficial materials of the slope stable by binding the loose materials together.

Mass movements can be classified by their speed of movement. Creep and solifluction are the two types of slow mass movement, which are measured in fractions of inches per year. Creep is the slowest mass movement process, where unconsolidated materials at the surface of a slope move slowly downslope. The materials move slightly faster at the surface than below, so evidence of creep appears in the form of slanted telephone poles. During solifluction, the warm sun of the brief summer season in cold regions thaws the upper few feet of the earth. This waterlogged soil flows downslope over the underlying permafrost.

Rapid mass movement processes occur at feet per second or miles per hour. Falls occur when loose rock or sediment is dislodged and drops from a steep slope, such as along sea cliffs where waves erode the cliff base. Topples occur when there is an overturning movement of the mass. A topple can turn into a fall or a slide. A slide is a mass of rock or sediment that becomes dislodged and moves along a plane of weakness, such as a fracture. A slump is a slide that separates along a concave surface. Lateral spreads occur when a fractured earth mass spreads out at the sides.

A flow occurs when a mass of wet or dry rock fragments or sediment moves downslope as a highly viscous fluid. There are several different flow types. A debris flow is a mass of relatively dry, broken pieces of earth material that suddenly has water added. The debris flow occurs on steeper slopes and moves at speeds of 1-25 miles (2-40 km.) per hour. A debris avalanche occurs when an entire area of soil and underlying weathered bedrock becomes detached from the underlying bedrock and moves quickly down the slope. This flow type is often triggered by heavy rains in areas where vegetation has been removed. An earthflow is a dry mass of clayey or silty material that moves relatively slowly down the slope. A mudflow is a mass of earth material mixed with water that moves quickly down the slope.

A quick clay can occur when partially saturated, solid, clayey sediments are subjected to an earthquake, explosion, or loud noise and become liquid instantly.

Sherry L. Eaton

FLUVIAL AND KARST PROCESSES

Earth's landscape has been sculptured into an almost infinite variety of forms. The earth's surface has been modified by various processes for thousands, even hundreds of millions, of years to arrive at the modern configuration of landscapes.

Each process that transforms the surface is classified as either endogenic or exogenic. Endogenic processes are driven by the earth's internal heat and energy and are responsible for major crustal deformation. Endogenic processes are considered constructional, because they build up the earth's surface and create new landforms, such as mountain systems. Conversely, exogenic processes are considered destructional because they result in the wearing away of landforms created by endogenic processes. Exogenic processes are driven by solar energy putting into motion the earth's atmosphere and water, resulting in the lowering of features originally created by endogenic processes.

The most effective exogenic processes for wearing away the landscape are those that involve the action of flowing water, commonly referred to as fluvial processes. Water flows over the surface as runoff, after it evaporates into the atmosphere and infiltrates into the soil. The water that is left over flows down under the influence of gravity and has tremendous energy for sculpting the earth's surface. Although flowing water is the most effective agent for modifying the landscape, it represents less than 0.01 percent of all the water on Earth's surface. By comparison, nearly 75 percent of the earth's surface water is stored within glaciers.

Drainage Basins

Fluvial processes can be considered from a variety of spatial scales. The largest scale is the drainage basin. A drainage basin is the area defined by topographic divides that diverts all water and material within the basin to a single outlet. Every stream of any size has its own drainage basin, and every portion of the earth's land surfaces are located within a drainage basin. Drainage basins vary tremendously in size, de-pending on the size of the river considered. For example, the largest drainage basin on earth is the Amazon, which drains about 2.25 million square miles (5.83 million sq. km.) of South America.

The Amazon Basin is so large that it could contain nearly the entire continent of Australia. By comparison, the Mississippi River drainage basin, the largest in North America, drains an area of about 1,235,000 square miles (3,200,000 sq. km.). Smaller rivers have much smaller basins, with many draining only an area roughly the size of a football field. While basins vary tremendously in size, they are spatially organized, with larger basins receiving the drainage from smaller basins, and eventually draining into the ocean. Because drainage basins receive water and material from the landscape within the basin, they are sensitive to environmental change that occurs within the basin. For example, during the twentieth century, the Mississippi River was influenced by many human-imposed changes that occurred either within the basin or directly within the channel, such as agriculture, dams and reservoirs, and levees.

Drainage Networks and Surface Erosion

Drainage basins can be subdivided into drainage networks by the arrangement of their valleys and interfluves. Interfluves are the ridges of higher elevation that separate adjacent valleys. Where an interfluve represents a natural boundary between two or more basins, it is referred to as a drainage divide. Valleys contain the larger rivers and are easily distinguished from interfluves by their relatively low, flat surfaces. Interfluves have relatively steep slopes and, for this reason, are eroded by runoff. The term erosion refers to the transport of material, in this case sediment that is dislodged from the surface.

Runoff starts as a broad sheet of slow-moving water that is not very erosive. As it continues to flow downslope, it speeds up and concentrates into rills, which are narrow, fast-moving lines of water. Because the runoff is concentrated within rills, the water travels faster and has more energy for erosion. Thus, rills are responsible for transporting sediment from higher points of elevation within the basin to the valleys, which are at a lower elevation. Rills can become powerful enough to scour deeply into the surface, developing into permanent channels called gullies.

The presence of many gullies indicates significant erosion on the landscape and represents an expensive and long-lasting problem if it is not remedied after initial development. The formations of gullies is often associated with human manipulation of the earth. For example, gullies can develop after improper land management, particularly intensive agricultural and grazing practices. A change in land use from natural vegetation, such as forests or prairie, can result in a type of land cover that is not suited for preventing erosion. Such land surfaces become susceptible to the formation of gullies during heavy, prolonged rains.

At a smaller scale, fluvial processes can be considered from the perspective of the river channel. River channels are located within the valleys of basins, offering a permanent conduit for drainage. Higher in the basin, river channels and valleys are relatively narrow, but grow larger toward the mouth of the basin as they receive drainage from smaller rivers within the basin. River channels may be categorized by their planform pattern, which refers to their overhead appearance, such as would be viewed from the window of an airplane.

The two major types of rivers are meandering and braided. Meandering rivers have a single channel that is sinuous and winding. These rivers are characterized as having orderly and symmetrical bends, causing the river to alternate directions as it flows across its valley. In contrast, braided rivers contain numerous channels divided by small islands, which results in a disorganized pattern. The islands within a braided river channel are not permanent. Instead, they erode and form over the course of a few years, or even during large flood events. Meandering channels usually have narrow and deep channels, but braided river channels are shallow and wide.

Sediment and Floodplains

Another distinction between braided and meandering river channels is the types of sediment they transport. Braided rivers transport a great amount of sediment that is deposited into midchannel islands within the river. Also, because braided rivers are frequently located higher in the drainage basin, they may have larger sediments from the erosion of adjacent slopes. In contrast, meandering river channels are located closer to the mouth of the basin and transport fine-grained sediment that is easily stored within point bars, which results in symmetrical bends within the river.

The sediments of both meandering and braided rivers are deposited within the valleys onto floodplains. Floodplains are wide, flat surfaces formed from the accumulation of alluvium, which is a term for sediment that is deposited by water. Floodplain sediments are deposited with seasonal flooding. When a river floods, it transports a large amount of sediment from the channel to the adjacent floodplain. After the water escapes the channel, it loses energy and can no longer transport the sediment. As a result, the sediment falls out of suspension and is deposited onto the floodplain. Because flooding occurs seasonally, floodplain deposits are layered and may accumulate into very thick alluvial deposits over thousands of years.

Karst Processes and Landforms

A specialized type of exogenic process that is also related to the presence of water is karst. Karst processes and topography are characterized by the solution of limestone by acidic groundwater into a number of distinctive landforms. While fluvial processes lower the landscape from the surface, karst processes lower the landscape from beneath the surface. Because limestone is a very permeable sedimentary rock, it allows for a large amount of groundwater flow. The primary areas for solution of the limestone occur along bedding planes and joints. This creates a positive feedback by increasing the amount of water flowing through the rock, thereby further increasing solution of the limestone. The result is a complex maze of underground conduits and caverns, and a surface with few rivers because of the high degree of infiltration.

The surface topography of karst regions often is characterized as undulating. A closer inspection reveals numerous depressions that lack surface outlets. Where this is best developed, it is referred to as cockpit karst. It occurs in areas underlain by extensive limestone and receiving high amounts of precipitation, for example, southern Illinois and Indiana in the midwestern United States, and in Puerto Rico and Jamaica.

Sinkholes are also common to karstic regions. Sinkholes are circular depressions having steep-sided vertical walls. Sinkholes can form either from the sudden collapse of the ceiling of an underground cavern or as a result of the gradual solution and lowering of the surface. Sinkholes can fill with sediments washed in from surface runoff. This reduces infiltration and results in the development of small circular lakes, particularly common in central Florida. Over time, erosion causes the vertical walls to retreat, resulting in uvalas, which are much larger flat-floored depressions.

Where there are numerous adjacent sinkholes, the retreat and expansion of the depressions causes them to coalesce, resulting in the formation of poljes. Unlike uvalas, poljes have an irregular shape, and the floor of the basin is not flat because of differences between the coalescing sinkholes.

Caves are among the most characteristic features of karst regions, but can only be seen beneath the surface. Caves can traverse the subsurface for miles, developing into a complex network of interconnected passages. Some caves develop spectacular formations as a result of the high amount of dissolved limestone transported by the groundwater. The evaporation of water results in the accumulation of carbonate deposits, which may grow for thousands of years. Some of the most common deposits are stalactites, which grow downward from the ceiling of the cave, and stalagmites, which grow upward and occasionally connect with stalactites to form large vertical columns.

Paul F. Hudson

GLACIATION

In areas where more snow accumulates each winter than can thaw in summer, glaciers form. Glacier ice, called firn, looks like rock but is not as strong as most rocks and is subject to intermittent thawing and freezing. Glacier ice can be brittle and fracture readily into crevasses, while other ice behaves as a plastic substance. A glacier is thickest in the area receiving the most snow, called the zone of accumulation. As the thickness piles up, it settles down and squeezes the limit of the ice outward in all directions. Eventually, the ice reaches a climate where the ice begins to melt and evaporate. This is called the zone of ablation.

Alpine Glaciation

Varied topographic evidence throughout the alpine environment attests to the sculpturing ability of glacial ice. The world's most spectacular mountain scenery has been produced by alpine glaciation, including the Matterhorn, Yosemite Valley, Glacier National Park, Mount Blanc, the Tetons, and Rocky

A FUTURE ICE AGE

If past history is an indicator, some time in the future conditions again will become favorable for the growth of glaciers. As recently as 1300 to 1600 CE, a cold period known as the Little Ice Age settled over Northern Europe and Eastern North America. Viking colonies perished as agriculture became unfeasible, and previously ice-free rivers in Europe froze over.

Another ice age would probably develop rapidly and be impossible to stop. Active mountain glaciers would bury living forests. Great ice caps would again cover Europe and North America, and move at a rate of 100 feet (30 meters) per day. Major cities and populations would shift to the subtropics and the topics.

Mountain National Park, all of which are visited by large numbers of people annually. Although alpine glaciation is still an active process of land sculpture in the high mountain ranges of the world, it is much less active than it was in the Ice Age of the Pleistocene epoch.

The prerequisites for alpine, or mountain, glaciation to become active are a mountainous terrain with Arctic climatic conditions in the higher elevations, and sufficient moisture to help snow and ice develop into glacial ice. As glaciers move out from their points of origin, they erode into the sides of mountains and increase the local relief in the higher elevations. The erosional features produced by alpine glaciation dominate mountain topography and usually are the most visible features on topographic maps. The eroded material is transported downvalley and deposited in a variety of landforms.

One kind of an erosional feature is a cirque, a hollow bowl-shaped depression. The bowl of the cirque commonly contains a small round lake or tarn. A steep-walled mountain ridge called an arête forms between two cirques. A high pyramidal peak, called horn, is formed by the intersecting walls of three or more cirques.

Erosion is particularly rapid at the head of a glacier. In valleys, moving glaciers press rock fragments against the sides, widening and deepening them by abrasion and forming broad U-shaped valleys. When glaciers recede, tributary streams become higher than the floor of the U-shaped valley and waterfalls occur over these hanging valleys. As the ice continues to melt, residual sediments called moraines may be deposited. Moraines are made up of glacier till, a collection of sediment of all sizes. Bands of sediment along the side of a valley glacier are lateral moraines; those crossing the valley are end or recessional moraines; where two glaciers join, a medial moraine is formed. Meltwater may also sort out the finer materials, transport them downvalley, and deposit them in beds as outwash.

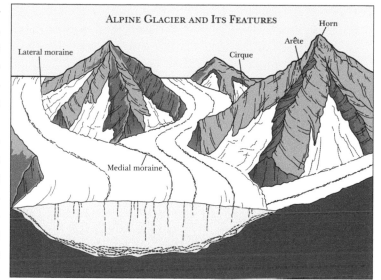

ALPINE GLACIER AND ITS FEATURES

Horn
Arête
Cirque
Lateral moraine
Medial moraine

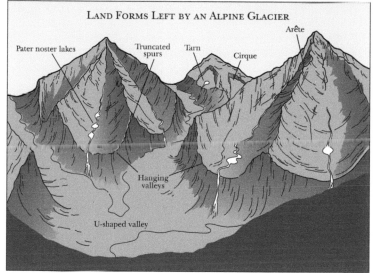

LAND FORMS LEFT BY AN ALPINE GLACIER

Arête
Pater noster lakes
Truncated spurs
Tarn
Cirque
Hanging valleys
U-shaped valley

Continental Glaciation

In the modern world, continental glaciation operates on a large scale only in Greenland and Antarctica. However, its existence in previous geologic ages is evidenced by strata of tillite (a compacted rock formed of glacial deposits) or, more frequently, by surficial deposits of glacial materials.

Much of the geomorphology of the northeastern quadrant of North America and the northwestern portion of Europe was formed during the Ice Age. During that time, great masses of ice accumulated on the continents and moved out from centers near the Hudson Bay and the Fenno-Scandian Shield, extending over the continents in great advancing and retreating lobes. In North America, the four

345

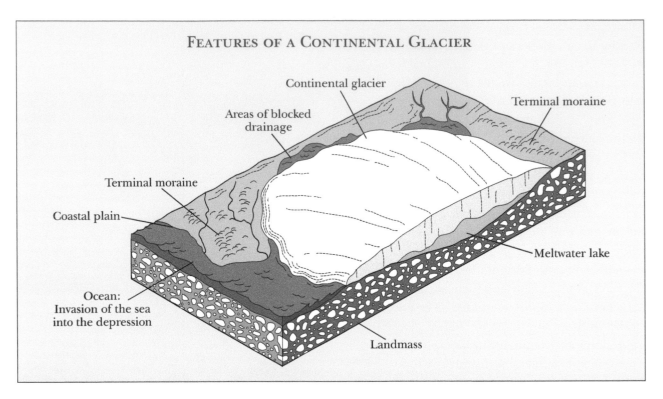

FEATURES OF A CONTINENTAL GLACIER

Continental glacier

Areas of blocked drainage

Terminal moraine

Terminal moraine

Coastal plain

Meltwater lake

Ocean: Invasion of the sea into the depression

Landmass

major stages of lobe advance were the Wisconsin (the most recent), the Illinoian, the Kansan, and the Nebraskan (the oldest). Between each of these major advances were pluvial periods in which the ice melted and great quantities of water rushed over or stood on the continents, creating distinctive features which can still be detected today.

The two major functions of gradation are accomplished by the processes of scour (degradation) in the areas close to the centers and deposition (aggradation) adjacent to the terminal or peripheral areas of the lobes. Thus, the overall effect of continental glaciation is to reduce relief—to scour high areas and fill in lower regions—unlike the changes caused by alpine glaciation.

Although continental glaciation usually does not result in the spectacular scenery of alpine glaciation, it was responsible for creating most of the Great Lakes and the lakes of Wisconsin, Michigan, Minnesota, Finland, and Canada; for gravel deposits; and for the rich agricultural lands of the Midwest, to mention just a few of its effects.

While glaciers were leveling hilly sections of North America and Europe by scraping them bare

of soil and cutting into the ice itself, they acquired a tremendous load of material. As a glacier warms and melts, there is a tremendous outflow of water, and the streams thus formed carry with them the debris of the glacier. The material deposited by glaciers is called drift or outwash. Glaciofluvial drift can be recognized by its separation into layers of finer sands and coarser gravels.

Kettles and kames are the most common features of the end moraines found at the outermost edges of a glacier. A kettle is a depression left when a block of ice, partially or completely buried in deposits of drift, melts away. Most of the lakes in the upper Great Lakes of the United States are kettle lakes. A kame is a round, cone-shaped hill. Kames are produced by deposition from glacial meltwater. Sometimes, the outwash material poured into a long and deep crevasse, rather than a hole. These tunnels have had their courses choked by debris, revealed today by long, narrow ridges, generally referred to as eskers.

Ron Janke

DESERT LANDFORMS

Deserts are often striking in color, form, or both. The underlying lack of water in deserts produces unique features not found in humid regions. Arid lands cover approximately 30 percent of the earth's land surface, an area of about 15.4 million square miles (40 million sq. km.). Arid lands include deserts and surrounding steppes, semiarid regions that act as transition zones between arid and humid lands.

Many of the world's largest and driest deserts are found between 20 and 40 degrees north and south latitude. These include the Mojave and Sonoran Deserts of the United States, the Sahara in northern Africa, and the Great Sandy Desert in Australia. In these deserts, the subtropical high prevents cloud formation and precipitation while increasing rates of surface evaporation.

Some arid lands, like China's Gobi Desert, form because they are far from oceans that are the dominant source for atmospheric water vapor and precipitation. Others, like California's Death Valley, are arid because mountain ranges block moisture from coming from the sea. The combination of mountain barriers and very low elevations makes Death Valley the hottest, driest desert in North America.

Sand Dunes

Many people envision deserts as vast expanses of blowing sand. Although wind plays a more important role in deserts than it does elsewhere, only about 25 percent of arid lands are covered by sand. Broad regions that are covered entirely in sand (such as portions of northwestern Africa, Arabia, and Australia) are referred to as sand seas. Why is wind more effective here than elsewhere?

The lack of soil water and vegetation, both of which act to bind grains together, allows enhanced eolian (wind) erosion. Very small particles are picked up and suspended within the moving air mass, while sand grains bounce along the surface. Removal of material often leaves behind depressions called blowouts or deflation hollows. Moving grains abrade cobbles and boulders at the surface, creating uniquely sculpted and smoothed rocks known as ventifacts. Bedrock outcrops can be streamlined as they are blasted by wind-borne grains to form features called yardangs. As these rocks are ground away, they contribute additional sediment to the wind.

Desert sand dunes are not stationary features—instead, they represent accumulations of moving sand. Wind blows sand along the desert floor. Where it collects, it forms dunes. Typically, dunes have relatively shallow windward faces and steeper slip faces. Sand grains bounce up the windward face and then eventually cascade down the slip face, the movement of individual grains driving movement of the entire dune in a downwind direction.

Four major dune types are found within arid regions. Barchan dunes are crescent-shaped features, with arms that point downwind. They may occur as isolated structures or within fields. They form where winds blow in a single direction and where the supply of sand is limited. With a larger supply of sand, barchan dunes can join with one another to form a transverse dune field.

There, ridges are perpendicular to the predominant wind direction. With quartering winds (that is, winds that vary in direction throughout a range of about 45 degrees) dune ridges form that are parallel to the average wind direction. These so-called longitudinal dunes have no clearly defined windward and slip faces. Where winds blow sand from all directions, star dunes form. Sand collects in the middle of the feature to form a peaked center with arms that spiral outward.

Badlands, Mesas, and Buttes

As scarce as it may be, water is still the dominant force in shaping desert landscapes. Annual precipitation may be low, but the amount of precipitation in a single storm may be a large fraction of the yearly total. An arid landscape that is underlain by poorly cemented rock or sediment, such as that found in western South Dakota, may form badlands as a result of the erosive ability of storm-water run-off. Overall aridity prevents vegetation from establishing the interconnected root system that holds soil particles together in more humid regions.

Cloudbursts cause rapid erosion that forms numerous gullies, deeply incised washes, and hoodoos, which are created when rock or sediment that is more resistant protects underlying material from erosion. Over time, protected sections stand as prominent spires while surrounding material is removed. Landscapes like those found in Badlands National Park in South Dakota are devoid of vegetation and erode rapidly during storms.

Arid regions that are underlain by flat-lying rock units can form mesas and buttes. Water follows fractures and other lines of weakness, forming ever-widening canyons. Over time, these grow into broad valleys. In northern Arizona's Monument Valley, remnants of original bedrock stand as isolated, flat-topped structures. Broad mesas are marked by their flat tops (made of a resistant rock like sandstone or basalt) and steep sides. Buttes are

DEATH VALLEY PLAYA

California's Death Valley is the driest desert in the United States, with an average rainfall of only 1.5 inches (38 millimeters) per year at the town of Furnace Creek. It is also consistently one of the hottest places on Earth, with a record high of 134 degrees Fahrenheit (57 degrees Celsius). In the distant past, however, Death Valley held lakes that formed in response to global cooling. Over 120,000 years ago, Death Valley hosted a 295-foot-deep (90 meters) body of water called Lake Manley. Evidence of this lake remains in evaporite deposits that make up the playa in the valley's center, in wave-cut shorelines, and in beach bars.

much narrower, with a small resistant cap, but are often as tall and steep as neighboring mesas.

Desert Pavement and Desert Varnish

Much of the desert floor is covered by desert pavement, an accumulation of gravel and cobbles that forms a surface fabric that can interconnect tightly. Fine material has been removed by wind and water, leaving behind larger fragments that inhibit further erosion. In many areas, desert pavements have been stable for long periods of time, as evidenced by their surface patina of desert varnish. Desert varnish is a thin outer coating of wind-deposited clay mixed with iron and manganese oxides. Varying in color from light brown to black, these coatings are thought to adhere to rocks by the action of single-celled microorganisms. Under a microscope, desert varnish can be seen to be made up of very fine layers. A thick, dark patina means that a rock has been exposed for a long time.

Playas

Where neither dunes nor rocky pavements cover the desert floor, one may find an accumulation of saline minerals. A playa is a flat surface that is often blindingly white in color. Playas are usually found in the centers of desert valleys and contain material that mineralized during the evaporation of a lake. Dry lake beds are a common feature of the Great Basin in the western United States. During glacial stages, the last of which occurred about 20,000 years ago, lakes grew in what are now arid, closed valleys. As the climate warmed, these lakes shrank, and many dried completely. As a lake evaporates, minerals that were held in solution crystallize, forming salts, including halite (table salt). These salt deposits frequently are mined for useful household and industrial chemicals.

Richard L. Orndorff

OCEAN MARGINS

Ocean margins are the areas where land borders the sea. Although often referred to as coastlines or beaches, ocean margins cover far greater territory than beaches. An ocean margin extends from the coastal plain—the fertile farming belt of land along the seacoast—to the edge of the gently sloping land submerged in water, called the continental shelf.

Ocean margin constitutes 8 percent of the world's surface. It is rich in minerals, both above and below water, and is home to 25 percent of Earth's people, along with 90 percent of the marine life. This fringe of land at the border of the ocean is ever changing. Tides wash sediment in and leave it behind, just below sea level. This process, called deposition, builds up land in some areas of the coastline. At the same time, ocean waves, winds, and storms wear away or erode parts of the shoreline. As land is worn away or built up, the amount of land above sea level changes. Factors such as climate, erosion, deposition, changes in sea level, and the effects of humans constantly change the shape of the ocean margin on Earth.

Beach Dynamics

The two types of coasts or land formations at the ocean margin are primary coasts and secondary coasts. Primary coasts are formed by systems on land, such as the melting of glaciers, wind or water erosion, and sediment deposited by rivers. Deltas and fjords are examples of primary coasts. Secondary coasts are formed by ocean patterns, such as erosion by waves or currents, sediment deposition by waves or currents, or changes by marine plants or animals. Beaches, coral reefs, salt marshes, and mangrove swamps are examples of secondary coasts.

Sediment carried by rivers to the sea is deposited to form deltas at the mouths of the rivers. Some of the sediment can wash out to sea, causing formations to build up at a distance from the shore. These formations eventually become barrier islands, which are often little more than 10 feet (3 km.) above sea level. As

a consequence, heavy storms, such as hurricanes, can cause great damage to barrier islands. Barrier islands naturally protect the coastline from erosion, however, especially during heavy coastal storms.

Sea level changes also affect the shape of the coastline. As oceans slowly rise, land is slowly consumed by the ocean. Barrier islands, having low sea levels, may slowly be covered with water. The melting of continental glaciers increased the sea level 0.06 inch (0.15 centimeter) per year during the twentieth century. As ocean waters warm, they expand, eating away at sea levels. Global warming caused by carbon dioxide levels in the atmosphere could cause sea levels to rise as much as 0.24 inch (0.6 centimeter) per year as a result of the warming of the water and glacial melting.

Human Influence

The shape of the ocean margin also changes radically as a result of human influence. According to the United States Geological Survey, 39 percent of the people living in the United States live directly on the the coasts. According to UN Atlas of Oceans, about 44 percent of the world's population lives within 93 miles (150 kilometers) of the coast. Pollution from toxins, dredging, recreational boating, and waste disposal kills plants and animals along the ocean margin. This changes the coastal shape, as mangrove forests, coral reefs, and other coastal lifeforms die.

A greater concern along the coastal fringe, however, is human development. Not only are people drawn to the fertile soil along the coastal zone of the continent, but they also develop islands and coves into resort communities. To protect homes and hotels along the coastal zone from coastal erosion, people build breakwalls, jetties, and sand and stone bars called groins.

These human-made barriers disrupt the natural method by which the ocean carries material along the coast. Longshore drift, a zigzag movement, deposits sediment from one area of the beach farther along the shoreline. Breakwalls, jetties, and groins disrupt this flow. As the ocean smashes against a breakwall, the property behind it may be safe for the present, but the coastline neighboring the breakwall takes a greater beating. The silt and sediment from upshore, which would replace that carried downshore, never arrives. Eventually, the breakwall will break down under the impact of the ocean force. Areas with breakwalls and jetties often suffer greater damage in coastal storms than areas that remain naturally open to the changing forces of the ocean.

To compensate for the destructive nature of human-made barriers, many recreational beaches replace lost sand with dredgings or deposit truckloads of sand from inland sources. For example, Virginia Beach in the United States spends between US$2 million and US$3 millon annually to restore beaches for the tourist season in this way.

Despite the changes in the shape of the ocean margin, it continues to provide a stable supply of resources—fish, seafood, minerals, sponges, and other marine plants and animals. Offshore drilling of oil and natural gas often takes place within 200 miles (322 km.) of shorelines.

Lisa A. Wroble

EARTH'S CLIMATES

THE ATMOSPHERE

The thin layer of gases that envelops the earth is the atmosphere. This layer is so thin that if the earth were the size of a desktop globe, more than 99 percent of its atmosphere would be contained within the thickness of an ordinary sheet of paper. Despite its thinness, the atmosphere sustains life on Earth, protecting it from the Sun's searing radiation and regulating the earth's temperature. Storms of the atmosphere carry water to the continents, and weathering by its wind and rain helps shape them.

Composition of the Atmosphere

The earth's atmosphere consists of gases, microscopic particles called aerosol, and clouds consisting of water droplets and ice particles. Its two principal gases are nitrogen and oxygen. In dry air, nitrogen occupies 78 percent, and oxygen 21 percent, of the atmosphere's volume. Argon, neon, xenon, helium, hydrogen, and other trace gases together equal less than 1 percent of the remaining volume.

These gases are distributed homogeneously in a layer called the homosphere, which occurs between the earth's surface and about 50 miles (80 km.) altitude. Above 50 miles altitude, in the heterosphere, the concentration of heavier gases decreases more rapidly than lighter gases.

The atmosphere has no firm top. It simply thins out until the concentration of its gas molecules approaches that of the gases in outer space. The concentration of nitrogen and oxygen remains essentially con-

stant in the atmosphere because a balance exists between the production and removal of these gases at the earth's surface. Decaying organic matter adds nitrogen to the atmosphere, while soil bacteria remove nitrogen. Oxygen enters the atmosphere primarily through photosynthesis and is removed through animal respiration, combustion, and decay of organic material, and by chemical reactions involving the creation of oxides.

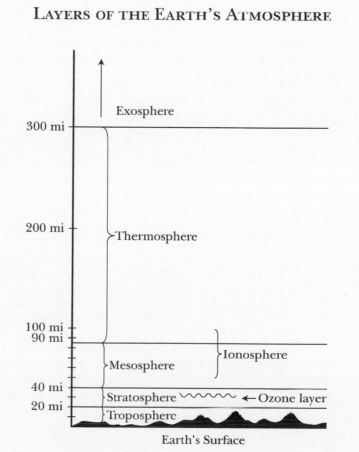

LAYERS OF THE EARTH'S ATMOSPHERE

THE GREENHOUSE EFFECT

Clouds and atmospheric gases such as water vapor, carbon dioxide, methane, and nitrous oxide absorb part of the infrared radiation emitted by the earth's surface and reradiate part of it back to the earth. This process effectively reduces the amount of energy escaping to space and is popularly called the "greenhouse effect" because of its role in warming the lower atmosphere. The greenhouse effect has drawn worldwide attention because increasing concentrations of carbon dioxide from the burning of fossil fuels result in a global warming of the atmosphere.

Scientists know that the greenhouse analogy is incorrect. A greenhouse traps warm air within a glass building where it cannot mix with cooler air outside. In a real greenhouse, the trapping of air is more important in maintaining the temperature than is the trapping of infrared energy. In the atmosphere, air is free to mix and move about.

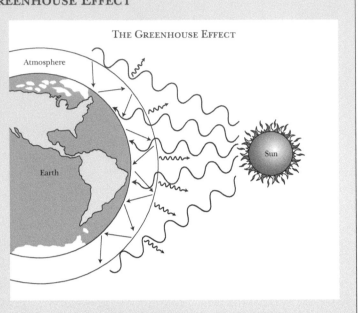

THE GREENHOUSE EFFECT

The atmosphere contains many gases that are present in small, variable concentrations. Three gases—water vapor, carbon dioxide and ozone—are vital to life on Earth. Water vapor enters the atmosphere through evaporation, primarily from the oceans, and through transpiration by plants. It condenses to form clouds, which provide the rain and snow that sustain life outside the oceans. The concentration of water vapor varies from about 4 percent by volume in tropical humid climates to a small fraction of a percent in polar dry climates. Water vapor plays an important role in regulating the temperature of the earth's surface and the atmosphere. Clouds reflect some of the incoming solar radiation, while water vapor and clouds both absorb earth's infrared radiation.

Carbon dioxide also absorbs the earth's infrared radiation. The global average atmospheric carbon dioxide in 2018 was 407.4 parts per million (ppm for short), with a range of uncertainty of plus or minus 0.1 ppm. Carbon dioxide levels today are higher than at any point in at least the past 800,000 years. The annual rate of increase in atmospheric carbon dioxide over the past 60 years is about 100 times faster than previous natural increases, such as those that occurred at the end of the last ice age 11,000–17,000 years ago.

Carbon dioxide enters the atmosphere as the result of decay of organic material, through respiration, during volcanic eruptions, and from the burning of fossil fuels. It is removed during photosynthesis and by dissolving in ocean water, where it is used by organisms and converted to carbonates. The increase in atmospheric carbon dioxide associated with the burning of fossil fuels has raised concerns that the earth's atmosphere may be warming through enhancement of the greenhouse effect.

Ozone, a gas consisting of molecules containing three oxygen atoms, forms in the upper atmosphere when oxygen atoms and oxygen molecules combine. Most ozone exists in the upper atmosphere between 6.2 and 31 miles (10 and 50 km.) in altitude, in concentrations of no more than 0.0015 percent by volume. This small amount of ozone sustains life outside the oceans by absorbing most of the Sun's ultraviolet radiation, thereby shielding the earth's surface from the radiation's harmful effects on living organisms. Paradoxically, ozone is an irritant near the earth's surface and is the major component of photochemical smog. Other gases that contribute to pollution include methane, nitrous oxide, hydrocarbons, and chlorofluorocarbons.

Aerosols represent another component of atmospheric pollution. Aerosols form in the atmosphere during chemical reactions between gases, through mechanical or chemical interactions between the earth, ocean surface, and atmosphere, and during evaporation of droplets containing dissolved or solid material. These microscopic particles are always present in air, with concentrations of about a few hundred per cubic centimeter in clean air to as many as a million per cubic centimeter in polluted air. Aerosols are essential to the formation of rain and snow, because they serve as centers upon which cloud droplets and ice particles form.

Energy Exchange in the Atmosphere

The Sun is the ultimate source of the energy in Earth's atmosphere. Its radiation, called electromagnetic radiation because it propagates as waves with electric and magnetic properties, travels to the surface of the earth's atmosphere at the speed of light. This energy spans many wavelengths, some of which the human eye perceives as colors. Visible wavelengths make up about 44 percent of the Sun's energy. The remainder of the Sun's radiant energy cannot be seen by human eyes. About 7 percent arrives as ultraviolet radiation, and most of the remaining energy is infrared radiation.

The Sun is not the only source of radiation. All objects emit and absorb radiation to some degree. Cooler objects such as the earth emit nearly all their energy at infrared wavelengths. Objects heat when they absorb radiation and cool when they emit radiation. The radiation emitted by the earth and atmosphere is called terrestrial radiation.

The balance between absorption of solar radiation and emission of terrestrial radiation ultimately determines the average temperature of the earth-atmosphere system. The vertical temperature distribution within the atmosphere also depends on the absorption and emission of radiation within the atmosphere, and the transfer of energy by the processes of conduction, convection, and latent heat exchange. Conduction is the direct transfer of heat from molecule to molecule. This process is most important in transferring heat from the earth's surface to the first few centimeters of the at-

---

THE OZONE HOLE

Since the 1970s, balloon-borne and satellite measurements of stratospheric ozone have shown rapidly declining stratospheric ozone concentrations over the continent of Antarctica, termed the "ozone hole." The lowest concentrations occur during the Antarctic spring, in September and October. The decrease in ozone has been associated with an increase in the concentration of chlorine, a gas introduced into the stratosphere through chemical reactions involving sunlight and chlorofluorocarbons—synthetic chemicals used primarily as refrigerants. The ozone hole over Antarctica has raised concern about possible worldwide reduction in the concentration of upper atmospheric ozone.

---

mosphere. Convection, the transfer of heat by rising or sinking air, transports heat energy vertically through the atmosphere.

Latent heat is the energy required to change the state of a substance, for example, from a liquid to a gas. Energy is transferred from the earth's surface to the atmosphere through latent heat exchange when water evaporates from the oceans and condenses to form rain in the atmosphere.

Only 48 percent of the solar energy reaching the top of the earth's atmosphere is absorbed by the earth's surface. The atmosphere absorbs another 23 percent. The remaining 30 percent is scattered back to space by atmospheric gases, clouds and the earth's surface. To understand the importance of terrestrial radiation and the greenhouse effect in the atmosphere's energy balance, consider the solar radiation arriving at the top of the earth to be 100 energy units, with 48 energy units absorbed by the earth's surface and 23 units by the atmosphere.

The earth's surface actually emits 117 units of energy upward as terrestrial radiation, more than twice as much energy as it receives from the Sun. Only 6 of these units are radiated to space—the atmosphere absorbs the remaining energy. Latent heat exchange, conduction, and convection account for another 30 units of energy transferred from the surface to the atmosphere. The atmosphere, in turn, radiates 96 units of energy back to the earth's surface (the greenhouse effect), and 64 units to space. The earth's and atmosphere's energy budget remains in balance, the atmosphere gaining and losing 160 units of energy, and the earth gaining and losing 147 units of energy.

Vertical Structure of the Atmosphere

Temperature decreases rapidly upward away from the earth's surface, to about –60 degrees Farenheit (–51 degrees Celsius) at an altitude of about 7.5 miles (12 km.). Above this altitude, temperature increases with height to about 32 degrees Farenheit (0 degrees Celsius) at an altitude of 31 miles (50 km.). The layer of air in the lower atmosphere where temperature decreases with height is called the troposphere. It contains about 75 percent of the atmosphere's mass. The layer of air above the troposphere, where temperature increases with height, is called the stratosphere. All but 0.1 percent of the remaining mass of the atmosphere resides in the stratosphere.

The stratosphere exists because ozone in the stratosphere absorbs ultraviolet light and converts it to heat. The boundary between the troposphere and stratosphere is called the tropopause. The tropopause is extremely important because it acts as a lid on the earth's weather. Storms can grow vertically in the troposphere, but cannot rise far, if at all, beyond the tropopause. In the polar regions, the tropopause can be as low as 5 miles (8 km.) above the surface, while in the tropics, the tropopause can be as high as 11 miles (18 km.). For this reason, tropical storms can extend to much higher altitudes than storms in cold regions.

The mesosphere extends from the top of the stratosphere, the stratopause, to an altitude of about 56 miles (90 km.). Temperature decreases with height within the mesosphere. The lowest average temperatures in the atmosphere occur at the mesopause, the top of the mesosphere, where the temperature is about –130 degrees Farenheit (–90 degrees Celsius). Only 0.0005 percent of the atmosphere's mass remains above the mesopause. In this uppermost layer, the thermosphere, there are few atoms and molecules. Oxygen molecules in the thermosphere absorb high-energy solar radiation. In this near vacuum, absorption of even small amounts of energy causes a large increase in temperature. As a result, temperature increases rapidly with height in the lower thermosphere, reaching about 1,300 degrees Farenheit (700 degrees Celsius) above 155 miles (250 km.) altitude.

The upper mesosphere and thermosphere also contain ions, electrically charged atoms or molecules. Ions are created in the atmosphere when air molecules collide with high-energy particles arriving from space or absorb high-energy solar radiation. Ions cannot exist very long in the lower atmosphere, because collisions between newly formed ions quickly restore ions to their uncharged state. However, above about 37 miles (60 km.) collisions are less frequent and ions can exist for longer times. This region of the atmosphere, called the ionosphere, is particularly important for amplitude-modulated (AM) radio communication because it reflects standard AM radio waves. At night, the lower ionosphere disappears as ions recombine, allowing AM radio waves to travel longer distances when reflected. For this reason, AM radio station signals can sometimes travel great distances at night.

The top of the atmosphere occurs at about 310 miles (500 km.). At this altitude, the distance between individual molecules is so great that energetic molecules can move into free space without colliding with neighbor molecules. In this uppermost layer, called the exosphere, the earth's atmosphere merges into space.

Robert M. Rauber

Global Climates

A region's climate is the sum of its long-term weather conditions. Most descriptions of climate emphasize temperature and precipitation characteristics, because these two climatic elements usually exert more impact on environmental conditions and human activities than do other elements, such as wind, humidity, and cloud cover. Climatic descriptions of a region generally cover both mean conditions and extremes. Climatic means are important because they represent average conditions that are frequently experienced; extreme conditions, such as severe storms, excessive heat and cold, and droughts, are important because of their adverse impact.

Important Climate Controls

A region's climate is largely determined by the interaction of six important natural controls: sun angle, elevation, ocean currents, land and water heating and cooling characteristics, air pressure and wind belts, and orographic influence.

Sun angle—the height of the Sun in degrees above the nearest horizon—largely controls the amount of solar heating that a site on Earth receives. It strongly influences the mean temperatures of most of the earth's surface, because the Sun is the ultimate energy source for nearly all the atmosphere's heat. The higher the angle of the Sun in the sky, the greater the concentration of energy, per unit area, on the earth's surface (assuming clear skies). From a global perspective, the Sun's mean angle is highest, on average, at the equator, and becomes progressively lower poleward. This causes a gradual decrease in mean temperatures with increasing latitude.

Sun angles also vary seasonally and daily. Each hemisphere is inclined toward the Sun during spring and summer, and away from the Sun during fall and winter. This changing inclination causes mean sun angles to be higher, and the length of daylight longer, during the spring and summer. Therefore, most locations, especially those outside the tropics, have warmer temperatures during these two seasons. The earth's rotation causes sun angles to be higher during midday than in the early morning and late afternoon, resulting in warmer temperatures at midday. Heating and cooling lags cause both seasonal and daily maximum and minimum temperatures typically to occur somewhat after the periods of maximum and minimum solar energy receipt.

Variations in elevation—the distance above sea level—can cause locations at similar latitudes to vary greatly in temperature. Temperatures decrease an average of about 3.5 degrees Fahrenheit per thousand feet (6.4 degrees Celsius per thousand meters). Therefore, high mountain and plateau stations are much colder than low-elevation stations at the same latitude.

Surface ocean currents can transport masses of warm or cold water great distances from their source regions, affecting both temperature and moisture conditions. Warm currents facilitate the evaporation of copious amounts of water into the atmosphere and add buoyancy to the air by heating it from below. This results in a general increase in precipitation totals. Cold currents evaporate water relatively slowly and chill the overlying air, thus stabilizing it and reducing its potential for precipitation.

The influence of ocean currents on land areas is greatest in coastal regions and decreases inland. The west coasts of continents (except for Europe) generally are paralleled by relatively cold currents, and the east coasts by relatively warm currents. For example, the warm Gulf Stream flows northward off the eastern United States, while the West Coast is cooled by the southward-flowing California Current.

Land can change temperature much more readily than water. As a result, the air over continents typically experiences larger annual temperature ranges (that is, larger temperature differences between summer and winter) and shorter heating and cooling lags than does the air over oceans. This same effect causes continental interiors and the leeward (downwind) coasts of continents typically to have larger temperature ranges than do windward (upwind) coasts. Climates that are dominated by air from landmasses are often described as continental climates. Conversely, climates dominated by air from oceans are described as maritime climates.

The seasonal heating and cooling of continents can also produce a monsoon influence, which has to do with annual shifts of wind patterns. Areas influenced by a monsoon, such as Southeast Asia, tend to have a predominantly onshore flow of moist maritime air during the summer. This often produces heavy rains. An offshore flow of dry air predominates in winter, producing fair weather.

Earth's atmosphere displays a banded, or beltlike, pattern of air pressure and wind systems. High pressure is associated with descending air and dry weather; low pressure is associated with rising air, which produces cloudiness and often precipitation. Wind is produced by differences in air pressure. The air blows outward from high-pressure systems and into low-pressure systems in a constant attempt to equalize air pressures.

The direction and speed of movement of weather systems, such as weather fronts and storms, are controlled by wind patterns, especially those several kilometers above the surface. The seasonal shift of global temperatures caused by the movement of the Sun's vertical rays between the Tropics of Cancer and Capricorn produces a latitudinal migration of both air pressure and wind belts. This shift affects the annual temperature and precipitation patterns of many regions.

Four air-pressure belts exist in each hemisphere. The intertropical convergence zone (ITCZ) is a broad belt of low pressure centered within a few degrees of latitude of the equator. The subtropical highs are high-pressure belts centered between 20 and 40 degrees north and south latitude, which are responsible for many of the world's deserts. The subpolar lows are low-pressure belts centered about 50 or 70 degrees north and south latitude. Finally, the polar highs are high-pressure centers located near the North and South Poles.

The air pressure gradient between these belts produces the earth's major wind belts. The regions between the ITCZ and the subtropical highs are dominated by the trade winds, a broad belt in each hemisphere of easterly (that is, moving east to west) winds. The middle latitudes are mostly situated between the subtropical highs and the subpolar lows and are within the westerly wind belt. This wind belt causes winds, and weather systems, to travel generally from west to east in the United States and Canada. Finally, the high-latitude zones between the subpolar lows and polar highs are situated within the polar easterlies.

The final factor affecting climate— orographic influence—is the lifting effect of mountain peaks or ranges on winds that pass over them. As air approaches a mountain barrier, it rises, typically producing clouds and precipitation on the windward (upwind) side of the mountains. After it crosses the crest, it descends the leeward (downwind) side of the mountains, generally producing dry weather. Most of the world's wettest locations are found on the windward sides of high mountain ranges; some deserts, such as those of the western interior United States, owe their aridity to their location on the leeward sides of orographic barriers.

World Climate Types

The global distribution of the world climate controls is responsible for the development of fourteen widely recognized climate types. In this section, the major characteristics of each of these climates will be briefly described. The climates are discussed in a rough poleward sequence.

Tropical Wet Climate

Sometimes called the tropical rain forest climate, the tropical wet climate exists chiefly in areas lying within 10 degrees of the equator. It is an almost seasonless climate, characterized by year-round warm, humid, rainy conditions that allow land areas to support a dense broadleaf forest cover. The warm temperatures, which for most locations average near 80 degrees Fahrenheit (27 degrees Celsius) throughout the year, result from the constantly high midday sun angles experienced at this low latitude.

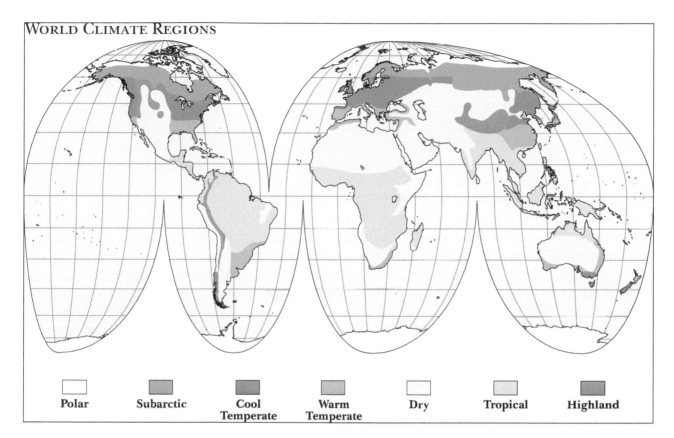

WORLD CLIMATE REGIONS

Polar   Subarctic   Cool Temperate   Warm Temperate   Dry   Tropical   Highland

The heavy precipitation totals result from the heating and subsequent rising of the warm moist air to form frequent showers and thunderstorms, especially during the afternoon hours. The dominance of the ITCZ enhances precipitation totals, helping make this climate type one of the world's rainiest.

Tropical Monsoonal Climate

The tropical monsoonal climate occurs in low-latitude areas, such as Southeast Asia, that have a warm, rainy climate with a short dry season. Temperatures are similar to those of the tropical wet climate, with the warmest weather often occurring during the drier period, when sunshine is more abundant. The heavy rainfalls result from the nearness of the ITCZ for much of the year, as well as the dominance of warm, moist air masses derived from tropical oceans. During the brief dry season, however, the ITCZ has usually shifted into the opposite hemisphere, and windflow patterns often have changed so as to bring in somewhat drier air derived from continental sources.

Tropical Savanna Climate

The tropical savanna climate, also referred to as the tropical wet-dry climate, occupies a large portion of the tropics between 5 and 20 degrees latitude in both hemispheres. It experiences a distinctive alternation of wet and dry seasons, caused chiefly by the seasonal shift in latitude of the subtropical highs and ITCZ. Summer is typically the rainy season because of the domination of the ITCZ. In many areas, an onshore windflow associated with the summer monsoon increases rainfalls at this time. In winter, however, the ITCZ shifts into the opposite hemisphere and is replaced by drier and more stable air associated with the subtropical high. In addition, the winter monsoon tendency often produces an outflow of continental air. The long dry season inhibits forest growth, so vegetation usually consists of a cover of drought-resistant shrubs or the tall savanna grasses after which the climate is named.

Subtropical Desert Climate

The subtropical desert climate has hot, arid conditions as a result of the year-round dominance of the subtropical highs. Summertime temperatures in this climate soar to the highest readings found anywhere on earth. The world's record high temperature was 134 degrees Fahrenheit (56.7 degrees Celsius), recorded in Furnace Creek Ranch, California (formerly Greenland Ranch) on July 10, 1913. Rainfall totals in this type of climate are generally less than 10 inches (25 centimeters) per year. What rainfall does occur often arrives as brief, sometimes violent, afternoon thunderstorms. Although summer temperatures are extremely hot, the dry air enables rapid cooling during the winter, so that temperatures are cool to mild at this time of year.

Subtropical Steppe Climate

The subtropical steppe climate is a semiarid climate, found mostly on the margins of the subtropical deserts. Precipitation usually ranges from 10 to 30 inches (25 to 75 centimeters), sufficient for a ground cover of shrubs or short steppe grasses. Areas on the equatorward margins of subtropical deserts typically receive their precipitation during a brief showery period in midsummer, associated with the poleward shift of the ITCZ. Areas on the poleward margins of the subtropical highs receive most of their rainfall during the winter, due to the penetration of cyclonic storms associated with the equatorward shift of the westerly wind belt.

Mediterranean Climate

The Mediterranean climate, also sometimes referred to as the dry summer subtropics, has a distinctive pattern of dry summers and more humid, moderately wet winters. This pattern is caused by the seasonal shift in latitude of the subtropical high and the westerlies. During the summer, the subtropical high shifts poleward into the Mediterranean climate regions, blanketing them with dry, warm, stable air. As winter approaches, this pressure center retreats equatorward, allowing the westerlies, with their eastward-traveling weather fronts and cyclonic storms, to overspread this region. The Mediterranean climate is found on the windward sides of continents, particularly the area surrounding the Mediterranean Sea and much of California. This results in the predomi-

nance of maritime air and relatively mild temperatures throughout the year.

Humid Subtropical Climate

The humid subtropical climate is found on the eastern, or leeward, sides of continents in the lower middle latitudes. The most extensive land area with this climate is the southeastern United States, but it is also seen in large areas in South America, Asia, and Australia. Temperature ranges are moderately large, with warm to hot summers and cool to mild winters. Mean temperatures for a given location are dictated largely by latitude, elevation, and proximity to the coast. Precipitation is moderate. Winter precipitation is usually associated with weather fronts and cyclonic storms that travel eastward within the westerly wind belt. During summer, most precipitation is in the form of brief, heavy afternoon and evening thunderstorms. Some coastal areas are subject to destructive hurricanes during the late summer and autumn.

Midlatitude Desert Climate

This type of climate consists of areas within the western United States, southern South America, and Central Asia that have arid conditions resulting from the moisture-blocking influence of mountain barriers. This climate is highly continental, with warm summers and cold winters. When precipitations occurs, it frequently comes in the form of winter snowfalls associated with weather fronts and cyclonic storms. Rainfall in summer typically occurs as afternoon thunderstorms.

Midlatitude Steppe

The midlatitude steppe climate is located in interior portions of continents in the middle latitudes, particularly in Asia and North America. This climate has semiarid conditions caused by a combination of continentality resulting from the large distance from oceanic moisture sources and the presence of mountain barriers. Like the midlatitude desert climate, this climate has large annual temperature ranges, with cold winters and warm summers. It also receives winter rains and snows chiefly from weather fronts and cyclonic storms; summer rains occur largely from afternoon convectional storms. In the Great Plains of the United States, spring can bring very turbulent conditions, with blizzards in early spring and hailstorms and tornadoes in mid to late spring.

Marine West Coast

This type of climate is typically located on the west coasts of continents just poleward of the Mediterranean climate. Its location in the heart of the westerly wind belt on the windward sides of continents produces highly maritime conditions. As a result, cloudy and humid weather is common, along with frequent periods of rainfall from passing weather fronts and cyclonic storms. These storms are often well developed in winter, resulting in extended periods of wet and windy weather. Precipitation amounts are largely controlled by the presence and strength of the orographic effect; mountainous coasts like the northwestern United States and the west coast of Canada are much wetter than are flatter areas like northern Europe. Temperatures are held at moderate levels by the onshore flow of maritime air. As a consequence, winters are relatively mild and summers relatively cool for the latitude.

Humid Continental Climate

The humid continental climate is found in the northern interiors of Eurasia (Europe and Asia) and North America. It does not occur in the Southern Hemisphere because of the absence of large land masses in the upper midlatitudes of that hemisphere. This climate type is characterized by low to moderate precipitation that is largely frontal and cyclonic in nature. Most precipitation occurs in summer, but cold winter temperatures typically cause the surface to be frozen and snow-covered for much of the late fall, winter, and early spring. Temperature ranges in this climate are the largest in the world. A town in Siberia, Verkhoyansk, holds the Guinness world record for the greatest temperature range at a single location is 221 degrees Fahrenheit (105 degrees Celsius), from -90 degrees Fahrenheit (-68 degrees Celsius) to 99 degrees Fahrenheit (37 degrees Celsius). Winter temperatures in parts of both North America and Siberia can fall well below -49 degrees Fahrenheit (-45 degrees Celsius), making these the coldest permanently settled sites in the world.

Tundra Climate

The tundra climate is a severely cold climate that exists mostly on the coastal margins of the Arctic Ocean in extreme northern North America and Eurasia, and along the coast of Greenland. The high-latitude location and proximity to icy water cause every month to have average temperatures below 50 degrees Fahrenheit (10 degrees Celsius), although a few months in summer have means above freezing. As a result of the cold temperatures, tundra areas are not forested, but instead typically have a sparse ground cover of grasses, sedges, flowers, and lichens. Even this vegetation is buried by a layer of snow during most of the year. Cold temperatures lower the water vapor holding capacity of the air, causing precipitation totals to be generally light. Most precipitation is associated with weather fronts and cyclonic storms and occurs during the summer half of the year.

Ice Cap Climate

The most poleward and coldest of the world's climates is called the ice cap climate. It is found on the continent of Antarctica, interior Greenland, and some high mountain peaks and plateaus. Because monthly mean temperatures are subfreezing throughout the year, areas with this climate are glaciated and have no permanent human inhabitants.

The coldest temperatures of all occur in interior Antarctica, where a Russian research station named Vostok recorded the world's coldest temperature of -128.6 degrees Fahrenheit (-89.2 degrees Celsius) on July 21, 1983. This climate receives little precipitation because the atmosphere can hold very little water vapor. A major moisture surplus exists, however, because of the lack of snowmelt and evaporation. This causes the build up of a surface snow cover that eventually compacts to form the icecaps that bury the surface. Snowstorms are often accompanied by high winds, producing blizzard conditions.

Global Warming

Though warming has not been uniform across the planet, the upward trend in the globally averaged temperature shows that more areas are warming than cooling. According to the National Oceanic and Atmospheric Administration (NOAA) 2018 Global Climate Summary, the combined land and ocean temperature has increased at an average rate of 0.13°F (0.07°C) per decade since 1880; however, the average rate of increase since 1981 (0.31°F/ 0.17°C) is more than twice as great. It is strongly suspected that human activities that increase the accumulation of greenhouse gases (heat-trapping gases) in the atmosphere may play a key role in the temperature rise.

Levels of carbon dioxide (CO_2) in the atmosphere are higher now than they have been at any time in the past 400,000 years. This gas is responsible for nearly two-thirds of the global-warming potential of all human-released gases. Levels surpassed 407 ppm in 2018 for the first time in recorded history. By comparison, during ice ages, CO_2 levels were around 200 parts per million (ppm), and during the warmer interglacial periods, they hovered around 280 ppm. The recent rise in CO_2 shows a remarkably constant relationship with fossil-fuel burning, which is understandable when one considers that about 60 percent of fossil-fuel emissions stay in the air. Atmospheric carbon dioxide concentrations are also increased by deforestation, which is occurring at a rapid rate in several tropical countries. Deforestation causes carbon dioxide levels to rise because trees remove large quantities of this gas from the atmosphere during the process of photosynthesis.

Research indicates that if atmospheric concentrations of greenhouse gases continue to increase at the 1990s pace, global temperatures could rise an additional 1.8 to 6.3 degrees Fahrenheit (1 to 3.5 degrees Celsius) during the twenty-first century. That level of temperature increase would produce major changes in global climates and plant and animal habitats and would cause sea levels to rise substantially.

Ralph C. Scott

CLOUD FORMATION

Clouds are visible manifestations of water in the air. Cloud patterns can provide even a casual observer with much information about air movements and the processes occurring in the atmosphere. The shapes and heights of the clouds and the directions from which they have come are valuable clues in understanding weather.

Importance of Cooling

Clouds are formed when water vapor in the air is transformed into either water droplets or ice crystals. Sometimes large amounts of moisture are added to the air, producing clouds, but clouds generally are formed when a large amount of air is cooled. The amount of water vapor that air can hold varies with temperature: Cold air can hold less water vapor than warmer air. If air is cooled to the point at which it can hold no more water vapor, the water vapor will condense into water droplets. The temperature at which condensation begins is called the dew point. At below freezing temperatures, the water vapor will turn or deposit into ice crystals.

Cloud droplets do not necessarily form even if the air is fully saturated, that is, holding as much water vapor as possible at a given temperature. Once formed, cloud droplets can evaporate again very easily. Two factors hasten the production and growth of cloud droplets. One is the presence of

CLOUD FORMATION

The hydrologic cycle is the continuous circulation of the earth's waters through evaporation, condensation, and precipitation. The cycle also moves water through runoff, infiltration, and transpiration.

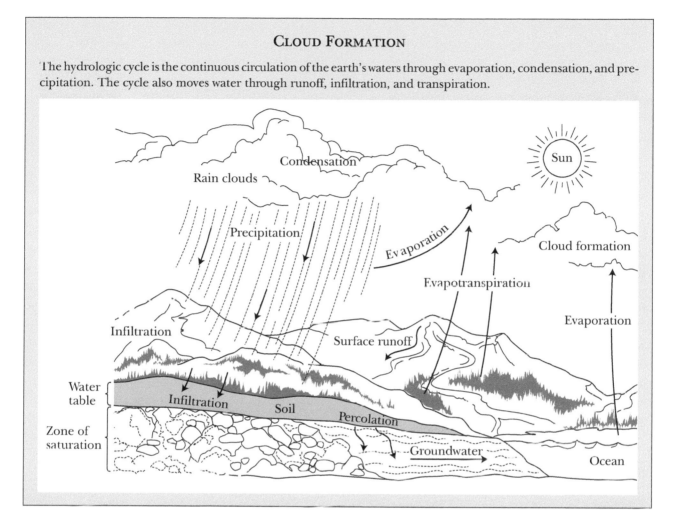

particles in the atmosphere that attract water. These are called hygroscopic particles or condensation nuclei. They include salt, dust, and pollen. Once water vapor condenses on these particles, more condensation can occur. Then the droplets can grow larger and bump into other droplets, growing even larger through this process, called coalescence.

Condensation and cloud droplet growth also is hastened when the air is very cold, at about -40 degrees Farenheit (which is also -40 degrees Celsius). At this temperature ice crystals form, but some water droplets can exist as liquid water. These water droplets are said to be supercooled. The water vapor is more likely to deposit on the ice crystals than on the supercooled water. Thus the ice crystals grow larger and the supercooled water droplets evaporate, resulting in more water vapor to deposit on ice crystals. Whether the cloud droplets start as hygro-

scopic particles or ice crystals, they eventually can grow in size to become a raindrop; around 1 million cloud droplets make one raindrop.

How and Why Rising Air Cools

In order for air to be cooled, it must rise or be lifted. When a volume of air, or an air parcel, is forced to rise through the surrounding air, the parcel expands in size as the pressure of the air around it declines with altitude. Close to the surface, the atmospheric pressure is relatively high because the density of the atmosphere is high. As altitude increases, the atmosphere declines in density, and the still air exerts less pressure. Thus, as an air parcel rises through the atmosphere, the pressure of the surrounding air declines, and the parcel takes up more space as it expands. Since work is done by the parcel as it expands, the parcel cools and its temperature declines.

An alternative explanation of the cooling is that the number of molecules in the air parcel remains the same, but when the volume is larger, the molecules produce less frictional heat because they do not bang into each other as much. The temperature of the air parcel declines, but no heat left the parcel—the change in temperature resulted from internal processes. The process of an air parcel rising, expanding, and cooling is called adiabatic cooling. Adiabatic means that no heat leaves the parcel. If the parcel rises far enough, it will cool sufficiently to reach its dewpoint temperature. With continued cooling, condensation will result—a cloud will be formed. At this height, which is called the lifting condensation level, an invisible parcel of air will turn into a cloud.

Uplift Mechanisms

An initial force is necessary to cause the air parcel to rise and then cool adiabatically. The three major processes are convection, orographic, and frontal or cyclonic.

With certain conditions, convection or vertical movement can cause clouds to form. On a sunny day, usually in the summer, the ground is heated unevenly. Some areas of the ground become warmer and heat the air above, making it warmer and less dense. A stream of air, called a thermal, may rise. As it rises, it cools adiabatically through expansion and may reach its dewpoint temperature. With continued cooling and rising, condensation will occur, forming a cloud. Since the cloud is formed by predominantly vertical motions, the cloud will be cumulus. With continued warming of the surface, the thermals may rise even higher, perhaps producing thunderstorm, or cumulonimbus, clouds. Thus, a sunny summer day can start off without a cloud in the sky, but can be stormy with many thunderstorms by afternoon.

Clouds also can form when air is forced to rise when it meets a mountain or other large vertical barrier. This type of lifting—orographic—is especially prevalent where air moves over the ocean and then is forced to rise up a mountain, as occurs on the west coast of North and South America. As the

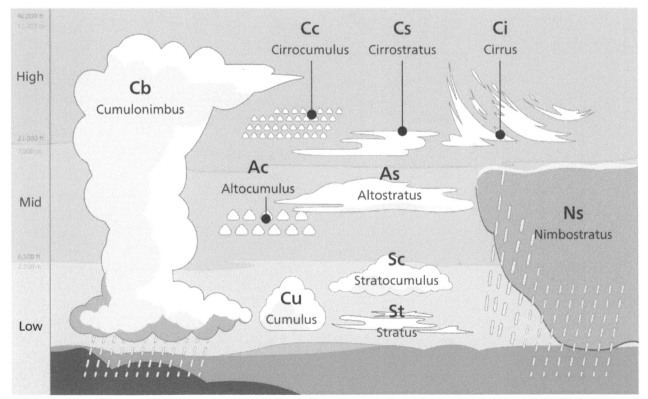

Cloud classification by altitude of occurrence. Multi-level and vertical genus-types not limited to a single altitude level include nimbostratus, cumulonimbus, and some of the larger cumulus species. (Illustration by Valentin de Bruyn)

air rises, it cools adiabatically and eventually becomes so cool that it cannot hold the water vapor. Condensation occurs and clouds form. The air continues to move up the mountain, producing clouds and precipitation on the side of the mountain from which the wind came, the windward side. However, the air eventually must fall down the other side of the mountain, the leeward side. That air is warmed and moisture evaporates, resulting in no clouds.

A third lifting mechanism is frontal, or cyclonic, action. This occurs when a large mass of cold air and a large mass of warm air—often hundreds of miles in area—meet. The warm air mass and the cold air mass will not mix freely, resulting in a border or front between the two air masses. The warm, less dense, air will always rise above the cold, denser, air mass. As the warm air rises, it cools, and when it reaches its dew point, clouds will form. If the warm air displaces the cold air, or a warm front occurs, the warm air will rise gradually, resulting in layered or stratiform clouds. The cloud types will change on an upward diagonal path, with the lowest being

stratus, and nimbostratus if rain occurs, followed by altostratus, then cirrostratus, and cirrus.

On the other hand, if the cold air displaces the warm air, the warm air will be forced to rise much more quickly. The clouds formed will be puffy or cumuliform—cumulus at the lowest levels, altocumulus and cirrocumulus at the highest altitudes. Sometimes cumulonimbus clouds will also form.

Sometimes when a cold front meets a warm front, the whole warm air mass is forced off the ground. This forms a cyclone—an area of low pressure—as the warm air rises. As this air rises, it cools. If it reaches its dew point, condensation and clouds will result. In oceanic tropical areas, a cyclone can form within warm, moist air. This air also will cool and, if it reaches its dew point, will condense and form clouds. Sometimes, these tropical cyclones are the precursors of hurricanes. The clouds associated with cyclones are usually cumulus, including cumulonimbus, as they are formed by rapidly rising air.

Margaret F. Boorstein

STORMS

A storm is an atmospheric disturbance that produces wind, is accompanied by some form of precipitation, and sometimes involves thunder and lightning. Storms that meet certain criteria are given specific names, such as hurricanes, blizzards, and tornadoes.

Stormy weather is associated with low atmospheric pressure, while clear, calm, dry weather is associated with high atmospheric pressure. Because of the way atmospheric pressure and wind direction are related, low-pressure areas are characterized by winds moving cyclonically (in a counterclockwise direction in the Northern Hemisphere; clockwise in the Southern Hemisphere) around the center of the low pressure. Storms of all kinds are associated with

cyclones, but two classes of cyclones—tropical and extratropical—produce most storms.

Tropical Cyclones
These storms develop during the summer and autumn in every tropical ocean except the South Atlantic and eastern South Pacific Oceans. Tropical cyclones that occur in the North Atlantic and eastern North Pacific Oceans are known as hurricanes; in the western North Pacific Ocean, as typhoons; and in the Indian and South Pacific Oceans, as cyclones.

All tropical cyclones develop in three stages. Arising from the formation of the initial atmospheric disturbance that is characterized by a cluster of thunderstorms, the first stage—tropical depres-

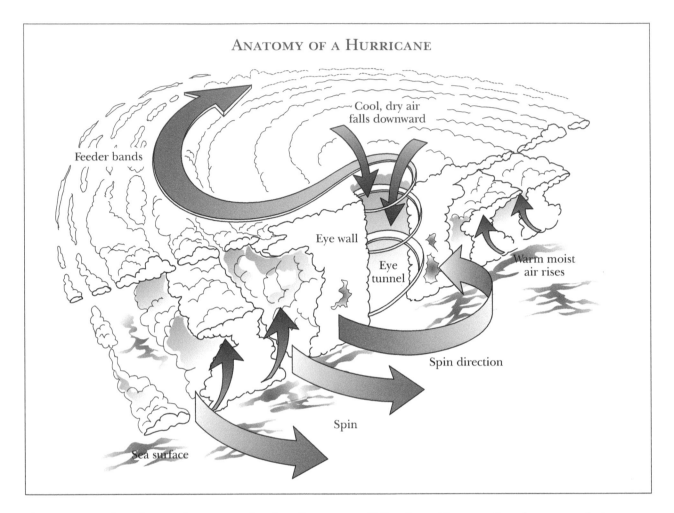

ANATOMY OF A HURRICANE

Feeder bands

Cool, dry air falls downward

Eye wall

Eye tunnel

Warm moist air rises

Spin direction

Spin

Sea surface

sion—occurs when the maximum sustained surface wind speeds (the average speed over one minute) range from 23–39 miles (37–61 km.) per hour. The second stage—tropical storm—occurs when sustained winds range from 40–73 miles (62–119 km.) per hour. At this stage, the storm is given a name. From 80 to 100 tropical storms develop each year across the world, with about half continuing to the final stage—hurricane—at which sustained wind speeds are 74 miles (120 km.) per hour or greater. Moving over land or into colder oceans initiates the end of the hurricane after a week or so by eliminating the hurricane's fuel—warm water.

A mature hurricane is a symmetrical storm, with the "eye" at the center; the eye develops as winds increase and become circular around the central core of low pressure. Within the eye, it is relatively warm, and there are light winds, no precipitation, and few clouds. This is caused by air descending in the center of the storm. Surrounding the eye is the "eye

wall," a ring of intense thunderstorms that can extend high into the atmosphere. Within the eye wall, the strongest winds and heaviest rainfall are found; this is also where warm, moist air, the hurricane's "fuel," flows into the storm. Spiraling bands of clouds, called "rain bands," surround the eye wall. Precipitation and wind speeds decrease from the eye wall out toward the edge of the rain bands, while atmospheric pressure is lowest in the eye and increases outward.

Hurricanes can be the most damaging storms because of their intensity and size. Damage is caused by high winds and the flying debris they carry, flooding from the tremendous amounts of rain a hurricane can produce, and storm surge. A storm surge, which accounts for most of the coastal property loss and 90 percent of hurricane deaths, is a dome of water that is pushed forward as the storm moves. This wall of water is lifted up onto the coast as the eye wall comes in contact with land. For exam-

NAMING HURRICANES

Hurricanes once were identified by their latitudes and longitudes, but this method of naming became confusing when two or more hurricanes developed at the same time in the same ocean. During World War II hurricanes were identified by radio code letters, such as Able and Baker. In 1953 the National Weather Service began using English female names in an alphabetized list. Male names and French and Spanish names were added in 1978. By 2000 six lists of names were used on a rotating basis. When a hurricane causes much death or destruction, as Hurricane Andrew did in August of 1992—its name is retired for at least ten years.

ple, a 25-foot (8-meter) storm surge created by Hurricane Camille in 1969 destroyed the Richelieu Apartments next to the ocean in Pass Christian, Mississippi. Ignoring advice to evacuate, twenty-five people had gathered there for a hurricane party; all but one was killed.

To help predict the damage that an approaching hurricane can cause, the Saffir-Simpson Scale was developed. A hurricane is rated from 1 (weak) to 5 (devastating), according to its central pressure, sustained wind speed, and storm surge height. Michael was a category 5 storm at the time of landfall on October 10, 2018, near Mexico Beach and Tyndall Air Force Base, Florida. Michael was the first hurricane to make landfall in the United States as a category 5 since Hurricane Andrew in 1992, and only the fourth on record. The others are the Labor Day Hurricane in 1935 and Hurricane Camille in 1969.

Extratropical Cyclones

Also known as midlatitude cyclones, these storms are traveling low-pressure systems that are seen on newspaper and television daily weather maps. They are created when a mass of moist, warm air from the south contacts a mass of drier, cool air from the north, causing a front to develop. At the front, the warmer air rides up over the colder air. This causes water vapor to condense and produces clouds and rain during most of the year, and snow in the winter.

Thunderstorms

Thunderstorms also develop in stages. During the cumulus stage, strong updrafts of warm air build the storm clouds. The storm moves into the mature stage when updrafts continue to feed the storm, but cool downdrafts are also occurring in a portion of the cloud where precipitation is falling. When the warm updrafts disappear, the storm's fuel is gone and the dissipating stage begins. Eventually, the cloud rains itself out and evaporates.

Thunderstorms can also form away from a frontal system, usually during summer. This formation is related to a relatively small area of warm, moist air

STORM CLASSIFICATIONS

Tropical Classification	Wind speed
Gale-force winds	>15 meters/second
Tropical depression	20-34 knots and a closed circulation
Tropical storm (named)	35-64 knots
Hurricane	65+ knots (74+ mph)
Saffir-Simpson Scale for Hurricanes	
Category 1	63-83 knots (74-95 mph)
Category 2	83-95 knots (96-110 mph)
Category 3	96-113 knots (111-130 mph)
Category 4	114-135 knots (131-155 mph)
Category 5	>135 knots (>155 mph)

Notes: 1 knot = 1 nautical mile/hour = 1.152 miles/hour = 1.85 kilometers/hour.
Source: National Aeronautics and Space Administration, Office of Space Science, Planetary Data System.
http:/atmos.nmsu.edu/jsdap/encyclopediawork.html

rising and creating a thunderstorm that is usually localized and short lived.

Wind, lightning, hail, and flooding from heavy rain are the main destructive forces of a thunderstorm. Lightning occurs in all mature thunderstorms as the positive and negative electrical charges in a cloud attempt to equal out, creating a giant spark. Most lightning stays within the clouds, but some finds its way to the surface. The lightning heats the air around it to incredible temperatures (54,000 degrees Farenheit/30,000 degrees Celsius), which causes the air to expand explosively, creating the shock wave called thunder. Since lightning travels at the speed of light and thunder at the speed of sound, one can estimate how many miles away the lightning is by counting the seconds between the lightning and thunder and dividing by five. People have been killed by lightning while boating, swimming, biking, golfing, standing under a tree, talking on the telephone, and riding on a lawnmower.

Hail is formed in towering cumulonimbus clouds with strong updrafts. It begins as small ice pellets that grow by collecting water droplets that freeze on contact as the pellets fall through the cloud. The strong updrafts push the pellets back into the cloud, where they continue collecting water droplets until they are too heavy to stay aloft and fall as hailstones. The more an ice pellet is pushed back into the cloud, the larger the hailstone becomes. The largest authenticated hailstone in the United States fell near Vivian, South Dakota, on July 23, 2010. It mea-sured 8.0 inches (20 cm) in diameter, 18 1/2 inches (47 cm.) in circumference, and weighed in at 1.9375 pounds (879 grams).

Tornadoes

For reasons not well understood, less than 1 percent of all thunderstorms spawn tornadoes. Called funnel clouds until they touch earth, tornadoes contain the highest wind speeds known.

Although tornadoes can occur anywhere in the world, the United States has the most, with an average of 1000 per year. Tornadoes have occurred in every state, but the greatest number hit a portion of the Great Plains from central Texas to Nebraska, known as "Tornado Alley." There cold Canadian air and warm Gulf Coast air often collide over the flat land, creating the wall cloud from which most tornadoes are spawned. May is the peak month for tornado activity, but they have been spotted in every month.

Because tornado winds cannot be measured directly, the tornado is ranked according to its damage, using the Fujita Intensity Scale. The scale ranges from an F0, with wind speeds less than 72 miles (116 km.) per hour, causing light damage, to an F5, with winds greater than 260 miles (419 km.) per hour, causing incredible damage. Most tornadoes are small, but the larger ones cause much damage and death.

Kay R. S. Williams

Earth's Biological Systems

Biomes

The major recognizable life zones of the continents, biomes are characterized by their plant communities. Temperature, precipitation, soil, and length of day affect the survival and distribution of biome species. Species diversity within a biome may increase its stability and capability to deliver natural services, including enhancing the quality of the atmosphere, forming and protecting the soil, controlling pests, and providing clean water, fuel, food, and drugs. Land biomes are the temperate, tropical, and boreal forests; tundra; desert; grasslands; and chaparral.

Temperate Forest

The temperate forest biome occupies the so-called temperate zones in the midlatitudes (from about 30 to 60 degrees north and south of the equator). Temperate forests are found mainly in Europe, eastern North America, and eastern China, and in narrow zones on the coasts of Australia, New Zealand, Tasmania, and the Pacific coasts of North and South America. Their climates are characterized by high rainfall and temperatures that vary from cold to mild.

Temperate forests contain primarily deciduous trees—including maple, oak, hickory, and beechwood—and, secondarily, evergreen trees—including pine, spruce, fir, and hemlock. Evergreen forests in some parts of the Southern Hemisphere contain eucalyptus trees.

The root systems of forest trees help keep the soil rich. The soil quality and color is due to the action of earthworms. Where these forests are frequently cut, soil runoff pollutes streams, which reduces fisheries because of the loss of spawning habitat.

Racoons, opposums, bats, and squirrels are found in the trees. Deer and black bear roam forest floors. During winter, small animals such as groundhogs and squirrels burrow in the ground.

Tropical Forest

Tropical forests are in frost-free areas between the Tropic of Cancer and the Tropic of Capricorn. Temperatures range from warm to hot year-round, because the Sun's rays shine nearly straight down around midday. These forests are found in northern Australia, the East Indies, southeastern Asia, equatorial Africa, and parts of Central America and northern South America.

Tropical forests have high biological diversity and contain about 15 percent of the world's plant species. Animal life lives at different layers of tropical forests. Nuts and fruits on the trees provide food for birds, monkeys, squirrels, and bats. Monkeys and sloths feed on tree leaves. Roots, seeds, leaves, and fruit on the forest floor feed deer, hogs, tapirs, antelopes, and rodents. The tropical forests produce rubber trees, mahogany, and rosewood. Large animals in these forests include the Asian tiger, the African bongo, the South American tapir, the Central and South American jaguar, the Asian and African leopard, and the Asian axis deer. Deforestation for agriculture and pastures has caused reduction in plant and animal diversity.

Boreal Forest

The boreal forest is a circumpolar Northern Hemisphere biome spread across Russia, Scandinavia, Canada, and Alaska. The region is very cold. Evergreen trees such as white spruce and black spruce

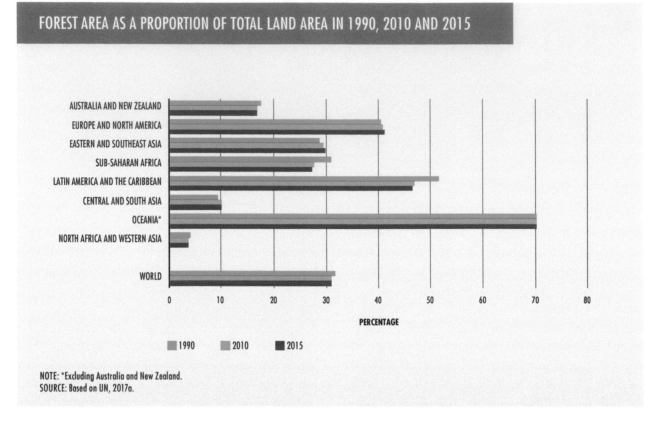

FOREST AREA AS A PROPORTION OF TOTAL LAND AREA IN 1990, 2010 AND 2015

NOTE: *Excluding Australia and New Zealand.
SOURCE: Based on UN, 2017a.

dominate this zone, which also contains larch, balsam, pine, and fir, and some deciduous hardwoods such as birch and aspen. The acidic needles from the evergreens make the leaf litter that is changed into soil humus. The acidic soil limits the plants that develop.

Animals in boreal forests include deer, caribou, bear, and wolves. Birds in this zone include goshawks, red-tailed hawks, sapsuckers, grouse, and nuthatches. Relatively few animals emigrate from this habitat during winter. Conifer seeds are the basic winter food. The disappearing aspen habitat of the beaver has decreased their numbers and has reduced the size of wetlands.

Tundra

About 5 percent of the earth's surface is covered with Arctic tundra, and 3 percent with alpine tundra. The Arctic tundra is the area of Europe, Asia, and North America north of the boreal coniferous forest zone, where the soils remain frozen most of the year. Arctic tundra has a permanent frozen subsoil, called permafrost. Deep snow and low temper-

atures slow the soil-forming process. The area is bounded by a 50 degrees Fahrenheit circumpolar isotherm, known as the summer isotherm. The cold temperature north of this line prevents normal tree growth.

The tundra landscape is covered by mosses, lichens, and low shrubs, which are eaten by caribou, reindeer, and musk oxen. Wolves eat these herbivores. Bear, fox, and lemming also live here. The larger mammals, including marine mammals and the overwintering birds, have large fat layers beneath the skin and long dense fur or dense feathers that provide protection. The small mammals burrow beneath the ground to avoid the harsh winter climate. The most common Arctic bird is the old squaw duck. Ptarmigans and eider ducks are also very common. Geese, falcons, and loons are some of the nesting birds of the area.

The alpine tundra, which exists at high altitude in all latitudes, is acted upon by winds, cold temperatures, and snow. The plant growth is mostly cushion and mat-forming plants.

BIOMES OF THE WORLD

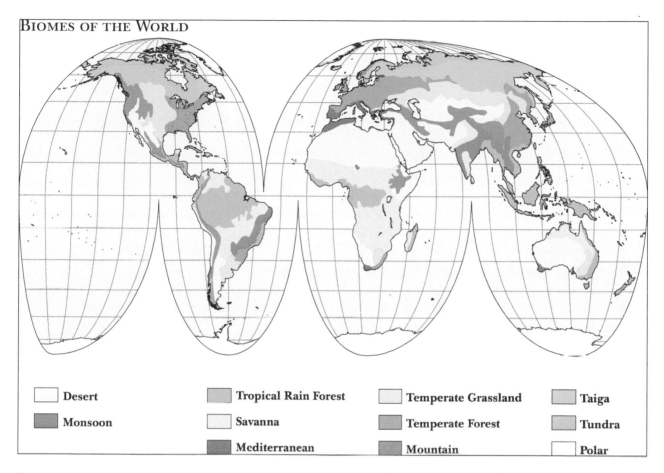

Desert	Tropical Rain Forest	Temperate Grassland	Taiga
Monsoon	Savanna	Temperate Forest	Tundra
	Mediterranean	Mountain	Polar

Desert

The desert biome covers about one-seventh of the earth's surface. Deserts typically receive no more than 10 inches (25 centimeters) of rainfall a year, but evaporation generally exceeds rainfall. Deserts are found around the Tropic of Cancer and the Tropic of Capricorn. As the warm air rises over the equator, it cools and loses its water content. This dry air descends in the two subtropical zones on each side of the equator; as it warms, it picks up moisture, resulting in drying the land.

Rainfall is a key agent in shaping the desert. The lack of sufficient plant cover removes the natural protection that prevents soil erosion during storms. High winds also cut away the ground.

Some desert plants obtain water from deep below the surface, for example, the mesquite tree, which has roots that are 40 feet (13 meters) deep. Other plants, such as the barrel cactus, store large amounts of water in their leaves, roots, or stems. Other plants slow the loss of water by having tiny leaves or shedding their leaves. Desert plants have very short growth periods, because they cannot grow during the long drought periods.

Desert animals protect themselves from the Sun's heat by eating at night, staying in the shade during the day, and digging burrows in the ground. Among the world's large desert animals are the camel, coyote, mule deer, Australian dingo, and Asian saiga. The digestive process of some desert animals produces water. A method used by some animals to conserve water is the reabsorption of water from their feces and urine.

Grassland

Grasslands cover about a quarter of the earth's surface, and can be found between forests and deserts. Treeless grasslands grow in parts of central North America, Central America, and eastern South America that have between 10 and 40 inches (250-1,000 millimeters) of erratic rainfall. The climate has a high rate of evaporation and periodic major droughts. The biome is also subject to fire.

Some grassland plants survive droughts by growing deep roots, while others survive by being dormant. Grass seeds feed the lizards and rodents that become the food for hawks and eagles. Large animals include bison, coyotes, mule deer, and wolves. The grasslands produce more food than any other biome. Poor grazing and agricultural practices and mining destroy the natural stability and fertility of these lands. The reduced carrying capacity of these lands causes an increase in water pollution and erosion of the soil. Diverse natural grasslands appear to be more capable of surviving drought than are simplified manipulated grass systems. This may be due to slower soil mineralization and nitrogen turnover of plant residues in the simplified system.

Savannas are open grasslands containing deciduous trees and shrubs. They are near the equator and are associated with deserts. Grasses grow in clumps and do not form a continuous layer. The northern savanna bushlands are inhabited by oryx and gazelles. The southern savanna supports springbuck and eland. Elephants, antelope, giraffe, zebras, and black rhinoceros are found on the savannas. Lions, leopards, cheetah, and hunting dogs are the primary predators here. Kangaroos are found in the savannas of Australia. Savannas cover South America north and south of the Amazon rain forest, where jaguar and deer can be found.

Chaparral

The chaparral or Mediterranean biome is found in the Mediterranean Basin, California, southern Australia, middle Chile, and Cape Province of South America. This region has a climate of wet winters and summer drought. The plants have tough leathery leaves and may contain thorns. Regional fires clear the area of dense and dead vegetation. Fire, heat, and drought shape the region. The vegetation dwarfing is due to the severe drought and extreme climate changes. The seeds from some plants, such as the California manzanita and South African fire lily, are protected by the soil during a fire and later germinate and rapidly grow to form new plants.

Ocean

The ocean biome covers more than 70 percent of the earth's surface and includes 90 percent of its volume. The ocean has four zones. The intertidal zone is shallow and lies at the land's edge. The continental shelf, which begins where the intertidal zone ends, is a plain that slopes gently seaward. The neritic zone (continental slope) begins at a depth of about 600 feet (180 meters), where the gradual slant of the continental shelf becomes a sharp tilt toward the ocean floor, plunging about 12,000 feet (3,660 meters) to the ocean bottom, which is known as the abyss. The abyssal zone is so deep that it does not have light.

Plankton are animals that float in the ocean. They include algae and copepods, which are microscopic crustaceans. Jellyfish and animal larva are also considered plankton. The nekton are animals that move freely through the water by means of their muscles. These include fish, whales, and squid. The benthos are animals that are attached to or crawl along the ocean's floor. Clams are examples of benthos. Bacteria decompose the dead organic materials on the ocean floor.

The circulation of materials from the ocean's floor to the surface is caused by winds and water temperature. Runoff from the land contains polluting chemicals such as pesticides, nitrogen fertilizers, and animal wastes. Rivers carry loose soil to the ocean, where it builds up the bottom areas. Overfishing has caused fisheries to collapse in every world sector. In some parts of the northwestern Altantic Ocean, there has been a shift from bony fish to cartilaginous fish dominating the fisheries.

Human Impact on Biomes

Human interaction with biomes has increased biotic invasions, reduced the numbers of species, changed the quality of land and water resources, and caused the proliferation of toxic compounds. Managed care of biomes may not be capable of undoing these problems.

Ronald J. Raven

NATURAL RESOURCES

SOILS

Soils are the loose masses of broken and chemically weathered rock mixed with organic matter that cover much of the world's land surface, except in polar regions and most deserts. The two major solid components of soil—minerals and organic matter—occupy about half the volume of a soil. Pore spaces filled with air and water account for the other half. A soil's organic material comes from the remains of dead plants and animals, its minerals from weathered fragments of bedrock. Soil is also an active, dynamic, ever-changing environment. Tiny pores in soil fill with air, water, bacteria, algae, and fungi working to alter the soil's chemistry and speed up the decay of organic material, making the soil a better living environment for larger plants and animals.

Soil Formation

The natural process of forming new soil is slow. Exactly how long it takes depends on how fast the bedrock below is weathered. This weathering process is a direct result of a region's climate and topography, because these factors influence the rate at which exposed bedrock erodes and vegetation is distributed. Global variations in these factors account for the worldwide differences in soil types.

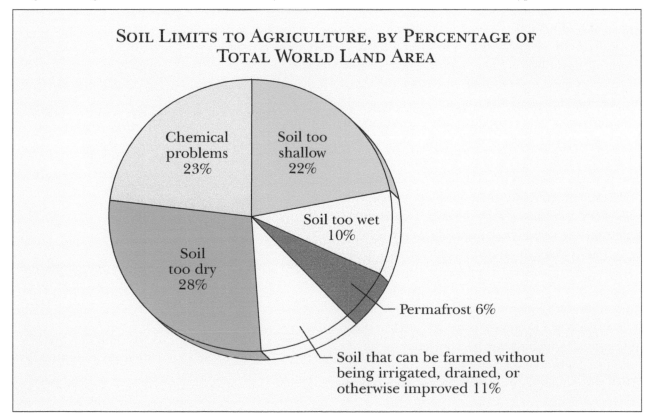

SOIL LIMITS TO AGRICULTURE, BY PERCENTAGE OF TOTAL WORLD LAND AREA

Chemical problems 23%

Soil too shallow 22%

Soil too wet 10%

Soil too dry 28%

Permafrost 6%

Soil that can be farmed without being irrigated, drained, or otherwise improved 11%

Climate is the principal factor in determining the type and rate of soil formation. Temperature and precipitation are the two main climatic factors that influence soil formation, and they vary with elevation and latitude. Water is the main agent of weathering, and the amount of water available depends on how much falls and how much runs off. The amount of precipitation and its distribution during the year influence the kind of soil formed and the rate at which it is formed. Increased precipitation usually results in increased rates of soil formation and deep soils. Temperature and precipitation also determine the kind and amount of vegetation in a region, which determines the amount of available organics.

Topography is a characteristic of the landscape involving slope angle and slope length. Topographic relief governs the amount of water that runs off or enters a soil. On flat or gently sloping land, soil tends to stay in place and may become thick, but as the slope increases so does the potential for erosion. On steep slopes, soil cover may be very thin, possibly only a few inches, because precipitation washes it downhill; on level plains, soil profiles may be several feet thick.

Types of Soil

Typically, bedrock first weathers to form regolith, a protosoil devoid of organic material. Rain, wind, snow, roots growing into cracks, freezing and thawing, uneven heating, abrasion, and shrinking and swelling break large rock particles into smaller ones. Weathered rock particles may range in size from clay to silt, sand, and gravel, with the texture and particle size depending largely on the type of bedrock. For example, shale yields finer-textured soils than sandstone. Soils formed from eroded limestone are rich in base minerals; others tend to be acidic. Generally, rates of soil formation are largely determined by the rates at which silicate minerals in the bedrock weather: the more silicates, the longer the formation time.

In regions where organic materials, such as plant and animal remains, may be deposited on top of regolith, rudimentary soils can begin to form. When waste material is excreted, or a plant or animal dies, the material usually ends up on the earth's surface. Organisms that cause decomposition, such as bacteria and fungi, begin breaking down the remains into a beneficial substance known as humus. Humus restores minerals and nutrients to the soil. It also improves the soil's structure, helping it to retain water. Over time, a skeletal soil of coarse, sandy material with trace amounts of organics gradually forms. Even in a region with good weathering rates and adequate organic material, it can take as long as fifty years to form 12 inches (30 centimeters) of soil. When new soil is formed from weathering bedrock, it can take from 100 to 1,000 years for less than an inch of soil to accumulate.

Water moves continually through most soils, transporting minerals and organics downward by a process called leaching. As these materials travel downward, they are filtered and deposited to form distinct soil horizons. Each soil horizon has its own color, texture, and mineral and humus content. The O-horizon is a thin layer of rotting organics covering the soil. The A-horizon, commonly called topsoil, is rich in humus and minerals. The B-horizon is a subsoil rich in minerals but poor in humus. The C-horizon consists of weathered bedrock; the D-horizon is the bedrock itself.

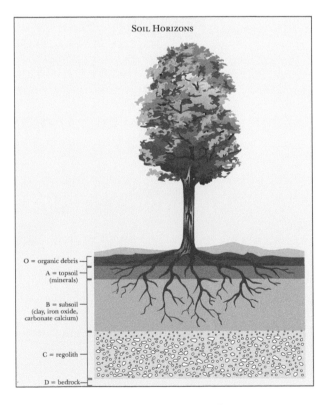

SOIL HORIZONS

O = organic debris
A = topsoil (minerals)
B = subsoil (clay, iron oxide, carbonate calcium)
C = regolith
D = bedrock

Because Earth's surface is made of many different rock types exposed at differing amounts and weathering at different rates at different locations, and because the availability of organic matter varies greatly around the planet due to climatic and seasonal conditions, soil is very diverse and fertile soil is unevenly distributed. Structure and composition are key factors in determining soil fertility. In a fertile soil, plant roots are able to penetrate easily to obtain water and dissolved nutrients. A loam is a naturally fertile soil, consisting of masses of particles from clays (less than 0.002 mm across), through silts (ten times larger) to sands (100 times larger), interspersed with pores, cracks, and crevices.

The Roles of Soil

In any ecosystem, soils play six key roles. First, soil serves as a medium for plant growth by mechanically supporting plant roots and supplying the eighteen nutrients essential for plants to survive. Different types of soil contain differing amounts of these eighteen nutrients; their combination often determines the types of vegetation present in a region, and as a result, influences the number and types of animals the vegetation can support, including humans. Humans rely on soil for crops necessary for food and fiber.

Second, the property of a particular soil is the controlling factor in how the hydrologic system in a region retains and transports water, how contaminants are stored or flushed, and at what rate water is naturally purified. Water enters the soil in the form of precipitation, irrigation, or snowmelt that falls or runs off soil. When it reaches the soil, it will either be surface water, which evaporates or runs into streams, or subsurface water, which soaks into the soil where it is either taken up by plant roots or percolates downward to enter the groundwater system. Passing through soil, organic and inorganic pollutants are filtered out, producing pure groundwater.

Soil also functions as an air-storage facility. Air is pushed into and drawn out of the soil by changes in barometric pressure, high winds, percolating water, and diffusion. Pore spaces within soil provide access to oxygen to organisms living underground as well as to plant roots. Soil pore spaces also contain carbon dioxide, which many bacteria use as a source of carbon.

Soil is nature's recycling system, through which organic waste products and decaying plants and animals are assimilated and their elements made available for reuse. The production and assimilation of humus within soil converts mineral nutrients into forms that can be used by plants and animals, who return carbon to the atmosphere as carbon dioxide. While dead organic matter amounts to only about 1 percent of the soil by weight, it is a vital component as a source of minerals.

Soil provides a habitat for many living things, from insects to burrowing animals, from single microscopic organisms to massive colonies of subterranean fungi. Soils contain much of the earth's genetic diversity, and a handful of soil may contain billions of organisms, belonging to thousands of species. Although living organisms only account for about 0.1 percent of soil by weight, 2.5 acres (one hectare) of good-quality soil can contain at least 300 million small invertebrates—mites, millipedes, insects, and worms. Just 1 ounce (30 grams) of fertile soil can contain 1 million bacteria of a single type, 100 million yeast cells, and 50,000 fungus mycelium. Without these, soil could not convert nitrogen, phosphorus, and sulphur to forms available to plants.

Finally, soil is an important factor in human culture and civilization. Soil is a building material used to make bricks, adobe, plaster, and pottery, and often provides the foundation for roads and buildings. Most important, soil resources are the basis for agriculture, providing people with their dietary needs.

Because the human use of soils has been haphazard and unchecked for millennia, soil resources in many parts of the world have been harmed severely. Human activities, such as overcultivation, inexpert irrigation, overgrazing of livestock, elimination of tree cover, and cultivating steep slopes, have caused natural erosion rates to increase many times over. As a result of mismanaged farm and forest lands, escalated erosional processes wash off or blow away an estimated 75 billion tons of soil annually, eroding away one of civilization's crucial resources.

Randall L. Milstein

WATER

Life on Earth requires water—without it, life on Earth would cease. As human populations grow, the freshwater resources of the world become scarcer and more polluted, while the need for clean water increases. Although nearly three-quarters of Earth's surface is covered with water, only about 0.3 percent of that water is freshwater suitable for consumption and irrigation. This is because more than 97 percent of Earth's water is ocean salt water, and most of the remaining freshwater is frozen in the Antarctic ice cap. Only the small amounts that remain in lakes, rivers, and groundwater is available for human use.

All of earth's water cycles between the ocean, land, atmosphere, plants, and animals over and over. On average, a molecule of surface water cycles from the ocean, to the atmosphere, to the land and back again in less than two weeks. Water consumed by plants or animals takes longer to return to the oceans, but eventually the cycle is completed.

Water's Uses

Water supports the lives of all living creatures. People use for drinking, cooking, cleaning, and bathing. Water also plays a key role in society since humans can travel on it, make electricity with it, fish in it, irrigate crops with it, and use it for recreation. Globally, more than 4 trillion cubic meters of freshwater is used each day. Agriculture accounts for about 70 percent, industry uses 20 percent, and domestic and municipal activities use 10 percent. To produce beef requires between 1,320 and 5,283 gallons (5,000 and 20,000 liters) of water for every 35 ounces (1 kg). A similar amount of wheat requires between 660 and 1056 gallons (2,500 and 4,000 liters) of water. Manufactured goods also require significant amounts of freshwater; a car consumes between 13,000 and 20,000 gallons (49,210 to 75,708 liters), a smartphone has a water footprint of nearly 3,100 gallons (11,734 liters) and a teeshirt uses around 660 gallons (2,498 liters).

The average American family uses more than 300 gallons of drinking quality water per day at home. Roughly 70 percent of this use occurs indoors for drinking, bathing and showering, flushing the toilet, and washing dishes. Outdoor water use for landscape watering, washing cars, and cleaning windows, etc., accounts for 30 percent of household use, although it can be much higher in drier parts of the country. For example, the arid West has some of the highest per capita residential water use because of landscape irrigation.

As the world's population grows, the demand for fresh water will also increase. A study by the World Bank concluded that approximately 80 percent of human illness results from insufficient water supplies and poor water quality caused by lack of sanitation, so careful management of water resources is essential for improving the health of people in the twenty-first century.

Groundwater Supply and Quality

The amount of groundwater in the Earth is seventy times greater than all of the freshwater lakes combined. Groundwater is held within the rocks below the ground surface and is the primary source of water in many parts of the world. In the United States, approximately 50 percent of the population uses some groundwater. However, problems with both groundwater supplies and its quality threaten its future use.

The U.S. Environmental Protection Agency (EPA) found that 45 percent of the large public water systems in the United States that use groundwater were contaminated with synthetic organic chemicals that posed potential health threats. Another major problem occurs when groundwater is used faster than it is replaced by precipitation infiltrating through the ground surface. Many of the arid regions of earth are already suffering from this problem. For example, one-third of the wells in Beijing, China, have gone dry due to overuse. In the United States, the Ogallala Aquifer of the Great Plains, the

The World and North America's Greatest Rivers and Lakes

Longest river	
Nile (North Africa)	4,130 miles (6,600 km.)
Missouri-Mississippi (United States)	3,740 miles ((6,000 km.)
Largest river by average discharge	
Amazon (South America)	6,181,000 cubic feet/second (175,000 cubic meters/second)
Missouri-Mississippi (United States)	600,440 cubic feet/second (17,000 cubic meters/second)
Largest freshwater lake by volume	
Lake Baikal (Russia)	5,280 cubic miles (22,000 cubic km.)
Lake Superior (United States)	3,000 cubic miles (12,500 cubic km.)

largest in North America, is being severely overused. This aquifer irrigates 30 percent of U.S. farmland, but some areas of the aquifer have declined by up to 60 percent. In one part of Texas, so much water has been pumped out that the aquifer has essentially dried up there. Once depleted, the aquifer will take over 6,000 years to replenish naturally through rainfall.

Surface Water Supply and Quality

Surface water is used for transportation, recreation, electrical generation, and consumption. Ships use rivers and lakes as transport routes, people fish and boat on rivers and lakes, and dams on rivers often are used to generate electricity. The largest river on earth is the Amazon in South America, which has an average flow of 212,500 cubic meters per second, more than twelve times greater than North America's Mississippi River. Earth's largest lake by volume—Lake Baikal in Russia—holds 5,521 cubic miles of water (23,013 cubic kilometers), or approximately 20 percent of Earth's fresh surface water. This is a volume of water approximately equivalent to all five of the North American Great Lakes combined.

Although surface water has more uses, it is more prone to pollution than groundwater. Almost every human activity affects surface water quality. For ex-

ample, water is used to create paper for books, and some of the chemicals used in the paper process are discharged into surface water sources. Most foods are grown with agricultural chemicals, which can contaminate water sources. According to 2018 surveys on national water quality from the U.S. Environmental Protection Agency, nearly half of U.S. rivers and streams and more than one-third of lakes are polluted and unfit for swimming, fishing, and drinking.

Earth's Future Water Supply

Inadequate water supplies and water quality problems threaten the lives of more than 1.5 million people worldwide. The World Health Organization estimates that polluted water causes the death of 361,000 children under five years of age each year and affects the health of 20 percent of Earth's population. As the world's population grows, these problems are likely to worsen.

The United Nations estimates that if current consumption patterns continue, 52 percent of the world's people will live in water-stressed conditions by 2050. Since access to clean freshwater is essential to health and a decent standard of living, efforts must be made to clean up and conserve the planet's freshwater.

Mark M. Van Steeter

EXPLORATION AND TRANSPORTATION

EXPLORATION AND HISTORICAL TRADE ROUTES

The world's exploration was shaped and influenced substantially by economic needs. Lacking certain resources and outlets for trade, many societies built ships, organized caravans, and conducted military expeditions to protect their frontiers and obtain new markets.

Over the last 5,000 years, the world evolved from a cluster of isolated communities into a firmly integrated global community and capitalist world system. By the beginning of the twentieth century, explorers had successfully navigated the oceans, seas, and landmasses and gathered many regional economies into the beginnings of a global economy.

Early Trade Systems

Trade and exploration accompanied the rise of civilization in the Middle East. Egyptian pharaohs, looking for timber for shipbuilding, established trade relations with Mediterranean merchants. Phoenicians probed for new markets off the coast of North Africa and built a permanent settlement at Carthage. By 513 BCE, the Persian Empire stretched from the Indus River in India to the Libyan coast, and it controlled the pivotal trade routes in Iran and Anatolia. A regional economy was taking shape, linking Africa, Asia, and Europe into a blended economic system.

Alexander the Great's victory against the Persian Empire in 330 BCE thrust Greece into a dominant position in the Middle Eastern economy. Trade between the Mediterranean and the Middle East increased, new roads and harbors were constructed, and merchants expanded into sub-Saharan Africa, Arabia, and India. The Romans later benefited from the Greek foundation. Through military and political conquest, Rome consolidated its control over such diverse areas as Arabia and Britain and built a system of roads and highways that facilitated the growth of an expanding world economy. At the apex of Roman power in 200 CE, trade routes provided the empire with Greek marble, Egyptian cloth, seafood from Black Sea fisheries, African slaves, and Chinese silk.

The emergence of a profitable Eurasian trade route linked people, customs, and economies from the South China Sea to the Roman Empire. Although some limited activity occurred during the Hellenistic period, East-West trade flourished following the rise of the Han Dynasty in China. With the opening of the Great Silk Road from 139 BCE to 200 CE, goods and services were exchanged between people from three different continents.

The Great Silk Road was an intricate network of middlemen stretching from China to the Mediterranean Sea. Eastern merchants sold their products at markets in Afghanistan, Iran, and even Syria, and exchanged a variety of commodities through the use of camel caravans. Chinese spices, perfumes, metals, and especially silk were in high demand. The Parthians from central Asia added their own sprinkling of merchandise, introducing both the East and the West to various exotic fruits, rare birds, and ostrich eggs.

Romans peddled glassware, statuettes, and acrobatic performing slaves. Since communication lines were virtually nonexistent during this period, trade routes were the only means by which ideas regarding art, religion, and culture could mix. The contacts and exchanges enacted along the Great Silk Road initiated a process of cultural diffusion among a diversity of cultures and increased each culture's knowledge of the vast frontiers of world geography.

The Atlantic Slave Trade

Beginning in the fifteenth century, European navigators explored the West African coastline seeking gold. Supplies were difficult to procure, because most of the gold mines were located in the interior along the Senegal River and in the Ashanti forests. Because mining required costly investments in time, labor, and security, the Europeans quickly shifted their focus toward the slave trade. Although slavery had existed since antiquity, the Atlantic slave trade generated one of the most significant movements of people in world history. It led to the forced migration of more than 10 million Africans to South America, the Caribbean islands, and North America. It ensured the success of several imperial conquests, and it transformed the demographic, cultural, and political landscape on four continents.

Originally driven by their quest to circumnavigate Africa and open a lucrative trade route with India, the Portuguese initiated a systematic exploration of the West African coastline. The architect of this system, Henry the Navigator, pioneered the use of military force and naval superiority to annex African islands and open up new trade routes, and he increased Portugal's southern frontier with every acquisition. In 1415 his ships captured Ceuta, a prosperous trade center located on the Mediterranean coast overlooking North African trade routes. Over the next four decades, Henry laid claim to the Madeira Islands, the Canary Islands, the Azores, and Cape Verde. After his death, other Portuguese explorers continued his pursuit of circumnavigation of Africa.

Diego Cão reached the Congo River in 1483 and sent several excursions up the river before returning to Lisbon. Two explorers completed the Portuguese mission at the end of the fifteenth century. Vasco da Gama, sailing from 1497 to 1499, and Bartholomeu Dias, from 1498 to 1499, sailed past the southern tip of Africa and eventually reached India. Since Muslims had already created a number of trade links between East Africa, Arabia, and India, Portuguese exploration furthered the integration of various regions into an emerging capitalist world system.

When the Portuguese shifted their trading from gold to slaves, the other European powers followed suit. The Netherlands, Spain, France, and England used their expanding naval technology to explore the Atlantic Ocean and ship millions of slaves across the ocean. A highly efficient and organized trade route quickly materialized. Since the Europeans were unwilling to venture beyond the walls of their coastal fortresses, merchants relied on African sources for slaves, supplying local kings and chiefs with the means to conduct profitable slave-raiding parties in the interior. In both the Congo and the Gold Coast region, many Africans became quite wealthy trading slaves.

In 1750 merchants paid King Tegbessou of Dahomey 250,000 pounds for 9,000 slaves, and his income exceeded the earnings of many in England's merchant and landowning class. After purchasing slaves, dealers sold them in the Americas to work in the mines or on plantations. Commodities such as coffee and sugar were exported back to Europe for home consumption. Merchants then sold alcohol, tobacco, textiles, and firearms to Africans in exchange for more slaves. This practice was abolished by the end of the nineteenth century, but not before more than 10 million Africans had been violently removed from their homeland. The Atlantic slave trade, however, joined port cities from the Gold Coast and Guinea in Africa with Rio de Janeiro, Hispaniola, Havana, Virginia, Charleston, and Liverpool, and constituted a pivotal step toward the rise of a unified global economy.

Magellan and Zheng He

The Portuguese explorer Ferdinand Magellan generated considerable interest in the Asian markets when he led an expedition that sailed around the world from 1519 to 1522. Looking for a quick route to Asia and the Spice Islands, he secured financial backing from the king of Spain. Magellan sailed from Spain in 1519, canvassed the eastern coastline of South America, and visited Argentina. He ultimately traversed the narrow straits along the southern tip of the continent and ventured into the uncharted waters of the Pacific Ocean.

Magellan explored the islands of Guam and the Philippines but was killed in a skirmish on Mactan in 1521. Some of his crew managed to return to Spain in 1522, and one member subsequently published a journal of the expedition that dramatically enhanced the world's understanding of the major sea lanes that connected the continents.

China also opened up new avenues of trade and exploration in Southeast Asia during the fifteenth century. Under the direction of Chinese emperor Yongle, explorer Zheng He organized seven overseas trips from 1405 to 1433 and investigated economic opportunities in Korea, Vietnam, the Indian Ocean, and Egypt. His first voyage consisted of more than 28,000 men and 400 ships and represented the largest naval force assembled prior to World War I.

Zheng's armada carried porcelains, silks, lacquerware, and artifacts to Malacca, the vital port city in Indonesia. He purchased an Arab medical text on drug therapy and had it translated into Chinese. He introduced giraffes and mahogany wood into the mainland's economy, and his efforts helped spread Chinese ideas, customs, diet, calendars, scales and measures, and music throughout the global economy. Zheng He's discoveries, coupled with all the material gathered by the European explorers, provided cartographers and geographers with a credible store of knowledge concerning world geography.

Emerging Global Trade Networks

From 1400 to 1900, several regional economic systems facilitated the exchange of goods and services throughout a growing world system. Building on the triangular relationships produced by the slave trade, the Atlantic region helped spread new food-

stuffs around the globe. Plants and plantation crops provided societies with a plentiful supply of sweet potatoes, squash, beans, and maize. This system, often referred to as the Columbian exchange, also assisted development in other regions by supplying the global economy with an ample money supply in gold and silver. Europeans sent textiles and other manufactures to the Americas. In return, they received minerals from Mexico; sugar and molasses from the Caribbean; money, rum, and tobacco from North America; and foodstuffs from South America. Trade routes also closed the distance between the Pacific coastline in the Americas and the Pacific Rim.

Additional thriving trade routes existed in the African-West Asian region. Linking Europe and Africa with Arabia and India, this area experienced a considerable amount of trade over land and through the sea lanes in the Persian Gulf and Red Sea. Europeans received grains, timber, furs, iron, and hemp from Russia in exchange for wool textiles and silver. Central Asians secured stores of cotton textiles, silk, wheat, rice, and tobacco from India and sold silver, horses, camel, and sheep to the Indians. Ivory, blankets, paper, saltpeter, fruits, dates, incense, coffee, and wine were regularly exchanged among merchants situated along the trade route connecting India, Persia, the Ottoman Empire, and Europe.

Finally, a Russian-Asian-Chinese market provided Russia's ruling czars with arms, sugar, tobacco, and grain, and a sufficient supply of drugs, medicines, livestock, paper money, and silver moved eastward. Overall, this system linked the economies of three continents and guaranteed that a nation could acquire essential foodstuffs, resources, and money from a variety of sources.

Several profitable trade routes existed in the Indian Ocean sector. After Malacca emerged as a key trading port in the sixteenth century, this territory served as an international clearinghouse for the global economy. Indians sent tin, elephants, and wood into Burma and Siam. Rice, silk, and sugar were sold to Bengal. Pepper and other spices were shipped westward across the Arabian Sea, while Ceylon furnished India with vast quantities of jewels, cinnamon, pearls, and elephants. The booming interregional trade routes positioned along the Indian coastline ensured that many of the vast commodities produced in the world system could be obtained in India.

The final region of crucial trade routes was between Southeast Asia and China. While the extent of Asian overseas trade prior to the twentieth century is usually downplayed, an abundance of products flowed across the Bay of Bengal and the South China Sea. Japan procured silver, copper, iron, swords, and sulphur from Cantonese merchants,

and Japanese-finished textiles, dyes, tea, lead, and manufactures were in high demand on the mainland. The Chinese also purchased silk and ceramics from the Philippines in exchange for silver. Burma and Siam traded pepper, sappan wood, tin, lead, and saltpeter to China for satin, velvet, thread, and labor. As goods increasingly moved from the Malabar coast in India to the northern boundaries of Korea and Japan, the Pacific Rim played a prominent role in the global economy.

Robert D. Ubriaco, Jr.

ROAD TRANSPORTATION

Roads—the most common surfaces on which people and vehicles move—are a key part of human and economic geography. Transportation activities form part of a nation's economic product: They strengthen regional economy, influence land and natural resource use, facilitate communication and commerce, expand choices, support industry, aid agriculture, and increase human mobility. The need for roads closely correlates with the relative location of centers of population, commerce, industry, and other transportation.

History of Road Making
The great highway systems of modern civilization have their origin in the remote past. The earliest travel was by foot on paths and trails. Later, pack animals and crude sleds were used. The development of the wheel opened new options. As various ancient civilizations reached a higher level, many of them realized the importance of improved roads.

The most advanced highway system of the ancient world was that of the Romans. When Roman civilization was at its peak, a great system of military roads reached to the limits of the empire. The typical Roman road was bold in conception and construction, built in a straight line when possible, with a deep multilayer foundation, perfect for wheeled vehicles.

After the decline of the Roman Empire, rural road building in Europe practically ceased, and roads fell into centuries of disrepair. Commerce traveled by water or on pack trains that could negotiate the badly maintained roads. Eventually, a commercial revival set in, and roads and wheeled vehicles increased.

Interest in the art of road building was revived in Europe in the late eighteenth century. P. Trésaguet, a noted French engineer, developed a new method of lightweight road building. The regime of French dictator Napoleon Bonaparte (1800–1814) encouraged road construction, chiefly for military purposes. At about the same time, two Scottish engineers, Thomas Telford and John McAdam, also developed road-building techniques.

Roads in the United States
Toward the end of the eighteenth century, public demand in the United States led to the improvement of some roads by private enterprise. These improvements generally took the form of toll roads,

called "turnpikes" because a pike was rotated in each road to allow entry after the fee was paid, and generally were located in areas adjacent to larger cities. In the early nineteenth century, the federal government paid for an 800-mile-long macadam road from Cumberland, Maryland, to Vandalia, Illinois.

With the development of railroads, interest in road building began to wane. By 1900, however, demand for better roads came from farmers, who wanted to move their agricultural products to market more easily. The bicycle craze of the 1890s and the advent of motorized vehicles also added to the demand for more and better roads. Asphalt and concrete technology was well developed by then; now, the problem was financing. Roads had been primarily a local issue, but the growing demand led to greater state and federal involvement in funding.

The Federal-Aid Highway Act of 1956 was a milestone in the development of highway transportation in the United States; it marked the beginning of the largest peacetime public works program in the history of the world, creating a 41,000-mile National System of Interstate and Defense Highways, built to high standards. Later legislation expanded funding, improved planning, addressed environmental concerns, and provided for more balanced transportation. Other developed countries also developed highway programs but were more restrained in construction.

Roads and Development

Transportation presents a severe challenge for sustainable development. The number of motor vehicles at the end of the second decade of the twenty-first century—estimated at more than 1.2 billion worldwide—is growing almost everywhere at higher rates than either population or the gross domestic product. Overall road traffic grows even more quickly. The tiny nation of San Marino has nearly 1.2 cars per person. Americans own one car for every 1.88 residents. In Great Britain, there is one car for every 5.3 people.

Highways around the world have been built to help strengthen national unity. The Trans-Canada Highway, the world's longest national road, for ex-

---

HIGHWAY CLASSIFICATION

Modern roads can be classified by roadway design or traffic function. The basic type of roadway is the conventional, undivided two-way road. Divided highways have median strips or other physical barriers separating the lanes going in opposite directions.

Another quality of a roadway is its right-of-way control. The least expensive type of system controls most side access and some minor at-grade intersections; the more expensive type has side access fully controlled and no at-grade intersections. The amount of traffic determines the number of lanes. Two or three lanes in each direction is typical, but some roads in Los Angeles have five lanes, while some sections of the Trans-Canada Highway have only one lane. Some highways are paid for entirely from public funds; if users pay directly when they use the road, they are called tollways or turnpikes.

Roads are classified as expressway, arterial, collector, and local in urban areas, with a similar hierarchy in rural areas. The highest level—expressway—is intended for long-distance travel.

---

ample, extends east-west across the breadth of the country. Completed in the 1960s, it had the same goal as the Canadian Pacific Railroad a century before, to improve east-west commerce within Canada.

Sometimes, existing highways need to be upgraded; in less-developed countries, this can simply mean paving a road for all-weather operation. An example of a late-1990s project of this nature was the Brazil-Venezuela Highway project, which had this description: Improve the Brazil-Venezuela highway link by completion of paving along the BR-174, which runs northward from Manaus in the Amazon, through Boa Vista and up to the frontier, so opening a route to the Caribbean. Besides the investment opportunities in building the road itself, the highway would result in investment opportunities in mining, tourism, telecommunications, soy and rice production, trade with Venezuela, manufacturing in the Manaus Free Trade Zone, ecotourism in the Amazon, and energy integration.

Growing road traffic has required increasingly significant national contributions to road construction. Beginning in the 1960s, the World Bank began to finance road construction in several countries. It required that projects be organized to the highest technical and economic standards, with private contracting and international competitive bidding rather than government workers. Still, there were questions as to whether these economic assessments had a road-sector bias and properly incorporated environmental costs. Sustainability was also a question—could the facilities be maintained once they were built?

In the 1990s, the World Bank financed a program to build an asphalt road network in Mozambique. Asphalt makes very smooth roads but is very maintenance-intensive, requiring expensive imported equipment and raw materials. By the end of the decade, the roads required resurfacing but the debt was still outstanding. Alternative materials would have given a rougher road, but it could have been built with local materials and labor.

The European Investment Bank has become a major player in the construction of highways linking Eastern and Western Europe to further European integration. Some of the fastest growth in the world in ownership of autos has been in Eastern Europe. There is a two-way feedback effect between highway construction and auto ownership.

Environment Consequences

Highways and highway vehicles have social, economic, and environmental consequences. Compromise is often necessary to balance transportation needs against these constraints. For example, in Israel, there has been a debate over construction of the Trans-Israel highway, a US$1.3 billon, six-lane highway stretching 180 miles (300 km.) from Galilee to the Negev.

Demand on resources for worldwide road infrastructure far exceeds available funds; governments increasingly are looking to external sources such as tolls. Private toll roads, common in the nineteenth century, are making a comeback. This has spread from the United States to Europe, where private and government-owned highway operators have begun to sell shares on the stock market. Private companies are not only operating and financing roads in Europe, they are also designing and building them. In Eastern Europe, where road construction languished under communism, private financing and toll collecting are seen as the means of supporting badly needed construction.

Industrial development in poor countries is adversely affected by limited transportation. Costs are high-unreliable delivery schedules make it necessary to maintain excessive inventories of raw materials and finished goods. Poor transport limits the radius of trade and makes it difficult for manufacturers to realize the economies of large-scale operations to compete internationally.

In more difficult terrain, roads become more expensive because of a need for cuts and fills, bridges, and tunnels. To save money, such roads often have steeper grades, sharper curves, and reduced width than might be desired. Severe weather changes also damage roads, further increasing maintenance costs.

Stephen B. Dobrow

RAILWAYS

Railroads were the first successful attempts by early industrial societies to develop integrated communication systems. Today, global societies are linked by Internet systems dependent upon communication satellites orbiting around Earth. The speed by which information and ideas can reach remote places breaks down isolation and aids in the developing of a world community. In the nineteenth century, railroads had a similar impact. Railroads were critical for the creation of an urban-industrial society: They linked regions and remote places together, were important contributors in developing nation-states, and revolutionized the way business was conducted through the creation of corporations. Although alternative forms of transportation exist, railroads remain important in the twenty-first century.

The Industrial Revolution and the Railroad

Development of the steam engine gave birth to the railroad. Late in the eighteenth century, James Watt perfected his steam engine in England. Water was superheated by a boiler and vaporized into steam, which was confined to a cylinder behind a piston. Pressure from expanding steam pushes the cylinder forward, causing it to do work if it is attached to wheels. Watt's engine was used in the manufacturing of textiles, thus beginning the Industrial Revolution whereby machine technology mass produced goods for mass consumption. Robert Fulton was the first innovator to commercially apply the steam engine to water transportation. His steamboat *Clermont* made its maiden voyage up the Hudson River in 1807.

Not until the 1820s was a steam engine used for land transportation. Rivers and lakes were natural features where no road needed to be built. Applying steam to land movement required some type of roadbed. In England, George Stephenson ran a locomotive over iron strips attached to wooden rails. Within a short time, England's forges were able to roll rails made completely of iron shaped like an inverted "U."

The amount of profit a manufacturer could make was determined partially by the cost of transportation. The lower the cost of moving cargo and people, the higher the profitability. Several alternatives existed before the emergence of railroads, although there were drawbacks compared to rail transportation. Toll roads were too slow. A loaded wagon pulled by four horses could average 15 miles (25 km.) a day. Canals were more efficient than early railroads, because barges pulled by mules moved faster over waterways. However, canals could not be built everywhere, especially over mountains.

The application of railroad technology, using steam as a power source, made it possible to overcome obstacles in moving goods and people over considerable distances and at profitable costs. Railroads transformed the way goods were purchased by reducing the costs for consumers, thus raising the living standards in industrial societies. Railroads transformed the human landscape by strengthening the link between farm and city, changed commercial cities into industrial centers, and started early forms of suburban growth well before automobiles arrived.

Financing Railroads

Constructing railroads was costly. Tunnels had to be blasted through mountains, and rivers had to be crossed by bridges. Early in the building of U.S. railroads, the nation's iron foundries could not meet the demands for rolled rails. Rails had to be imported from England until local forges developed more efficient technologies. Once a railway was completed, there was a constant need to maintain the right-of-way so that traffic flow would not be disrupted. Accidents were frequent, and it was an early practice to burn damaged cars because salvaging them was too expensive.

In some countries, railroads were built and operated by national governments. In the United States, railroads were privately owned; however, it was impossible for any single individual to finance and operate a rail system with miles of track. Businessmen raised money by selling stocks and bonds. Just as investors buy stocks in modern high-technology companies, investors purchased stocks and bonds in railroads.

Investing in railroads was good as long as they earned profits and returned money to their investors, but not all railroads made sufficient profits to reward their investors. Competition among railroads was heavy in the United States, and some railroads charged artificially low fares to attract as much business as they could. When ambitious investment schemes collapsed, railroads went bankrupt and were taken over by financiers.

Selling shares of common stock and bonds was made possible by creating corporations. Railroads were granted permission from state governments to organize a corporation. Every investor owned a portion of the railroad. Stockholders' interests were served by boards of directors, and all business transactions were opened for public inspection. One important factor of the corporation was that it relieved individuals from the responsibilities associated with accidents. The railroad, as a corporation, was held accountable, and any compensation for claims made against the company came out of corporate funds, not from individual pockets. This had an impact on the law profession, as law schools began specializing in legal matters relevant to railroads and interstate commerce.

The Success of Railroads

Railroads usually began by radiating outward from port cities where merchants engaged in transoceanic trade. A classic example, in the United States, is the country's first regional railroad—the Baltimore and Ohio. Construction commenced from Baltimore in 1828; by 1850, the railroad had crossed the Appalachian Mountains and was on the Ohio River at Wheeling, Virginia.

Once trunk lines were established, rail networks became more intensive as branch lines were built to link smaller cities and towns. Countries with extremely large continental dimensions developed interior articulating cities where railroads from all directions converged. Chicago and Atlanta are two such cities in the United States. Chicago was surrounded by three circular railroads (belts) whose only function was to interchange cars. Railroads from the Pacific Coast converged with lines from the Atlantic Coast as well as routes moving north from the Gulf Coast.

Mechanized farms and heavy industries developed within the network. Railroads made possible the extraction of fossil fuels and metallic ores, the necessary ingredients for industrial growth. Extension of railroads deep into Eastern Europe helped to generate massive waves of immigration into both North and South America, creating multicultural societies.

Building railroads in Africa and South Asia made it possible for Europe to increase its political control over native populations. The ultimate aim of the colonial railroad was to develop a colony's economy according to the needs of the mother country. Railroads were usually single-line routes transhipping commodities from interior centers to coastal ports for exportation. Nairobi, Kenya, began as a rail hub linking British interests in Uganda with Kenya's port city of Mombasa. Similar examples existed in Malaysia and Indonesia.

Railroads generated conflicts among colonial powers as nations attempted to acquire strategic resources. In 1904–1905 Russia and Japan fought a war in the Chinese province of Manchuria over railroad rights; Imperial Germany attempted to get around British interests in the Middle East by building a railroad linking Berlin with Baghdad to give Germany access to lucrative oil fields. India was a region of loosely connected provinces until British railroads helped establish unification. The resulting sense of national unity led to the termination of British rule in 1947 and independence for India and Pakistan.

In the United States, private railroads discontinued passenger service among cities early in the 1970s and the responsibility was assumed by the federal government (Amtrak). Most Americans riding trains do so as commuters traveling from the suburbs to jobs in the city. The U.S. has no true high-speed trains, aside from sections of Amtrak's Acela line in the Northeast Corridor, where it can reach 150 mph for only 34 miles of its 457-mile span. Passenger service remains popular in Japan and Europe. France, Germany, and Japan operate high-speed luxury trains with speeds averaging above 100 miles (160 km.) per hour.

Railroads are no longer the exclusive means of land transportation as they were early in the twentieth century. Although competition from motor vehicles and air freight provide alternate choices, railroads have remained important. France and England have direct rail linkage beneath the English Channel. In the United States, great railroad mergers and the application of computer technology have reduced operating costs while increasing profits. Transoceanic container traffic has been aided by railroads hauling trailers on flatcars. Railroads began the process of bringing regions within a nation together in the nineteenth century just as the computer and the World Wide Web began uniting nations throughout the world at the end of the twentieth century.

Sherman E. Silverman

AIR TRANSPORTATION

The movement of goods and people among places is an important field of geographic study. Transportation routes form part of an intricate global network through which commodities flow. Speed and cost determine the nature and volume of the materials transported, so air transportation has both advantages and disadvantages when compared with road, rail, or water transport.

Early Flying Machines

The transport of people and freight by air is less than a century old. Although hot-air balloons were used in the late eighteenth century for military purposes, aerial mapping, and even early photography, they were never commercially important as a means of transportation. In the late nineteenth century, the German count Ferdinand von Zeppelin began experimenting with dirigibles, which added self-propulsion to lighter-than-air craft. These aircraft were used for military purposes, such as the bombing of Paris in World War I. However, by the 1920s zeppelins had become a successful means of passenger transportation. They carried thousands of passengers on trips in Europe or across the Atlantic Ocean and also were used for exploration. Nevertheless, they had major problems and were soon superseded by flying machines heavier than air. The early term for such a machine was an aeroplane, which is still the word used for airplane in Great Britain.

Following pioneering advances with the internal combustion engine and in aerodynamic theory using gliders, the development of powered flight in a heavier-than-air machine was achieved by Wilbur and Orville Wright in December, 1903. From that time, the United States moved to the forefront of aviation, with Great Britain and Germany also making significant contributions to air transport. World War I saw the further development of aviation for military purposes, evidenced by the infamous bombing of Guernica.

Early Commercial Service

Two decades after the Wright brothers' brief flight, the world's first commercial air service began, covering the short distance from Tampa to St. Petersburg in Florida. The introduction of airmail service by the U.S. Post Office provided a new, regular source of income for commercial airlines in the United States, and from these beginnings arose the modern Boeing Company, United Airlines, and American Airlines. Europe, however, was the home of the world's first commercial airlines. These include the Deutsche Luftreederie in Germany, which connected Berlin, Leipzig, and Weimar in 1919; Farman in France, which flew from Paris to London; and KLM in the Netherlands (Amsterdam to London), followed by Qantas—the Queensland and Northern Territory Aerial Services, Limited—in Australia. The last two are the world's oldest still operating airlines.

Aircraft played a vital role in World War II, as a means of attacking enemy territory, defending territory, and transporting people and equipment. A humanitarian use of air power was the Berlin Airlift of 1948, when Western nations used airplanes to deliver food and medical supplies to the people of West Berlin, which the Soviet Union briefly blockaded on the ground.

Cargo and Passenger Service

The jet engine was developed and used for fighter aircraft during World War II by the Germans, the British, and the United States. Further research led to civil jet transport, and by the 1970s, jet planes accounted for most of the world's air transportation. Air travel in the early days was extremely expensive, but technological advances enabled longer flights with heavier loads, so commercial air travel became both faster and more economical.

Most air travel is made for business purposes. Of business trips between 750 and 1,500 miles (1,207 and 2,414 km.), air captures almost 85 percent, and of trips more than 1,500 miles (2,414 km.), 90 percent are made by air. The United States had 5,087 public airports in 2018, a slight decrease from the 5,145 public airports operating in 2014. Conversely, the number of private airports increased over this period from 13,863 to 14,549.

The biggest air cargo carriers in 2019 were Federal Express, which carried more than 15.71 billion freight tonne kilometres (FTK), Emirates Skycargo (12.27 billion FTK), and United Parcel Service (11.26 billion FTK).

The first commercial supersonic airliner, the British-French Concorde, which could fly at more than twice the speed of sound, began regular service in early 1976. However, the fleet was grounded after a Concorde crash in France in mid-2000. The first space shuttle flew in 1981. There have been 135 shuttle missions since then, ending with the successful landing of Space Shuttle Orbiter Atlantis on July 21, 2011. The shuttles have transported 600 people and 3 million pounds (1.36 million kilograms) of cargo into space.

Health Problems Transported by Air

The high speed of intercontinental air travel and the increasing numbers of air travelers have increased the risk of exotic diseases being carried into destination countries, thereby globalizing diseases previously restricted to certain parts of the world. Passengers traveling by air might be unaware that they are carrying infections or viruses. The worldwide spread of HIV/AIDS after the 1980s was accelerated by international air travel.

Disease vectors such as flies or mosquitoes can also make air journeys unnoticed inside airplanes. At some airports, both airplane interiors and passengers are subjected to spraying with insecticide upon arrival and before deplaning. The West Nile virus (West Nile encephalitis) was previously found only in Africa, Eastern Europe, and West Asia, but in the 1990s it appeared in the northeastern United States, transported there by birds, mosquitos, or people.

It was feared in the mid-1990s that the highly infectious and deadly Ebola virus, which originated in tropical Africa, might spread to Europe and the United States, by air passengers or through the importing of monkeys. The devastation of native bird communities on the island of Guam has been traced to the emergence there of a large population of brown tree snakes, whose ancestors are thought to have arrived as accidental stowaways on a military airplane in the late 1940s.

Ray Sumner

ENERGY AND ENGINEERING

ENERGY SOURCES

Energy is essential for powering the processes of modern industrial society: refining ores, manufacturing products, moving vehicles, heating buildings, and powering appliances. In 1999 energy costs were half a trillion dollars in the United States alone. All technological progress has been based on harnessing more energy and using it more effectively. Energy use has been shaped by geography and also has shaped economic and political geography.

Ancient to Modern Energy

Energy use in traditional tribal societies illustrates all aspects of energy use that apply in modern human societies. Early Stone Age peoples had only their own muscle power, fueled by meat and raw vegetable matter. Warmth for living came from tropical or subtropical climates. Then a new energy source, fire, came into use. It made cold climates livable. It enabled the cooking of roots, grains, and heavy animal bones, vastly increasing the edible food supply. Its heat also hardened wood tools, cured pottery, and eventually allowed metalworking.

Nearly as important as fire was the domestication of animals, which multiplied available muscle energy. Domestic animals carried and pulled heavy loads. Domesticated horses could move as fast as the game to be hunted or large animals to be herded.

Increased energy efficiency was as important as new energy sources in making tribal societies more successful. Cured animal hides and woven cloth were additional factors enabling people to move to cooler climates. Cooking fires also allowed drying meat into jerky to preserve it against times of limited supply. Fire-cured pottery helped protect food against pests and kept water close by. However, energy benefits had costs. Fire drives for hunting may have caused major animal extinctions. Periodic burning of areas for primitive agriculture caused erosion. Trees became scarce near the best campsites because they had been used for camp fires—the first fuel shortage.

Energy Fundamentals

Human use of energy revolves about four interrelated factors: energy sources, methods of harnessing the sources, means of transporting or storing energy, and methods of using energy. The potential energies and energy flows that might be harnessed are many times greater than present use.

The Sun is the primary source of most energy on Earth. Sunlight warms the planet. Plants use photosynthesis to transform water and carbon dioxide into the sugars that power their growth and indirectly power plant-eating and meat-eating animals. Many other energies come indirectly from the Sun. Remains of plants and animals become fossil fuels. Solar heat evaporates water, which then falls as rain, causing water flow in rivers. Regional differences in the amount of sunlight received and reflected cause temperature differences that generate winds, ocean currents, and temperature differences between different ocean layers. Food for muscle power of humans and animals is the most basic energy system.

Energy Sources

Biomass—wood or other vegetable matter that can be burned—is still the most important energy source in much of the world. Its basic use is to provide heat for cooking and warmth. Biomass fuels are often agricultural or forestry wastes. The advantage of biomass is that it is grown, so it can be replaced. However, it has several limitations. Its low energy content per unit volume and unit mass makes it unprofitable to ship, so its use is limited to the amount nearby. Collecting and processing biomass fuels costs energy, so the net energy is less. Biomass energy production may compete with food production, since both come from the soil. Finally, other fuels can be cheaper.

Greater concentration of biomass energy or more efficient use would enable it to better compete against other energy sources. For example, fermenting sugars into fuel alcohol is one means of concentrating energy, but energy losses in processing make it expensive.

Fossil fuels have more concentrated chemical energy than biomass. Underground heat and pressure compacts trees and swampy brush into the progressively more energy-concentrated peat, lignite coal, bituminous coal, and anthracite or black coal, which is mostly carbon. Industrializing regions turned to coal when they had exhausted their firewood. Like wood, coal could be stored and shoveled into the fire box as needed. Large deposits of coal are still available, but growth in the use of coal slowed by the mid-twentieth century because of two competing fossil fuels, petroleum and natural gas.

Petroleum includes gasoline, diesel fuel, and fuel oil. It forms from remains of one-celled plants and animals in the ocean that decompose from sugars into simpler hydrogen and carbon compounds (hydrocarbons). Petroleum yields more energy per unit than coal, and it is pumped rather than shoveled. These advantages mean that an oil-fired vehicle can be cheaper and have greater range than a coal-fired vehicle.

There are also hydrocarbon gases associated with petroleum and coal. The most common is the natural gas methane. Methane does not have the energy density of hydrocarbon liquids, but it burns cleanly and is a fuel of choice for end uses such as homes and businesses.

Petroleum and natural gas deposits are widely scattered throughout the world, but the greatest known deposits are in an area extending from Saudi Arabia north through the Caucasus Mountains. Deposits extend out to sea in areas such as the Persian Gulf, the North Sea, and the Gulf of Mexico. Other sources, such as oil tar sands and shale oil, are currently seen as a potentially important source of energy, but controversies surrounding the extraction, refining, and delivery processes make these energy sources a matter of significant debate and concern.

Heat engines transform the potential of chemical energies. James Watt's steam engine (1782) takes heat from burning wood or coal (external combustion), boils water to steam, and expands it through pistons to make mechanical motion. In the twentieth century, propeller-like steam turbines were developed to increase efficiency and decrease complexity. Auto and diesel engines burn fuel inside the engine (internal combustion), and the hot gases expand through pistons to make mechanical motion. Expanding them through a gas turbine is a jet engine. Heat engines can create energy from other sources, such as concentrated sunlight, nuclear fission, or nuclear fusion. The electrical generator transforms mechanical motion into electricity that can move by wire to uses far away. Such transportation (or wheeling) of electricity means that one power plant can serve many customers in different locations.

Flowing water and wind are two of the oldest sources of industrial power. The Industrial Revolution began with water power and wind power, but they could only be used in certain locations, and they were not as dependable as steam engines. In the early twentieth century, electricity made river power practical again. Large dams along river valleys with adequate water and steep enough slopes enabled areas like the Tennessee Valley to be industrial centers. In the 1970s wind power began to be used again, this time for generating electricity.

Solar energy can be tapped directly for heat or to make electricity. Although sunlight is free, it is not concentrated energy, so getting usable energy requires more equipment cost. Consequently, fossil-fueled heat is cheaper than solar heat, and power from the conventional utility grid has been much less expensive than solar-generated electricity. However, prices of solar equipment continue to drop as technologies improve.

Future Energy Sources

Possible future energy sources are nuclear fission, nuclear fusion, geothermal heat, and tides. Fission reactors contain a critical mass of radioactive heavy elements that sustains a chain reaction of atoms splitting (fissioning) into lighter elements—releasing heat to run a steam turbine. Tremendous amounts of fission energy are available, but reactor costs and safety issues have kept nuclear prices higher than that of coal.

Nuclear fusion involves the same reaction that powers the Sun: four hydrogen atoms fusing into one helium atom. However, duplicating the Sun's heat in a small area without damaging the surrounding reactor may be too expensive to allow profitable fusion reactors.

Geothermal power plants, tapping heat energy from within the earth, have operated since 1904, but widespread use depends on cheaper drilling to make them practical in more than highly volcanic areas. Tidal power is limited to the few bays that concentrate tidal energy.

Energy and Warfare

Much of ancient energy use revolved about herding animals and conducting warfare. Horse riders moved faster and hit harder than warriors on foot. The bow and arrow did not change appreciably for thousands of years. Herders on the plains rode horses and used the bow and arrow as part of tending their flocks, and the small amounts of metal needed for weapons was easily acquired. Consequently, the herders could invade and plunder much more advanced peoples. From Scythians to Parthians to Mongols, these people consistently destroyed the more advanced civilizations.

The geographical effect was that ancient civilizations generally developed only if they had physical barriers separating them from the flat plains of herding peoples. Egypt had deserts and seas. The Greeks and Romans lived on mountainous peninsulas, safe from easy attack. The Chinese built the Great Wall along their northern frontier to block invasions.

Nomadic riders dominated until the advent of an energy system of gunpowder and steel barrels began delivering lead bullets. With them, the Russians broke the power of the Tartars in Eurasia in the late fifteenth century, and various peoples from Europe conquered most of the world. Energy and industrial might became progressively more important in war with automatic weapons, high explosives, aircraft, rockets, and nuclear weapons.

By World War II, oil had become a reason for war and a crucial input for war. The Germans attempted to seize petroleum fields around Baku on the Caspian. Later in the war, major Allied attacks targeted oil fields in Romania and plants in Germany synthesizing liquid fuels. During the Arab-Israeli War of 1973, Arabs countered Western support of Israel with an oil boycott that rocked Western economies. In 1990 Iraq attempted to solve a border dispute with its oil-rich neighbor, Kuwait, by seizing all of Kuwait. An alliance, led by the United States, ejected the Iraqis.

Other wars occur over petroleum deposits that extend out to sea. European nations bordering on the North Sea negotiated a complete demarcation of economic rights throughout that body. Tensions between China and other Asian countries continue to mount over rights to the resources available in the South China Sea. Current estimates suggest that there may be 90 trillion cubic feet of natural gas and 11 billion barrels of oil in proved and probable reserves, with much more potentially undiscovered. The area is claimed by China, Vietnam, Malaysia, and the Philippines. Turkey and Greece have not resolved ownership division of Aegean waters that might have oil deposits.

Energy, Development, and Energy Efficiency

Ancient civilizations tended to grow and use locally available food and firewood. Soils and wood supplies often were depleted at the same time, which often coincided with declines in those civilizations.

The Industrial Revolution caused development to concentrate in new wooded areas where rivers suitable for power, iron ore, and coal were close together, for example, England, Silesia, and the Pittsburgh area. The iron ore of Alsace in France combined with nearby coal from the Ruhr in Germany fueled tremendous growth, not always peacefully.

By the late nineteenth century, the development of Birmingham, Alabama, demonstrated that railroads enabled a wider spread between coal deposits, iron ore deposits, and existing population centers. By the 1920s, the Soviet Union developed entirely new cities to connect with resources. By the 1970s, unit trains and ore-carrying ships transported coal from the thick coal beds in Montana and Wyoming to the United States' East Coast and to countries in Asia.

The mechanized transport of electrical distribution and distribution of natural gas in pipelines also changed settlement patterns. Trains and subway trains allowed cities to spread along rail corridors in the late nineteenth century and early twentieth century. By the 1940s, cars and trucks enabled cities such as Los Angeles and Phoenix to spread into suburbs. The trend continues with independent solar power that allows houses to be sited anywhere.

Advances in technology have allowed people to get more while using less energy. For example, early peoples stampeded herds of animals over cliffs for food, which was mostly wasted. Horseback hunting was vastly more efficient. Likewise, fireplaces in colonial North America were inefficient, sending most of their heat up the chimney. In the late eighteenth century, inventor and statesman Benjamin Franklin developed a metallic cylinder radiating heat in all directions, which saved firewood.

The ancient Greeks and others pioneered the use of passive solar energy and efficiency after they exhausted available firewood. They sited buildings to absorb as much low winter sun as possible and constructed overhanging roofs to shade buildings from the high summer sun. That siting was augmented by heavy masonry building materials that buffered the buildings from extremes of heat and cold. Later, metal pipes and glass meant that solar energy could be used for water and space heating.

The first seven decades of the twentieth century saw major declines in energy prices, and cars and appliances became less efficient. That changed abruptly with the energy crises and high prices of the 1970s. Since then, countries such as Japan, with few local energy resources, have worked to increase efficiency so they will be less sensitive to energy shocks and be able to thrive with minimal energy inputs. This trend could lead eventually to economies functioning on only solar and biomass inputs.

Solid-state electronics, use of light emitting diode (LED) or compact fluorescent lamps (CFLs) rather than incandescent bulbs, and fuel cells, which convert fuel directly into electricity more efficiently than combustion engines, all could lead to less energy use. The speed of their adoption depends on the price of competing energies. According to the U.S. Energy Information Administration's (EIA) International Energy Outlook 2019 (IEO2019), the global supply of crude oil, other liquid hydrocarbons, and biofuels is expected to be adequate to meet world demand through 2050. However, many have noted that continuing to burn fossil fuels at our current rate is not sustainable, not because reserves will disappear, but because the damage to the climate would be unacceptable.

Energy and Environment

Energy affects the environment in three major ways. First, firewood gathering in underdeveloped countries contributes to deforestation and resulting erosion. Although more efficient stoves and small solar cookers have been designed, efficiency increases are competing against population increases.

Energy production also frequently causes toxic pollutant by-products. Sulfur dioxide (from sulfur impurities in coal and oil) and nitrogen oxides (from nitrogen being formed during combustion) damage lungs and corrode the surfaces of buildings. Lead additives in gasoline make internal combustion engines run more efficiently, but they cause low-grade lead poisoning. Spent radioactive fuel from nuclear fission reactors is so poisonous that it must be guarded for centuries.

Finally, carbon dioxide from the burning of fossil fuels may be accelerating the greenhouse effect, whereby atmospheric carbon dioxide slows the planetary loss of heat. If the effect is as strong as some research suggests, global temperatures may increase several degrees on average in the twenty-first century, with unknown effects on climate and sea level.

Roger V. Carlson

ALTERNATIVE ENERGIES

The energy that lights homes and powers industry is indispensable in modern societies. This energy usually comes from mechanical energy that is converted into electrical energy by means of generators—complex machines that harness basic energy captured when such sources as coal, oil, or wood are burned under controlled conditions. This energy, in turn, provides the thermal energy used for heating, cooling, and lighting and for powering automobiles, locomotives, steamships, and airplanes. Because such natural resources as coal, oil, and wood are being used up, it is vital that these nonrenewable sources of energy be replaced by sources that are renewable and abundant. It is also desirable that alternative sources of energy be developed in order to cut down on the pollution that results from the combustion of the hydrocarbons that make the nonrenewable fuels burn.

The Sun as an Energy Source
Energy is heat. The Sun provides the heat that makes Earth habitable. As today's commonly used fuel resources are used less, solar energy will be used increasingly to provide the power that societies need in order to function and flourish.

There are two forms of solar energy: passive and active. Humankind has long employed passive solar energy, which requires no special equipment. Ancient cave dwellers soon realized that if they inhabited caves that faced the Sun, those caves would be warmer than those that faced away from the Sun. They also observed that dark surfaces retained heat and that dark rocks heated by the Sun would radiate the heat they contained after the Sun had set. Modern builders often capitalize on this same knowledge by constructing structures that face south in the Northern Hemisphere and north in the Southern Hemisphere. The windows that face the Sun are often large and unobstructed by draperies and curtains. Sunlight beats through the glass and, in passive solar houses, usually heats a dark stone or brick floor that will emit heat during the hours when there is no sunlight. Just as an automobile parked in the sunlight will become hot and retain its heat, so do passive solar buildings become hot and retain their heat.

Active solar energy is derived by placing specially designed panels so that they face the Sun. These panels, called flat plate collectors, have a flat glass top beneath which is a panel, often made of copper with a black overlay of paint, that retains heat. These panels are constructed so that heat cannot escape from them easily. When water circulated through pipes in the panels becomes hot, it is either pumped into tanks where it can be stored or circulated through a central heating system.

Some active solar devices are quite complex and best suited to industrial use. Among these is the focusing collector, a saucer-shaped mirror that centers the Sun's rays on a small area that becomes extremely hot. A power plant at Odeillo in the French Pyrenees Mountains uses such a system to concentrate the Sun's rays on a concave mirror. The mirror directs its incredible heat to an enormous, confined body of water that the heat turns to steam, which is then used to generate electricity.

Another active solar device is the solar or photovoltaic cell, which gathers heat from the Sun and turns it into energy directly. Such cells help to power spacecraft that cannot carry enough conventional fuel to sustain them through long missions in outer space.

Geothermal Heating

The earth's core is incredibly hot. Its heat extends far into the lower surfaces of the planet, at times causing eruptions in the form of geysers or volcanoes. Many places on Earth have springs that are warmed by heat from the earth's core.

In some countries, such as Iceland, warm springs are so abundant that people throughout the country bathe in them through the coldest of winters. In Iceland, geothermal energy is used to heat and light homes, making the use of fossil fuels unnecessary.

Hot areas exist beneath every acre of land on Earth. When such areas are near the surface, it is easy to use them to produce the energy that humans require. As dependence on fossil fuels decreases, means will increasingly be found of drawing on Earth's subterranean heat as a major source of energy.

Wind Power

Anyone who has watched a sailboat move effortlessly through the water has observed how the wind can be used as a source of kinetic energy—the kind of energy that involves motion—whose movement is transferred to objects that it touches. Wind power has been used throughout human history. In its more refined aspects, it has been employed to power windmills that cause turbines to rotate, providing generators with the power they require to produce electricity.

Windmills typically have from two to twenty blades made of wood or of heavy cloth such as canvas. Windmills are most effective when they are located in places where the wind regularly blows with considerable velocity. As their blades turn, they cause the shafts of turbines to rotate, thus powering generators. The electricity created is usually trans-

mitted over metal cables for immediate use or for storage.

Modern vertical-axis wind turbines have two or three strips of curved metal that are attached at both ends to a vertical pole. They can operate efficiently even if they are not turned toward the wind. These windmills are a great improvement over the old horizontal axis windmills that have been in use for many years. From 2000 to 2015, cumulative wind capacity around the world increased from 17,000 megawatts to more than 430,000 megawatts. In 2015, China also surpassed the EU in the number of installed wind turbines and continues to lead installation efforts. Production of wind electricity in 2016 accounted for 16 percent of the electricity generated by renewables.

Oceans as Energy Sources

Seventy percent of the earth's surface is covered by oceans. Their tides, which rise and fall with predictable regularity twice a day, would offer a ready source of energy once it becomes economically feasible to harness them and store the electrical energy they can provide. The most promising spots to build facilities to create electrical energy from the tides are places where the tides are regularly quite dramatic, such as Nova Scotia's Bay of Fundy, where the difference between high and low tides averages about 55 feet (17 meters).

Some tidal power stations that currently exist were created by building dams across estuaries. The sluices of these dams are opened when the tide comes in and closed after the resulting reservoir fills. The water captured in the reservoir is held for several hours until the tide is low enough to create a considerable difference between the level of the wa-

---

OCEAN ENERGY

The oceans have tremendous untapped energy flows in currents and tremendous potential energy in the temperature differences between warmer tropical surface waters and colder deep waters, known as ocean thermal energy conversion. In both cases, the insurmountable cost has been in transporting energy to users on shore.

ter in the reservoir and that outside it. Then the sluice gates are opened and, as the water rushes out at a high rate of speed, it turns turbines that generate electricity.

The world's first large-scale tidal power plant was the Rance Tidal Power Station in France, which became operational in 1966. It was the largest tidal power station in terms of output until Sihwa Lake Tidal Power Station opened in South Korea in August 2011.

Future of Renewable Energy

As pollution becomes a huge problem throughout the world, the race to find nonpolluting sources of energy is accelerating rapidly. Scientists are working on unlocking the potential of the electricity generated by microbes as a fuel source, for example. New technologies are making renewable energy sources economically practical. As supplies of fossil fuels have diminished, pressure to become less dependent on them has grown worldwide. Alternative energy sources are the wave of the future.

R. Baird Shuman

ENGINEERING PROJECTS

Human beings attempt to overcome the physical landscape by building forms and structures on the earth. Most structures are small-scale, like houses, telephone poles, and schools. Other structures are great engineering works, such as hydroelectric projects, dams, canals, tunnels, bridges, and buildings.

Hydroelectric Projects

The potential for hydroelectricity generation is greatest in rapidly flowing rivers in mountainous or hilly terrain. The moving water turns turbines that, in turn, generate electricity. Hydroelectric power projects also can be built on escarpments and fall lines, where there is tremendous untapped energy in the falling water.

Most of the potential for hydroelectricity remains untapped. Only about one-sixth of the suitable rivers and falls are used for hydroelectric power. Certain areas of the world have used more of their potential than others. The percent of potential hydropower capacity that has not been developed is 71 percent in Europe, 75 percent in North America, 79 percent in South America, 95 percent in Africa, 95 percent in the Middle East, and 82 percent in Asia-Pacific. China, Brazil, Canada, and the United States currently produce the most hydroelectric power.

In Africa, only Zambia, Zimbabwe, and Ghana produce significant hydroelectricity. The region's total generating capacity needs to increase by 6 percent per year to 2040 from the current total of 125 GW to keep pace with rising electricity demand. In Southeast Asia, countries continue to grapple with the need to build up their hydroelectric plants without causing harm to the rivers that are used to supply food, water, and transportation.

Dams

Dams serve several purposes. One purpose is the generation of hydroelectric power, as discussed above. Dams also provide flood control and irrigation. Rivers in their natural state tend to rise and fall with the seasons. This can cause serious problems for people living in downstream valleys. Flood-control dams also can be used to regulate the flow of water used for irrigation and other projects. A final reason to build dams is to reduce swampland, in order to control insects and the diseases they carry.

Famous dams are found in all regions of the world. In North America, two of the most notable dams are Hoover Dam, completed in 1936, on the Colorado River between Arizona and Nevada; and the Grand Coulee Dam, completed in 1942, on the Columbia River in Washington State.

In South America, the most famous dam is the Itaipu Dam, completed in 1983, on the Paraná River between Brazil and Paraguay. In Africa, the Aswan High Dam was completed in 1970, on the Nile River in Egypt, and the Kariba Dam was completed in 1958, on the Zambezi River between Zambia and Zimbabwe. In Asia, the Three Gorges Dam spans the Yangtze River by the town of Sandouping in Hubei province, China. The Three Gorges Dam has been the world's largest power station in terms of installed capacity (22,500 MW) since 2012.

Bridges

Bridges are built to span low-lying land between two high places. Most commonly, there is a river or other body of water in the way, but other features that might be spanned include ravines, deep valleys and trenches, and swamps. A related engineering

---

ENGINEERING WORKS AND ENVIRONMENTAL PROBLEMS

Although engineering allows humans to overcome natural obstacles, works of engineering often have unintended consequences. Many engineering projects have caused unanticipated environmental problems.

Dams, for instance, create large lakes behind them by trapping water that is released slowly. This water typically contains silt and other material that eventually would have formed soil downstream had the water been allowed to flow naturally. Instead, the silt builds up behind the dam, eventually diminishing the lake's usefulness. As an additional consequence, there is less silt available for soil-building downstream.

Canals also can cause environmental harm by diverting water from its natural course. The river from which water is diverted may dry up, negatively affecting fish, animals, and the people who live downstream.

The benefits of engineering works must be weighed against the damage they do to the environment. They may be worthwhile, but they are neither all good nor all bad: There are benefits and drawbacks in building any engineering project.

---

project is the causeway, in which land in a low-lying area is built up and a road is then constructed on it.

The longest bridge in the world is the Akashi Kaikyo in Japan near Osaka. It was built in 1998 and spans 6,529 feet (1,990 meters), connecting the island of Hōnshū to the small island of Awaji. The Storebælt Bridge in Denmark, also completed in 1998, spans 5,328 feet (1,624 meters), connecting the island of Sjaelland, on which Copenhagen is situated, with the rest of Denmark. Another bridge spanning more than 5,300 feet is the Osman Gazi Bridge in Turkey. The bridge was opened on 1 July 1, 2016, ad to become the longest bridge in Turkey and the fourth-longest suspension bridge in the world by the length of its central span. The length of the bridge is expected to be surpassed by the Çanakkale 1915 Bridge, which is currently under construction across the Dardanelles strait.

Other long bridges can be found across the Humber River in Hull, England; across the Chiang Jiang (Yangtze River) in China; in Hong Kong, Norway, Sweden, and Turkey and elsewhere in Japan.

The longest bridge in the United States is the Lake Pontchartrain Causeway, Louisiana, which spans 24 miles (38.5 km), the Verrazano-Narrows Bridge in New York City between Staten Island and Brooklyn was once the longest suspension bridge in the world. Completed in 1964, its main span measures 4,260 feet (1,298 meters).

Canals

Moving goods and people by water is generally cheaper and easier, if a bit slower, than moving them by land. Before the twentieth century, that cost savings overwhelmed the advantages of land travel—speed and versatility. Therefore, human beings have wanted to move things by water whenever possible. To do so, they had two choices: locate factories and people near water, such as rivers, lakes, and oceans, or bring water to where the factories and people are, by digging canals.

One of the most famous canals in the world is the Erie Canal, which runs from Albany to Buffalo in New York State. Built in 1825 and running a length of 363 miles (584 km.), the Erie Canal opened up the Great Lakes region of North America to development and led to the rise of New York City as one of the world's dominant cities.

Two other important canals in world history are the Panama Canal and the Suez Canal. The Panama Canal connects the Atlantic and Pacific Oceans over a length of 50.7 miles (81.6 km.) on the isthmus of Panama in Central America. Completed in 1914, the Panama Canal eliminated the long and dangerous sea journey around the tip of South America. The Suez Canal in Egypt, which runs for 100 miles (162 km.) and was completed in 1856, eliminates a similar journey around the Cape of Good Hope in South Africa.

The longest canal in the world is the Grand Canal in China, which was built in the seventh century and stretches a length of 1,085 miles (2,904 km.). It connects Tianjin, near Beijing in the north of China, with Nanjing on the Chang Jiang (Yangtze River) in Central China. The Karakum Canal runs across the Central Asian desert in Turkmenistan from the Amu Darya River westward to Ashkhabad. It was begun in the 1954, and completed in 1988 and is navigable over much of its 854-mile (1,375-km.) length. The Karakum Canal and carries 13 cubic kilometres (3.1 cu mi) of water annually from the Amu-Darya River across the Karakum Desert to irrigate the dry lands of Turkmenistan.

Many canals are found in Europe, particularly in England, France, Belgium, the Netherlands, and Germany, and in the United States and Canada, especially connecting the Great Lakes to each other and to the Ohio and Mississippi Rivers.

Tunnels

Tunnels connect two places separated by physical features that would make it extremely difficult, if not impossible, for them to be connected without cutting directly through them. Tunnels can be used in place of bridges over water bodies so that water traffic is not impeded by a bridge span. Tunnels of this type are often found in port cities, and cities with them include Montreal, Quebec; New York City; Hampton Roads, Virginia; Liverpool, England; or Rio de Janeiro, Brazil.

Tunnels are often used to go through mountains that might be too tall to climb over. Trains especially

are sensitive to changes in slope, and train tunnels are found all over the world. Less common are automobile and truck tunnels, although these are also found in many places. Train and automotive tunnels through mountains are common in the Appalachian Mountains in Pennsylvania, the Rockies in the United States and Canada, Japan, and the Alps in Italy, France, Switzerland, and Austria.

The Chunnel

Arguably the most famous—and one of the most ambitious—tunnels in the world goes by the name Chunnel. Completed in 1994, it connects Dover, England, to Calais, France, and runs 31 miles (50 km.). "Chunnel" is short for the Channel Tunnel, named for the English Channel, the body of water that it goes under. It was built as a train tunnel, but cars and trucks can be carried through it on trains. In the year 2000 plans were underway to cut a second tunnel, to carry automobiles and trucks, that would run parallel to the first Chunnel.

The Seikan Tunnel in Japan, connects the large island of Hōnshū with the northern island of Hōkkaidō. The Seikan Tunnel is nearly 2.4 miles (4 km.) longer than Europe's Chunnel; however, the undersea portion of the tunnel is not as long as that of the Chunnel.

Buildings

Historically, North America has been home to the tallest buildings in the world. Chicago has been called the birthplace of the skyscraper and was at one time home to the world's tallest building. In 1998, however, the two Petronas Towers (each 1,483 feet/452 meters tall) were completed in Kuala Lumpur, Malaysia, surpassing the height of the world's tallest building, Chicago's Sears Tower (1,450 feet/442 meters), which had been completed in 1974. In 2019, the tallest completed building in the world is the 2,717-foot (828-metre) tall Burj Khalifa in Dubai, the tallest building since 2008.

Of the twenty tallest buildings standing in the year 2019, China is home to ten (Shanghai Tower, Ping An Finance Center, Goldin Finance 117, Guangzhou CTF Finance Center, Tianjin CFT Finance Center, China Zun, Shanghai World Financial Center, International Commerce Center, Wuhan Greenland Center, Changsha); Malaysia (the Petronas towers) and the United States (One World Trade Center and Central Park Towers) boast two each; Vietnam has one (Landmark 81 in Ho Chi Minh City), as does Russia (Lakhta Center), Taiwan (Taipei 101), South Korea (Lotte World Trade Center), Saudi Aragia (Abraj Al-Bait Clock Tower in Mecca).

Timothy C. Pitts

INDUSTRY AND TRADE

MANUFACTURING

Manufacturing is the process by which value is added to materials by changing their physical form—shape, function, or composition. For example, an automobile is manufactured by piecing together thousands of different component parts, such as seats, bumpers, and tires. The component parts in unassembled form have little or no utility, but pieced together to produce a fully functional automobile, the resulting product has significant utility. The more utility something has, the greater its value. In other words, the value of the component parts increases when they are combined with the other parts to produce a useful product.

Employment in Manufacturing

On a global scale, 28 percent of the world's working population had jobs in the manufacturing sector in the third decade of the century. The rest worked in agriculture (28 percent) and services (49 percent). The importance of each of these sectors varies from country to country and from time period to time period. High-income countries have a higher percentage of their labor force employed in manufacturing than low-income countries do. For example, in the United States 19 percent of the labor force worked in manufacturing by 2019, whereas the African country of Tanzania had only 7 percent of its labor force employed in the manufacturing sector at that time.

At the end of the twentieth century, the vast majority of the U.S. labor force (74 percent) worked in services, a sector that includes jobs such as computer programmers, lawyers, and teachers. By the end of the second decade of the twenty-first century, the percentage has risen to slightly more than 79 percent. Only 1 percent worked in agriculture and mining. This employment structure is typical for a high-income country. In low-income countries, in contrast, the majority of the labor force have agricultural jobs. In Tanzania, for example, 66 percent of the labor force worked in agriculture, while services accounted for 27 percent of the jobs.

The importance of manufacturing as an employer changes over time. In 1950 manufacturing accounted for 38 percent of all jobs in the United States. The percentage of jobs accounted for by the manufacturing sector in high-income countries has decreased in the post-World War II period. The decreasing share of manufacturing jobs in high-income countries is partly attributable to the fact that many manufacturing companies have replaced people with machines on assembly lines. Because one machine can do the work of many people, manufacturing has become less labor-intensive (uses fewer people to perform a particular task) and more capital-intensive (uses machines to perform tasks formerly done by people). In the future, manufacturing in high-income countries is expected to become increasingly capital-intensive. It is not inconceivable that manufacturing's share of the U.S. labor force could fall below 10 percent over the course of the twenty-first century.

Geography of Manufacturing

Every country produces manufactured goods, but the vast bulk of manufacturing activity is concentrated geographically. Four countries—China, the United States, Japan, and Germany—produce almost 60 percent of the world's manufactured goods. The concentration of manufacturing activity in a small number of regions means that there are other regions where very little manufacturing occurs. Africa is a prime example of a region with little manufacturing.

Different countries tend to specialize in the production of different products. For example, 50 percent of the automobiles that were produced in that late 1990s were produced in three countries—Germany, Japan, and the United States. In the production of television sets, the top three countries were China, Japan, and South Korea, which together produced 48 percent of the world's television sets. It is important to note that these patterns change over time. For example, in 1960 the top three automobile-producing countries were Germany, the United Kingdom, and the United States, which together produced 76 percent of the world's automobiles.

Multinational Corporations

A multinational corporation is a corporation that is headquartered in one country but owns business facilities, for example, manufacturing plants, in other countries. Some examples of multinational corporations from the manufacturing sector include the automobile maker Ford, whose headquarters are the in the United States, the pharmaceutical company Bayer, whose headquarters are in Germany, and the candy manufacturer Nestlé, whose headquarters are in Switzerland. Since the end of World War II, multinational corporations have become increasingly important in the world economy. Most multinational corporations are headquartered in high-income countries, such as Japan, the United Kingdom, and the United States.

Companies open manufacturing plants in other countries for a variety of reasons. One of the most common reasons is that it allows them to circumvent barriers to trade that are imposed by foreign governments, especially tariffs and quotas. A tariff is an import tax that is imposed upon foreign-manufactured goods as they enter a country. A quota is a limitation imposed on the volume of a particular good that a particular country can export to another country. The net effect of tariffs and quotas is to increase the cost of imported goods for consumers.

Governments impose tariffs and quotas partly to raise revenue and partly to encourage consumers to purchase goods manufactured in their own country. Foreign manufacturers faced with tariffs and quotas often begin manufacturing their product in the country imposing the tariffs and quotas. As tariffs and quotas apply to imported goods only, producing in the country imposing the quotas or tariffs effectively makes these trade barriers obsolete.

Companies also open manufacturing plants in other countries because of differences in labor costs among countries. While most manufacturing takes place in high-income countries, some low-income countries have become increasingly attractive as production locations because their workers can be hired much more cheaply than in high-income countries. For example, in late 2019, the average manufacturing job in the United States paid more than US$22.50 per hour. By comparison, manufacturing employees in the Philippines earned a few cents more than US$2.50 per hour.

This dramatic differences in labor costs have prompted some companies to close down their manufacturing plants in high-income countries and open up new plants in low-income countries. This has resulted in high-income countries purchasing more manufactured goods from low-income countries.

More than half the clothing imported into the United States came from Asian countries, for example, China, Taiwan, and South Korea, where labor costs were much lower than in the United States. Much of this clothing was made in factories where workers were paid by companies headquartered in the United States. For example, most of the Nike sports shoes that were sold in the United States were made in China, Indonesia, Vietnam, and Pakistan.

Transportation and Communications Technology

The ability of companies to have manufacturing plants in other countries stems from the fact that the world has a sophisticated and efficient transportation and communications system. An advanced transportation and communications system makes it relatively easy and relatively cheap to transfer information and goods between geographically distant locations. Thus, Nike can manufacture soccer balls in Pakistan and transport them quickly and cheaply to customers in the United States.

The extent to which transportation and communications systems have improved during the last two centuries can be illustrated by a few simple examples. In 1800, when the stagecoach was the primary method of overland transportation, it took twenty hours to travel the ninety miles from Lansing, Michigan, to Detroit, Michigan. Today, with the automobile, the same journey takes approximately ninety minutes. In 1800 sailing ships traveling at an average speed of ten miles per hour were used to transport people and goods between geographically distant countries. In the year 2019 jet-engine aircraft could traverse the globe at speeds in excess of 600 miles per hour. Communications technology has also improved over time.

In 1930 a three-minute telephone call between New York and London, England, cost more than US$385 in 2018 dollars. In the year 2019 the same telephone call could be made for less than a dime.

In addition to modern telephones, there are fax machines, email, videoconferencing capabilities, and a host of other technologies that make communication with other parts of the world both inexpensive and swift.

Future Prospects
The global economy of the twenty-first century presents a wide variety of opportunities and challenges. Sophisticated communications and transportation networks provide increasing numbers of manufacturing companies with more choices as to where to locate their factories. However, high-income countries like the United States are increasingly in competition with other countries (both high- and low-income) to maintain existing and manufacturing investments and attract new ones. Persuading existing companies to keep their U.S. factories open and not move overseas has been a major challenge. Likewise, making the United States as an attractive place for foreign companies to locate their manufacturing plants is an equally challenging task.

Neil Reid

GLOBALIZATION OF MANUFACTURING AND TRADE

Why are most of the patents issued worldwide assigned to Asian corporations? How did a Taiwanese earthquake prevent millions of Americans from purchasing memory upgrades for their computers? Why have personal incomes in Beijing nearly doubled in less than a decade?

Answers to these questions can be found in the geography of globalization. Globalization is an economic, political, and social process characterized by the integration of the world's many systems of manufacturing and trade into a single and increasingly seamless marketplace. The result: a new world geography.

This new geography is associated with the expansion of manufacturing and trade as capitalist principles replace old ideologies and state-controlled economies. With expanded free markets, the process of manufacturing and trading is constantly changing. Globalization delivers economic growth through improved manufacturing processes, newly developed goods, foreign investment in overseas manufacturing, and expanded employment.

The economies of developing countries are slowly transitioning from agricultural to industrial activities. Nevertheless, more than 65 percent of workers in these countries continue to work in agriculture. Meanwhile, developed countries, such as Australia and Germany, are experiencing high-technology service sector growth and reduced manufacturing employment. In the United States, nearly 30 percent of all workers were employed in manufacturing during the 1950s, but by 2019, less than 8.5 percent were.

In between these extremes, former state-controlled economies, like Romania, are adopting more efficient economic development strategies. Other nations and economic models, such as Indonesia and China, are pulled into the global marketplace by the growth and expansion of market economies. Despite the different economic paths of developing, transitioning, and developed nations, manufacturing and trade link all nations together and represent an economic convergence with important implications for political, business, and labor leaders—as well as all the world's citizens.

The geographies of manufacturing and trade can be examined as the distribution and location of economic activities in response to technological change and political and economic change.

Distribution and Location

Questions about where people live, work, and spend their money can be answered by reading product labels in any shopping mall, supermarket, or automobile dealership. They reveal the fact that manufacturing is a multistage process of component fabrication and final product assembly that can occur continents apart. For example, a shirt may be designed in New Jersey, assembled in Costa Rica from North Carolina fabric, and sold in British Columbia. To understand how goods produced in faraway locations are sold at neighborhood stores, geographers investigate the spatial, or geographic, distribution of natural resources, manufacturing plants, trading patterns, and consumption.

Historically, the geography of manufacturing and trade has been closely linked to the distribution of raw materials, workers, and buyers. In earlier times, this meant that manufacturing and trade were highly localized functions. In the eighteenth century, every North American town had cobblers or blacksmiths who produced goods from local resources for sale in local markets. By the start of the Industrial Revolution, improved transportation and manufacturing techniques had significantly enlarged the geography of manufacturing and trade. As distances increased, new manufacturing and trading centers developed. The location of these centers was contingent upon site and situation. Site and situation refer to a physical location, or site, relative to needed materials, transportation networks, and markets. For example, Pittsburgh, Pennsylvania, became the site of a major steel industry because it was near coal and iron resources. Pittsburgh also benefited from its historical role as a port town on a major river system that provided access to both western and eastern markets.

While relative location and transportation costs continue to be important factors, the geographic distribution of production and movement of goods across space is more complex than the simple calculus of site and situation. New global and local geographies of manufacturing and trade have been fueled by two major factors: technology and political change.

Technological Change

The old saying that time is money partially explains where goods are manufactured and traded. By compressing time and space, technology has enabled people, goods, and information to go farther more quickly. In the process, technology has reduced interaction costs, such as telecommunications. Just as steel enabled railroads to push farther westward, new technologies reduce the distance between places and people.

By increasing physical and virtual access to people, places, and things, technology has eliminated many barriers to global trade. However, improved telecommunications and transportation are only part of technology's contribution to globalization. If time is money, new efficient manufacturing processes also have reduced costs and facilitated globalization.

Armed with more efficient production processes, reliable telecommunications infrastructures, and transportation improvements, businesses can increase profits and remain competitive by seeking out lower-cost labor markets thousands of miles from consumers. As trade and manufacturing are increasingly spatially separate activities, the geographic distribution of manufacturing promotes an uneven distribution of income. The global distribution of manufacturing plants is closely related to industry-specific skill and wage requirements. For example, low-wage and low-skill jobs tend to concentrate in the developing regions of Asia, South America, and Africa. Alternately, high-technology and high-wage manufacturing activities concentrate in more developed regions.

In some cases, high wages and global competition force corporations to move their manufacturing plants to save costs and remain competitive. During the early 1990s, this byproduct of globalization was a major issue during the U.S. and Canadian debates to ratify the North American Free Trade Agreement (NAFTA). Focusing on primarily U.S. and Canadian companies that moved jobs to Mexico, the debate contributed to growing anxiety over job security as plants relocate to low-cost labor markets in South America and around the world.

As global competition increases, the geography of manufacturing and trade is increasingly global and rapidly changing. One company that has adapted to the shifting nature of global trade and manufacturing is Nike. Based in Beaverton, Oregon, Nike designs and develops new products at its Oregon world headquarters. However, Nike has internationalized much of its manufacturing capacity to compete in an aggressive athletic apparel industry. Over the last twenty-five years, Nike's strategy has meant shifts in production from high-wage U.S. locations to numerous low-wage labor markets around Pacific Rim.

Political and Economic Change: A New World Order

In order for companies such as Nike to successfully adapt to changing global dynamics, a stable international, or multilateral, trading system must be in place. In 1948 the General Agreement on Tariffs and Trade (GATT) was the first major step toward developing this stable global trading infrastructure. During that same period, the World Bank and International Monetary Fund were created to stabilize and standardize financial markets and practices. However, Cold War politics postponed complete economic integration for nearly half a century. Since the collapse of communism, globalization has accelerated as economies coalesce around the principles of free markets and capitalism. These important changes have become institutionalized through multilateral trade agreements and international trading organizations.

International trading organizations try to minimize or eliminate barriers to free and fair trade between nations. Trade barriers include tariffs (taxes levied on imported goods), product quotas, government subsidies to domestic industry, domestic content rules, and other regulations. Barriers prevent competitive access to domestic markets by artificially raising the prices of imported goods too high or preventing foreign firms from achieving economies of scale. In some cases, tariffs can also be used to promote fair trade by effectively leveling the playing field.

Because tariffs can be used both to promote fair trade and to unfairly protect markets, trading organizations are responsible for distinguishing between the two. For example, the Asian Pacific Economic Cooperation (APEC) forum has established guidelines to promote fair trade and attract foreign investment. APEC initiatives include a public Web-based database of member state tariff schedules and related links. Through programs such as the APEC information-sharing project, trading organizations are streamlining the international business process and promoting the overall stability of international markets.

The Future

As the globalization of manufacturing and trade continues, a new world geography is emerging. Unlike the Cold War's east-west geography and politics of ideology, an economic politics divides the developed and developing world along a north-south axis. While the types of conflicts associated with these new politics and the rules of engagement are unclear, it is evident that a new hierarchy of nations is emerging.

Globalization will raise the economic standard of living in most nations, but it has also widened the gap between richer and poorer countries. A small group of nations generates and controls most of the world's wealth. Conversely, the poorest countries account for roughly two-thirds of the world's population and less than 10 percent of its wealth.

This fundamental question of economic justice was a motive behind globalization's first major political clash. During the 1999 World Trade Organization (WTO) meetings in Seattle, Washington, approximately 50,000 environmentalists, labor unionists, and human and animal rights activists protested against numerous issues, including cultural intolerance, economic injustice, environmental degradation, political repression, and unfair labor practices they attribute to free trade. While the protesters managed to cancel the opening ceremonies, the United Nations secretary-general, Kofi Annan, expressed the general sentiment of most WTO member states. Agreeing that the protesters' concerns were important, Annan also asserted that the globalization of manufacturing and trade should not be used as a scapegoat for domestic failures to protect individual rights. More important, the secretary-general feared that those issues could

be little more than a pretext for a return to unilateral trade policies, or protectionism.

Like the Seattle protesters, supporters of multilateral trade advocate political and economic reforms. Proponents emphasize that open markets promote open societies. Free traders earnestly believe economic engagement encourages rogue nations to improve poor human rights, environmental, and labor records. It is argued that economic engagement raises the expectations of citizens, thereby promoting change.

Conclusion

Technological and political change have made global labor and consumer markets more accessible and established an economic world hierarchy. At the top, one-fifth of the world's population consumes the vast majority of produced goods and controls more than 80 percent of the wealth. At the bottom of this hierarchy, poor nations are industrializing but possess less than 10 percent of the world's wealth. In political, social, and cultural terms, this global economic reality defines the contours and cleavages of a changing world geography. Whether geographers calculate the economic and political costs of a widening gap between rich and poor or chart the flow of funds from Tokyo to Toronto, the globalization of manufacturing and trade will remain central to the study of geography well into the twenty-first century.

Jay D. Gatrell

MODERN WORLD TRADE PATTERNS

Trade, its routes, and its patterns are an integral part of modern society. Trade is primarily based on need. People trade the goods that they have, including money, to obtain the goods that they don't have. Some nations are very rich in agriculture or natural resources, while others are centers of industrial or technical activity. Because nations' needs change

only slowly, trade routes and trading patterns develop that last for long periods of time.

Types of Trade

The movement of goods can occur among neighboring countries, such as the United States and Mexico, or across the globe, as between Japan and

Italy. Some trade routes are well established with regularly scheduled service connecting points. Such service is called liner service. Liners may also serve intermediate points along a trade route to increase their revenue.

Some trade occurs only seasonally, such as the movement of fresh fruits from Chile to California. Some trade occurs only when certain goods are demanded, such as special orders of industrial goods. This type of service is provided by operators called tramps. They go where the business of trade takes them, rather than along fixed liner schedules and routes.

Many people think of international trade as being carried on great ships plying the oceans of the world. Such trade is important; however, a considerable amount of trade is carried by other modes of transportation. Ships and airplanes carry large volumes of freight over large distances, while trucks, trains, barges, and even animal transport are used to move goods over trade routes among neighboring or landlocked countries.

Trade Routes

Through much of human history, trade routes were limited. Shipping trade carried on sailing vessels, for example, was limited by the prevailing winds that powered the ships. Land routes were limited by the location of water, mountain ranges, and the slow development of roads through thick forests and difficult terrain. The mechanization of transportation eventually freed ships and other forms of transport to follow more direct trade routes. Also, the development of canals and transcontinental highway systems allowed trade routes to develop based solely upon economic requirements.

Other changes in trade routes have occurred with industrialization of transport systems. The world began to have a great need for coal. Trade routes ran to the countries in which coal was mined. Ships and trains delivered coal to the power industry worldwide. Later, trade shifted to locations where oil (petroleum) was drilled. Now, oil is delivered to those same powerplants and industrial sites around the world.

Noneconomic Factors

Some trade is not purely economic in nature. Political relationships among countries can play an important part in their trade relations. For example, many national governments try to protect their countries' automobile and electronics industries from outside competition by not allowing foreign goods to be imported easily. Governments control imports by assessing duties, or tariffs, on selected imports.

Some national governments use the concept of cabotage to protect their home transportation industries by requiring that certain percentages of imported and exported trade goods be carried by their own carriers. For example, the U.S. government might require that 50 percent of its trade use American ships, planes, or trucks. The government might also require that all American carriers employ only American citizens.

Nations also can exert pressure on their trading partners by limiting access to port or airport facilities. Stronger nations may force weaker nations into accepting unequal trade agreements. For example, the United States once had an agreement with Germany concerning air passenger service between the two countries. The agreement allowed United States carriers to carry 80 percent of the passengers, while German carriers were permitted to carry only 20 percent of the passengers.

Multilateral Trade

In situations in which pairs of trading nations do not have direct diplomatic contact with each other, they make their trade arrangements through other nations. Such trade is referred to as multilateral. Certain carriers cater to this type of trade. They operate their ships or planes in around-the-world service. They literally travel around the globe picking up and depositing cargo along the way for a variety of nations.

Trade Patterns

For many years, world populations were coast centered. This means that most of the people in the country lived close to the coast. This was due primarily to the availability of water transportation systems to move both goods and people. At this time, major railroad, highway and airline systems did not exist. As railway and highway systems pushed into the interiors of nations, the population followed, and goods were needed as well as produced in these areas. Thus, over the years many inland population centers have developed that require transportation systems to move goods into and away from this area.

In these cases, international trade to these inland centers required the use of a number of different modes of transportation. Each of the different modes required additional paperwork and time for repackaging and securing of the cargo. For example, cargo coming off ships from overseas was unloaded and placed in warehouse storage. At some later time, it was loaded onto trucks that carried it to railyards. There it would be unloaded, stored, and then loaded onto railcars. At the destination, the cargo would once again be shifted to trucks for the final delivery. During the course of the trip, the cargo would have been handled a number of times, with the possibility of damage or loss occurring each time.

Containerization

As more goods began to move in international trade, the systems for packaging and securing of cargo became more standardized. In the 1960s, shipments began to move in containers. These are highway truck trailers which have been removed from the chassis leaving only the box. Container packaging has become the standard for most cargos moving today in both domestic and international trade. With the advent of containerization of cargo in international trade, cargo movements could quickly move intermodally. Intermodal shipping involves the movement of cargo by using more than a single mode of transportation.

Land, water, and air carriers have attempted to make the intermodal movement of cargo in international trade as seamless as possible. They have not only standardized the box for carrying cargo, but they have also standardized the handling equipment, so that containers move quickly from one mode to another. Advances in communications and

---

THE WORLD TRADE ORGANIZATION AND GLOBAL TRADING

In 1998 domestic political pressures and an expected domestic surplus of rice prompted the Japanese government to unilaterally implement a 355-percent tariff on foreign rice, violating the United Nations' General Agreement on Tariffs and Trade (GATT). On April 1, 1999, Japan agreed to return to GATT import levels and imposed new over-quota tariffs. While domestic Japanese politics could have prompted a trade war with rice-exporting countries, the crisis demonstrates how multilateral trading initiatives promote stability. Without an agreement, rice exporters might not have gained access to Japanese markets. By returning to GATT minimum quotas and implementing over-quota taxes, the compromise addressed the interests of both domestic and foreign rice growers.

---

electronic banking allow the paperwork and payments also to be completed and transferred rapidly.

As the demands for products have grown and as the size of industrial plants has grown, the size of movements of raw materials and containerized cargo has also grown. Thus, the sizes of the ships and trains required to move these large volumes of cargo have also increased.

The development of VLCC's (very large crude carriers) has allowed shippers to move large volumes of oil products. The development of large bulk carriers has allowed for the carriage of large volumes of dry raw materials such as grains or iron ore. These large vessels take advantage of what is known as economies of scale. Goods can be moved more cheaply when large volumes of them are moved at the same time. This is because the doubling of the volume of cargo moved does not double the cost to build or operate the vessels in which it is carried. This savings reduces the cost to move large volumes of cargo.

Intermodal Transportation

Intermodal transportation has allowed cargo to move seamlessly across both international boundaries and through different modes of transportation. This seamless movement has changed ocean trade routes over recent years.

The development of the Pacific Rim nations created a demand for trade between East Asia and both the United States and Europe. This trade has usually taken the all-water routes between Asia and Europe. Ships moving from East Asia across the Pacific Ocean pass through the Panama Canal and cross the Atlantic Ocean to reach Western Europe. This journey is in excess of 10,000 miles (16,000 km.) and usually takes about thirty days for most ships to complete. The all-water route from Asia to New York is similar. The distance is almost as great as that to Europe and requires about twenty-one to twenty-four days to complete.

Intermodal transportation has given shippers alternatives to all-water routes. A great volume of Asian goods is now shipped to such western U.S. ports as Seattle, Oakland, and Los Angeles, from which these goods are carried by trains across the United States to New York. The overall lengths of these routes to New York are only about 7,400 miles (12,000 km.) and take between only fifteen and nineteen days to complete. Cargos continuing to Europe are put back on ships in New York and complete their journeys in an additional seven to ten days. Such intermodal shipping can save as much as a week in delivery time.

Airfreight

Another changing trend in trade patterns is the development of airfreight as an international competitor. Modern aircraft have improved dramatically both in their ability to lift large weights of cargo as well as their ability to carry cargos over long distances. Because of the speed at which aircraft travel in comparison to other modes of transportation, goods can be moved quickly over large distances. Thus, high-value cargos or very fragile cargos can move very quickly by aircraft.

The drawback to airfreight movement of cargo is that it is more expensive than other modes of travel. However, for businesses that need to move perishable commodities, such as flowers of the Netherlands, or expensive commodities, such as Paris fashions or Singapore-made computer chips, airfreight has become both economic and essential.

Robert J. Stewart

POLITICAL GEOGRAPHY

FORMS OF GOVERNMENT

Philosophers and political scientists have studied forms of government for many centuries. Ancient Greek philosophers such as Plato and Aristotle wrote about what they believed to be good and bad forms of government. According to Plato's famous work, *The Republic*, the best form of government was one ruled by philosopher-kings. Aristotle wrote that good governments, whether headed by one person (a kingship), a few people (an aristocracy), or many people (a polity), were those that ruled for the benefit of all. Those that were based on narrow, selfish interests were considered bad forms of government, whether ruled by an individual (a tyranny), a few people (an oligarchy), or many people (a democracy). Thus, democracy was not always considered a good form of government.

Constitutions and Political Institutions

All governments have certain things in common: institutions that carry out legislative, executive, and judicial functions. How these institutions are supposed to function is usually spelled out in a country's constitution, which is a guide to organizing a country's political system. Most, but not all, countries have written constitutions. Great Britain, for example, has an unwritten constitution based on documents such as the Magna Carta, the English Bill of Rights, and the Treaty of Rome, and on unwritten codes of behavior expected of politicians and members of the royal family.

The world's oldest written constitution still in use is that of the United States. All countries have written or unwritten constitutions, and most follow them most of the time. Some countries do not follow their constitutions—for example, the Soviet Union did not; other countries, for example France, change their constitutions frequently.

Constitutions usually first specify if the country is to be a monarchy or a republic. Few countries still have monarchies, and those that do usually grant the monarch only ceremonial powers and duties. Countries with monarchies at the beginning of the twenty-first century included Spain, Great Britain, Lesotho, Swaziland, Sweden, Saudi Arabia, and Jordan. Most countries that do not have monarchies are republics.

Constitutions also specify if power is to be concentrated in the hands of a strong national government, which is a unitary system; if it is to be divided between a national and various subnational governments such as states, provinces, or territories, which is a federal system; or if it is to be spread among various subnational governments that might delegate some power to a weak national government, which is a confederate system.

Examples of countries with unitary systems include Great Britain, France, and China; federal systems include the United States, Germany, Russia, Canada, India, and Brazil. There were no confederate systems by the third decade of the twenty-first century, although there are examples from history as well as confederations of various groups and nations. The United States under its eighteenth-century Articles of Confederation and the nineteenth-century Confederate States of America, made up of the rebelling Southern states, were confederate systems. Switzerland was a confederation for much of the nineteenth century. The concept of dividing power between the national and subnational governments is called the vertical axis of power.

MONARCHIES OF THE WORLD

Realm/Kingdom	Monarch	Type
Principality of Andorra	Co-Prince Emmanuel Macron; Co-Prince Archbishop Joan Enric Vives Sicília	Constitutional
Antigua and Barbuda	Queen Elizabeth II	Constitutional
Commonwealth of Australia	Queen Elizabeth II	Constitutional
Commonwealth of the Bahamas	Queen Elizabeth II	Constitutional
Barbados	Queen Elizabeth II	Constitutional
Belize	Queen Elizabeth II	Constitutional
Canada	Queen Elizabeth II	Constitutional
Grenada	Queen Elizabeth II	Constitutional
Jamaica	Queen Elizabeth II	Constitutional
New Zealand	Queen Elizabeth II	Constitutional
Independent State of Papua New Guinea	Queen Elizabeth II	Constitutional
Federation of Saint Kitts and Nevis	Queen Elizabeth II	Constitutional
Saint Lucia	Queen Elizabeth II	Constitutional
Saint Vincent and the Grenadines	Queen Elizabeth II	Constitutional
Solomon Islands	Queen Elizabeth II	Constitutional
Tuvalu	Queen Elizabeth II	Constitutional
United Kingdom of Great Britain and Northern Ireland	Queen Elizabeth II	Constitutional
Kingdom of Bahrain	King Hamad bin Isa	Mixed
Kingdom of Belgium	King Philippe	Constitutional
Kingdom of Bhutan	King Jigme Khesar Namgyel	Constitutional
Brunei Darussalam	Sultan Hassanal Bolkiah	Absolute
Kingdom of Cambodia	King Norodom Sihamoni	Constitutional
Kingdom of Denmark	Queen Margrethe II	Constitutional
Kingdom of Eswatini	King Mswati III	Absolute
Japan	Emperor Naruhito	Constitutional
Hashemite Kingdom of Jordan	King Abdullah II	Constitutional
State of Kuwait	Emir Sabah al-Ahmad	Constitutional
Kingdom of Lesotho	King Letsie III	Constitutional
Principality of Liechtenstein	Prince Regnant Hans-Adam II (Regent: The Hereditary Prince Alois)	Constitutional
Grand Duchy of Luxembourg	Grand Duke Henri	Constitutional
Malaysia	Yang di-Pertuan Agong Abdullah	Constitutional
Principality of Monaco	Sovereign Prince Albert II	Constitutional

MONARCHIES OF THE WORLD (continued)

Realm/Kingdom	Monarch	Type
Kingdom of Morocco	King Mohammed VI	Constitutional
Kingdom of the Netherlands	King Willem-Alexander	Constitutional
Kingdom of Norway	King Harald V	Constitutional
Sultanate of Oman	Sultan Haitham bin Tariq	Absolute
State of Qatar	Emir Tamim bin Hamad	Mixed
Kingdom of Saudi Arabia	King Salman bin Abdulaziz	Absolute theocracy
Kingdom of Spain	King Felipe VI	Constitutional
Kingdom of Sweden	King Carl XVI Gustaf	Constitutional
Kingdom of Thailand	King Vajiralongkorn	Constitutional
Kingdom of Tonga	King Tupou VI	Constitutional
United Arab Emirates	President Khalifa bin Zayed	Mixed
Vatican City State	Pope Francis	Absolute theocracy

Whether governments share power with subnational governments or not, there must be institutions to make laws, enforce laws, and interpret laws: the legislative, executive, and judicial branches of government. How these branches interact is what determines whether governments are parliamentary, presidential, or mixed parliamentary-presidential. In a presidential system, such as in the United States, the three branches—legislative, executive, and judicial—are separate, independent, and designed to check and balance each other according to a constitution. In a parliamentary system, the three branches are not entirely separate, and the legislative branch is much more powerful than the executive and judicial branches.

Great Britain is a good example of a parliamentary system. Some countries, such as France and Russia, have created a mixed parliamentary-presidential system, wherein the three branches are separate but are not designed to check and balance each other. In a mixed parliamentary-presidential system, the executive (led by a president) is the most powerful branch of government.

Looking at political systems in this way—how the legislative, executive, and judicial branches of government interact—is to examine the horizontal axis of power. All governments are unitary, federal, or confederate, and all are parliamentary, presidential, or mixed parliamentary-presidential. One can find examples of different combinations. Great Britain is unitary and parliamentary. Germany is federal and parliamentary. The United States is federal and presidential. France is unitary and mixed parliamentary-presidential. Russia is federal and mixed parliamentary-presidential. Furthermore, virtually all countries are either republics or monarchies.

Types of Government

Constitutions describe how the country's political institutions are supposed to interact and provide a guide to the relationship between the government and its citizens. Thus, while governments may have similar political institutions—for example, Germany and India are both federal, parliamentary republics—how the leaders treat their citizens can vary widely. However, governments may have political systems that function similarly although they have different forms of constitutions and institutions. For example, Great Britain, a unitary, parliamentary monarchy with an unwritten constitution, treats its citizens very similarly to the United States, which is a federal, presidential republic with a written constitution.

The three most common terms used to describe the relationships between those who govern and those who are governed are democratic, authoritarian, and totalitarian. Characteristics of democracies are free, fair, and meaningfully contested elections; majority rule and respect for minority rights and opinions; a willingness to hand power to the opposition after an election; the rule of law; and civil rights and liberties, including freedom of speech and press, freedom of association, and freedom to travel. The United States, Canada, Japan, and most European countries are democratic.

An authoritarian system is one that curtails some or all of the characteristics of a democratic regime. For example, authoritarian regimes might permit token electoral opposition by allowing other political parties to run in elections, but they do not allow the opposition to win those elections. If the opposition did win, the authoritarian regime would not hand over power. Authoritarian regimes do not respect the rule of law, the rights of minorities to dissent, or freedom of the press, speech, or association. Authoritarian governments use the police, courts, prisons, and the military to intimidate and threaten their citizens, thus preventing people from uniting to challenge the existing political rulers. Afghanistan, Cuba, Iran, Uzbekistan, Saudi Arabia, Chad, Syria, Libya, Sudan, Belarus, and China are examples of countries with authoritarian regimes.

Totalitarian regimes are similar to authoritarian regimes but are even more extreme. Under a totalitarian regime, there is no legal opposition, no freedom of speech, and no rule of law whatsoever. Totalitarian regimes attempt to control totally all members of the society to the point where everyone always must actively demonstrate their loyalty to and support for the regime. Nazi Germany under Adolf Hitler's rule (1933-1945) and the Soviet Union under Joseph Stalin's rule (1928-1953) are examples of totalitarian regimes. As of 2019, only Eritrea and North Korea are the still have governments classified as totalitarian dictatorships.

Forms of Government: Putting it All Together

In *The Republic*, Plato asserts that people have varied dispositions, and, therefore, there are various types of governments. In recent years, regimes have been created that some call mafiacracies (rule by criminal mafias), narcocracies (rule by narcotics gangs), gerontocracies (rule by very old people), theocracies (rule by religious leaders), and so forth. Such variations show the ingenuity of the human mind in devising forms of government.

Whatever labels that are given to a political system, there remain basic questions to be asked about that regime: Is it a monarchy or a republic? Is all power concentrated in the hands of a national government, or is power shared between a national government and the states or provinces? Are its institutions those of a parliamentary, presidential, or mixed parliamentary-presidential system? Is it democratic, authoritarian, or totalitarian? Finally, does it live up to its constitution, both in terms of how power is supposed to be distributed among institutions and in its relationship between the government and the people? To paraphrase Aristotle, how many rulers are there, and in whose interests do they rule?

Nathaniel Richmond

POLITICAL GEOGRAPHY

Students of politics have been aware that there is a significant relationship between physical and political geography since the time of ancient Greece.

The ancient Greek philosopher Plato argued that a *polis* (politically organized society) must be of limited geographical size and limited population or it would lack cohesion. The ideal *polis* would be only as geographically large as required to feed about 5,000 people, its maximum population.

Plato's illustrious pupil, Aristotle, agreed that stable states must be small. "One can build a wall around the Hellespont," the main territory of ancient Greece, he wrote in his treatise *Politics*, "but that will not make it a polis." Today human ideas differ about the maximum area of a successful state or nation-state, but the close influence of physical geography on political geography and their profound mutual effects on politics itself are not in question.

Geographical Influences on Politics

The physical shape and contours of states may be called their physical geography; the political shape and contours of states, starting with their basic structure as unified state, federation, or confederation, are primary features of their political geography. The idea of "political geography" also can refer to variations in a population's political attitudes and behavior that are influenced by geographical features. Thus, the combination of plentiful land and sparse population tend toward an independent spirit, especially where the economy is agriculturally based. This has historically been the case in the western United States; in the Pampas region of Argentina, where cattle are raised by independent-minded gauchos (cowboys); and on the Brazilian frontier, where government regulation is routinely resisted.

Likewise, where physical geography presents significant difficulties for inhabitants in earning a living or associating, as where there is rough terrain and poor soil or inhospitable climate, the populace is likely to exhibit a hardy, self-reliant character that strongly influences political preferences. Thus, physical geography helps to shape national character, including aspects of a nation's politics.

Furthermore, it is well known that where physical geography isolates one part of a country's population from the rest, political radicalism may take root. This tendency is found in coastal cities and remote regions, where labor union radicalism has often been pronounced. Populations in coastal loca-

tions with access to foreign trade often show a more liberal, tolerant, and outgoing spirit, as reflected in their political opinions. In ancient Greece, the coastal access enjoyed by Athens through a nearby port in the fifth century BCE had a strong influence on its liberal and democratic political order. In modern times, China's coastal cities, such as Tientsin, and North American cities such as San Francisco, show similar influences.

The Geographical Imperative

In many instances, political geography is shaped by what may be called the "geographical imperative." Physical geography in these instances demands, or at least strongly suggests, that political geography follow its course. The numerous valleys of mountainous Greece strongly influenced the emergence of the small, often fiercely independent, polis of ancient times. The formation and borders of Asian states such as Bhutan, Nepal, and Tibet have been strongly influenced by the Himalaya Mountains, and the Alps, which shape Switzerland.

As another example, physical geography demands that the land between the Pacific Ocean and the Andes Mountains along the western edge of South America be organized as a separate country—Chile. Island geography often plays a decisive role in its political geography. The qualified political unity of Great Britain can be directly traced to its insular status. Small islands often find themselves combined into larger units, such as the Hawaiian Islands.

The absence of the geographical imperative, however, leaves political geography an open question. For example, Indonesia comprises some 1,300 islands stretching 3,000 miles in bodies of water such as the Indian Ocean and the Celebes Sea. With so many islands, Indonesia lacks a geographical imperative to be a unified state. It also lacks the imperative of ethnic and cultural homogeneity and cohesion, a circumstance mirrored in its political life, since it has remained unified only through military force. As control by the military waned after the fall of the authoritarian General Suharto in 1998, conflicts among the nation's diverse peoples have threatened its breakup. No such threat, however, confronts Australia, an immense island continent where a European majority dominates a fragmented and primitive aboriginal minority. In Australia, the geographical imperative suggests a unity supported by the cultural unity of the majority.

As many examples show, the geographical imperative is not absolute. For example, mountainous Greece is politically united in the twenty-first century. Although long shielded geographically, Tibet lost its political independence after it was successfully invaded by China. The formerly independent Himalayan state Sikkim was taken over by India. Thus, political will trumps physical geography.

The frequency of exceptions to the geographical imperative illustrates that human freedom, while not unlimited, often plays a key role in shaping political geography. As one example, the Baltic Republics—Lithuania, Latvia, and Estonia—historically have been dominated, or largely swallowed up, by neighboring Russia. By the start of the twenty-first century, however, they had regained their independence through the political will to self-rule and the drive for cultural survival.

Strategically Significant Locations

Locations of great economic or military significance become focal points of political attention and, potentially, of military conflict. There are innumerable such places in the world, but several stand out as models of how important physical geography can be for political geography in the context of international politics.

One significant example is the Panama Canal, without which ships must sail around South America. The Suez Canal, which connects European and Asian shipping, is a similar waterway, saving passage around Africa. The canal's significance was reduced after 1956, however, when its blockage after the Arab-Israeli war of that year led to the building of supertankers too large to traverse it. Another example is Gibraltar, whose fortifications command the entrance to the Mediterranean Sea from the Atlantic Ocean. A final example is the Bosporus, the tiny entrance from the Black Sea to waters leading to the Mediterranean Sea. It is the only warm-water route to and from Eastern Russia and therefore is of great military and economic importance for regional and world power politics.

Charles F. Bahmueller

GEOPOLITICS

Geopolitics is a concept pertaining to the role of purely geographical features in the relations among states in international politics. Geopolitics is especially concerned with the geographical locations of the states in relationship to one another. Geopolitical relationships incorporate social, economic, political, and historical features of the states that interact with purely geographical elements to influence the strategic thinking and behavior of nations in the international sphere.

Coined in 1899 by the Swedish theorist Rudolf Kjellen, the term "geopolitics" combines the logic of the search for security and competition for dominance among states with geographical methodology. *Geopolitics* must not, however, be confused with *political geography*, which focuses on individual states' territorial sizes, boundaries, resources, internal political relations, and relations with other states.

Geopolitical is a term frequently used by military and political strategists, politicians and diplomats, political scientists, journalists, statesmen, and a variety of other government officials, such as policy planners and intelligence analysts.

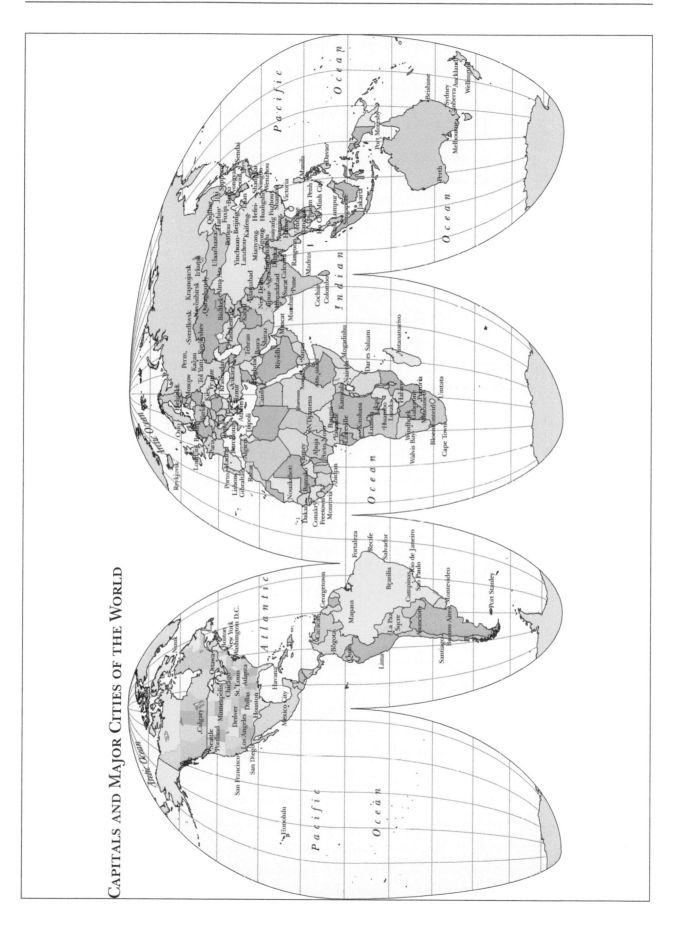

CAPITALS AND MAJOR CITIES OF THE WORLD

Power Struggles Among States

The idea of geopolitics arises in the course of what might be considered the universal struggle for power among the world's most powerful nations, which compete for political and military leadership. How one state can threaten another, for example, is often influenced by geographical factors in combination with technological, social, economic and other factors. The extent to which individual states can threaten each other depends in no small measure on purely geographical considerations.

By the close of twentieth century the Cold War that had dominated world security concerns was over. Nevertheless, the United States still worried about the danger of being attacked by nuclear missiles fired, not by the former Soviet Union, but by irresponsible, fanatical, or suicidal states. American political leaders and military planners were concerned with the geographical position of so-called "rogue states." or "states of concern." In 1994, North Korea, Cuba, Iran, Libya under Muammar Gaddafi, and Ba'athist Iraq were listed as states of concern. By 2019, a list of state sponsors of terrorism included Iran, North Korea, Sudan, and Syria.

Geographical factors play prominent roles in assessments of the different threats that those states presented to American interests. How far those states are located from American territory determines whether their missiles might pose a serious threat. A missile may be able to reach only the periphery of U.S. soil, or it might be able to carry only a small payload. Similar considerations determine the threat such states pose for U.S. forces stationed abroad, as well as for such important U.S. allies as Japan, Western Europe, or Israel. Such questions are thus said to constitute geopolitical, or geostrategic, considerations.

There are many examples of the influence of geopolitical factors on international relations among nations in the past. For example, the Bosporus, the narrow sea lane linking the Black Sea and the Mediterranean where Istanbul is situated, has long been considered of great strategic importance. In the nineteenth century, the Bosporus was the only direct route through which the Russian

> ### A PEACEFULLY RESOLVED BORDER DISPUTE
>
> The peaceful resolution of the border dispute between the Southern African states of Botswana and Namibia was hailed by observers of African politics. Instead of resorting to the armed warfare that so often has marked similar disputes on the continent, the two states chose a different course in 1996, when they found negotiations stalemated. They submitted their claims to the International Court of Justice in The Hague and agreed to accept the court's ruling. Late in 1999, by an eleven-to-four vote, the court ruled for Botswana, and Namibia kept its word to embrace the decision. At issue was a tiny island in the Chobe River on Botswana's northern border. An 1890 treaty between colonial rulers Great Britain and Germany had described the border at the disputed point vaguely, as the river's "main channel." The court took the course of the deepest channel to mark the agreed boundary, giving Botswana title to the 1.4-square-mile (3.5-sq. km.) territory.

navy could reach southern Europe and the Mediterranean Sea.

Because of Russia's nineteenth century history of expansionism and its integration into the pre-World War I European state system, with its networks of competing military alliances, the Bosporus took on added geopolitical meaning. It was the congested (and therefore vulnerable) space through which Russian naval power had to pass to reach the Mediterranean.

Historical Origins of Geopolitics

Although political geography was a well-established field by the late nineteenth century, geopolitics was just beginning to emerge as a field of study and political analysis at the end of the century. In 1896 the German theorist Friedrich Ratzel published his *Political Geography*, which put forward the idea of the state as territory occupied by a people bound together by an idea of the state. Ratzel's theory embraced Social Darwinist notions that justified the current boundaries of nations. Ratzel viewed the state as a biological organism in competition for land with other states. The ethical implication of his theory seemed to be that "might makes right."

That theme set the stage for later German geopolitical thought, especially the notion of the need for *Lebensraum* (living room)—space into which the people of a nation could expand. German dictator Adolf Hitler justified his attack on Russia during World War II partly upon his claim that the German people needed more *Lebensraum* to the east. To some modern geographers, the use of geopolitical theories to serve German fascism and to justify other instances of military aggression tarnished geopolitics itself as a field of study.

Historical Development of Geopolitics

Modern geopolitics has further origins in the work of the Scottish geographer Sir Halford John Mackinder. In 1904 he published a seminal article, "The Geographical Pivot of History," in which he argued that the world is made up of a Eurasian "heartland" and a secondary hinterland (the remainder of the world), which he called the "marginal crescent." According to his theory, international politics is the struggle to gain control of the heartland. Any state that managed that feat would dominate the world.

A major proposition of Mackinder's theory was that geographical factors are not merely causative factors, but coercive. He tried to describe the physical features of the world that he believed directed human actions. In his view, "Man and not nature initiates, but nature in large measure controls." Geopolitical factors were therefore to a great extent determinants of the behavior of states. If this were true, geopolitics as a science could have deep relevance and corresponding influence among governments.

After Mackinder's time, the concept of geopolitics had a double significance. On the one hand it was a purely descriptive theory of geographic causation in history. On the other hand, its purveyors also believed, as Mackinder argued in 1904, that geopolitics has "a practical value as setting into perspective some of the competing forces in current international politics." Mackinder sought to promote this field of study as a companion to British statecraft, a tool to further Britain's national interest. By extension, geopolitical theory could assist any government in forming its political/military strategy.

As applied to the early twentieth-century world of international politics, however, Mackinder's theory had major weaknesses. Among his most glaring oversights were his failure to appreciate the rise of the United States, which attained considerable naval power after the turn of the century. Also, he failed to foresee the crucial strategic role that air power would play in warfare—and with it the immense change that air power could make in geopolitical considerations. Air power moves continents closer together, revolutionizing their geopolitical relationships.

One of Mackinder's chief critics was Nicolas John Spykman. Spykman argued that Mackinder had overvalued the potential economic, and therefore political, power of the Eurasian heartland, which could never reach its full potential because it could not overcome the obstacles to internal transportation. Moreover, the weaknesses of the remainder of the world—in effect, northern, western and southern Europe—could be overcome through forging alliances.

The dark side of geopolitical thought as handmaiden to political and military strategy became apparent in the Germany of the 1920s. At that time German theorists sought the resurrection of a German state broken by failure in World War I, the harsh terms of the Versailles Treaty that ended the war, and the hyperinflation that followed, wiping out the German middle class. In his 1925 article "Why Geopolitik?" Karl Haushofer urged the practical applications of *Geopolitik*. He urged that this form of analysis had not only "come to stay" but could also form important services for German political leaders, who should use all available tools "to carry on the fight for Germany's existence."

Haushofer ominously suggested that the "struggle" for German existence was becoming increasingly difficult because of the growth of the country's population. A people, he wrote, should study the living spaces of other nations so it could be prepared to "seize any possibility to recover lost ground." This discussion clearly implied that, from geopolitical necessity, Germany should seek additional territory to feed itself—a view carried into effect by Hitler in his quest for *Lebensraum* in attacking the Soviet Union, including its wheat-producing breadbasket, the Ukraine.

After World War II, a chastened Haushofer sought to soft-pedal both the direction and influence of his prewar writings. However, Hitler's morally heinous use of *Geopolitik* left geopolitical theorizing permanently tainted, in some eyes. Nevertheless, there is no necessary connection between geopolitics as a purely analytic description and geopolitics as the basis for a selfish search for power and advantage.

Geopolitics in the Twenty-first Century

Geopolitical considerations were unquestionably of profound relevance to the principal states of the post-World War II Cold War period. After the fall of the Berlin Wall in 1989, however, some theorists thought that the age of geopolitics had passed. In 1990 American strategic theorist Edward N. Luttwak, for example, argued that the importance of military power in international affairs had declined precipitously with the winding down of the Cold War. Military power had been overtaken in significance by economic prowess. Consequently, geopolitics had been eclipsed by what Luttwak called "geoeconomics," the waging of geopolitical struggle by economic means.

The view of Luttwak and various geographers of the declining significance of military power and geopolitical analysis, however, was soon proved to be overdrawn by events. As early as the first months of 1991, before the Soviet Union was officially dismantled, military power asserted itself as a key determinant on the international scene. Led by the United States, a far-flung alliance of nations participated in a war to remove Iraqi dictator Saddam Hussein's forces from neighboring Kuwait, which Iraq had illegally occupied. The decisive and successful use of military power in that war dramatically disproved assertions of its growing irrelevance.

Similarly, in the first three decades of the twenty-first century, military power retained its pre-eminence in the dynamics of international politics, even as economic forces were seen to gather momentum. To states throughout Asia and the West (especially Western Europe and the United States), the relative military capability of potential adversaries, and therefore geopolitics, remained a vital feature of the international order. Central to this view of the world scene is the growing military rivalry of the United States and China in East Asia. As China modernizes and expands its nuclear and conventional forces, it may feel itself capable of challenging America's predominant military power and prestige in East Asia. This possibility heightens the use of geopolitical thinking, giving it currency in analyzing this emerging situation.

Geopolitics as Civilizational Clash

A sometimes controversial expression of geopolitical analysis has been offered by Samuel Huntington of Harvard University. In his *The Clash of Civilizations and the Remaking of World Order* (1996) Huntington constructs a theory to explain certain tendencies of international behavior. He divides the world into a number of cultural groupings, or "civilizations," and argues that the character of various international conflicts can best be explained as conflicts or clashes of civilizations. In his view, Western civilization differs from the civilization of Orthodox Christianity, with a variety of conflicts erupting between the two. An example is the attack by the North Atlantic Treaty Organization (NATO), the bastion of the West, on Serbia, which is part of the Orthodox East.

Huntington's other civilizations include Islamic, Jewish, Eastern Caribbean, Hindu, Sinic (Chinese), and Japanese. The clash between Israel and its neighbors, the struggle between Pakistan and India over Kashmir, the rivalries between the United States and China and between China and India, for example, can be viewed as civilizational conflicts. Huntington has stated, however, that his theory is not intended to explain all of the historical past, and he does not expect it to remain valid long into the future.

Charles F. Bahmueller

NATIONAL PARK SYSTEMS

The world's first national parks were established as a response to the exploitation of natural resources, disappearance of wildlife, and destruction of natural landscapes that took place during the late nineteenth century. Government efforts to preserve natural areas as parks began with the establishment of Yellowstone National Park in the United States in 1872 and were soon adopted in other countries, including Australia, Canada, and New Zealand.

While the preservation of nature continues to be an important benefit provided by national parks, worldwide increases in population and the pressures of urban living have raised public interest in setting aside places that provide opportunities for solitude and interaction with nature.

Because national parks have been established by nations with diverse cultural values, land resources, and management philosophies, there is no single definition of what constitutes a national park. In some countries, areas used principally for recreational purposes are designated as national parks; other countries emphasize preservation of outstanding scenic, geologic, or biological resources. The terminology used for national parks also varies among countries. For example, protected areas that are similar to national parks may be called reserves, preserves, or sanctuaries.

Diverse landscapes are protected within national parks, including swamps, river deltas, dune areas, mountains, prairies, tropical rain forests, temperate forests, arid lands, and marine environments. Individual parks within nations form networks that vary with respect to size, accessibility, function, and the type of natural landscapes preserved. Some national park areas are isolated and sparsely populated, such as Greenland National Park; others, such as Peak District National Park in Great Britain, contain numerous small towns and are easily accessible to urban populations.

The functions of national parks include the preservation of scenic landscapes, geological features, wilderness, and plants and animals within their natural habitats. National parks also serve as outdoor laboratories for education and scientific research and as reservoirs for genetic information. Many are components of the United Nations International Biosphere Reserve Program.

National parks also play important roles in preserving cultures, by protecting archaeological, cultural, and historical sites. The United Nations recognizes several national parks that possess important cultural attributes as World Heritage Sites. Tourism to national parks has become important to the economies of many developing nations, especially in Eastern and Southern Africa, India, Nepal, Ecuador, and Indonesia. Parks are sources of local employment and can stimulate improvements to transportation and other types of infrastructure while encouraging productive use of lands that are of marginal agricultural use.

The International Union for Conservation of Nature has developed a system for classifying the world's protected areas, with Category II areas designated as national parks. Using this definition, there are 3,044 national parks in the world, with a mean average size of 457 square miles (1,183 sq. km.) each. Together, they cover an area of about 1.5 million square miles (4 million sq. km.), accounting for about 2.7 percent of the total land area on Earth.

STEPHEN T. MATHER AND THE U.S. NATIONAL PARK SERVICE

In 1914 businessman and conservationist Stephen T. Mather wrote to Secretary of the Interior Franklin K. Lane about the poor condition of California's Yosemite and Sequoia National Parks. Lane wrote back, "if you don't like the way the national parks are being run, come on down to Washington and run them yourself." Mather accepted the challenge and became an assistant to Lane and later the first director of the U.S. National Park Service, from 1917 to 1929.

North America

In 1916 management of U.S. national parks and monuments was shifted from the U.S. Army to the newly established National Park Service (NPS). The system has since grown in size to protect sixty-one national parks, as well as other natural areas including national monuments, seashores, and preserves.

North America's second-largest system of national parks is Parks Canada, created in 1930. Among the best-known Canadian parks is Banff, established in southern Alberta in 1885. Preserved within this area are glacially carved valleys, evergreen forests, and turquoise lakes. Parks Canada has the goal of protecting representative examples of each of Canada's vegetation and physiographic regions.

Mexico began providing protection for natural areas in the late nineteenth century. Among its system of sixty-seven national parks is Dzibilchaltún, an important Mayan archaeological site on the Yucatán Peninsula. With fewer resources available for park management, the emphasis in Mexico remains the preservation of scenic beauty for public use.

South America

Two of South America's best-known national parks are located within Argentina's park system. Nahuel Huapi National Park preserves two rare deer species of the Andes, while Iguazú National Park, located on the border with Brazil, is home to tapir, ocelot, and jaguar.

Located on a plateau of the western slope of the Andes Mountains in Chile, Lauca National Park is one of the world's highest parks, with an average elevation of more than 14,000 feet (4,267 meters)-an altitude nearly as high as the tallest mountains in the continental United States. Huascarán, another mountain park located in western Peru, boasts twenty peaks that exceed 19,000 feet (5,791 meters) in elevation. The volcanic islands of Galapagos Islands National Park, managed by Ecuador, have been of interest to biologists since British naturalist Charles Darwin studied variation and adaptation in animal species there in 1835.

Australia and New Zealand

Established in 1886, Royal was Australia's first national park. Perhaps better known to tourists, Uluru National Park in Australia's Northern Territory protects two rock domes, Ayer's Rock and Mount Olga, that rise above the plains 15 miles (40 km.) apart.

Along with Australia and other former colonies of Great Britain, New Zealand was a leader in establishing early national parks. The first of these was Tongagiro, created in 1887 to protect sacred lands of the Maori people on the North Island. New Zealand's South Island features several national parks including Fiordland, created in 1904 to preserve high mountains, forests, rivers, waterfalls, and other spectacular features of glacial origin.

Africa

Game poaching continues to be a severe problem in Africa, where animals are slaughtered for ivory, meat, and hides. Many African national parks were established to protect large game. South Africa's national park system began in 1926, when the Sabie Game Preserve of the eastern Transvaal region became Kruger National Park. Among South Africa's greatest attractions to foreign visitors, Kruger is famous for its population of lions and elephants.

East Africa is also known for outstanding game sanctuaries, such as Serengeti National Park, created prior to Tanzania's independence from Great Britain. Another national park in Tanzania, Kilimanjaro, protects Africa's highest and best-known mountain. Other African countries with well-developed park systems include Kenya, the Democratic Republic of the Congo (formerly Zaire), and Zambia. Although there is now a network of national parks in Africa that protects a wide range of habitats in various regions, there remains a need to protect additional areas in the arid northern part of the continent that includes the Sahara Desert.

Europe

In comparison with the United States, the national park concept spread more slowly within Europe. In 1910 Germany set aside Luneburger Heide National Park near the Elbe River, and in 1913, Sweden established Sarek, Stora Sjöfallet, Peljekasje, and Abisko National Parks. Swiss National Park was founded in Switzerland in 1914, in the Lower Engadine region. Great Britain has several national parks, including Lake District, a favorite recreation destination for English poet William Wordsworth. Spain's Doñana National Park, located on its southwestern coast, preserves the largest dune area on the European continent.

Asia

The system of land tenure and rural economy in many Asian countries has made it difficult for national governments to set aside large areas free from human exploitation. Many national parks established by colonial powers prior to World War II were maintained or expanded by countries following independence. For example, Kaziranga National Park is a refuge for the largest heard of rhinocerous in India. Established in 1962, Thailand's Khao Yai National Park protects a sample of the country's wildlife, while Indonesia's Komodo Island National Park preserves the habitat for the large lizards known as Komodo dragons.

In Japan, high population density has made it difficult to limit human activities within large areas. Some Japanese national parks are principally recreation areas rather than wildlife sanctuaries and may contain cultural features such as Shinto shrines. One of the best known national parks in Japan is Fuji-Hakone-Izu, which contains world-famous Mount Fuji, a volcano with a nearly symmetrical shape.

The Future

National parks serve as relatively undisturbed enclaves that protect examples of the world's most outstanding natural and cultural resources. The movement to establish these areas is a relatively recent attempt to achieve an improved balance between human activities and the earth. In recent years, rising incomes and lower costs for international travel have improved the accessibility of national parks to a larger number of persons, meaning that park visitation is likely to continue to rise.

Thomas A. Wikle

BOUNDARIES AND TIME ZONES

INTERNATIONAL BOUNDARIES

International boundaries are the marked or imaginary lines traversing natural terrain of land or water that mark off the territory of one politically organized society—a state or nation-state—from other states. In addition, states claim "air boundaries." While satellites circumnavigate the earth without nations' permission, airplanes and other air vessels that fly much lower must gain the permission of states over whose territory they travel.

The existence of international boundaries is a consequence of the "territoriality" that is a feature of modern human societies. All politically organized societies, except for nomadic tribes, claim to rule some exactly defined geographical territory. International boundaries provide the limits that define this territory.

International boundaries have ancient origins. For example, the oldest sections of the Great Wall of China date back to the Ch'in Dynasty of the second century BCE. The Roman Empire also maintained boundaries to its territories, such as Hadrian's Wall in the north of England, built by the Romans in 122 CE as a defensive barrier against marauders. In these and other ancient instances, however, there was little thought that borders must be exact.

The existence of precisely drawn boundaries among states is relatively recent. The modern state has existed for no more than a few hundred years. In addition, means to determine many boundaries have come into existence only in the nineteenth and twentieth centuries, with the invention of scientific methods and instruments, along with accompanying vocabulary, for determining exact boundaries. The most basic terms of this vocabulary begin with "latitude" and "longitude" and their subdivi-

sions into the "minutes" and "seconds" used in determining boundaries. In modern times, a new attitude toward states' territory was born, especially with the nineteenth century forms of nationalism, which tend to regard every acre of territory as sacred.

Types of Boundaries

There are several types of international boundaries. Some are geographical features, including rivers, lakes, oceans, and seas. Thus boundaries of the United States include the Great Lakes, which border Canada to the north; the Rio Grande, a river that forms part of the U.S. boundary with Mexico to the south; the Atlantic and Pacific Oceans, to the east and west, respectively; and the Gulf of Mexico, to the south. In Africa, Lake Victoria bounds parts of Tanzania, Uganda, and Kenya; and rivers, such as sections of the Congo and the Zambezi, form natural boundaries among many of the continent's states.

Other geographical features, such as mountains, often form international boundaries. The Pyrenees, for example, separate France and Spain and cradle the tiny state of Andorra. In South America, the Andes frequently serve as a boundary, such as between Argentina and Chile. The Himalayas in South Central Asia create a number of borders, such as between India, China, and Tibet and between Nepal, Butan, and their neighbors. When there are no clear geographical barriers between states, boundaries must be decided by mutual consent or the threat of force. In the 2016 presidential campaign, Donald Trump repeatedly called for a wall to be built between the United States and Mexico, claim-

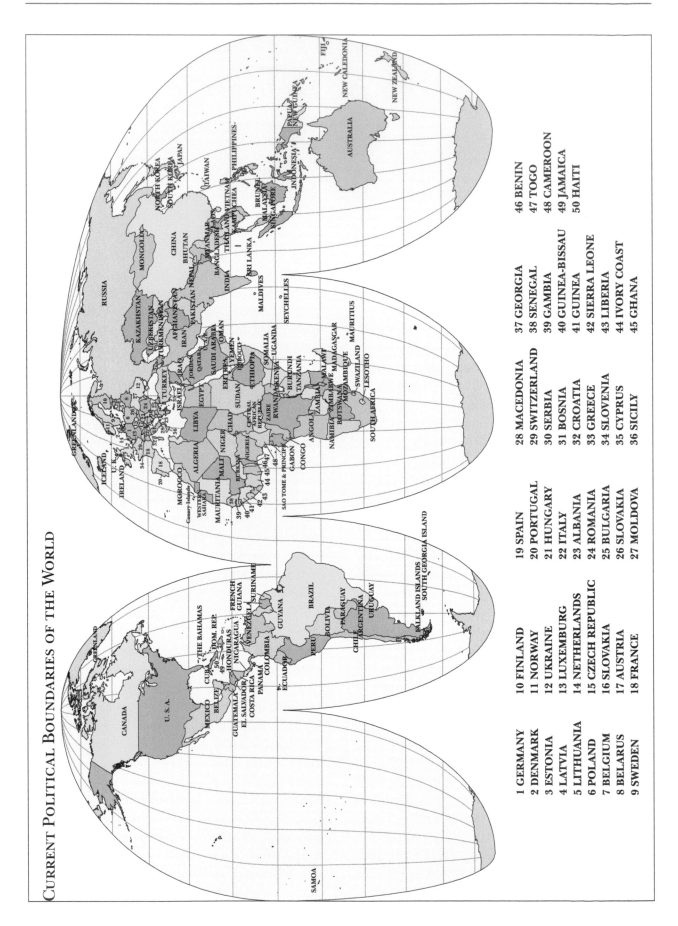

CURRENT POLITICAL BOUNDARIES OF THE WORLD

1 GERMANY
2 DENMARK
3 ESTONIA
4 LATVIA
5 LITHUANIA
6 POLAND
7 BELGIUM
8 BELARUS
9 SWEDEN

10 FINLAND
11 NORWAY
12 UKRAINE
13 LUXEMBURG
14 NETHERLANDS
15 CZECH REPUBLIC
16 SLOVAKIA
17 AUSTRIA
18 FRANCE

19 SPAIN
20 PORTUGAL
21 HUNGARY
22 ITALY
23 ALBANIA
24 ROMANIA
25 BULGARIA
26 SLOVAKIA
27 MOLDOVA

28 MACEDONIA
29 SWITZERLAND
30 SERBIA
31 BOSNIA
32 CROATIA
33 GREECE
34 SLOVENIA
35 CYPRUS
36 SICILY

37 GEORGIA
38 SENEGAL
39 GAMBIA
40 GUINEA-BISSAU
41 GUINEA
42 SIERRA LEONE
43 LIBERIA
44 IVORY COAST
45 GHANA

46 BENIN
47 TOGO
48 CAMEROON
49 JAMAICA
50 HAITI

ing that Mexico would pay for it. As of 2019, the wall has not been completed, however.

Creation and Change of International Boundaries

War and conquest often have been used to determine borders. Such wars, however, historically have created hostility among losers. Political pressures to recover lost lands build up among aggrieved losers, and such irredentist claims provide fuel for future wars. A classic example is the fate of the regions of Alsace and Lorraine between France and Germany. Although natural resources in the form of coal played a substantial role in the dispute over this area, national pride was also a potent element.

Whether boundaries are fixed through compelling geographical imperatives or in their absence, states typically sign treaties agreeing to their location. These may be treaties that conclude wars, or boundary commissions set up by those involved may draw up borders to which states give formal agreement. In 1846, for example, negotiators for Great Britain and the United States settled on the forty-ninth parallel as the boundary between the western United States and Canada, although in the United States, "Fifty-four [degrees latitude] Forty [minutes] or Fight" had been a popular motto in the presidential election campaign of 1844.

Sometimes no accepted borders exist because of chronic hostility between states. Thus, maps of the Kashmir region between India and Pakistan, claimed by both countries, show only a "line of control" or cease-fire line to divide the two warring states. Similarly, only a cease-fire line, drawn at the armistice of the Korean War of 1950-1953, divides North and South Korea; a mutually agreed-upon border remains unfixed.

In rare instances, no true boundary exists to mark where a state's territory begins and ends. Classic cases are found on the Arabian Peninsula, where the land borders of principalities, known as the Gulf Sheikdoms, are vague lines in the sand. Such circumstances usually create no difficulties where nothing is at stake, but when oil is discovered, states must come to agreement or risk coming to blows.

In other instances, negotiations and international arbitration have been effective for determining borders. Perhaps the most important principle for determining the borders of newly created states is found in the Latin phrase, *Uti possidetis iurus*. This principle is used when states become independent after having been colonies or constituent parts of a larger state that has broken up. The principle holds that states shall respect the borders in place when they were colonies. *Uti possidetis* was first extensively used in South America in the nineteenth century, when European colonial powers withdrew, leaving several newly born states to determine their own boundaries. The principle may be used as a basis for border agreements among the fifteen states of the former Soviet Union.

Besides war and negotiation, purchase has sometimes been a means of creating international boundaries. For example, in 1853 the United States purchased territory from Mexico in the southwest; in 1867, it purchased Alaska from Russia.

In rare cases, natural boundaries may change naturally or be changed deliberately by one side, incurring resentment among victims. An example occurred in 1997, when Vietnam complained that China had built an embankment on a border river embankment that caused the river to change its course; China countered that Vietnam had built a dam altering the river's course.

Other border difficulties among states include conflicts over water that flows from one country to another. In the 1990s, for example, Mexico complained of excessive U.S. use of Colorado River waters and demanded adjustment.

Border Disputes

Border disputes among states in the past two centuries have been numerous and lethal. In the twentieth century, numerous such controversies degenerated into violence. In Asia, India and Pakistan fought over Kashmir, beginning in 1947-1949 and recurring in 1965 and 1999. China has been involved in violent border disputes with India, especially in 1962; Vietnam in 1979; and Russia in 1969. In South America, border wars between Ecuador and Peru broke out in 1941, 1981, and 1995. This dispute was settled by negotiation in 1998. In Africa, among numerous recent armed conflicts, the bloody border conflict between Eritrea and Ethiopia in the 1990s was notable.

Other recent disputes have ended peacefully. Eritrea avoided violence with Yemen over several Red Sea islands by accepting arbitration by an international tribunal. In 1995 Saudi Arabia and the United Arab Emirates negotiated a peaceful agreement to their border dispute involving oil rights.

As of 2019, there are four ongoing border conflicts: Israeli-Syrian ceasefire line incidents during the Syrian Civil War, the War in Donbass, India-Pakistan military confrontation, and the 2019 Turkish offensive into north-eastern Syria, code-named by Turkey as Operation Peace Spring. Many unresolved boundary disputes might yet lead to conflicts. Among the most complex is the multinational dispute over the 600 tiny Spratly Islands in the South China Sea. Uninhabited but potentially valuable because of oil, the Spratlys are claimed by China, Brunei, Malaysia, Indonesia, the Philippines, Taiwan, and Vietnam.

Border Policies

Problems with international borders are not limited to territorial disputes. Policies regarding how borders should be operated—including the key questions of who and what should be allowed entrance and exit under what conditions—can be expected to continue as long as independent states exist. While the members of the European Union have agreed to allow free passage of people and goods among themselves, this policy does not extent to nonmembers.

The most important purpose of states is to protect the lives and property of their citizens. One of the principal purposes of international boundaries is to further this purpose. Most states insist on controlling their borders, although borders seem increasingly porous. Given the imperatives of control and the increasing difficulties of maintaining it, issues surrounding international borders are expected to continue indefinitely in the twenty-first century.

Charles F. Bahmueller

GLOBAL TIME AND TIME ZONES

Before the nineteenth century, people kept time by local reckoning of the position of the Sun; consequently, thousands of local times existed. In medieval Europe, "hours" varied in length, depending upon the seasons: Each hour was determined by the Roman Catholic Church. In the sixteenth century, Holy Roman emperor Charles V was the first secular ruler to decree hours to be of equal length. As the industrial and scientific revolutions swept Europe, North America, and other areas, some form of time standardization became necessary as communities and regions increasingly interacted. In 1780 Geneva, Switzerland, was the first locality known to employ a standard time, set by the

town-hall clockkeeper, throughout the town and its immediate vicinity.

The growth and expansion of railroads, providing the first relatively fast movement of people and goods from city to city, underscored the need for a standard system in Great Britain. As early as 1828, Sir John Herschel, Astronomer Royal, called for a national standard time system based on instruments at the Royal Observatory at Greenwich. That practice began in 1852, when the British telegraph system had developed sufficiently for the Greenwich time signals to be sent instantly to any point in the country.

As railroads expanded through North America, they exposed a problem of local time variation similar to that in Great Britain but on a far larger scale, since the distances between the East and West Coasts were much greater than in Great Britain. In order for long-distance train schedules to work, different parts of the country had to coordinate their clocks. The first to suggest a standard time framework for the United States was Charles F. Dowd, president of Temple Grove Seminary for Women in Saratoga Springs, New York. Initially, Dowd proposed putting all U.S. railroads on a single standard time, based on the time in Washington, D.C. When he realized that the time in California would be behind such a standard by almost four hours, he produced a revised system, establishing four time zones in the United States. Dowd's plan, published in 1870, included the first known map of a time zone system for the country.

Not everyone was happy with the designation of Washington, D.C., as the administrative center of time in the United States. Northeastern railroad executives urged that New York, the commercial capital of the nation, be used instead: Many cities and towns in the region already had standardized to New York time out of practical necessity. Dowd proposed a compromise: to set the entire national time zone system in the United States using the Greenwich prime meridian, already in use in many parts of the world for maritime and scientific purposes. In 1873 the American Association of Railways (AAR) flatly rejected the proposal.

In the end, Dowd proved to be a visionary. In 1878 Sandford Fleming, chief engineer of the government of Canada, proposed a worldwide system of twenty-four time zones, each fifteen degrees of longitude in width, and each bisected by a meridian, beginning with the prime meridian of Greenwich. William F. Allen, general secretary of the AAR and armed with a deep knowledge of railroad practices and politics, took up the crusade and persuaded the railroads to agree to a system. At noon on Sunday, November 18, 1883, most of the more than six hundred U.S. railroad lines dropped the fifty-three arbitrary times they had been using and adopted Greenwich-indexed meridians that defined the times in each of four times zones: eastern,

central, mountain, and Pacific. Most major cities in the United States and Canada followed suit.

Time System for the World

Almost at the same time that American railroads adopted a standard time zone system, the State Department, authorized by the United States Congress, invited governments from around the world to assemble delegates in Washington, D.C., to adopt a global system. The International Meridian Conference assembled in the autumn of 1884, attended by representatives of twenty-five countries. Led by Great Britain and the United States, most favored adoption of Greenwich as the official prime meridian and Greenwich mean time as universal time.

There were other contenders: The French wanted the prime meridian to be set in Paris, and the Germans wanted it in Berlin; others proposed a mountaintop in the Azores or the tip of the Great Pyramid in Egypt. Greenwich won handily. The conference also agreed officially to start the universal day at midnight, rather than at noon or at sunrise, as practiced in many parts of the world. Each time zone in the world eventually came to have a local name, although technically, each goes by a letter in the alphabet in order eastward from Greenwich.

Once a global system was in place, there was a new issue: Many jurisdictions wanted to adjust their clocks for part of the year to account for differences in the number of hours of daylight between summer and winter months. In 1918 Congress decreed a system of daylight saving time for the United States but almost immediately abolished it, leaving state governments and communities to their local options. Daylight saving time, or a form of it, returned in the United States and many Allied nations during World War II. In the Uniform Time Act of 1966, Congress finally established a national system of daylight saving time, although with an option for states to abstain.

To the extent that it indicates how human communities want to manipulate time for social, political, or economic reasons, the issue of daylight saving time, rather than the establishment of a system of world time zones, is a better clue to the geographical issues involved in time administration. Both the history and the present format of the world time zone system show that the mathematically precise arrangement envisioned by many of the pioneers of time zones is not as important as things on the ground.

In the United States, the railroad time system adopted in 1883 drew the boundary between eastern time and central time more or less between the thirteen original states and the trans-Appalachian West: The entire Midwest, including Ohio, Indiana, and Michigan, fell in the central time zone. As the center of population migrated westward, train speeds increased, highways developed, and New York emerged as the center of mass media in the United States, the boundary between the eastern and central time zones marched steadily westward. In 1918 it ran down the middle of Ohio; by the 1960s, it was at the outskirts of Chicago.

One of the principal reasons for the popularity of Greenwich as the site of the prime meridian (zero degrees longitude), is that it places the international date line (180 degrees longitude)—where, in effect, time has to move forward to the next day rather than the next hour—far out in the Pacific Ocean where few people are affected by what otherwise would be an awkward arrangement. However, even this line is somewhat irregular, to avoid placing a small section of eastern Russia and some of the Aleutian Islands of the United States in different days.

Coordinated Universal Time Coordinated Universal Time (or UTC) is the primary time standard by which the world regulates clocks and time. It is within about 1 second of mean solar time at 0° longitude, and is not adjusted for daylight saving time. In some countries, the term Greenwich Mean Time is used. The co-ordination of time and frequency transmissions around the world began on January 1, 1960. UTC was first officially adopted as CCIR Recommendation 374, Standard-Frequency and Time-Signal Emissions, in 1963, but the official abbreviation of UTC and the official English name of Coordinated Universal Time (along with the French equivalent) were not adopted until 1967. UTC uses a *slightly* different second called the *SI second*. That is based on *atomic clocks*. Atomic clocks are more regular than the slightly variable Earth's rotation period. Hence, the essential difference between GMT and UTC is that they use different definitions of exactly how long one second of time is.

By 1950 most nations had adopted the universal time zone system, although a few followed later: Saudi Arabia in 1962, Liberia in 1972. Despite adhering to the system in principle, many nations take considerable liberties with the zones, especially if their territory spans several. All of Western Europe, despite covering an area equivalent to two zones, remains on a single standard. The People's Republic of China, which stretches across five different time zones, arbitrarily sets the entire country officially on Beijing time, eight hours behind Greenwich. Iran, Afghanistan, India, and Myanmar, each of which straddle time zone boundaries, operate on half-hour compromise systems as their time standards (as does Newfoundland). As late as 1978, Guyana's standard time was three hours, forty-five minutes in advance of Greenwich.

It can be argued that adoption of a worldwide system of time zones in the late nineteenth century was one of the earliest manifestations of the emergence of a global economy and society, and has been a crucial factor in the unfolding of this process throughout the twentieth century and beyond.

Ronald W. Davis

GLOBAL EDUCATION

THEMES AND STANDARDS IN GEOGRAPHY EDUCATION

Many people believe that the study of geography consists of little more than knowing the locations of places. Indeed, in the past, whole generations of students grew up memorizing states, capitals, rivers, seas, mountains, and countries. Most students found that approach boring and irrelevant. During the 1990s, however, geography education in the United States underwent a remarkable transformation.

While it remains important to know the locations of places, geography educators know that place name recognition is just the beginning of geographic understanding. Geography classes now place greater emphasis on understanding the characteristics of and the connections between places. Three things have led to the renewal of geography education: the five themes of geography, the national geography standards, and the establishment of a network of geographic alliances.

The Five Themes of Geography
One of the first efforts to move geography education beyond simple memorization was the National Geographic Society's publication of five themes of geography in 1984: location, place, human-environment interactions, movement, and regions. Not intended to be a checklist or recipe for understanding the world, these themes merely provided a framework for teachers—many of whom did not have a background in the subject—to incorporate geography throughout a social studies curriculum. The five themes were promoted widely by the National Geographic Society and are still used by some teachers to organize their classes.

Location is about knowing where things are. Both the absolute location (where a place is on earth's surface) and relative location (the connections between places) are important. The concept of place involves the physical and human characteristics that distinguish one place from another. The theme of human/environment interaction recognizes that people have relationships within defined places and are influenced by their surroundings. For example, many different types of housing have been created as adaptations to the world's diverse climates. The theme of movement involves the flow of people, goods, and ideas around the world. Finally, regions are human creations to help organize and understand Earth, and geography studies how they form and change.

The National Geography Standards
Geography was one of six subjects identified by President George H. W. Bush and the governors of the U.S. states when they formulated the National Education Goals in 1989. While the goals themselves foundered amid the political debate that followed their adoption, one tangible result of the initiative was the creation of Geography for Life: The National Geography Standards. More than 1,000 teachers, professors, business people, and government officials were involved in the writing of Geography for Life. The project was supported by four geography organizations: the American Geographical Society, the Association of American Geographers, the National Council for Geographic Education, and the National Geographic Society. The resulting book defines what every U.S. student

---

GEOGRAPHY STANDARDS

The geographically informed person knows and understands the following:

- how to use maps and other geographic representations, tools, and technologies to acquire, process, and report information from a spatial perspective;
- how to use mental maps to organize information about people, places, and environments in a spatial context;
- how to analyze the spatial organization of people, places, and environments on Earth's surface;
- the physical and human characteristics of places;
- that people create regions to interpret Earth's complexity;
- how culture and experience influence people's perceptions of places and regions;
- the physical processes that shape the patterns of Earth's surface;
- the characteristics and spatial distribution of ecosystems on Earth's surface;
- the characteristics, distribution, and migration of human populations on Earth's surface;
- the characteristics, distribution, and complexity of Earth's cultural mosaics;
- the patterns and networks of economic interdependence on Earth's surface;
- the processes, patterns, and functions of human settlement;
- how the forces of cooperation and conflict among people influence the division and control of Earth's surface;
- how human actions modify the physical environment;
- how physical systems affect human systems;
- the changes that occur in the meaning, use, distribution, and importance of resources;
- how to apply geography to interpret the past;
- how to apply geography to interpret the present and plan for the future.

Source: National Geography Standards Project. Geography for Life: National Geography Standards, Second Edition. Washington, D.C.: National Geographics Research and Exploration, 2012.

---

should know and be able to accomplish in geography.

Each of the eighteen standards is designed to develop students' geographic skills, including asking geographic questions; acquiring, organizing, and analyzing geographic information; and answering the questions. Each standard features explanations, examples, and specific requirements for students in grades four, eight, and twelve.

Geography Alliances and the Future of Geography Education

To publicize efforts in geography education, a network of geography alliances was established between 1986 and 1993. Today, each U.S. state has a geography alliance that links teachers and organizations such as the National Geographic Society and the National Council for Geographic Education to sponsor workshops, teacher training sessions, field experiences, and other ways of sharing the best in geographic teaching and learning.

A 2013 executive summary prepared by the National Geographic Society for the *Road Map for 21st Century Geography Education Project* continues to champion the goal of better geography education in K–12 schools. The Road Map Project represents the collaborative effort of four national organizations: the American Geographical Society (AGS), the Association of American Geographers (AAG), the National Council for Geographic Education (NCGE), and the National Geographic Society (NGS). The project partners share belief that geography education is essential for student success in all aspects of their adult lives—careers, civic lives, and personal decision making. It also is essential for the education of specialists who can help society addressing critical issues in the areas of social welfare, economic stability, environmental health, and international relations.

Eric J. Fournier

GLOBAL DATA

WORLD GAZETTEER OF OCEANS AND CONTINENTS

Places whose names are printed in SMALL CAPS *are subjects of their own entries in this gazetteer.*

Aden, Gulf of. Deep-water area between the RED and ARABIAN SEAS, bounded by Somalia, Africa, on the south and Yemen on the north. Water is warmer and saltier in the Gulf of Aden than in the Red and Arabian Seas, because little water enters from rain or land runoff.

Africa. Second-largest continent, connected to ASIA by the narrow isthmus of Suez. Bounded on the east by the INDIAN OCEAN and on the west by the ATLANTIC OCEAN. Countries of Africa are Algeria, Angola, Benin, Botswana, Burkina Faso, Burundi, Cameroon, Central African Republic, Chad, Congo, Côte d'Ivoire (Ivory Coast), the Democratic Republic of Congo, Egypt, Ethio-

pia, Gabon, Gambia, Ghana, Guinea, Kenya, Liberia, Libya, Madagascar, Malawi, Mali, Mauritania, Morocco, Mozambique, Namibia, Niger, Nigeria, Rio Muni (Mbini), Rwanda, Senegal, Sierra Leone, Somalia, South Africa, Sudan, Tanzania, Togo, Tunisia, Uganda, Western Sahara, Zambia, and Zimbabwe. Climate ranges from hot and rainy near the equator, to hot and dry in the huge Sahara Desert in the north and the Kalahari Desert in the south, to warm and fairly mild at the northern and southern extremes. Paleontological evidence indicates that humans originally evolved in Africa.

Agulhas Current. Warm, swift ocean current moving south along East AFRICA's coast. Part moves between AFRICA and MADAGASCAR to form the Mozambique Current. The warm water of the

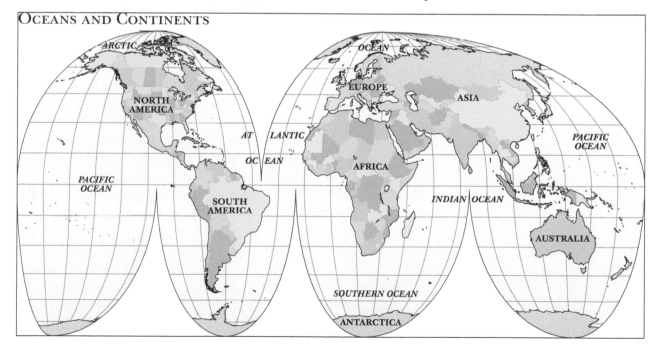

OCEANS AND CONTINENTS

Agulhas Current increases the average temperatures in the eastern part of South Africa.

Agulhas Plateau. Relatively small ocean-bottom plateau that lies south of South AFRICA, at the area where the INDIAN and ATLANTIC OCEANS meet.

Aleutian Islands. Chain of volcanic islands that extends 1,100 miles (1,770 km.) from the tip of the Alaska Peninsula to the Kamchatka Peninsula in Russia and forms the boundary between the North PACIFIC OCEAN and the BERING SEA. The area is hazardous to navigation and has been called the "Home of Storms."

Aleutian Trench. Located on the northern margin of the PACIFIC OCEAN, stretching 3,666 miles (5,900 km.) from the western edge of the Aleutian Island chain to Prince William Sound, Alaska. Depth is 25,263 feet (7,700 meters).

American Highlands. Elevated region on the ANTARCTIC coast between Enderby Land and Wilkes Land, located far south of India. The Lambert and Fisher glaciers originate in the American Highlands and move down to feed the AMERY ICE SHELF.

Amery Ice Shelf. Year-round shelf of relatively flat ice in a bay of ANTARCTICA, located at approximately longitude 70 degrees east, between MAC. ROBERTSON LAND and the AMERICAN HIGHLANDS. The ice shelf is fed by the Lambert and Fisher glaciers.

Amundsen Sea. Portion of the southernmost PACIFIC OCEAN off the Wahlgreen Coast of ANTARCTICA, approximately longitude 100 to 120 degrees west. Named for the Norwegian explorer Roald Amundsen, who became the first person to reach the SOUTH POLE in 1911.

Antarctic Circle. Latitude of 66.3 degrees south. South of this line, the Sun does not set on the day of the summer solstice, about December 22 in the SOUTHERN HEMISPHERE, and does not rise on the day of the winter solstice, about June 21.

Antarctic Circumpolar Current. Eastward-flowing current that circles ANTARCTICA and extends from the surface to the deep ocean floor. The largest-volume current in the oceans. Extends northward to approximately 40 degrees south latitude and is driven by westerly winds.

Antarctic Convergence. Meeting place where cold Antarctic water sinks below the warmer sub-Antarctic water.

Antarctic Ocean. See SOUTHERN OCEAN.

Antarctica. Fifth-largest continent, located at the southernmost part of the world. There are two major regions; western Antarctica, which includes the mountainous Antarctic peninsula, and eastern Antarctica, which is mostly a low continental shield area. An ice cap up to 13,000 feet (4,000 meters) thick covers 95 percent of the continent's surface. Temperatures in the austral summer (December, January, and February) rarely rise above 0 degrees Fahrenheit (-18 degrees Celsius) except on the peninsula. By international treaty, the continent is not owned by any single country, and human access is largely regulated. There has never been a self-supporting human habitation on Antarctica.

Arabian Sea. Portion of the INDIAN OCEAN bounded by India on the east, Pakistan on the north, and Oman and Yemen of the Arabian Peninsula on the west.

Arctic Circle. Latitude of 66.3 degrees north. North of this line, the Sun does not set on the day of the summer solstice, about June 21 in the NORTHERN HEMISPHERE, and does not rise on the day of the winter solstice, about December 22.

Arctic Ocean. World's smallest ocean. It centers on the geographic NORTH POLE and connects to the PACIFIC OCEAN through the BERING SEA, and to the ATLANTIC OCEAN through the GREENLAND SEA. The Arctic Ocean is covered with ice up to 13 feet (4 meters) thick all year, except at its edges. Norwegian explorers on the ship *Fram* stayed locked in the icepack from 1893 to 1896, in order to study the movement of polar ice. They drifted in the ice a total of 1,028 miles (1,658 km.), from the Bering Sea to the Greenland Sea, proving that there was no land mass under the Arctic ice at the top of the world. Also known as Arctic Sea or Arctic Mediterranean Sea.

Argentine Basin. Basin on the floor of the western ATLANTIC OCEAN, off the coast of Argentina in SOUTH AMERICA. Among ocean basins, this one is unusually circular.

Ascension Island. Isolated volcanic island in the South ATLANTIC OCEAN, about midway between SOUTH AMERICA and AFRICA. One of the islands visited by British biologist Charles Darwin during his five-year voyage on the *Beagle*.

Asia. Largest continent; joins with EUROPE to form the great Eurasian landmass. Asia is bounded by the ARCTIC OCEAN on the north, the western PACIFIC OCEAN on the east, and the INDIAN OCEAN on the south. Its countries include Afghanistan, Bahrain, Bangladesh, Bhutan, Cambodia, China, India, Iran, Iraq, Irian Jaya, Israel, Japan, Jordan, Kalimantan, Kazakhstan, North and South Korea, Kyrgyzstan, Laos, Lebanon, Malaysia, Myanmar, Mongolia, Nepal, Oman, Pakistan, the Philippines, Russia, Sarawak, Saudi Arabia, Sri Lanka, Sumatra, Syria, Tajikistan, Thailand, Asian Turkey, Turkmenistan, United Arab Emirates, Uzbekistan, Vietnam, and Yemen. Climates include virtually all types on earth, from arctic to tropical, desert to rain forest. Asia has the highest (Mount Everest) and lowest (Dead Sea) surface points in the world. Nearly 60 percent of the world's people live in Asia.

Atlantic Ocean. Second-largest body of water in the world, covering more than 25 percent of Earth's surface. Bordered by NORTH and SOUTH AMERICA on the west, and EUROPE and East AFRICA on the east. The widest part (5,500 miles/8,800 km.) lies between West AFRICA and Mexico, along 20 degrees latitude. Scientists disagree on the north-south boundaries of the Atlantic; if one includes the ARCTIC OCEAN and the SOUTHERN OCEAN, the Atlantic Ocean extends about 13,300 miles (21,400 km.). The deepest spot (28,374 feet/8,648 meters) is found in the PUERTO RICO TRENCH. The Atlantic Ocean has been a major route for trade and communications, especially between North America and Europe, for hundreds of years. This is because of its relatively narrow size and favorable currents, such as the GULF STREAM.

Australasia. Loosely defined term for the region, which, at the least, includes AUSTRALIA and New Zealand; at the most, it also includes other South Pacific Islands in the region.

Australia. Smallest continent, sometimes called the "island continent." Located between the INDIAN and PACIFIC OCEANS. It is the only continent occupied by a single nation, the Commonwealth of Australia. Australia is the flattest and driest continent; two-thirds is either desert or semiarid. Geologically, it is the oldest and most isolated continent. Unlike any other place on Earth, large mammals never evolved in Australia. Marsupials (pouched, warm-blooded animals) and unusual birds developed in their place.

Azores. Archipelago (group of islands) in the eastern ATLANTIC OCEAN lying about 994 miles (1,600 km.) west of Portugal. The islands are of volcanic origin and have been known, fought over, and used by the Europeans since before the fourteenth century. Spanish explorer Christopher Columbus stopped in the Azores to wait for favorable winds before his first trip across the ATLANTIC OCEAN.

Barents Sea. Partially enclosed section of the ARCTIC OCEAN, bounded by Russia and Norway on the south and the Russian island of Navaya Zemlaya on the east. The Barents Sea was important in World War II because Allied convoys had to cross it, through storms and submarine patrols, to deliver war supplies to Murmansk, the only ice-free port in western Russia. It was named for the Dutch explorer Willem Barents.

Bays. See under individual names.

Beaufort Sea. Area of the ARCTIC OCEAN located off the northern coast of Alaska and western Canada. It is usually frozen over and has no islands. Named for British admiral Sir Francis Beaufort, who devised the Beaufort Wind Scale as a means of classifying wind force at sea.

Bengal, Bay of. Northeast arm of the INDIAN OCEAN, bounded by India on the west and Myanmar on the east. The Ganges River empties into the Bay of Bengal. The great ports of Calcutta and Madras in India, and Rangoon in Myanmar lie in the bay, making it a busy and important area for shipping for centuries.

Benguela Current. Northward-flowing current along the western coast of Southern AFRICA. Normally, the Benguela Current carries cold, rich water that wells up from the ocean depths and supports a large fishing industry. A change in winds can reduce the oxygen supply and kill huge numbers of fish, similar to what may happen off the coast of Peru during El Niño weather conditions.

Bering Sea. Portion of the northernmost PACIFIC OCEAN that is bounded by the state of Alaska on the east, Russia and the Kamchatka Peninsula on the west, and the BERING STRAIT on the north. It is a valuable fishing ground, rich in shrimp, crabs, and fish. Whales, fur seals, sea otters, and walrus are also found there.

Bering Strait. Narrowest point of connection between the BERING SEA and the ARCTIC OCEAN, located between the easternmost point of Siberia on the west and Alaska on the east. The Bering Strait is 52 miles (84 km.) wide. During the Ice Age, when the sea level was lower, humans and animals were able to walk from the Asian continent across a land bridge—now known as Beringia—to the North American continent across the frozen strait, providing the first human access to the Americas.

Bikini Atoll. Small atoll in the Marshall Islands group in the western PACIFIC OCEAN. In the 1940s, the United States began testing nuclear bombs on Bikini and neighboring atolls. The U.S. Army removed the inhabitants of Bikini, and testing occurred from 1946 to 1958. The Bikini inhabitants were allowed to return in 1969, then removed again in 1978 when high levels of radioactivity were found to remain.

Black Sea. Large inland sea situated where southeastern EUROPE meets ASIA; connected to the MEDITERRANEAN SEA through Turkey's Bosporus strait. The sea covers an area of about 178,000 square miles (461,000 sq. km.), with a maximum depth of more than 7,250 feet (2,210 meters).

Brazil Current. Extension of part of the warm, westward-flowing South EQUATORIAL CURRENT, which turns south to the coast of Brazil. The Brazil Current has very salty water because of its long flow across the equator. It joins the WEST WIND DRIFT and moves eastward across the South ATLANTIC OCEAN as part of the SOUTH ATLANTIC GYRE.

California, Gulf of. Branch of the eastern PACIFIC OCEAN that separates Baja California from mainland Mexico. Warm, nutrient-rich water supports a variety of fish, oysters, and sponges. California gray whales migrate to the gulf to give birth and breed, January through March. Fisheries and tourism are important industries in the Gulf of California. Also known as the Sea of Cortés.

California Current. Cool water that flows southeast along the western coast of NORTH AMERICA from Washington State to Baja California. The eastern portion of the NORTH PACIFIC GYRE.

Canada Basin. Part of the ocean floor that lies north of northeastern Canada and Alaska. The BEAUFORT SEA lies above the Canada Basin.

Cape Horn. Southernmost tip of SOUTH AMERICA. It is the site of notoriously severe storms and is hazardous to shipping.

Cape Verde Plateau. ATLANTIC OCEAN plateau lying off the western bulge of the AFRICAN continent. The volcanic Cape Verde Islands lie on the plateau.

Caribbean Sea. Portion of the western ATLANTIC OCEAN bounded by CENTRAL and SOUTH AMERICA to the west and south, and the islands of the Antilles chain on the north and east. Mostly tropical in climate, the Caribbean Sea supports a large variety of plant and animal life. Its islands, including Puerto Rico, the Cayman Islands, and the Virgin Islands, are popular tourist sites.

Caspian Sea. World's largest inland sea. Located east of the Caucasus Mountains at EUROPE's southeasternmost extremity, it dominates the expanses of western Central ASIA. Its basin is 750 miles (1,200 kilometers) long, and its average width is 200 miles (320 kilometers). It covers 149,200 square miles (386,400 sq. km.).

Central America. Region generally understood to constitute the irregularly shaped neck of land linking North and South America, containing Belize, Guatemala, Honduras, El Salvador, Nicaragua, Costa Rica, and Panama.

Chukchi Sea. Portion of the Arctic Ocean, bounded by the Bering Strait on the south, Siberia on the southwest, and Alaska on the southeast. The Chukchi Sea is the area of exchange between waters and sea life of the Pacific and Arctic Oceans, and so is an area of interest to oceanographers and fishermen.

Clarion Fracture Zone. East-west-running fracture zone that begins off the west coast of Mexico and extends approximately 2.500 miles (4,023 km.) to the southwest.

Cocos Basin. Relatively small ocean basin located off the west coast of Sumatra in the northeast Indian Ocean.

Coral Sea. Area of the Pacific Ocean off the northeast coast of Australia, between Australia on the southwest, Papua New Guinea and the Solomon Islands on the northeast, and New Caledonia on the east. Site of a naval battle in 1942 that prevented the Japanese invasion of Australia.

Cortés, Sea of. See California, Gulf of.

Denmark Strait. Channel that separates Greenland and Iceland and connects the North Atlantic Ocean with the Arctic Ocean.

Dover, Strait of. Body of water between England and the European continent, separating the North Sea from the English Channel. It is 33 miles (53 km.) wide at its narrowest point. The tunnel between England and France (known as the "Chunnel") was cut into the rock under the Strait of Dover.

Drake Passage. Narrow part of the Southern Ocean that connects the Atlantic and Pacific Oceans between the southern tip of South America and the Antarctic peninsula. Named for sixteenth-century English navigator and explorer Sir Francis Drake, who discovered the passage when his ship was blown into it during a violent storm. Also called Drake Strait.

East China Sea. Area of the western Pacific Ocean bounded by China on the west, the Yellow Sea on the north, and Japan on the northeast. Large oil deposits were found under the East China Sea floor in the 1980s.

East Pacific Rise. Broad, nearly continuous undersea mountain range that extends from the southern end of Baja California southward, then curves east near Antarctica. It is formed along the southeast side of the Pacific Plate and is part of the Ring of Fire, a nearly continuous ring of volcanic and tectonic activity around the rim of the Pacific Ocean. Also called East Pacific Ridge.

East Siberian Sea. Portion of the Arctic Ocean bounded by the Chukchi Sea on the east, Siberia on the south, and the Laptev Sea on the west. Much of the East Siberian Sea is covered with ice year-round.

Eastern Hemisphere. The half of the earth containing Europe, Asia, and Africa; generally understood to fall between longitudes 20 degrees west and 160 degrees east.

El Niño. Conditions—also known as El Niño-Southern Oscillation (ENSO) events—that occur every two to ten years and cause weather and ocean temperature changes off the coast of Ecuador and Peru. Most of the time, the Peru Current causes cold, nutrient-rich water to well up off the coast of Ecuador and Peru. During ENSO years, the cold upwelling is replaced by warmer surface water that does not support plankton and fish. Fisheries decline and seabirds starve. Climatic changes of El Niño can bring floods to normally dry areas and drought to wet areas. Effects can extend across North and South America, and to the western Pacific Ocean. During the 1990s, the ENSO event fluctuated but did not go completely away, which caused tremendous damage to fisheries and agriculture, storms and droughts in North America, and numerous hurricanes.

Emperor Seamount Chain. Largest known example of submerged underwater volcanic ridges, located in the northern Pacific Ocean and extending southward from the Kamchatka Peninsula in Russia for about 2,500 miles (4,023 km.).

Enderby Land. Section of Antarctica that lies between the Indian Ocean and the South Polar Plateau, east of Queen Maud Land. Enderby Land lies between approximately longitude 45 and 60 degrees east.

English Channel. Strait water separating continental France from Great Britain. Runs for roughly 350 miles (560 km.), from the Atlantic Ocean in the west to the Strait of Dover in the east.

Equatorial Current. Currents just north and south of the equator that flow from east to west. Equatorial currents are found in the Pacific and Atlantic Oceans. The equatorial currents and the trade winds, which move in the same direction, greatly aid oceangoing traffic.

Eurasia. Term for the combined landmass of Europe and Asia.

Europe. Sixth-largest continent, actually a large peninsula of the Eurasian landmass. Europe is densely populated and includes the countries of Albania, Andorra, Austria, Belarus, Belgium, Bulgaria, Bosnia-Herzegovina, Croatia, the Czech Republic, Denmark, Estonia, Finland, France, Germany, Greece, Hungary, Iceland, Ireland, Italy, Latvia, Lithuania, Macedonia, Malta, Moldova, Monaco, the Netherlands, Norway, Poland, Portugal, Romania, Slovakia, Spain, Switzerland, Turkey, and the United Kingdom (England, Northern Ireland, Scotland, and Wales). Climate ranges from near arctic in the north, to temperate and Mediterranean in the south.

Florida Current. Water moving northward along the east coast of Florida to Cape Hatteras, North Carolina, where it joins the Gulf Stream.

Fundy, Bay of. Large inlet on the North American Atlantic coast, northwest of Maine, separating New Brunswick and Nova Scotia in Canada. Renowned for having the largest tidal change in the world, more than 56 feet (17 meters).

Galápagos Islands. Located directly on the equator, 600 miles (965 km.) west of Ecuador. The islands are volcanic in origin and sit directly in the cold Peru Current, which cools the islands and creates unusual microclimates and fogs. The extreme isolation of the islands allowed unique species to develop. Biologist Charles Darwin visited the Galápagos in the 1830s, and the unusual organisms he observed helped him to conceive the theory of evolution.

Galápagos Rift. Divergent plate boundary extending between the Galápagos Islands and South America. The first hydrothermal vent community was discovered in 1977 in the Galápagos Rift. This unusual type of biological habitat is based on energy from bacteria that use heat and chemicals to make food, instead of sunlight.

Grand Banks. Portion of the northwest Atlantic Ocean southeast of Nova Scotia and Newfoundland. The Grand Banks are extremely rich fishing grounds, although in the 1980s and 1990s catches of cod, flounder, and many other fish dropped dramatically due to overfishing and pollution.

Great Barrier Reef. Largest coral reef in the world, lying in the Coral Sea off the east coast of Australia. The reef system and its small islands stretch for more than 1,100 miles (1,750 km.) and is difficult to navigate through. The reefs are home to an incredible variety of tropical marine life, including large numbers of sharks.

Greenland. Largest island in the world that is not rated as a continent; lies between the northernmost part of the Atlantic Ocean and the Arctic Ocean, northeast of the North American continent. About 90 percent of Greenland is permanently covered with an ice sheet and glaciers. Residents engage in limited agriculture, growing potatoes, turnips, and cabbages. Most people live along the southwest coast, where the climate is warmed by the North Atlantic current.

Greenland Sea. Body of water bounded by Greenland on the west, Iceland on the north, and Spitsbergen on the east. It is often ice-covered.

Guinea, Gulf of. Arm of the North Atlantic Ocean below the great bulge of West Africa.

Gulf Stream. Current of westward-moving warm water originating along the equator in the Atlantic Ocean. The mass of water moves along the east coast of Florida as the FLORIDA CURRENT, then turns in a northeasterly direction off North Carolina to become the Gulf Stream. The Gulf Stream flows northeast past Newfoundland and the western edge of the British Isles. The warmer water of the Gulf Stream moderates the climate of northwestern EUROPE, causing temperatures in winter to be several degrees warmer than in areas of NORTH AMERICA at the same latitudes. The Gulf Stream decreases the time required for ships to travel from North America to Europe. This was an important factor in trade and communication in American Colonial times and has continued to be significant.

Gulfs. See under individual names.

Hatteras Abyssal Plain. Part of the floor of the northwest ATLANTIC OCEAN Basin, east of North Carolina. It rises to form shallow sandbars around Cape Hatteras, which are a notorious navigational hazard. In the seventeenth and eighteenth centuries, so many ships were lost in the area that Cape Hatteras became known as "The Graveyard of the Atlantic."

horse latitudes. Latitude belts between 30 and 35 degrees north and south latitude, where winds are usually light and variable and the climate mostly hot and dry.

Humboldt Current. See PERU CURRENT

Iceland. Island country bounded by the GREENLAND SEA on the north, the NORWEGIAN SEA on the east, and the ATLANTIC OCEAN on the south and west. Total area of 39,768 square miles (103,000 sq. km.). The nearest land mass is GREENLAND, 200 miles (320 km.) to the northwest. Situated on top of the northern part of the Atlantic Mid-Oceanic Ridge, it is characterized by major volcanic activities, geothermal springs, and glaciers.

Idzu-Bonin Trench. Ocean trench in the western PACIFIC OCEAN, about 6,082 miles (9,810 km.) long and 2,624 feet (800 meters) deep.

Indian Ocean. Third-largest of the world's oceans, bounded by the continents of AFRICA to the west, ASIA to the north, AUSTRALIA to the east, and ANTARCTICA to the south. Most of the Indian Ocean lies below the equator. It has an approximate area of 33 million square miles (76 million sq. km.) and an average depth of about 13,120 feet (4,000 meters). Its deepest point is 24,442 feet (7,450 meters), in the JAVA TRENCH. The Indian Ocean was the first major ocean to be used as a trade route, particularly by the Egyptians. About 600 BCE, the Egyptian ruler Necho sent an expedition into the Indian Ocean, and the ship circumnavigated Africa, probably the first time this feat was accomplished. Warm winds blowing over the northern part of the Indian Ocean from May to September pick up huge amounts of moisture, which falls on India and Sri Lanka as monsoons. Fishing is important and mostly is done by small, family boats. About 40 percent of the world's offshore oil production comes from the Indian Ocean.

Indonesian Trench. See JAVA TRENCH.

Intracoastal Waterway. Series of bays, sounds, and channels, part natural and part human-made, that extends along the eastern coast of the United States from the Delaware River in New Jersey, south to the tip of Florida, then around the west coast of Florida. It extends around the Gulf Coast to the Rio Grande in Texas. It runs 2,455 miles (3,951 km.) and is an important, protected route for commercial and pleasure boat traffic.

Japan, Sea of. Marginal sea of the western Pacific Ocean that is bounded by Japan on the east and the Russian mainland on the west. Its surface area is approximately 377,600 square miles (978,000 sq. km.). It has an average depth of 5,750 feet (1,750 meters) and a maximum depth of 12,276 feet (3,742 meters).

Japan Trench. Ocean trench approximately 497 miles (800 km.) long, beginning at the eastern edge of the Japanese islands and stretching southward toward the MARIANA TRENCH. Depth is 27,560 feet (8,400 meters).

Java Sea. Portion of the western PACIFIC OCEAN between the islands of Java and Borneo. The sea has a total surface area of 167,000 square miles (433,000 sq. km.) and a comparatively shallow average depth of 151 feet (46 meters).

Java Trench. Ocean trench in the INDIAN OCEAN, 2,790 miles (4,500 km.) long and 24,443 feet (7,450 meters) deep. Also called the Indonesian Trench.

Kermadec Trench. Ocean trench approximately 930 miles (1,500 km.) long, located in the southwest PACIFIC OCEAN, beginning northeast of New Zealand. It has a depth of 32,800 feet (10,000 meters). Its northern end connects with the TONGA TRENCH.

Kurile Trench. Ocean trench approximately 1,367 miles (2,200 km.) long along the northeast rim of the PACIFIC OCEAN, beginning at the north end of the Japanese island chain and extending northeastward. Depth is 34,451 feet (10,500 meters).

Labrador Current. Cold current that begins in Baffin Bay between GREENLAND and northeastern Canada and flows southward. The Labrador Current sometimes carries icebergs into North Atlantic shipping channels; such an iceberg caused the famous sinking of the great passenger ship *Titanic* in 1912.

Laptev Sea. Marginal sea of the ARCTIC OCEAN off the coast of northern Siberia. The Taymyr Peninsula bounds it on the west and the New Siberian Islands on the east. Its area is about 276,000 square miles (714,000 sq. km.). Its average depth is 1,896 feet (578 meters), and the greatest depth is 9,774 feet (2,980 meters).

Lord Howe Rise. Elevation of the floor of the western PACIFIC OCEAN that lies between AUSTRALIA and New Guinea and under the TASMAN SEA.

Mac. Robertson Land. Land near the coast of ANTARCTICA, located between the INDIAN OCEAN and the south Polar Plateau, east of ENDERBY LAND. Mac.Robertson Land lies between approximately longitude 60 and 65 degrees east.

Macronesia. Loose grouping of islands in the ATLANTIC OCEAN that includes the Azores, Madeira, the Canary Islands and Cape Verde. The term is derived from Greek words meaning "large" and "island" and should not be confused with MICRONESIA, small islands in the central and North PACIFIC OCEAN.

Madagascar. Large island nation, officially called the Malagasy Republic, located in the INDIAN OCEAN about 200 miles from the southeast coast of AFRICA. Although geographically tied to the African continent, it has a culture more closely tied to those of France and Southeast Asia. Area is 226,657 square miles (587,042 sq. km.).

Magellan, Strait of. Waterway connecting the south ATLANTIC OCEAN with the South Pacific. Ships passing through the strait, north of Tierra del Fuego Island, avoid some of the world's roughest seas around CAPE HORN.

magnetic poles. The two points on the earth, one in the NORTHERN HEMISPHERE and one in the SOUTHERN HEMISPHERE, which are defined by the internal magnetism of the earth. Each point attracts one end of a compass needle and repels the opposite end.

Malacca, Strait of. Relatively narrow passage (200 miles/322 kilometers wide) bordered by Malaysia and Sumatra and linking the SOUTH CHINA SEA and the JAVA SEA. It is one of the most heavily traveled waterways in the world, with more than one thousand ships every week.

Mariana Trench. Lowest point on Earth's surface, with a maximum depth of 36,150 feet (11,022 meters) in the Challenger Deep. The Mariana Trench is located on the western margin of the PACIFIC OCEAN southeast of Japan, and is approximately 1,584 miles (2,550 km.) long.

Marie Byrd Land. Section of ANTARCTICA located at the base of the Antarctic peninsula and shaped like a large peninsula itself. It is bounded at its base by the ROSS ICE SHELF and the Ronne Ice Shelf.

Mediterranean Sea. Large sea that separates the continents of EUROPE, AFRICA, and ASIA. It takes its name from Latin words meaning "in the middle of land"—a reference to its nearly land-locked nature. Covers about 969,100 square miles (2.5 million sq. km.) and extends 2,200 miles (3,540 km.) from west to east and about 1,000 miles (1,600 km.) from north to south at its widest. Its greatest depth is 16,897 feet (5,150 meters).

Melanesia. One of three divisions of the Pacific Islands, along with MICRONESIA and POLYNESIA; located in the western Pacific. The name Melanesia, for "dark islands," was given to the area because of its inhabitants' dark skins

Mexico, Gulf of. Nearly enclosed arm of the western ATLANTIC OCEAN, bounded by the states of Florida, Alabama, Mississippi, Louisiana, and Texas, and Mexico and the Yucatan Peninsula. Cuba is located in the gap between the Yucatan Peninsula and Florida. Most ocean water enters through the Yucatan passage and exits the Gulf of Mexico around the tip of Florida, becoming the FLORIDA CURRENT. Fisheries, tourism, and oil production are important activities.

Micronesia. One of three divisions of the Pacific Islands, along with MELANESIA and POLYNESIA. Micronesia means "small islands." Micronesia's islands are mostly atolls and coral islands, but some are of volcanic origin. The more than 2,000 islands of Micronesia are located in the Pacific Ocean east of the Philippines, mostly north of the EQUATOR.

Mid-Atlantic Ridge. Steep-sided, underwater mountain range running down the middle of the ATLANTIC OCEAN. Formed by the divergent boundaries, or region where tectonic plates are separating.

Mozambique Current. See AGULHAS CURRENT.

New Britain Trench. Ocean trench in the southwest PACIFIC OCEAN, about 5,158 miles (8,320 km.) long and 2,460 feet (750 meters) deep.

New Hebrides Basin. Part of the CORAL SEA, located east of AUSTRALIA and west of the New Hebrides island chain. The basin contains volcanic islands, both old and recent.

New Hebrides Trench. Ocean trench in the southwest PACIFIC OCEAN, about 5,682 miles (9,165 km.) long and 3,936 feet (1,200 meters) deep.

North America. Third-largest continent, usually considered to contain all land and nearby islands in the WESTERN HEMISPHERE north of the Isthmus of Panama, which connects it to SOUTH AMERICA. The major mainland countries are Canada, the United States, Mexico, Guatemala, El Salvador, Honduras, Nicaragua, Costa Rica, and Panama. Island countries include the islands of the CARIBBEAN SEA and GREENLAND. Climate ranges from arctic to tropical.

North Atlantic Current. Continuation of the GULF STREAM, originating near the GRAND BANKS off Newfoundland. It curves eastward and divides into a northern branch, which flows into the NORWEGIAN SEA, a southern branch, which flows eastward, and a branch that forms the Canary Current and flows south along the coast of EUROPE.

North Atlantic Gyre. Large mass of water, located in the ATLANTIC OCEAN in the NORTHERN HEMISPHERE, that rotates clockwise. Warm water moves toward the pole and cold water moves toward the equator.

North Pacific Current. Eastward flow of water in the PACIFIC OCEAN in the NORTHERN HEMISPHERE. It originates as the Kuroshio Current and moves from Japan toward NORTH AMERICA.

North Pacific Gyre. Large mass of water, located in the PACIFIC OCEAN in the NORTHERN HEMISPHERE, that rotates clockwise. Warm water moves toward the pole and cold water moves toward the equator.

North Pole. Northern end of the earth's geographic axis, located at 90 degrees north latitude and longitude zero degrees. The North Pole itself is located on the Polar Abyssal Plain, about 14,000 feet (4,000 meters) deep in the ARCTIC OCEAN. U.S. explorer Robert Edwin is credited with being the first person to reach the North Pole, in 1909, although there is historical dispute over the claim. The North Pole is different from the North MAGNETIC POLE.

North Sea. Arm of the northeastern ATLANTIC OCEAN, bounded by Great Britain on the west and Norway, Denmark, and Germany on the east and south. The North Sea is one of the great fishing areas of the world and an important source of oil.

Northern Hemisphere. The half of the earth above the equator.

Norwegian Sea. Section of the North Atlantic Ocean. Norway borders it on the east and Iceland on the west. A submarine ridge linking

GREENLAND, ICELAND, the Faroe Islands, and northern Scotland separates the Norwegian Sea from the open ATLANTIC OCEAN. Cut by the ARCTIC CIRCLE, the sea is often associated with the ARCTIC OCEAN to the north. Reaches a maximum depth of about 13,020 feet (3,970 meters).

Oceania. Loosely applied term for the large island groups of the central and South Pacific; sometimes used to include AUSTRALIA and New Zealand.

Okhotsk, Sea of. Nearly enclosed area of the northwestern PACIFIC OCEAN bounded by Russia's Kamchatka Peninsula on the east and Siberia on the west. It is open to the Pacific Ocean on the south side only through Japan and the Kuril Islands, a string of islands belonging to Russia.

Pacific Ocean. Largest body of water in the world, covering more than one-third of Earth's surface—an area of about 70 million square miles (181 million sq. km.), more than the entire land area of the world. At its widest point, between Panama in CENTRAL AMERICA and the Philippines, it stretches 10,700 miles (17,200 km.). It runs 9,600 miles (15,450 km.) from the BERING STRAIT in the north to ANTARCTICA in the south. Bordered by NORTH and SOUTH AMERICA in the east, and ASIA and AUSTRALIA in the west. The average depth is about 12,900 feet (3,900 meters). It contains the deepest point on Earth (36,150 feet/11,022 meters), in the Challenger Deep of the MARIANA TRENCH, southwest of Japan. The Pacific Ocean bottom is more geologically varied than the INDIAN or ATLANTIC OCEANS; it has more volcanoes, ridges, trenches, seamounts, and islands. The vast size of the Pacific Ocean was a formidable barrier to travel, communications, and trade well into the nineteenth century. However, evidence shows that people crossed the Pacific Ocean in rafts or canoes as early as 3,000 BCE.

Pacific Rim. Modern term for the nations of ASIA and NORTH and SOUTH AMERICA that border, or are in, the PACIFIC OCEAN. Used mostly in discussions of economic growth.

Palau Trench. Ocean trench in the western PACIFIC OCEAN, about 250 miles (400 km.) long and 26,425 feet (8,054 meters) deep.

Palmer Land. Section of ANTARCTICA that occupies the base of the Antarctic peninsula.

Panama, Isthmus of. Narrow neck of land that joins CENTRAL and SOUTH AMERICA. In 1914 the Panama Canal was opened through the isthmus, creating a direct sea link between the PACIFIC OCEAN and the CARIBBEAN SEA. The canal stretches about 50 miles (80 km.) from Panama City on the Pacific to Colón on the Caribbean. More than 12,000 ships pass through the canal annually.

Persian Gulf. Large extension of the ARABIAN SEA that separates Iran from the Arabian Peninsula in the Middle East. It covers about 88,000 square miles (226,000 sq. km.) and is about 620 miles (1,000 km.) long and 125–185 miles (200–300 km.) wide.

Peru-Chile Trench. Ocean trench that runs along the eastern boundary of the PACIFIC OCEAN, off the western edge of SOUTH AMERICA. It is 3,666 miles (5,900 km.) long and 26,576 feet (8,100 meters) deep.

Peru Current. Cold, broad current that originates in the southernmost part of the SOUTH PACIFIC GYRE and flows up the west coast of SOUTH AMERICA. Off the coast of Peru, prevailing winds usually push the warmer surface water to the west. This causes the nutrient-rich, colder water of the Peru Current to well up to the surface, which provides excellent feeding for fish. At times, the upwelling ceases and biological, economic, and climatic catastrophe can result in EL NIÑO weather conditions. Also known as the Humboldt Current.

Philippine Trench. Ocean trench located on the western rim of the PACIFIC OCEAN, at the eastern margin of the PHILIPPINE ISLANDS. It is about 870 miles (1,400 km.) long and 34,451 feet (10,500 meters) deep.

Polynesia. One of three main divisions of the Pacific Islands, along with MELANESIA and MICRONESIA. The islands are spread through the central and South Pacific. Polynesia means "many islands." Mostly small, the islands are predominantly coral atolls, but some are of volcanic origin.

Puerto Rico Trench. Ocean trench in the western Atlantic Ocean, about 27,500 feet (8,385 meters) deep and 963 miles (1,550 km.) long.

Queen Maud Land. Section of Antarctica that lies between the Atlantic Ocean and the south Polar Plateau, between approximately longitude 15 and 45 degrees east.

Red Sea. Narrow arm of water separating Africa from the Arabian Peninsula. One of the saltiest bodies of ocean water on Earth, as a result of high evaporation and little freshwater input. It was used as a trade route for Mediterranean, Indian, and Chinese peoples for centuries before the Europeans discovered it in the fifteenth century. The Suez Canal was opened in 1869 between the Mediterranean Sea and the Red Sea, cutting the distance from the northern Indian Ocean to northern Europe by about 5,590 miles (9,000 km.). This greatly increased the economic and military importance of the Red Sea.

Ring of Fire. Nearly continuous ring of volcanic and tectonic activity around the margins of the Pacific Ocean.

Ross Ice Shelf. Thick layer of ice in the Ross Sea off the coast of Antarctica. The relatively flat ice is attached to and nourished by a continental glacier.

Ross Sea. Bay in the Southern Ocean off the coast of Antarctica, located south of New Zealand. Named for English explorer James Clark Ross, the first person to break through the Antarctic ice pack in a ship, in 1841.

St. Peter and St. Paul. Cluster of rocks showing above the surface of the Atlantic Ocean between Brazil and West Africa. Important landmarks in the days of slave ships.

Sargasso Sea. Warm, salty area of water located in the Atlantic Ocean south and east of Bermuda, formed from water that circulates around the center of the North Atlantic Gyre. Named for the seaweed, *Sargassum*, that floats on the surface in large amounts.

Scotia Sea. Area of the southernmost Atlantic Ocean between the southern tip of South America and the Antarctic peninsula. The area is known for severe storms.

Seas. See under individual names.

Siam, Gulf of. See Thailand, Gulf of.

South America. Fourth-largest continent, usually considered to contain all land and nearby islands in the Western Hemisphere south of the Isthmus of Panama, which connects it to North America. Countries are Argentina, Bolivia, Brazil, Chile, Colombia, Ecuador, French Guiana, Guyana, Paraguay, Peru, Suriname, Uruguay, and Venezuela. Climate ranges from tropical to cold, nearly sub-Antarctic.

South Atlantic Gyre. Large mass of water, located in the Atlantic Ocean in the Southern Hemisphere, that rotates counterclockwise. Warm water moves toward the pole and cold water moves toward the equator.

South China Sea. Portion of the western Pacific Ocean that lies along the east coast of China, Vietnam, and the southeastern part of the Gulf of Thailand. The eastern and southern edges are defined by the Philippine and Indonesian Islands.

South Equatorial Current. Part of the South Atlantic Gyre that is split in two by the eastern prominence of Brazil. One part moves along the northeastern coast of South America toward the Caribbean Sea and the North Atlantic Ocean; the other turns southward and forms the Brazil Current.

South Pacific Gyre. Large mass of water, located in the Pacific Ocean in the Southern Hemisphere, that rotates counterclockwise. Warm water moves toward the pole and cold water moves toward the equator.

South Pole. Southern end of the earth's geographic axis, located at 90 degrees south latitude and longitude zero degrees. The first person to reach the South Pole was Norwegian explorer Roald Amundsen, in 1911. The South Pole is different from the South Magnetic Pole.

Southeastern Pacific Plateau. Portion of the Pacific Ocean floor closest to South America.

Southern Hemisphere. The half of the earth below the equator.

Southern Ocean. Not officially recognized as one of the major oceans, but a commonly used term for water surrounding ANTARCTICA and extending northward to 50 degrees south latitude. Also known as the Antarctic Ocean.

Straits. See under individual names.

Sunda Shelf. One of the largest continental shelves in the world, nearly 772,000 square miles (2 million sq. km.). Located in the JAVA SEA, SOUTH CHINA SEA, and Gulf of THAILAND. The area was above water in the Quaternary period, enabling large animals such as elephants and rhinoceros to migrate to Sumatra, Java, and Borneo.

Surtsey Island. Island formed by a volcanic explosion off the coast of ICELAND in 1963. It is valuable to scientists studying how island flora and fauna develop and is a popular tourist site.

Tashima Current. See TSUSHIMA CURRENT.

Tasman Sea. Area of the PACIFIC OCEAN off the southeast coast of AUSTRALIA, between Australia and Tasmania on the west and New Zealand on the east. First crossed by the Morioris people sometime before 1300 CE Also called the Tasmanian Sea.

Tasmanian Sea. See TASMAN SEA.

Thailand, Gulf of. Also known as the Gulf of Siam, inlet of the South China Sea, located between the Malay Archipelago and the Southeast Asian mainland. Bounded by Thailand, Cambodia, and Vietnam.

Tonga Trench. Ocean trench in the PACIFIC OCEAN, northeast of New Zealand. It stretches for 870 miles (1,400 km.), beginning at the northern end of the KERMADEC TRENCH. Depth is 32,810 feet (10,000 meters).

Tsushima Current. Warm current in the western PACIFIC OCEAN that flows out of the YELLOW SEA into the Sea of JAPAN in the spring and summer. Also called Tashima Current.

Walvis Ridge (Walfisch Ridge). Long, narrow undersea elevation near the southwestern coast of AFRICA, which extends about 1,900 miles (3,000 km.) in a southwesterly direction under the ATLANTIC OCEAN.

Weddell Sea. Bay in the SOUTHERN OCEAN bounded by the ANTARCTIC peninsula on the west and a northward bulge of ANTARCTICA on the east, stretching from approximately longitude 60 to 10 degrees west. One of the harshest environments on Earth; surface water temperatures stay near 32 degrees Fahrenheit (0 degrees Celsius) all year. The Weddell Sea was the site of much whaling and seal hunting in the nineteenth and twentieth centuries.

West Caroline Trench. See YAP TRENCH.

West Wind Drift. Surface portion of the ANTARCTIC CIRCUMPOLAR CURRENT, driven by westerly winds. Often extremely rough; seas as high as 98 feet (30 meters) have been reported.

Western Hemisphere. The half of the earth containing NORTH and SOUTH AMERICA; generally understood to fall between longitudes 160 degrees east and 20 degrees west.

Wilkes Land. Broad section near the coast of ANTARCTICA, which lies south of AUSTRALIA and east of the AMERICAN HIGHLANDS. Wilkes Land is the nearest landmass to the South MAGNETIC POLE.

Yap Trench. Ocean trench in the western PACIFIC OCEAN, about 435 miles (700 km.) long and 27,900 feet (8,527 meters) deep. Also called the West Caroline Trench.

Yellow Sea. Area of the PACIFIC OCEAN bounded by China on the north and west and Korea on the east. Named for the large amounts of yellow dust carried into it from central China by winds and by the Yangtze, Yalu, and Yellow Rivers. Parts of the sea often show a yellow color from the dust.

Kelly Howard

WORLD'S OCEANS AND SEAS

Name	Approximate Area		Average Depth	
	Sq. Miles	Sq. Km.	Feet	Meters
Pacific Ocean	64,000,000	165,760,000	13,215	4,028
Atlantic Ocean	31,815,000	82,400,000	12,880	3,926
Indian Ocean	25,300,000	65,526,700	13,002	3,963
Arctic Ocean	5,440,200	14,090,000	3,953	1,205
Mediterranean & Black Seas	1,145,100	2,965,800	4,688	1,429
Caribbean Sea	1,049,500	2,718,200	8,685	2,647
South China Sea	895,400	2,319,000	5,419	1,652
Bering Sea	884,900	2,291,900	5,075	1,547
Gulf of Mexico	615,000	1,592,800	4,874	1,486
Okhotsk Sea	613,800	1,589,700	2,749	838
East China Sea	482,300	1,249,200	617	188
Hudson Bay	475,800	1,232,300	420	128
Japan Sea	389,100	1,007,800	4,429	1,350
Andaman Sea	308,100	797,700	2,854	870
North Sea	222,100	575,200	308	94
Red Sea	169,100	438,000	1,611	491
Baltic Sea	163,000	422,200	180	55

MAJOR LAND AREAS OF THE WORLD

Area	Approximate Land Area		Percent of World Total
	Sq. Mi.	Sq. Km.	
World	57,308,738	148,429,000	100.0
Asia (including Middle East)	17,212,041	44,579,000	30.0
Africa	11,608,156	30,065,000	20.3
North America	9,365,290	24,256,000	16.3
Central America, South America, & Caribbean	6,879,952	17,819,000	8.9
Antarctica	5,100,021	13,209,000	8.9
Europe	3,837,082	9,938,000	6.7
Oceania, including Australia	2,967,966	7,687,000	5.2

MAJOR ISLANDS OF THE WORLD

		Area	
Island	Location	Sq. Mi.	Sq. Km.
Greenland	North Atlantic Ocean	839,999	2,175,597
New Guinea	Western Pacific Ocean	316,615	820,033
Borneo	Western Pacific Ocean	286,914	743,107
Madagascar	Western Indian Ocean	226,657	587,042
Baffin	Canada, North Atlantic Ocean	183,810	476,068
Sumatra	Indonesia, northeast Indian Ocean	182,859	473,605
Hōnshū	Japan, western Pacific Ocean	88,925	230,316
Great Britain	North Atlantic Ocean	88,758	229,883
Ellesmere	Canada, Arctic Ocean	82,119	212,688
Victoria	Canada, Arctic Ocean	81,930	212,199
Sulawesi (Celebes)	Indonesia, western Pacific Ocean	72,986	189,034
South Island	New Zealand, South Pacific Ocean	58,093	150,461
Java	Indonesia, Indian Ocean	48,990	126,884
North Island	New Zealand, South Pacific Ocean	44,281	114,688
Cuba	Caribbean Sea	44,218	114,525
Newfoundland	Canada, North Atlantic Ocean	42,734	110,681
Luzon	Philippines, western Pacific Ocean	40,420	104,688
Iceland	North Atlantic Ocean	39,768	102,999
Mindanao	Philippines, western Pacific Ocean	36,537	94,631
Ireland	North Atlantic Ocean	32,597	84,426
Hōkkaidō	Japan, western Pacific Ocean	30,372	78,663
Hispaniola	Caribbean Sea	29,355	76,029
Tasmania	Australia, South Pacific Ocean	26,215	67,897
Sri Lanka	Indian Ocean	25,332	65,610
Sakhalin (Karafuto)	Russia, western Pacific Ocean	24,560	63,610
Banks	Canada, Arctic Ocean	23,230	60,166
Devon	Canada, Arctic Ocean	20,861	54,030
Tierra del Fuego	Southern tip of South America	18,605	48,187
Kyūshū	Japan, western Pacific Ocean	16,223	42,018
Melville	Canada, Arctic Ocean	16,141	41,805
Axel Heiberg	Canada, Arctic Ocean	15,779	40,868
Southampton	Hudson Bay, Canada	15,700	40,663

COUNTRIES OF THE WORLD

Country	Region	Population	Area Square Miles	Area Square Kilometers	Population Density Persons/ Sq. Mi.	Population Density Persons/ Sq. Km.
Afghanistan	Asia	31,575,018	249,347	645,807	127	49
Albania	Europe	2,862,427	11,082	28,703	259	100
Algeria	Africa	42,545,964	919,595	2,381,741	47	18
Andorra	Europe	76,177	179	464	425	164
Angola	Africa	29,250,009	481,354	1,246,700	60	23
Antigua and Barbuda	Caribbean	104,084	171	442	609	235
Argentina	South America	44,938,712	1,073,518	2,780,400	41	16
Armenia	Europe	2,962,100	11,484	29,743	259	100
Australia	Australia	25,576,880	2,969,907	7,692,024	9	3
Austria	Europe	8,877,036	32,386	83,879	275	106
Azerbaijan	Asia	10,027,874	33,436	86,600	300	116
Bahamas	Caribbean	386,870	5,382	13,940	73	28
Bahrain	Asia	1,543,300	300	778	5,136	1,983
Bangladesh	Asia	167,888,084	55,598	143,998	3,020	1,166
Barbados	Caribbean	287,025	166	430	1,730	668
Belarus	Europe	9,465,300	80,155	207,600	119	46
Belgium	Europe	11,515,793	11,787	30,528	976	377
Belize	Central America	398,050	8,867	22,965	44	17
Benin	Africa	11,733,059	43,484	112,622	269	104
Bhutan	Asia	821,592	14,824	38,394	55	21
Bolivia	South America	11,307,314	424,164	1,098,581	26	10
Bosnia and Herzegovina	Europe	3,511,372	19,772	51,209	179	69
Botswana	Africa	2,302,878	224,607	581,730	10.4	4
Brazil	South America	210,951,255	3,287,956	8,515,767	64	25
Brunei	Asia	421,300	2,226	5,765	189	73
Bulgaria	Europe	7,000,039	42,858	111,002	163	63
Burkina Faso	Africa	20,244,080	104,543	270,764	194	75
Burundi	Africa	10,953,317	10,740	27,816	1,020	394
Cambodia	Asia	16,289,270	69,898	181,035	233	90
Cameroon	Africa	24,348,251	179,943	466,050	135	52
Canada	North America	37,878,499	3,855,103	9,984,670	10	4
Cape Verde	Africa	550,483	1,557	4,033	352	136
Central African Republic	Africa	4,737,423	240,324	622,436	21	8
Chad	Africa	15,353,184	495,755	1,284,000	31	12
Chile	South America	17,373,831	291,930	756,096	60	23
China, People's Republic of	Asia	1,400,781,440	3,722,342	9,640,821	376	145
Colombia	South America	46,103,400	440,831	1,141,748	105	40
Comoros	Africa	873,724	719	1,861	1,215	469

| | | | Area | | Population Density | |
| | | | Square Miles | Square Kilometers | Persons/ Sq. Mi. | Persons/ Sq. Km. |
Country	Region	Population				
Costa Rica	Central America	5,058,007	19,730	51,100	256	99
Côte d'Ivoire (Ivory Coast)	Africa	25,823,071	124,680	322,921	207	80
Croatia	Europe	4,087,843	21,831	56,542	186	72
Cuba	Caribbean	11,209,628	42,426	109,884	264	102
Cyprus	Europe	864,200	2,276	5,896	381	147
Czech Republic	Europe	10,681,161	30,451	78,867	350	135
Dem. Republic of the Congo	Africa	86,790,567	905,446	2,345,095	96	37
Denmark	Europe	5,814,461	16,640	43,098	350	135
Djibouti	Africa	1,078,373	8,880	23,000	122	47
Dominica	Caribbean	71,808	285	739	251	97
Dominican Republic	Caribbean	10,358,320	18,485	47,875	559	216
Ecuador	South America	17,398,588	106,889	276,841	163	63
Egypt	Africa	99,873,587	387,048	1,002,450	258	100
El Salvador	Central America	6,704,864	8,124	21,040	826	319
Equatorial Guinea	Africa	1,358,276	10,831	28,051	124	48
Eritrea	Africa	3,497,117	46,757	121,100	75	29
Estonia	Europe	1,324,820	17,505	45,339	75	29
Eswatini (Swaziland)	Africa	1,159,250	6,704	17,364	174	67
Ethiopia	Africa	107,534,882	410,678	1,063,652	262	101
Fed. States of Micronesia	Pacific Islands	105,300	271	701	388	150
Fiji	Pacific Islands	884,887	7,078	18,333	124	48
Finland	Europe	5,527,405	130,666	338,424	41	16
France	Europe	67,022,000	210,026	543,965	319	123
Gabon	Africa	2,067,561	103,347	267,667	21	8
Gambia	Africa	2,228,075	4,127	10,690	539	208
Georgia	Europe	3,729,600	26,911	69,700	140	54
Germany	Europe	83,073,100	137,903	357,168	603	233
Ghana	Africa	30,280,811	92,098	238,533	329	127
Greece	Europe	10,724,599	50,949	131,957	210	81
Grenada	Caribbean	108,825	133	344	818	316
Guatemala	Central America	17,679,735	42,042	108,889	420	162
Guinea	Africa	12,218,357	94,926	245,857	129	50
Guinea-Bissau	Africa	1,604,528	13,948	36,125	114	44
Guyana	South America	782,225	83,012	214,999	9.3	3.6
Haiti	Caribbean	11,263,077	10,450	27,065	1,077	416
Honduras	Central America	9,158,345	43,433	112,492	210	81
Hungary	Europe	9,764,000	35,919	93,029	272	105
Iceland	Europe	360,390	39,682	102,775	9.1	3.5
India	Asia	1,357,041,500	1,269,211	3,287,240	1,069	413
Indonesia	Asia	268,074,600	735,358	1,904,569	365	141
Iran	Asia	83,096,438	636,372	1,648,195	131	50

Country	Region	Population	Area		Population Density	
			Square Miles	Square Kilometers	Persons/ Sq. Mi.	Persons/ Sq. Km.
Iraq	Asia	39,309,783	169,235	438,317	233	90
Ireland	Europe	4,921,500	27,133	70,273	181	70
Israel	Asia	9,141,680	8,522	22,072	1,073	414
Italy	Europe	60,252,824	116,336	301,308	518	200
Jamaica	Caribbean	2,726,667	4,244	10,991	642	248
Japan	Asia	126,140,000	145,937	377,975	865	334
Jordan	Asia	10,587,132	34,495	89,342	307	119
Kazakhstan	Asia	18,592,700	1,052,090	2,724,900	18	7
Kenya	Africa	47,564,296	224,647	581,834	212	82
Kiribati	Pacific Islands	120,100	313	811	383	148
Kuwait	Asia	4,420,110	6,880	17,818	642	248
Kyrgyzstan	Asia	6,309,300	77,199	199,945	83	32
Laos	Asia	6,492,400	91,429	236,800	70	27
Latvia	Europe	1,910,400	24,928	64,562	78	30
Lebanon	Asia	6,855,713	4,036	10,452	1,740	672
Lesotho	Africa	2,263,010	11,720	30,355	194	75
Liberia	Africa	4,475,353	37,466	97,036	119	46
Libya	Africa	6,470,956	683,424	1,770,060	9.6	3.7
Liechtenstein	Europe	38,380	62	160	622	240
Lithuania	Europe	2,793,466	25,212	65,300	111	43
Luxembourg	Europe	613,894	998	2,586	614	237
Madagascar	Africa	25,680,342	226,658	587,041	114	44
Malawi	Africa	17,563,749	45,747	118,484	383	148
Malaysia	Asia	32,715,210	127,724	330,803	256	99
Maldives	Asia	378,114	115	298	3,287	1,269
Mali	Africa	19,107,706	482,077	1,248,574	39	15
Malta	Europe	493,559	122	315	3,911	1,510
Marshall Islands	Pacific Islands	55,500	70	181	795	307
Mauritania	Africa	3,984,233	397,955	1,030,700	10.4	4
Mauritius	Africa	1,265,577	788	2,040	1,606	620
Mexico	North America	126,577,691	759,516	1,967,138	166	64
Moldova	Europe	2,681,735	13,067	33,843	205	79
Monaco	Europe	38,300	0.78	2.02	49,106	18,960
Mongolia	Asia	3,000,000	603,902	1,564,100	4.9	1.9
Montenegro	Europe	622,182	5,333	13,812	117	45
Morocco	Africa	35,773,773	172,414	446,550	207	80
Mozambique	Africa	28,571,310	308,642	799,380	93	36
Myanmar (Burma)	Asia	54,339,766	261,228	676,577	207	80
Namibia	Africa	2,413,643	318,580	825,118	7.5	2.9
Nauru	Pacific Islands	11,000	8	21	1,357	524
Nepal	Asia	29,609,623	56,827	147,181	521	201

Country	Region	Population	Area Square Miles	Area Square Kilometers	Population Density Persons/ Sq. Mi.	Population Density Persons/ Sq. Km.
Netherlands	Europe	17,370,348	16,033	41,526	1,083	418
New Zealand	Pacific Islands	4,952,186	104,428	270,467	47	18
Nicaragua	Central America	6,393,824	46,884	121,428	137	53
Niger	Africa	21,466,863	458,075	1,186,408	47	18
Nigeria	Africa	200,962,000	356,669	923,768	565	218
North Korea	Asia	25,450,000	46,541	120,540	546	211
North Macedonia	Europe	2,077,132	9,928	25,713	210	81
Norway	Europe	5,328,212	125,013	323,782	41	16
Oman	Asia	4,183,841	119,499	309,500	36	14
Pakistan	Asia	218,198,000	310,403	803,940	703	271
Palau	Pacific Islands	17,900	171	444	104	40
Panama	Central America	4,158,783	28,640	74,177	145	56
Papua New Guinea	Pacific Islands	8,558,800	178,704	462,840	47	18
Paraguay	South America	7,052,983	157,048	406,752	44	17
Peru	South America	32,162,184	496,225	1,285,216	65	25
Philippines	Asia	108,785,760	115,831	300,000	939	363
Poland	Europe	38,386,000	120,728	312,685	319	123
Portugal	Europe	10,276,617	35,556	92,090	290	112
Qatar	Asia	2,740,479	4,468	11,571	614	237
Republic of the Congo	Africa	5,380,895	132,047	342,000	41	16
Romania	Europe	19,405,156	92,043	238,391	210	81
Russia[1]	Europe/Asia	146,877,088	6,612,093	17,125,242	23	9
Rwanda	Africa	12,374,397	10,169	26,338	1,217	470
St Kitts and Nevis	Caribbean	56,345	104	270	541	209
St Lucia	Caribbean	180,454	238	617	756	292
St Vincent and Grenadines	Caribbean	110,520	150	389	736	284
Samoa	Pacific Islands	199,052	1,093	2,831	181	70
San Marino	Europe	34,641	24	61	1,471	568
Sahrawi Arab Dem. Rep.[2]	Africa	567,421	97,344	252,120	6	2.3
São Tomé and Príncipe	Africa	201,784	386	1,001	523	202
Saudi Arabia	Asia	34,218,169	830,000	2,149,690	41	16
Senegal	Africa	16,209,125	75,955	196,722	212	82
Serbia	Europe	6,901,188	29,913	77,474	231	89
Seychelles	Africa	96,762	176	455	552	213
Sierra Leone	Africa	7,901,454	27,699	71,740	285	110
Singapore	Asia	5,638,700	279	722.5	20,212	7,804
Slovakia	Europe	5,450,421	18,933	49,036	287	111
Slovenia	Europe	2,084,301	7,827	20,273	267	103
Solomon Islands	Pacific Islands	682,500	10,954	28,370	62	24
Somalia	Africa	15,181,925	246,201	637,657	62	24
South Africa	Africa	58,775,022	471,359	1,220,813	124	48

Country	Region	Population	Area Square Miles	Area Square Kilometers	Population Density Persons/ Sq. Mi.	Population Density Persons/ Sq. Km.
South Korea	Asia	51,811,167	38,691	100,210	1,339	517
South Sudan	Africa	12,575,714	248,777	644,329	52	20
Spain	Europe	46,934,632	195,364	505,990	241	93
Sri Lanka	Asia	21,803,000	25,332	65,610	860	332
Sudan	Africa	40,782,742	710,251	1,839,542	57	22
Suriname	South America	568,301	63,251	163,820	9.1	3.5
Sweden	Europe	10,344,405	173,860	450,295	59	23
Switzerland	Europe	8,586,550	15,940	41,285	539	208
Syria	Asia	17,070,135	71,498	185,180	238	92
Taiwan	Asia	23,596,266	13,976	36,197	1,689	652
Tajikistan	Asia	9,127,000	55,251	143,100	166	64
Tanzania	Africa	55,890,747	364,900	945,087	153	59
Thailand	Asia	66,455,280	198,117	513,120	335	130
Timor-Leste	Asia	1,167,242	5,760	14,919	202	78
Togo	Africa	7,538,000	21,853	56,600	344	133
Tonga	Pacific Islands	100,651	278	720	362	140
Trinidad and Tobago	Caribbean	1,359,193	1,990	5,155	683	264
Tunisia	Africa	11,551,448	63,170	163,610	183	71
Turkey	Europe/Asia	82,003,882	302,535	783,562	271	105
Turkmenistan	Asia	5,851,466	189,657	491,210	31	12
Tuvalu	Pacific Islands	10,200	10	26	1,020	392
Uganda	Africa	40,006,700	93,263	241,551	429	166
Ukraine[3]	Europe	41,990,278	232,820	603,000	180	70
United Arab Emirates	Asia	9,770,529	32,278	83,600	303	117
United Kingdom	Europe	66,435,600	93,788	242,910	708	273
United States	North America	330,546,475	3,796,742	9,833,517	87	34
Uruguay	South America	3,518,553	68,037	176,215	52	20
Uzbekistan	Asia	32,653,900	172,742	447,400	189	73
Vanuatu	Pacific Islands	304,500	4,742	12,281	65	25
Vatican City	Europe	1,000	0.17	0.44	5,887	2,273
Venezuela	South America	32,219,521	353,841	916,445	91	35
Vietnam	Asia	96,208,984	127,882	331,212	751	290
Yemen	Asia	28,915,284	175,676	455,000	166	64
Zambia	Africa	16,405,229	290,585	752,612	57	22
Zimbabwe	Africa	15,159,624	150,872	390,757	101	39

Notes: (1) Including the population and area of Autonomous Republic of Crimea and City of Sevastopol, Ukraine's administrative areas on the Crimean Peninsula, which are claimed by Russia; (2) Administration is split between Morocco and the Sahrawi Arab Democratic Republic (Western Sahara), both of which claim the entire territory; (3) Excludes Crimea.
Source: U.S. Census Bureau, International Data Base

PAST AND PROJECTED WORLD POPULATION GROWTH, 1950-2050

Year	Approximate World Population	Ten-Year Growth Rate (%)
1950	2,556,000,053	18.9
1960	3,039,451,023	22.0
1970	3,706,618,163	20.2
1980	4,453,831,714	18.5
1990	5,278,639,789	15.2
2000	6,082,966,429	12.6
2010	6,848,932,929	10.7
2020	7,584,821,144	8.7
2030	8,246,619,341	7.3
2040	8,850,045,889	5.6
2050	9,346,399,468	—

Note: The listed years are the baselines for the estimated ten-year growth rate figures; for example, the rate for 1950-1960 was 18.9%.
Source: U.S. Census Bureau, International Data Base

WORLD'S LARGEST COUNTRIES BY AREA

			Area	
Rank	Country	Region	Sq. Miles	Sq. Km.
1	Russia	Europe/Asia	6,612,093	17,125,242
2	Canada	North America	3,855,103	9,984,670
3	United States	North America	3,796,742	9,833,517
4	China	Asia	3,722,342	9,640,821
5	Brazil	South America	3,287,956	8,515,767
6	Australia	Australia	2,969,907	7,692,024
7	India	Asia	1,269,211	3,287,240
8	Argentina	South America	1,073,518	2,780,400
9	Kazakhstan	Asia	1,052,090	2,724,900
10	Algeria	Africa	919,595	2,381,741
11	Democratic Rep. of the Congo	Africa	905,446	2,345,095
12	Saudi Arabia	Asia	830,000	2,149,690
13	Mexico	North America	759,516	1,967,138
14	Indonesia	Asia	735,358	1,904,569
15	Sudan	Africa	710,251	1,839,542
16	Libya	Africa	683,424	1,770,060
17	Iran	Asia	636,372	1,648,195
18	Mongolia	Asia	603,902	1,564,100
19	Peru	South America	496,225	1,285,216
20	Chad	Africa	495,755	1,284,000
21	Mali	Africa	482,077	1,248,574
22	Angola	Africa	481,354	1,246,700
23	South Africa	Africa	471,359	1,220,813
24	Niger	Africa	458,075	1,186,408
25	Colombia	South America	440,831	1,141,748
26	Bolivia	South America	424,164	1,098,581
27	Ethiopia	Africa	410,678	1,063,652
28	Mauritania	Africa	397,955	1,030,700
29	Egypt	Africa	387,048	1,002,450
30	Tanzania	Africa	364,900	945,087

Source: U.S. Census Bureau, International Data Base

World's Smallest Countries by Area

Rank	Country	Region	Area Sq. Miles	Area Sq. Km.
1	Vatican City*	Europe	0.17	0.44
2	Monaco*	Europe	0.78	2.02
3	Nauru	Pacific Islands	8	21
4	Tuvalu	Pacific Islands	10	26
5	San Marino*	Europe	24	61
6	Liechtenstein*	Europe	62	160
7	Marshall Islands	Pacific Islands	70	181
8	Saint Kitts and Nevis	Central America	104	270
9	Maldives	Asia	115	298
10	Malta	Europe	122	315
11	Grenada	Caribbean	133	344
12	Saint Vincent and the Grenadines	Central America	150	389
13	Barbados	Caribbean	166	430
14	Antigua and Barbuda	Caribbean	171	442
15	Palau	Pacific Islands	171	444
16	Seychelles	Africa	176	455
17	Andorra*	Europe	179	464
18	Saint Lucia	Central America	238	617
19	Federated States of Micronesia	Pacific Islands	271	701
20	Tonga	Pacific Islands	278	720

Note: Asterisks (*) denote countries on continents; all other countries are islands or island groups.
Source: U.S. Census Bureau, International Data Base

WORLD'S LARGEST COUNTRIES BY POPULATION

Rank	Country	Region	Population
1	China	Asia	1,401,028,280
2	India	Asia	1,357,746,150
3	United States	North America	329,229,067
4	Indonesia	Asia	268,074,600
5	Pakistan	Asia	218,385,000
6	Brazil	South America	211,032,216
7	Nigeria	Africa	200,962,000
8	Bangladesh	Asia	167,983,726
9	Russia	Europe/Asia	146,877,088
10	Mexico	North America	126,577,691
11	Japan	Asia	126,020,000
12	Philippines	Asia	108,210,625
13	Ethiopia	Africa	107,534,882
14	Egypt	Africa	99,930,038
15	Vietnam	Asia	96,208,984
16	Democratic Rep. of the Congo	Africa	86,790,567
17	Germany	Europe	83,149,300
18	Iran	Asia	83,142,818
19	Turkey	Europe/Asia	82,003,882
20	France	Europe	67,060,000
21	Thailand	Asia	66,461,867
22	United Kingdom	Europe	66,435,600
23	Italy	Europe	60,252,824
24	South Africa	Africa	58,775,022
25	Tanzania	Africa	55,890,747
26	Myanmar	Asia	54,339,766
27	South Korea	Asia	51,811,167
28	Kenya	Africa	47,564,296
29	Spain	Europe	46,934,632
30	Colombia	South America	46,127,200

Source: U.S. Census Bureau, International Data Base

WORLD'S SMALLEST COUNTRIES BY POPULATION

Rank	Country	Region	Population
1	Vatican City	Europe	1,000
2	Tuvalu	Pacific Islands	10,200
3	Nauru	Pacific Islands	11,000
4	Palau	Pacific Islands	17,900
5	San Marino	Europe	34,641
6	Monaco	Europe	38,300
7	Liechtenstein	Europe	38,380
8	Marshall Islands	Pacific Islands	55,500
9	Saint Kitts and Nevis	Central America	56,345
10	Dominica	Caribbean	71,808
11	Andorra	Europe	76,177
12	Seychelles	Africa	96,762
13	Tonga	Pacific Islands	100,651
14	Antigua and Barbuda	Caribbean	104,084
15	Federated States of Micronesia	Pacific Islands	105,300
16	Grenada	Caribbean	108,825
17	Saint Vincent and the Grenadines	Central America	110,520
18	Kiribati	Pacific Islands	120,100
19	Saint Lucia	Central America	180,454
20	Samoa	Pacific Islands	199,052
21	São Tomé and Príncipe	Africa	201,784
22	Barbados	Caribbean	287,025
23	Vanuatu	Pacific Islands	304,500
24	Iceland	Europe	360,390
25	Maldives	Asia	378,114
26	Bahamas	Caribbean	386,870
27	Belize	Central America	398,050
28	Brunei	Asia	421,300
29	Malta	Europe	493,559
30	Cape Verde	Africa	550,483

Source: U.S. Census Bureau, International Data Base.

WORLD'S MOST DENSELY POPULATED COUNTRIES

				Area		Persons Per Square	
Rank	Country	Region	Population	Sq. Miles	Sq. Km.	Mile	Km.
1	Monaco	Europe	38,300	0.78	2.02	49,106	18,960
2	Singapore	Asia	5,638,700	279	722.5	20,212	7,804
3	Vatican City	Europe	1,000	0.17	0.44	5,887	2,273
4	Bahrain	Asia	1,543,300	300	778	5,136	1,983
5	Malta	Europe	493,559	122	315	3,911	1,510
6	Maldives	Asia	378,114	115	298	3,287	1,269
7	Bangladesh	Asia	167,983,726	55,598	143,998	3,021	1,167
8	Lebanon	Asia	6,855,713	4,036	10,452	1,740	672
9	Barbados	Caribbean	287,025	166	430	1,730	668
10	Taiwan	Asia	23,596,266	13,976	36,197	1,689	652
11	Mauritius	Africa	1,265,577	788	2,040	1,606	620
12	San Marino	Europe	34,641	24	61	1,471	568
13	Nauru	Pacific Islands	11,000	8	21	1,344	519
14	South Korea	Asia	51,811,167	38,691	100,210	1,339	517
15	Rwanda	Africa	12,374,397	10,169	26,338	1,217	470
16	Comoros	Africa	873,724	719	1,861	1,215	469
17	Netherlands	Europe	17,426,881	16,033	41,526	1,087	420
18	Haiti	Caribbean	11,263,077	10,450	27,065	1,077	416
19	Israel	Asia	9,149,500	8,522	22,072	1,074	415
20	India	Asia	1,357,746,150	1,269,211	3,287,240	1,070	413
21	Burundi	Africa	10,953,317	10,740	27,816	1,020	394
22	Tuvalu	Pacific Islands	10,200	10	26	1,015	392
23	Belgium	Europe	11,515,793	11,849	30,689	974	376
24	Philippines	Asia	108,210,625	115,831	300,000	934	361
25	Japan	Asia	126,020,000	145,937	377,975	862	333
26	Sri Lanka	Asia	21,803,000	25,332	65,610	860	332
27	El Salvador	Central America	6,704,864	8,124	21,040	826	319
28	Grenada	Caribbean	108,825	133	344	818	316
29	Marshall Islands	Pacific Islands	55,500	70	181	795	307
30	Saint Lucia	Central America	180,454	238	617	756	292

Source: U.S. Census Bureau, International Data Base

World's Least Densely Populated Countries

Rank	Country	Region	Population	Area		Persons Per Square	
				Sq. Miles	Sq. Km.	Mile	Km.
1	Mongolia	Asia	3,000,000	603,902	1,564,100	4.9	1.9
2	Western Sahara	Africa	567,421	97,344	252,120	6	2.3
3	Namibia	Africa	2,413,643	318,580	825,118	7.5	2.9
4	Australia	Australia	25,594,366	2,969,907	7,692,024	9	3
5	Suriname	South America	568,301	63,251	163,820	9.1	3.5
6	Iceland	Europe	360,390	39,682	102,775	9.1	3.5
7	Guyana	South America	782,225	83,012	214,999	9.3	3.6
8	Libya	Africa	6,470,956	683,424	1,770,060	9.6	3.7
9	Canada	North America	37,898,384	3,855,103	9,984,670	10	4
10	Mauritania	Africa	3,984,233	397,955	1,030,700	10.4	4
11	Botswana	Africa	2,302,878	224,607	581,730	10.4	4
12	Kazakhstan	Asia	18,592,700	1,052,090	2,724,900	18	7
13	Central African Republic	Africa	4,737,423	240,324	622,436	21	8
14	Gabon	Africa	2,067,561	103,347	267,667	21	8
15	Russia	Europe/Asia	146,877,088	6,612,093	17,125,242	23	9
16	Bolivia	South America	11,307,314	424,164	1,098,581	26	10
17	Chad	Africa	15,353,184	495,755	1,284,000	31	12
18	Turkmenistan	Asia	5,851,466	189,657	491,210	31	12
19	Oman	Asia	4,183,841	119,499	309,500	36	14
20	Mali	Africa	19,107,706	482,077	1,248,574	39	15
21	Argentina	South America	44,938,712	1,073,518	2,780,400	41	16
22	Saudi Arabia	Asia	34,218,169	830,000	2,149,690	41	16
23	Republic of the Congo	Africa	5,399,895	132,047	342,000	41	16
24	Finland	Europe	5,527,405	130,666	338,424	41	16
25	Norway	Europe	5,328,212	125,013	323,782	41	16
26	Paraguay	South America	7,052,983	157,048	406,752	44	17
27	Belize	Central America	398,050	8,867	22,965	44	17
28	Algeria	Africa	42,545,964	919,595	2,381,741	47	18
29	Niger	Africa	21,466,863	458,075	1,186,408	47	18
30	Papua New Guinea	Pacific Islands	8,558,800	178,704	462,840	47	18

Source: U.S. Census Bureau, International Data Base

WORLD'S MOST POPULOUS CITIES

Rank	City	Country	Region	Population
1	Chongqing	China	Asia	30,484,300
2	Shanghai	China	Asia	24,256,800
3	Beijing	China	Asia	21,516,000
4	Chengdu	China	Asia	16,044,700
5	Karachi	Pakistan	Asia	14,910,352
6	Guangzhou	China	Asia	14,043,500
7	Istanbul	Turkey	Europe	14,025,000
8	Tokyo	Japan	Asia	13,839,910
9	Tianjin	China	Asia	12,784,000
10	Mumbai	India	Asia	12,478,447
11	São Paulo	Brazil	South America	12,252,023
12	Moscow	Russia	Europe/Asia	12,197,596
13	Kinshasa	Dem. Rep. of Congo	Africa	11,855,000
14	Baoding	China	Asia	11,194,372
15	Lahore	Pakistan	Asia	11,126,285
16	Wuhan	China	Asia	11,081,000
17	Delhi	India	Asia	11,034,555
18	Harbin	China	Asia	10,635,971
19	Suzhou	China	Asia	10,459,890
20	Cairo	Egypt	Africa	10,230,350
21	Seoul	South Korea	Asia	10,197,604
22	Jakarta	Indonesia	Asia	10,075,310
23	Lima	Peru	South America	9,174,855
24	Mexico City	Mexico	North America	9,041,395
25	Ho Chi Minh City	Vietnam	Asia	8,993,082
26	Dhaka	Bangladesh	Africa	8,906,039
27	London	United Kingdom	Europe	8,825,001
28	Bangkok	Thailand	Asia	8,750,600
29	Xi'an	China	Asia	8,705,600
30	New York	United States	North America	8,622,698
31	Bangalore	India	Asia	8,425,970
32	Shenzhen	China	Asia	8,378,900
33	Nanjing	China	Asia	8,230,000
34	Tehran	Iran	Asia	8,154,051
35	Rio de Janeiro	Brazil	South America	6,718,903
36	Shantou	China	Asia	5,391,028
37	Kolkata	India	Asia	4,486,679
38	Shijiazhuang	China	Asia	4,303,700
39	Los Angeles	United States	North America	3,884,307
40	Buenos Aires	Argentina	South America	3,054,300

MAJOR LAKES OF THE WORLD

Lake	Location	Surface Area		Maximum Depth	
		Sq. Mi.	Sq. Km.	Feet	Meters
Caspian Sea	Central Asia	152,239	394,299	3,104	946
Superior	North America	31,820	82,414	1,333	406
Victoria	East Africa	26,828	69,485	270	82
Huron	North America	23,010	59,596	750	229
Michigan	North America	22,400	58,016	923	281
Aral	Central Asia	13,000	33,800	223	68
Tanganyika	East Africa	12,700	32,893	4,708	1,435
Baikal	Russia	12,162	31,500	5,712	1,741
Great Bear	North America	12,000	31,080	270	82
Nyasa	East Africa	11,600	30,044	2,316	706
Great Slave	North America	11,170	28,930	2,015	614
Chad	West Africa	9,946	25,760	23	7
Erie	North America	9,930	25,719	210	64
Winnipeg	North America	9,094	23,553	204	62
Ontario	North America	7,520	19,477	778	237
Balkhash	Central Asia	7,115	18,428	87	27
Ladoga	Russia	7,000	18,130	738	225
Onega	Russia	3,819	9,891	361	110
Titicaca	South America	3,141	8,135	1,214	370
Nicaragua	Central America	3,089	8,001	230	70
Athabasca	North America	3,058	7,920	407	124
Rudolf	Kenya, East Africa	2,473	6,405	—	—
Reindeer	North America	2,444	6,330	—	—
Eyre	South Australia	2,400	6,216	varies	varies
Issyk-Kul	Central Asia	2,394	6,200	2,297	700
Urmia	Southwest Asia	2,317	6,001	49	15
Torrens	Australia	2,200	5,698	—	—
Vänern	Sweden	2,141	5,545	322	98
Winnipegosis	North America	2,086	5,403	59	18
Mobutu Sese Seko	East Africa	2,046	5,299	180	55
Nettilling	North America	1,950	5,051	—	—

Note: The sizes of some lakes vary with the seasons.

MAJOR RIVERS OF THE WORLD

| River | Region | Source | Outflow | Approximate Length | |
				Miles	Km.
Nile	N. Africa	Tributaries of Lake Victoria	Mediterranean Sea	4,180	6,690
Mississippi-Missouri-Red Rock	N. America	Montana	Gulf of Mexico	3,710	5,970
Yangtze Kiang	East Asia	Tibetan Plateau	China Sea	3,602	5,797
Ob	Russia	Altai Mountains	Gulf of Ob	3,459	5,567
Yellow (Huang He)	East Asia	Kunlun Mountains, west China	Gulf of Chihli	2,900	4,667
Yenisei	Russia	Tannu-Ola Mountains, western Tuva, Russia	Arctic Ocean	2,800	4,506
Paraná	S. America	Confluence of Paranaiba and Grande Rivers	Río de la Plata	2,795	4,498
Irtysh	Russia	Altai Mountains, Russia	Ob River	2,758	4,438
Congo	Africa	Confluence of Lualaba and Luapula Rivers, Congo	Atlantic Ocean	2,716	4,371
Heilong (Amur)	East Asia	Confluence of Shilka and Argun Rivers	Tatar Strait	2,704	4,352
Lena	Russia	Baikal Mountains, Russia	Arctic Ocean	2,652	4,268
Mackenzie	N. America	Head of Finlay River, British Columbia, Canada	Beaufort Sea	2,635	4,241
Niger	West Africa	Guinea	Gulf of Guinea	2,600	4,184
Mekong	Asia	Tibetan Plateau	South China Sea	2,500	4,023
Mississippi	N. America	Lake Itasca, Minnesota	Gulf of Mexico	2,348	3,779
Missouri	N. America	Confluence of Jefferson, Gallatin, and Madison Rivers, Montana	Mississippi River	2,315	3,726
Volga	Russia	Valdai Plateau, Russia	Caspian Sea	2,291	3,687
Madeira	S. America	Confluence of Beni and Maumoré Rivers, Bolivia-Brazil boundary	Amazon River	2,012	3,238
Purus	S. America	Peruvian Andes	Amazon River	1,993	3,207
São Francisco	S. America	S.W. Minas Gerais, Brazil	Atlantic Ocean	1,987	3,198

River	Region	Source	Outflow	Approximate Length	
				Miles	Km.
Yukon	N. America	Junction of Lewes and Pelly Rivers, Yukon Terr., Canada	Bering Sea	1,979	3,185
St. Lawrence	N. America	Lake Ontario	Gulf of St. Lawrence	1,900	3,058
Rio Grande	N. America	San Juan Mountains, Colorado	Gulf of Mexico	1,885	3,034
Brahmaputra	Asia	Himalayas	Ganges River	1,800	2,897
Indus	Asia	Himalayas	Arabian Sea	1,800	2,897
Danube	Europe	Black Forest, Germany	Black Sea	1,766	2,842
Euphrates	Asia	Confluence of Murat Nehri and Kara Su Rivers, Turkey	Shatt-al-Arab	1,739	2,799
Darling	Australia	Eastern Highlands, Australia	Murray River	1,702	2,739
Zambezi	Africa	Western Zambia	Mozambique Channel	1,700	2,736
Tocantins	S. America	Goiás, Brazil	Pará River	1,677	2,699
Murray	Australia	Australian Alps, New S. Wales	Indian Ocean	1,609	2,589
Nelson	N. America	Head of Bow River, western Alberta, Canada	Hudson Bay	1,600	2,575
Paraguay	S. America	Mato Grosso, Brazil	Paraná River	1,584	2,549
Ural	Russia	Southern Ural Mountains, Russia	Caspian Sea	1,574	2,533
Ganges	Asia	Himalayas	Bay of Bengal	1,557	2,506
Amu Darya (Oxus)	Asia	Nicholas Range, Pamir Mountains, Turkmenistan	Aral Sea	1,500	2,414
Japurá	S. America	Andes, Colombia	Amazon River	1,500	2,414
Salween	Asia	Tibet, south of Kunlun Mountains	Gulf of Martaban	1,500	2,414
Arkansas	N. America	Central Colorado	Mississippi River	1,459	2,348
Colorado	N. America	Grand County, Colorado	Gulf of California	1,450	2,333
Dnieper	Russia	Valdai Hills, Russia	Black Sea	1,419	2,284
Ohio-Allegheny	N. America	Potter County, Pennsylvania	Mississippi River	1,306	2,102
Irrawaddy	Asia	Confluence of Nmai and Mali rivers, northeast Burma	Bay of Bengal	1,300	2,092
Orange	Africa	Lesotho	Atlantic Ocean	1,300	2,092

| River | Region | Source | Outflow | Approximate Length | |
				Miles	Km.
Orinoco	S. America	Serra Parima Mountains, Venezuela	Atlantic Ocean	1,281	2,062
Pilcomayo	S. America	Andes Mountains, Bolivia	Paraguay River	1,242	1,999
Xi Jiang	East Asia	Eastern Yunnan Province, China	China Sea	1,236	1,989
Columbia	N. America	Columbia Lake, British Columbia, Canada	Pacific Ocean	1,232	1,983
Don	Russia	Tula, Russia	Sea of Azov	1,223	1,968
Sungari	East Asia	China-North Korea boundary	Amur River	1,215	1,955
Saskatchewan	N. America	Canadian Rocky Mountains	Lake Winnipeg	1,205	1,939
Peace	N. America	Stikine Mountains, British Columbia, Canada	Great Slave River	1,195	1,923
Tigris	Asia	Taurus Mountains, Turkey	Shatt-al-Arab	1,180	1,899

Highest Peaks in Each Continent

| Continent | Mountain | Location | Height | |
			Feet	Meters
Asia	Everest	Tibet & Nepal	29,028	8,848
South America	Aconcagua	Argentina	22,834	6,960
North America	McKinley	Alaska	20,320	6,194
Africa	Kilimanjaro	Tanzania	19,340	5,895
Europe	Elbrus	Russia & Georgia	18,510	5,642
Antarctica	Vinson Massif	Ellsworth Mountains	16,066	4,897
Australia	Kosciusko	New South Wales	7,316	2,228

Note: The world's highest sixty-six mountains are all in Asia.

MAJOR DESERTS OF THE WORLD

Desert	Location	Approximate Area		Type
		Sq. Miles	Sq. Km.	
Antarctic	Antarctica	5,400,000	14,002,200	polar
Sahara	North Africa	3,500,000	9,075,500	subtropical
Arabian	Southwest Asia	1,000,000	2,593,000	subtropical
Great Western (Gibson, Great Sandy, and Great Victoria)	Australia	520,000	1,348,360	subtropical
Gobi	East Asia	500,000	1,296,500	cold winter
Patagonian	Argentina, South America	260,000	674,180	cold winter
Kalahari	Southern Africa	220,000	570,460	subtropical
Great Basin	Western United States	190,000	492,670	cold winter
Thar	South Asia	175,000	453,775	subtropical
Chihuahuan	Mexico	175,000	453,775	subtropical
Karakum	Central Asia	135,000	350,055	cold winter
Colorado Plateau	Southwestern United States	130,000	337,090	cold winter
Sonoran	United States and Mexico	120,000	311,160	subtropical
Kyzylkum	Central Asia	115,000	298,195	cold winter
Taklimakan	China	105,000	272,265	cold winter
Iranian	Iran	100,000	259,300	cold winter
Simpson	Eastern Australia	56,000	145,208	subtropical
Mojave	Western United States	54,000	140,022	subtropical
Atacama	Chile, South America	54,000	140,022	cold coastal
Namib	Southern Africa	13,000	33,709	cold coastal
Arctic	Arctic Circle			polar

Highest Waterfalls of the World

Waterfall	Location	Source	Height Feet	Meters
Angel	Canaima National Park, Venezuela	Rio Caroni	3,212	979
Tugela	Natal National Park, South Africa	Tugela River	3,110	948
Utigord	Norway	glacier	2,625	800
Monge	Marstein, Norway	Mongebeck	2,540	774
Mutarazi	Nyanga National Park, Zimbabwe	Mutarazi River	2,499	762
Yosemite	Yosemite National Park, California, U.S.	Yosemite Creek	2,425	739
Espelands	Hardanger Fjord, Norway	Opo River	2,307	703
Lower Mar Valley	Eikesdal, Norway	Mardals Stream	2,151	655
Tyssestrengene	Odda, Norway	Tyssa River	2,123	647
Cuquenan	Kukenan Tepuy, Venezuela	Cuquenan River	2,000	610
Sutherland	Milford Sound, New Zealand	Arthur River	1,904	580
Kjell	Gudvanger, Norway	Gudvangen Glacier	1,841	561
Takkakaw	Yoho Natl Park, British Columbia, Canada	Takkakaw Creek	1,650	503
Ribbon	Yosemite National Park, California, U.S.	Ribbon Stream	1,612	491
Upper Mar Valley	near Eikesdal, Norway	Mardals Stream	1,536	468
Gavarnie	near Lourdes, France	Gave de Pau	1,388	423
Vettis	Jotunheimen, Norway	Utla River	1,215	370
Hunlen	British Columbia, Canada	Hunlen River	1,198	365
Tin Mine	Kosciusko National Park, Australia	Tin Mine Creek	1,182	360
Silver Strand	Yosemite National Park, California, U.S.	Silver Strand Creek	1,170	357
Basaseachic	Baranca del Cobre, Mexico	Piedra Volada Creek	1,120	311
Spray Stream	Lauterburnnental, Switzerland	Staubbach Brook	985	300
Fachoda	Tahiti, French Polynesia	Fautaua River	985	300
King Edward VIII	Guyana	Courantyne River	850	259
Wallaman	near Ingham, Australia	Wallaman Creek	844	257
Gersoppa	Western Ghats, India	Sharavati River	828	253
Kaieteur	Guyana	Rio Potaro	822	251
Montezuma	near Rosebery, Tasmania	Montezuma River	800	240
Wollomombi	near Armidale, Australia	Wollomombi River	722	2203

Source: Fifth Continent Australia Pty Limited

GLOSSARY

Places whose names are printed in SMALL CAPS *are subjects of their own entries in this glossary.*

Ablation. Loss of ice volume or mass by a GLACIER. Ablation includes melting of ice, SUBLIMATION, DEFLATION (removal by WIND), EVAPORATION, and CALVING. Ablation occurs in the lower portions of glaciers.

Abrasion. Wearing away of ROCKS in STREAMS by grinding, especially when rocks and SEDIMENT are carried along by stream water. The STREAMBED and VALLEY are carved out and eroded, and the rocks become rounded and smoothed by abrasion.

Absolute location. Position of any PLACE on the earth's surface. The absolute location can be given precisely in terms of DEGREES, MINUTES, and SECONDS of LATITUDE (0 to 90 degrees north or south) and of LONGITUDE (0 to 180 degrees east or west). The EQUATOR is 0 degrees latitude; the PRIME MERIDIAN, which runs through Greenwich in England, is 0 degrees longitude.

Abyss. Deepest part of the OCEAN. Modern TECHNOLOGY—especially sonar—has enabled accurate mapping of the ocean floors, showing that there are MOUNTAIN CHAINS, or RIDGES, in all the oceans, as well as deep CANYONS or TRENCHES closer to the edges of the oceans.

Acid rain. PRECIPITATION containing high levels of nitric or sulfuric acid; a major environmental problem in parts of North America, Europe, and Asia. Natural precipitation is slightly acidic (about 5.6 on the pH SCALE), because CARBON DIOXIDE—which occurs naturally in the ATMOSPHERE—is dissolved to form a weak carbonic acid.

Adiabatic. Change of TEMPERATURE within the ATMOSPHERE that is caused by compression or expansion without addition or loss of heat.

Advection. Horizontal movement of AIR from one PLACE to another in the ATMOSPHERE, associated with WINDS.

Advection fog. FOG that forms when a moist AIR mass moves over a colder surface. Commonly, warm moist air moves over a cool OCEAN CURRENT, so the air cools to SATURATION POINT and fog forms. This phenomenon, known as sea fog, occurs along subtropical west COASTS.

Aerosol. Substances held in SUSPENSION in the ATMOSPHERE, as solid particles or liquid droplets.

Aftershock. EARTHQUAKE that follows a larger earthquake and originates at or near the focus of the latter; many aftershocks may follow a major earthquake, decreasing in frequency and magnitude with time.

Agglomerate. Type of ROCK composed of volcanic fragments, usually of different sizes and rough or angular.

Aggradation. Accumulation of SEDIMENT in a STREAMBED. Aggradation often results from reduced flow in the channel during dry periods. It also occurs when the STREAM's load (BEDLOAD and SUSPENDED LOAD) is greater than the stream capacity. A BRAIDED STREAM pattern often results.

Air current. Air currents are caused by differential heating of the earth's surface, which causes heated air to rise. This causes WINDS at the surface as well as higher in the earth's ATMOSPHERE.

Air mass. Large body of air with distinctive homogeneous characteristics of TEMPERATURE, HUMIDITY, and stability. It forms when air remains stationary over a source REGION for a period of time, taking on the conditions of that region. An air mass can extend over a million square miles with a depth of more than a mile. Air masses are classified according to moisture content (*m* for maritime or *c* for continental) and temperature

465

(*A* for ARCTIC, *P* for polar, *T* for tropical, or *E* for equatorial). The air masses affecting North America are mP, cP, and mT. The interaction of AIR masses produces WEATHER. The line along which air masses meet is a FRONT.

Albedo. Measure of the reflective properties of a surface; the ratio of reflected ENERGY (INSOLATION) to the total incoming energy, expressed as a percentage. The albedo of Earth is 33 percent.

Alienation (land). Land alienation is the appropriation of land from its original owners by a more powerful force. In preindustrial societies, the ownership of agricultural land is of prime importance to subsistence farmers.

Alkali flat. Dry LAKEBED in an arid REGION, covered with a layer of SALTS. A well-known example is the Alkali Flat area of White Sands National Monument in New Mexico; it is the bed of a large lake that formed when the GLACIERS were melting. It is covered with a form of gypsum crystals called selenite. This material is blown off the surface into large SAND DUNES. Also called a salina. See also BITTER LAKE.

Allogenic sediment. SEDIMENT that originates outside the PLACE where it is finally deposited; SAND, SILT, and CLAY carried by a STREAM into a LAKE are examples.

Alluvial fan. Common LANDFORM at the mouth of a CANYON in arid REGIONS. Water flowing in a narrow canyon immediately slows as it leaves the canyon for the wider VALLEY floor, depositing the SEDIMENTS it was transporting. These spread out into a fan shape, usually with a BRAIDED STREAM pattern on its surface. When several alluvial fans grow side by side, they can merge into one continuous sloping surface between the HILLS and the valley. This is known by the Spanish word *bajada*, which means "slope."

Alluvial system. Any of various depositional systems, excluding DELTAS, that form from the activity of RIVERS and STREAMS. Much alluvial SEDIMENT is deposited when rivers top their BANKS and FLOOD the surrounding countryside. Buried alluvial sediments may be important water-bearing RESERVOIRS or may contain PETROLEUM.

Alluvium. Material deposited by running water. This includes not only fertile SOILS, but also CLAY, SILT, or SAND deposits resulting from FLUVIAL processes. FLOODPLAINS are covered in a thick layer of alluvium.

Altimeter. Instrument for measuring ALTITUDE, or height above the earth's surface, commonly used in airplanes. An altimeter is a type of ANEROID BAROMETER.

Altitudinal zonation. Existence of different ECOSYSTEMS at various ELEVATIONS above SEA LEVEL, due to TEMPERATURE and moisture differences. This is especially pronounced in Central America and South America. The hot and humid COASTAL PLAINS, where bananas and sugarcane thrive, is the *tierra caliente*. From about 2,500 to 6,000 feet (750–1,800 meters) is the *tierra templada*; crops grown here include coffee, wheat, and corn, and major cities are situated in this zone. From about 6,000 to 12,000 feet (1,800–3,600 meters) is the *tierra fria*; here only hardy crops such as potatoes and barley are grown, and large numbers of animals are kept. From about 12,000 to 15,000 feet (3,600 to 4,500 meters) lies the *tierra helada*, where hardy animals such as sheep and alpaca graze. Above 15,000 feet (4,500 meters) is the frozen *tierra nevada*; no permanent life is possible in the permanent SNOW and ICE FIELDS there.

Angle of repose. Maximum angle of steepness that a pile of loose materials such as SAND or ROCK can assume and remain stable; the angle varies with the size, shape, moisture, and angularity of the material.

Antecedent river. STREAM that was flowing before the land was uplifted and was able to erode at the pace of UPLIFT, thus creating a deep CANYON. Most deep canyons are attributed to antecedent rivers. In the Davisian CYCLE OF EROSION, this process was called REJUVENATION.

Anthropogeography. Branch of GEOGRAPHY founded in the late nineteenth century by German geographer Friedrich Ratzel. The field is closely related to human ECOLOGY—the study of humans, their DISTRIBUTION over the earth, and their interaction with their physical ENVIRONMENT.

Anticline. Area where land has been UPFOLDED symmetrically. Its center contains stratigraphically older ROCKS. See also SYNCLINE.

Anticyclone. High-pressure system of rotating WINDS, descending and diverging, shown on a WEATHER chart by a series of closed ISOBARS, with a high in the center. In the NORTHERN HEMISPHERE, the rotation is CLOCKWISE; in the SOUTHERN HEMISPHERE, the rotation is COUNTERCLOCKWISE. An anticyclone brings warm weather.

Antidune. Undulatory upstream-moving bed form produced in free-surface flow of water over a SAND bed in a certain RANGE of high flow speeds and shallow flow depths.

Antipodes. TEMPERATE ZONE of the SOUTHERN HEMISPHERE. The term is now usually applied to the countries of Australia and New Zealand. The ancient Greeks had believed that if humans existed there, they must walk upside down. This idea was supported by the Christian Church in the Middle Ages.

Antitrade winds. WINDS in the upper ATMOSPHERE, or GEOSTROPHIC winds, that blow in the opposite direction to the TRADE WINDS. Antitrade winds blow toward the northeast in the NORTHERN HEMISPHERE and toward the southeast in the SOUTHERN HEMISPHERE.

Aperiodic. Irregularly occurring interval, such as found in most WEATHER CYCLES, rendering them virtually unpredictable.

Aphelion. Point in the earth's 365-DAY REVOLUTION when it is at its greatest distance from the SUN. This is caused by Earth's elliptical ORBIT around the Sun. The distance at aphelion is 94,555,000 miles (152,171,500 km.) and usually falls on July 4. The opposite of PERIHELION.

Aposelene. Earth's farthest point from the MOON.

Aquifer. Underground body of POROUS ROCK that contains water and allows water PERCOLATION through it. The largest aquifer in the United States is the Ogallala Aquifer, which extends south from South Dakota to Texas.

Arête. Serrated or saw-toothed ridge, produced in glaciated MOUNTAIN areas by CIRQUES eroding on either side of a RIDGE or mountain RANGE. From the French word for knife-edge.

Arroyo. Spanish word for a dry STREAMBED in an arid area. Called a WADI in Arabic and a WASH in English.

Artesian well. WELL from which GROUNDWATER flows without mechanical pumping, because the water comes from a CONFINED AQUIFER, and is therefore under pressure. The Great Artesian Basin of Australia has hundreds of artesian wells, called BORES, that provide drinking water for sheep and cattle. The name comes from the Artois REGION of France, where the phenomenon is common. A subartesian well is sunk into an UNCONFINED AQUIFER and requires a pump to raise water to the surface.

Asteroid belt. REGION between the ORBITS of Mars and Jupiter containing the majority of ASTEROIDS.

Asthenosphere. Part of the earth's UPPER MANTLE, beneath the LITHOSPHERE, in which PLATE movement takes place. Also known as the low-velocity zone.

Astrobleme. Remnant of a large IMPACT CRATER on Earth.

Astronomical unit (AU). Unit of measure used by astronomers that is equivalent to the average distance from the SUN to Earth (93 million miles/150 million km.).

Atmospheric pressure. Weight of the earth's ATMOSPHERE, equally distributed over earth's surface and pressing down as a result of GRAVITY. On average, the atmosphere has a force of 14.7 pounds per square inch (1 kilogram per centimeter) squared at SEA LEVEL, also expressed as 1013.2 millibars. Variations in atmospheric pressure, high or low, cause WINDS and WEATHER changes that affect CLIMATE. Pressure decreases rapidly with ALTITUDE or distance from the surface: Half of the total atmosphere is found below 18,000 feet (5,500 meters); more than 99 percent of the atmosphere is below 30 miles (50 km.) of the surface. Atmospheric pressure is measured with a BAROMETER.

Atoll. Ring-shaped growth of CORAL REEF, with a LA-GOON in the middle. Charles Darwin, who observed many Pacific atolls during his voyage on the *Beagle* in the nineteenth century, suggested that they were created from FRINGING REEFS around volcanic ISLANDS. As such islands sank beneath the water (or as SEA LEVELS rose), the coral continued growing upward. SAND resting atop an atoll enables plants to grow, and small human societies have arisen on some atolls. The world's largest atoll, Kwajalein in the Marshall Islands, measures about 40 by 18 miles (65 by 30 km.), but perhaps the most famous atoll is Bikini Atoll—the SITE of nuclear-bomb testing during the 1950s.

Aurora. Glowing and shimmering displays of colored lights in the upper ATMOSPHERE, caused by interaction of the SOLAR WIND and the charged particles of the IONOSPHERE. Auroras occur at high LATITUDES. Near the North Pole they are called aurora borealis or northern lights; near the South Pole, aurora australis or southern lights.

Austral. Referring to an object or occurrence that is located in the SOUTHERN HEMISPHERE or related to Australia.

Australopithecines. Erect-walking early human ancestors with a cranial capacity and body size within the RANGE of modern apes rather than of humans.

Avalanche. Mass of SNOW and ice falling suddenly down a MOUNTAIN slope, often taking with it earth, ROCKS, and trees.

Bank. Elevated area of land beneath the surface of the OCEAN. The term is also used for elevated ground lining a body of water.

Bar (climate). Measure of ATMOSPHERIC PRESSURE per unit surface area of 1 million dynes per square centimeter. Millibars (thousandths of a bar) are the MEASUREMENT used in the United States. Other countries use kilopascals (kPa); one kilopascal is ten millibars.

Bar (land). RIDGE or long deposit of SAND or gravel formed by DEPOSITION in a RIVER or at the COAST. Offshore bars and baymouth bars are common coastal features.

Barometer. Instrument used for measuring ATMOSPHERIC PRESSURE. In the seventeenth century, Evangelista Torricelli devised the first barometer—a glass tube sealed at one end, filled with mercury, and upended into a bowl of mercury. He noticed how the height of the mercury column changed and realized this was a result of the pressure of air on the mercury in the bowl. Early MEASUREMENTS of atmospheric pressure were, therefore, expressed as centimeters of mercury, with average pressure at SEA LEVEL being 29.92 inches (760 millimeters). This cumbersome barometer was replaced with the ANEROID BAROMETER—a sealed and partially evacuated box connected to a needle and dial, which shows changes in atmospheric pressure. See also ALTIMETER.

Barrier island. Long chain of SAND islands that forms offshore, close to the COAST. LAGOONS or shallower MARSHES separate the barrier islands from the mainland. Such LOCATIONS are hazardous for SETTLEMENTS because they are easily swept away in STORMS and HURRICANES.

Basalt. IGNEOUS EXTRUSIVE ROCK formed when LAVA cools; often black in color. Sometimes basalt occurs in tall hexagonal columns, such as the Giant's Causeway in Ireland, or the Devils Postpile at Mammoth, California.

Basement. Crystalline, usually PRECAMBRIAN, IGNEOUS and METAMORPHIC ROCKS that occur beneath the SEDIMENTARY ROCK on the CONTINENTS.

Basin order. Approximate measure of the size of a STREAM BASIN, based on a numbering scheme applied to RIVER channels as they join together in their progress downstream.

Batholith. Large LANDFORM produced by IGNEOUS INTRUSION, composed of CRYSTALLINE ROCK, such as GRANITE; a large PLUTON with a surface area greater than 40 square miles (100 sq. km.). Most mountain RANGES have a batholith underneath.

Bathymetric contour. Line on a MAP of the OCEAN floor that connects points of equal depth.

Beaufort scale. SCALE that measures WIND force, expressed in numbers from 0 to 12. The original Beaufort scale was based on descriptions of the state of the SEA. It was adapted to land conditions, using descriptions of chimney smoke, leaves of trees, and similar factors. The scale was devised in the early nineteenth century by Sir Francis Beaufort, a British naval officer.

Belt. Geographical REGION that is distinctive in some way.

Bergeron process. PRECIPITATION formation in COLD CLOUDS whereby ice crystals grow at the expense of supercooled water droplets.

Bight. Wide or open BAY formed by a curve in the COASTLINE, such as the Great Australian Bight.

Biogenic sediment. SEDIMENT particles formed from skeletons or shells of microscopic plants and animals living in seawater.

Biostratigraphy. Identification and organization of STRATA based on their FOSSIL content and the use of fossils in stratigraphic correlation.

Bitter lake. Saline or BRACKISH LAKE in an arid area, which may dry up in the summer or in periods of DROUGHT. The water is not suitable for drinking. Another name for this feature is "salina." See also ALKALI FLAT.

Block lava. LAVA flows whose surfaces are composed of large, angular blocks; these blocks are generally larger than those of AA flows and have smooth, not jagged, faces.

Block mountain. MOUNTAIN or mountain RANGE with one side having a gentle slope to the crest, while the other slope, which is the exposed FAULT SCARP, is quite steep. It is formed when a large block of the earth's CRUST is thrust upward on one side only, while the opposite side remains in place. The Sierra Nevada in California are a good example of block mountains. Also known as fault-block mountain.

Blowhole. SEA CAVE or tunnel formed on some rocky, rugged COASTLINES. The pressure of the seawater rushing into the opening can force a jet of seawater to rise or spout through an opening in the roof of the cave. Blowholes are found in Scotland, Tasmania, and Mexico, and on the Hawaiian ISLANDS of Kauai and Maui.

Bluff. Steep slope that marks the farthest edge of a FLOODPLAIN.

Bog. Damp, spongy ground surface covered with decayed or decaying VEGETATION. Bogs usually are formed in cool CLIMATES through the in-filling, or silting up, of a LAKE. Moss and other plants grow outward toward the edge of the lake, which gradually becomes shallower, until the surface is completely covered. Bogs also can form on cold, damp MOUNTAIN surfaces. Many bogs are filled with PEAT.

Bore. Standing WAVE, or wall, of water created in a narrow ESTUARY when the strong incoming, or FLOOD, TIDE meets the RIVER water flowing outward; it moves upstream with the advancing tide, and downstream with the EBB TIDE. South America's Amazon River and Asia's Mekong River have large bores. In North America, the bore in the Bay of Fundy is visited by many tourists each year. Its St. Andrew's wharf is designed to handle changes in water level of as much as 53 feet (15 meters) in one DAY.

Boreal. Alluding to an item or event that is in the NORTHERN HEMISPHERE.

Bottom current. Deep-sea current that flows parallel to BATHYMETRIC CONTOURS.

Brackish water. Water with SALT content between that of SALT WATER and FRESH WATER; it is common in arid areas on the surface, in coastal MARSHES, and in salt-contaminated GROUNDWATER.

Braided stream. STREAM having a CHANNEL consisting of a maze of interconnected small channels within a broader STREAMBED. Braiding occurs when the stream's load exceeds its capacity, usually because of reduced flow.

Breaker. WAVE that becomes oversteepened as it approaches the SHORE, reaching a point at which it cannot maintain its vertical shape. It then breaks, and the water washes toward the shore.

Breakwater. Large structure, usually of ROCK, built offshore and parallel to the COAST, to absorb WAVE ENERGY and thus protect the SHORE. Between the breakwater and the shore is an area of calm water, often used as a boat anchorage or

HARBOR. A similar but smaller structure is a seawall.

Breeze. Gentle WIND with a speed of 4 to 31 miles (6 to 50 km.) per hour. On the BEAUFORT SCALE, the numbers 2 through 6 represent breezes of increasing strength.

Butte. Flat-topped HILL, smaller than a MESA, found in arid REGIONS.

Caldera. Large circular depression with steep sides, formed when a VOLCANO explodes, blowing away its top. The ERUPTION of Mount St. Helens produced a caldera. Crater Lake in Oregon is a caldera that has filled with water. From the Spanish word for kettle.

Calms of Cancer. Subtropical BELT of high pressure and light WINDS, located over the OCEAN near 25 DEGREES north LATITUDE. Also known as the HORSE LATITUDES.

Calms of Capricorn. Subtropical BELT of high pressure and light WINDS, located over the OCEAN near 25 DEGREES south LATITUDE.

Calving. Loss of glacial mass when GLACIERS reach the SEA and large blocks of ice break off, forming ICEBERGS.

Cancer, tropic of. PARALLEL of LATITUDE at 23.5 DEGREES north; this line is the latitude farthest north on the earth where the noon SUN is ever directly overhead. The REGION between it and the tropic of CAPRICORN is known as the TROPICS.

Capricorn, tropic of. Line of LATITUDE at 23.5 DEGREES south; this line is the latitude farthest south on the earth where the noon SUN is ever directly overhead. The REGION between it and the tropic of CANCER is known as the TROPICS.

Carbon dating. Method employed by physicists to determine the age of organic matter—such as a piece of wood or animal tissue—to determine the age of an archaeological or paleontological SITE. The method works on the principle that the amount of radioactive carbon in living matter diminishes at a steady, and measurable, rate after the matter dies. Technique is also known as carbon-14 dating, after the radioactive car-

bon-14 isotope it uses. Also known as radiocarbon dating.

Carrying capacity. Number of animals that a given area of land can support, without additional feed being necessary. Lush GRASSLAND may have a carrying capacity of twenty sheep per acre, while more arid, SEMIDESERT land may support only two sheep per acre. The term sometimes is used to refer to the number of humans who can be supported in a given area.

Catastrophism. Theory, popular in the eighteenth and nineteenth centuries, that explained the shape of LANDFORMS and CONTINENTS and the EXTINCTION of species as the results of intense or catastrophic events. The biblical FLOOD of Noah was one such event, which supposedly explained many extinctions. Catastrophism is linked closely to the belief that the earth is only about 6,000 years old, and therefore tremendous forces must have acted swiftly to create present LANDSCAPES. An alternative or contrasting theory is UNIFORMITARIANISM.

Catchment basin. Area of land receiving the PRECIPITATION that flows into a STREAM. Also called catchment or catchment area.

Central place theory. Theory that explains why some SETTLEMENTS remain small while others grow to be middle-sized TOWNS, and a few become large cities or METROPOLISES. The explanation is based on the provision of goods and services and how far people will travel to acquire these. The German geographer Walter Christaller developed this theory in the 1930s.

Centrality. Measure of the number of functions, or services, offered by any CITY in a hierarchy of cities within a COUNTRY or a REGION. See also CENTRAL PLACE THEORY.

Chain, mountain. Another term for mountain RANGE.

Chemical farming. Application of artificial FERTILIZERS to the SOIL and the use of chemical products such as insecticides, fungicides, and herbicides to ensure crop success. Chemical farming is practiced mainly in high-income countries, because the cost of the chemical products is high. Farmers in low-income economies rely more on natural organic fertilizers such as animal waste.

Chemical weathering. Chemical decomposition of solid ROCK by processes involving water that change its original materials into new chemical combinations.

Chlorofluorocarbons (CFCs). Manufactured compounds, not occurring in nature, consisting of chlorine, fluorine, and carbon. CFCs are stable and have heat-absorbing properties, so they have been used extensively for cooling in refrigeration and air-conditioning units. Previously, they were used as propellants for aerosol products. CFCs rise into the STRATOSPHERE where ULTRAVIOLET RADIATION causes them to react with OZONE, changing it to oxygen and exposing the earth to higher levels of ultraviolet (UV) radiation. Therefore, the manufacture and use of CFCs was banned in many countries. The commercial name for CFCs is Freon.

Chorology. Description or mapping of a REGION. Also known as chorography.

Chronometer. Highly accurate CLOCK or timekeeping device. The first accurate and effective chronometers were constructed in the mid-eighteenth century by John Harrison, who realized that accurate timekeeping was the secret to NAVIGATION at SEA.

Cinder cone. Small conical HILL produced by PYROCLASTIC materials from a VOLCANO. The material of the cone is loose SCORIA.

Circle of illumination. Line separating the sunlit part of the earth from the part in darkness. The circle of illumination moves around the earth once in every approximately 24 hours. At the VERNAL and autumnal EQUINOXES, the circle of illumination passes through the POLES.

Cirque. Circular BASIN at the head of an ALPINE GLACIER, shaped like an armchair. Many cirques can be seen in MOUNTAIN areas where glaciers have completely melted since the last ICE AGE.

City Beautiful movement. Planning and architectural movement that was at its height from around 1890 to the 1920s in the United States. It was believed that classical architecture, wide and carefully laid-out streets, parks, and urban monuments would reflect the higher values of the society and be a civilizing, even uplifting, experience for the citizens of such cities. Civic pride was fostered through remodeling or modernizing older URBAN AREAS. Chicago, Illinois, and Pasadena, California, are cities where the planners of the City Beautiful movement left their imprint.

Clastic. ROCK or sedimentary matter formed from fragments of older rocks.

Climatology. Study of Earth CLIMATES by analysis of long-term WEATHER patterns over a minimum of thirty years of statistical records. Climatologists—scientists who study climate—seek similarities to enable grouping into climatic REGIONS. Climate patterns are closely related to natural VEGETATION. Computer TECHNOLOGY has enabled investigation of phenomena such as the EL NIÑO effect and global climate change. The KÖPPEN CLIMATE CLASSIFICATION system is the most commonly used scheme for climate classification.

Climograph. Graph that plots TEMPERATURE and PRECIPITATION for a selected LOCATION. The most commonly used climographs plot monthly temperatures and monthly precipitation, as used in the KÖPPEN CLIMATE CLASSIFICATION. Also spelled "climagraph." The term climagram is rarely used.

Clinometer. Instrument used by surveyors to measure the ELEVATION of land or the inclination (slope) of the land surface.

Cloud seeding. Injection of CLOUD-nucleating particles into likely clouds to enhance PRECIPITATION.

Cloudburst. Heavy rain that falls suddenly.

Coal. One of the FOSSIL FUELS. Coal was formed from fossilized plant material, which was originally FOREST. It was then buried and compacted, which led to chemical changes. Most coal was formed during the CARBONIFEROUS PERIOD (286 million to 360 million years ago) when the earth's CLIMATE was wetter and warmer than at present.

Coastal plain. Large area of flat land near the OCEAN. Coastal plains can form in various ways, but FLUVIAL DEPOSITION is an important process. In the United States, the coastal plain extends from Texas to North Carolina.

Coastal wetlands. Shallow, wet, or flooded shelves that extend back from the freshwater-saltwater interface and may consist of MARSHES, BAYS, LAGOONS, tidal flats, or MANGROVE SWAMPS.

Cognitive map. Mental image that each person has of the world, which includes LOCATIONS and connections. These maps expand as children mature, from plans of their rooms, to their houses, to their neighborhoods. Adults know certain parts of the CITY and the streets connecting them.

Coke. Type of fuel produced by heating COAL.

Col. Lower section of a RIDGE, usually formed by the headward EROSION of two CIRQUE GLACIERS at an ARÊTE. Sometimes called a saddle.

Colonialism. Control of one COUNTRY over another STATE and its people. Many European countries have created colonial empires, including Great Britain, France, Spain, Portugal, the Netherlands, and Russia.

Columbian exchange. Interaction that occurred between the Americas and Europe after the voyages of Christopher Columbus. Food crops from the New World transformed the diet of many European countries.

Comet. Small body in the SOLAR SYSTEM, consisting of a solid head with a long gaseous tail. The elliptical ORBIT of a comet causes it to range from very close to the SUN to very far away. In ancient times, the appearance of a comet in the sky was thought to be an omen of great events or changes, such as war or the death of a king.

Comfort index. Number that expresses the combined effects of TEMPERATURE and HUMIDITY on human bodily comfort. The index number is obtained by measuring ambient conditions and comparing these to a chart.

Commodity chain. Network linking labor, production, delivery, and sale for any product. The chain begins with the production of the raw material, such as the extraction of MINERALS by miners, and extends to the acquisition of the finished product by a consumer.

Complex crater. IMPACT CRATER of large diameter and low depth-to-diameter ratio caused by the presence of a central UPLIFT or ring structure.

Composite cone. Cone or VOLCANO formed by volcanic explosions in which the LAVA is of different composition, sometimes fluid, sometimes PYROCLASTS such as cinders. The alternation of layers allows a concave shape for the cone. These are generally regarded as the world's most beautiful volcanoes. Composite volcanoes are sometimes called STRATOVOLCANOES.

Condensation nuclei. Microscopic particles that may have originated as DUST, soot, ASH from fires or VOLCANOES, or even SEA SALT; an essential part of CLOUD formation. When AIR rises and cools to the DEW POINT (saturation), the moisture droplets condense around the nuclei, leading to the creation of raindrops or snowflakes. A typical air mass might contain 10 billion condensation nuclei in a single cubic yard (1 cubic meter) of air.

Cone of depression. Cone-shaped depression produced in the WATER TABLE by pumping from a WELL.

Confined aquifer. AQUIFER that is completely filled with water and whose upper BOUNDARY is a CONFINING BED; it is also called an artesian aquifer.

Confining bed. Impermeable layer in the earth that inhibits vertical water movement.

Confluence. PLACE where two STREAMS or RIVERS flow together and join. The smaller of the two streams is called a TRIBUTARY.

Conglomerate. Type of SEDIMENTARY ROCK consisting of smaller rounded fragments naturally cemented together by another MINERAL. If the cemented fragments are jagged or angular, the rock is called breccia.

Conical projection. MAP PROJECTION that can be imagined as a cone of paper resting like a witch's hat on a globe with a light source at its center; the images of the CONTINENTS would be projected onto the paper. In reality, maps are constructed mathematically. A conic projection can show only part of one HEMISPHERE. This projection is suitable for constructing a MAP of the United States, as a good EQUAL-AREA representation can be achieved. Also called conic projection.

Consequent river. RIVER that flows across a LANDSCAPE because of GRAVITY. Its direction is determined by the original slope of the land. TRIBUTARY streams, which develop later as EROSION proceeds, are called subsequent streams.

Continental climate. CLIMATE experienced over the central REGIONS of large LANDMASSES; drier and subject to greater seasonal extremes of TEMPERATURE than at the CONTINENTAL MARGINS.

Continental rift zones. Continental rift zones are PLACES where the CONTINENTAL CRUST is stretched and thinned. Distinctive features include active VOLCANOES and long, straight VALLEY systems formed by normal FAULTS. Continental rifting in some cases has evolved into the breaking apart of a CONTINENT by SEAFLOOR SPREADING to form a new OCEAN.

Continental shelf. Shallow, gently sloping part of the seafloor adjacent to the mainland. The continental shelf is geologically part of the CONTINENT and is made of CONTINENTAL CRUST, whereas the OCEAN floor is OCEANIC CRUST. Although continental shelves vary greatly in width, on average they are about 45 miles (75 km.) wide and have slopes of 7 minutes (about one-tenth of a degree). The average depth of a continental shelf is about 200 feet (60 meters). The outer edge of the continental shelf is marked by a sharp change in angle where the CONTINENTAL SLOPE begins. Most continental shelves were exposed above current SEA LEVEL during the PLEISTOCENE EPOCH and have been submerged by rising sea levels over the past 18,000 years.

Continental shield. Area of a CONTINENT that contains the oldest ROCKS on Earth, called CRATONS. These are areas of granitic rocks, part of the CONTINENTAL CRUST, where there are ancient MOUNTAINS. The Canadian Shield in North America is an example.

Convectional rain. Type of PRECIPITATION caused when AIR over a warm surface is warmed and rises, leading to ADIABATIC cooling, CONDENSATION, and, if the air is moist enough, rain.

Convective overturn. Renewal of the bottom waters caused by the sinking of SURFACE WATERS that have become denser, usually because of decreased TEMPERATURE.

Convergence (climate). AIR flowing in toward a central point.

Convergence (physiography). Process that occurs during the second half of a SUPERCONTINENT CYCLE, whereby crustal PLATES collide and intervening OCEANS disappear as a result of plate SUBDUCTION.

Convergent plate boundary. Compressional PLATE BOUNDARY at which an oceanic PLATE is subducted or two continental plates collide.

Convergent plate margin. Area where the earth's LITHOSPHERE is returned to the MANTLE at a SUBDUCTION ZONE, forming volcanic "ISLAND ARCS" and associated HYDROTHERMAL activity.

Conveyor belt current. Large CYCLE of water movement that carries warm water from the north Pacific westward across the Indian Ocean, around Southern Africa, and into the Atlantic, where it warms the ATMOSPHERE, then returns at a deeper OCEAN level to rise and begin the process again.

Coordinated universal time (UTC). International basis of time, introduced to the world in 1964. The basis for UTC is a small number of ATOMIC CLOCKS. Leap seconds are occasionally added to UTC to keep it synchronized with universal time.

Core-mantle boundary. SEISMIC discontinuity 1,790 miles (2,890 km.) below the earth's surface that separates the MANTLE from the OUTER CORE.

Core region. Area, generally around a COUNTRY's CAPITAL CITY, that has a large, dense POPULATION and is the center of TRADE, financial services, and production. The rest of the country is referred to as the PERIPHERY. On a larger scale, the CONTINENT of Europe has a core region, which includes London, Paris, and Berlin; Iceland, Portugal, and Greece are peripheral LOCATIONS.

Coriolis effect. Apparent deflection of moving objects above the earth because of the earth's ROTA-

tion. The deflection is to the right in the NORTH-
ERN HEMISPHERE and to the left in the SOUTH-
ERN HEMISPHERE. The deflection is inversely
proportional to the speed of the earth's rotation,
being negligible at the EQUATOR but at its maximum near the POLES. The Coriolis effect is a major influence on the direction of surface WINDS.
Sometimes called Coriolis force.

Corrasion. EROSION and lowering of a STREAMBED
by FLUVIAL action, especially by ABRASION of the
bedload (material transported by the STREAM)
but also including SOLUTION by the water.

Cosmogony. Study of the origin and nature of the
SOLAR SYSTEM.

Cotton Belt. Part of the United States extending
from South Carolina through Georgia, Alabama, Mississippi, Tennessee, Louisiana, Arkansas, Texas, and Oklahoma, where cotton was
grown on PLANTATIONS using slave labor before
the Civil War. After that war, the South stagnated
for almost a century. Racial SEGREGATION contributed to cultural isolation from the rest of the
United States. Cotton is still produced in this RE-
GION, but California has overtaken the Southern
STATES as a cotton producer, and other agricultural products, such as soybeans and poultry,
have become dominant crops in the old Cotton
Belt. In-migration, due to the SUN BELT
attraction, has led to rapid urban growth.

Counterurbanization. Out-migration of people
from URBAN AREAS to smaller TOWNS or RURAL
areas. As large modern cities are perceived to be
overcrowded, stressful, polluted, and dangerous, many of their residents move to areas they
regard as more favorable. Such moves are often
related to individuals' retirements; however,
younger workers and families are also part of
counterurbanization.

Crater morphology. Structure or form of CRATERS
and the related processes that developed them.

Craton. Large, geologically old, relatively stable
CORE of a continental LITHOSPHERIC PLATE,
sometimes termed a CONTINENTAL SHIELD.

Creep. Slow, gradual downslope movement of SOIL
materials under gravitational stress. Creep tests
are experiments conducted to assess the effects

of time on ROCK properties, in which environmental conditions (surrounding pressure,
TEMPERATURE) and the deforming stress are
held constant.

Crestal plane. Plane or surface that goes through
the highest points of all beds in a fold; it is coin-
cident with the axial plane when the axial plane
is vertical.

Cross-bedding. Layers of ROCK or SAND that lie at
an angle to horizontal bedding or to the ground.

Crown land. Land belonging to a NATION'S MONAR-
CHY.

Crude oil. Unrefined OIL, as it occurs naturally.
Also called PETROLEUM.

Crustal movements. PLATE TECTONICS theorizes
that Earth's CRUST is not a single rigid shell, but
comprises a number of large pieces that are in
motion, separating or colliding. There are two
types of crust—the older continental and the
much younger OCEANIC CRUST. When PLATES diverge, at SEAFLOOR SPREADING zones, new (oceanic) crust is created from the MAGMA that flows
out at the MID-OCEAN RIDGES. When plates converge and collide, denser oceanic crust is
SUBDUCTED under the lighter CONTINENTAL
CRUST. The boundaries at the areas where plates
slide laterally, neither diverging nor converging,
are called TRANSFORM FAULTS. The San Andreas
Fault represents the world's best-known transform BOUNDARY. As a result of crustal movements, the earth can be deformed in several
ways. Where PLATE BOUNDARIES converge, compression can occur, leading to FOLDING and the
creation of SYNCLINES and ANTICLINES. Other
stresses of the crust can lead to fracture, or faulting, and accompanying EARTHQUAKES. LAND-
FORMS created in this way include HORSTS,
GRABEN, and BLOCK MOUNTAINS.

Culture hearth. LOCATION in which a CULTURE has
developed; a CORE REGION from which the culture later spread or diffused outward through a
larger REGION. Mesopotamia, the Nile Valley,
and the Peruvian ALTIPLANO are examples of
culture hearths.

Curie point. TEMPERATURE at which a magnetic MINERAL locks in its magnetization. Also known as Curie temperature.

Cycle of erosion. Influential MODEL of LANDSCAPE change proposed by William Morris Davis near the end of the nineteenth century. The UPLIFT of a relatively flat surface, or PLAIN, in an area of moderate RAINFALL and TEMPERATURE, led to gradual EROSION of the initial surface in a sequence Davis categorized as Youth, Maturity, and Old Age. The final landscape was called PENEPLAIN. Davis also recognized the stage of REJUVENATION, when a new uplift could give new ENERGY to the cycle, leading to further downcutting and erosion. The model also was used to explain the sequence of LANDFORMS developed in REGIONS of ALPINE GLACIERS. The model has been criticized as misleading, since CRUSTAL MOVEMENT is continuous and more frequent than Davis perhaps envisaged, but remained useful as a description of TOPOGRAPHY. Also known as the Davisian cycle or geomorphic cycle.

Cyclonic rain. In the NORTHERN HEMISPHERE winter, two low-pressure systems or CYCLONES—the Aleutian Low and the Icelandic Low—develop over the OCEAN near 60 DEGREES north LATITUDE. The polar FRONT forms where the cold and relatively dry ARCTIC AIR meets the warmer, moist air carried by westerly WINDS. The warm air is forced upward, cools, and condenses. These cyclonic STORMS often move south, bringing winter PRECIPITATION to North America, especially to the STATES of Washington and Oregon.

Cylindrical projection. MAP PROJECTION that represents the earth's surface as a rectangle. It can be imagined as a cylinder of paper wrapped around a globe with a light source at its center; the images of the CONTINENTS would be projected onto the paper. In reality, MAPS are constructed mathematically. It is impossible to show the North Pole or South Pole on a cylindrical projection. Although the map is conformal, distortion of area is extreme beyond 50 DEGREES north and south LATITUDES. The Mercator projection, developed in the sixteenth century by the Flemish cartographer Gerardus Mercator, is the best-known cylindrical projection. It has been popular with seamen because the shortest route between two PORTS (the GREAT CIRCLE route) can be plotted as straight lines that show the COMPASS direction that should be followed. Use of this projection for other purposes, however, can lead to misunderstandings about size; for example, compare Greenland on a globe and on a Mercator map.

Datum level. Baseline or level from which other heights are measured, above or below. MEAN SEA LEVEL is the datum commonly used in surveying and in the construction of TOPOGRAPHIC MAPS.

Daylight saving time. System of seasonal adjustments in CLOCK settings designed to increase hours of evening sunlight during summer months. In the spring, clocks are set ahead one hour; in the fall, they are put back to standard time. In North America, these changes are made on the first Sunday in April and the last Sunday in October. The U.S. Congress standardized daylight saving time in 1966; however, parts of Arizona, Indiana, and Hawaii do not follow the system.

Débâcle. In a scientific context, this French word means the sudden breaking up of ice in a RIVER in the spring, which can lead to serious, sudden flooding.

Debris avalanche. Large mass of SOIL and ROCK that falls and then slides on a cushion of AIR downhill rapidly as a unit.

Debris flow. Flowing mass consisting of water and a high concentration of SEDIMENT with a wide RANGE of size, from fine muds to coarse gravels.

Declination, magnetic. Measure of the difference, in DEGREES, between the earth's NORTH MAGNETIC POLE and the North Pole on a MAP; this difference changes slightly each year. The needle of a magnetic COMPASS points to the earth's geomagnetic pole, which is not exactly the same as the North Pole of the geographic GRID or the set of lines of LATITUDE and LONGITUDE. The geomagnetic poles, north and south, mark the ends of the AXIS of the earth's MAGNETIC FIELD, but this field is not stationary. In fact, the geomagnetic poles have completely reversed hundreds of times throughout earth history. Lines of equal magnetic declination are called ISOGONIC LINES.

Declination of the Sun. LATITUDE of the SUBSOLAR POINT, the PLACE on the earth's surface where the SUN is directly overhead. In the course of a year, the declination of the Sun migrates from 23.5 DEGREES north LATITUDE, at the (northern) summer SOLSTICE, to 23.5 degrees south latitude, at the (northern) WINTER SOLSTICE. Hawaii is the only part of the United States that experiences the Sun directly overhead twice a year.

Deep-focus earthquakes. EARTHQUAKES occurring at depths ranging from 40 to 400 miles (70–700 km.) below the earth's surface. This RANGE of depths represents the zone from the base of the earth's CRUST to approximately one-quarter of the distance into Earth's MANTLE. Deep-focus earthquakes provide scientists information about the PLANET's interior structure, its composition, and SEISMICITY. Observation of deep-focus earthquakes has played a fundamental role in the discovery and understanding of PLATE TECTONICS.

Deep-ocean currents. Deep-ocean currents involve significant vertical and horizontal movements of seawater. They distribute oxygen- and nutrient-rich waters throughout the world's OCEANS, thereby enhancing biological productivity.

Defile. Narrow MOUNTAIN PASS or GORGE through which troops could march only in single file.

Deflation. EROSION by WIND, resulting in the removal of fine particles. The LANDFORM that typically results is a deflation hollow.

Deforestation. Removal or destruction of FORESTS. In the late twentieth century, there was widespread concern about tropical deforestation—destruction of the tropical RAIN FOREST—especially that of Brazil. Forest clearing in the TROPICS is uneconomic because of low SOIL fertility. Deforestation causes severe EROSION and environmental damage; it also destroys habitat, which leads to the EXTINCTION of both plant and animal species.

Degradation. Process of CRATER EROSION from all processes, including WIND and other meteorological mechanisms.

Degree (geography). Unit of LATITUDE or LONGITUDE in the geographic GRID, used to determine ABSOLUTE LOCATION. One degree of latitude is about 69 miles (111 km.) on the earth's surface. It is not exactly the same everywhere, because the earth is not a perfect sphere. One degree of longitude varies greatly in length, because the MERIDIANS converge at the POLES. At the EQUATOR, it is 69 miles (111 km.), but at the North or South Pole it is zero.

Degree (temperature). Unit of MEASUREMENT of TEMPERATURE, based on the CELSIUS SCALE, except in the United States, which uses the FAHRENHEIT SCALE. On the Celsius scale, one degree is one-hundredth of the difference between the freezing point of water and the boiling point of water.

Demographic measure. Statistical data relating to POPULATION.

Demographic transition. MODEL of POPULATION change that fits the experience of many European countries, showing changes in birth and death rates. In the first stage, in preindustrial countries, population size was stable because both BIRTH RATES and DEATH RATES were high. Agricultural reforms, together with the INDUSTRIAL REVOLUTION and subsequent medical advances, led to a rapid fall in the death rate, so that the second and third stages of the model were periods of rapid population growth, often called the POPULATION EXPLOSION. In the fourth stage of the model, birth rates fall markedly, leading again to stable population size.

Dendritic drainage. Most common pattern of STREAMS and their TRIBUTARIES, occurring in areas of uniform ROCK type and regular slope. A MAP, or aerial photograph, shows a pattern like the veins on a leaf—smaller streams join the main stream at an acute angle.

Denudation. General word for all LANDFORM processes that lead to a lowering of the LANDSCAPE, including WEATHERING, mass movement, EROSION, and transport.

Deposition. Laying down of SEDIMENTS that have been transported by water, WIND, or ice.

Deranged drainage. LANDSCAPE whose integrated drainage network has been destroyed by irregular glacial DEPOSITION, yielding numerous shallow LAKE BASINS.

Derivative maps. MAPS that are prepared or derived by combining information from several other maps.

Desalinization. Process of removing SALT and MINERALS from seawater or from saline water occurring in AQUIFERS beneath the land surface to render it fit for AGRICULTURE or other human use.

Desert climate. Low PRECIPITATION, low HUMIDITY, high daytime TEMPERATURES, and abundant sunlight are characteristics of desert climates. The hot DESERTS of the world generally are located on the western sides of CONTINENTS, at LATITUDES from fifteen to thirty DEGREES north or south of the EQUATOR. One definition, based on precipitation, defines deserts as areas that receive between 0 and 9 inches (0 to 250 millimeters) of precipitation per year. REGIONS receiving more precipitation are considered to have a SEMIDESERT climate, in which some AGRICULTURE is possible.

Desert pavement. Surface covered with smoothed PEBBLES and gravels, found in arid areas where DEFLATION (WIND EROSION) has removed the smaller particles. Called a "gibber plain" in Australia.

Desertification. Increase in DESERT areas worldwide, largely as a result of overgrazing or poor agricultural practices in semiarid and marginal CLIMATES. DEFORESTATION, DROUGHT, and POPULATION increase also contribute to desertification. The REGION of Africa just south of the Sahara Desert, known as the SAHEL, is the largest and most dramatic demonstration of desertification.

Detrital rock. SEDIMENTARY ROCK composed mainly of grains of silicate MINERALS as opposed to grains of calcite or CLAYS.

Devolution. Breaking up of a large COUNTRY into smaller independent political units is the final and most extreme form of devolution. The Soviet Union devolved from one single country into fifteen separate countries in 1991. At an intermediate level, devolution refers to the granting of political autonomy or self-government to a REGION, without a complete split. The reopening of the Scottish Parliament in 1999 and the Northern Ireland parliament in 2000 are examples of devolution; the Parliament of the United Kingdom had previously met only in London and made laws there for all parts of the country. Canada experienced devolution with the creation of the new territory of Nunavut, whose residents elect the members of their own legislative assembly.

Dew point. TEMPERATURE at which an AIR mass becomes saturated and can hold no more moisture. Further cooling leads to CONDENSATION. At ground level, this produces DEW.

Diagenesis. Conversion of unconsolidated SEDIMENT into consolidated ROCK after burial by the processes of compaction, cementation, recrystallization, and replacement.

Diaspora. Dispersion of a group of people from one CULTURE to a variety of other REGIONS or to other lands. A Greek word, used originally to refer to the Jewish diaspora. Jewish people now live in many countries, although they have Israel as a HOMELAND. Similar to this are the diasporas of the Irish and the overseas Chinese.

Diastrophism. Deformation of the earth's CRUST by faulting or FOLDING.

Diatom ooze. Deposit of soft mud on the OCEAN floor consisting of the shells of diatoms, which are microscopic single-celled creatures with SILICA-rich shells. Diatom ooze deposits are located in the southern Pacific around Antarctica and in the northern Pacific. Other PELAGIC, or deep-ocean, SEDIMENTS include CLAYS and calcareous ooze.

Dike (geology). LANDFORM created by IGNEOUS intrusion when MAGMA or molten material within the earth forces its way in a narrow band through overlying ROCK. The dike can be exposed at the surface through EROSION.

Dike (water). Earth wall or DAM built to prevent flooding; an EMBANKMENT or artificial LEVEE. Sometimes specifically associated with structures built in the Netherlands to prevent the entry of seawater. The land behind the dikes was reclaimed for AGRICULTURE; these new fields are called POLDERS.

Distance-decay function. Rate at which an activity diminishes with increasing distance. The effect that distance has as a deterrent on human activity is sometimes described as the FRICTION OF DISTANCE. It occurs because of the time and cost of overcoming distances between people and their desired activity. An example of the distance-decay function is the rate of visitors to a football stadium. The farther people have to travel, the less likely they are to make this journey.

Distributary. STREAM that takes waters away from the main CHANNEL of a RIVER. A DELTA usually comprises many distributaries. Also called distributary channel.

Diurnal range. Difference between the highest and lowest TEMPERATURES registered in one twenty-four-hour period.

Diurnal tide. Having only one high tide and one low tide each lunar DAY; TIDES on some parts of the Gulf of Mexico are diurnal.

Divergent boundary. BOUNDARY that results where two PLATES are moving apart from each other, as is the case along MID-OCEANIC RIDGES.

Divergent margin. Area where the earth's CRUST and LITHOSPHERE form by SEAFLOOR SPREADING.

Doline. Large SINKHOLE or circular depression formed in LIMESTONE areas through the CHEMICAL WEATHERING process of carbonation.

Dolomite. MINERAL consisting of calcium and magnesium carbonate compounds that often forms from PRECIPITATION from seawater; it is abundant in ancient ROCKS.

Downwelling. Sinking of OCEAN water.

Drainage basin. Area of the earth's surface that is drained by a STREAM. Drainage basins vary greatly in size, but each is separated from the next by RIDGES, or drainage DIVIDES. The CATCHMENT of the drainage basin is the WATERSHED.

Drift ice. ARCTIC or ANTARCTIC ice floating in the open SEA.

Drumlin. Low HILL, shaped like half an egg, formed by DEPOSITION by CONTINENTAL GLACIERS. A drumlin is composed of TILL, or mixed-size materials. The wider end faces upstream of the glacier's movement; the tapered end points in the direction of the ice movement. Drumlins usually occur in groups or swarms.

Dust devil. Whirling cloud of DUST and small debris, formed when a small patch of the earth's surface becomes heated, causing hot AIR to rise; cooler air then flows in and begins to spin. The resulting dust devil can grow to heights of 150 feet (50 meters) and reach speeds of 35 miles (60 km.) per hour.

Dust dome. Dome of AIR POLLUTION, composed of industrial gases and particles, covering every large CITY in the world. The pollution sometimes is carried downwind to outlying areas.

Earth pillar. Formation produced when a boulder or caprock prevents EROSION of the material directly beneath it, usually CLAY. The clay is easily eroded away by water during RAINFALL, except where the overlying ROCK protects it. The result is a tall, slender column, as high as 20 feet (6.5 meters) in exceptional cases.

Earth radiation. Portion of the electromagnetic spectrum, from about 4 to 80 microns, in which the earth emits about 99 percent of its RADIATION.

Earth tide. Slight deformation of Earth resulting from the same forces that cause OCEAN TIDES, those that are exerted by the MOON and the SUN.

Earthflow. Term applied to both the process and the LANDFORM characterized by fluid downslope movement of SOIL and ROCK over a discrete plane of failure; the landform has a HUMMOCKY surface and usually terminates in discrete lobes.

Earth's heat budget. Balance between the incoming SOLAR RADIATION and the outgoing terrestrial reradiation.

Eclipse, lunar. Obscuring of all or part of the light of the Moon by the shadow of the earth. A lunar eclipse occurs at the full moon up to three times a year. The surface of the Moon changes from gray to a reddish color, then back to gray. The sequence may last several hours.

Eclipse, solar. At least twice a year, the Sun, Moon, and Earth are aligned in one straight line. At that time, the Moon obscures all the light of the Sun along a narrow band of the earth's surface, causing a total eclipse; in regions of Earth adjoining that area, there is a partial eclipse. A corona (halo of light) can be seen around the Sun at the total eclipse. Viewing a solar eclipse with naked eyes is extremely dangerous and can cause blindness.

Ecliptic, plane of. Imaginary plane that would touch all points in the earth's orbit as it moves around the Sun. The angle between the plane of the ecliptic and the earth's axis is 66.5 degrees.

Edge cities. Forms of suburban downtown in which there are nodal concentrations of office space and shopping facilities. Edge cities are located close to major freeways or highway intersections, on the outer edges of metropolitan areas.

Effective temperature. Temperature of a planet based solely on the amount of solar radiation that the planet's surface receives; the effective temperature of a planet does not include the greenhouse temperature enhancement effect.

Ejecta. Material ejected from the crater made by a meteoric impact.

Ekman layer. Region of the sea, from the surface to about 100 meters down, in which the wind directly affects water movement.

Eluviation. Removal of materials from the upper layers of a soil by water. Fine material may be removed by suspension in the water; other material is removed by solution. The removal by solution is called leaching. Eluviation from an upper layer leads to illuviation in a lower layer.

Enclave. Piece of territory completely surrounded by another country. Two examples are Lesotho, which is surrounded by the Republic of South Africa, and the Nagorno-Karabakh region, populated by Armenians but surrounded by Azerbaijan. The term is also used for smaller regions, such as ethnic neighborhoods within larger cities. See also Exclave.

Endemic species. Species confined to a restricted area in a restricted environment.

Endogenic sediment. Sediment produced within the water column of the body in which it is deposited; for example, calcite precipitated in a lake in summer.

Environmental degradation. Situation that occurs in slum areas and squatter settlements because of poverty and inadequate infrastructure. Too-rapid human population growth can lead to the accumulation of human waste and garbage, the pollution of groundwater, and denudation of nearby forests. As a result, life expectancy in such degraded areas is lower than in the rural communities from which many of the settlers came. Infant mortality is particularly high. When people leave an area because of such environmental degradation, that is referred to as ecomigration.

Environmental determinism. Theory that the major influence on human behavior is the physical environment. Some evidence suggests that temperature, precipitation, sunlight, and topography influence human activities. Originally espoused by early German geographers, this theory has led to some extreme stances, however, by authors who have sought to explain the dominance of Europeans as a result of a cool temperate climate.

Eolian (aeolian). Relating to, or caused by, wind. In Greek mythology, Aeolus was the ruler of the winds. Erosion, transport, and deposition are common eolian processes that produce landforms in desert regions.

Eolian deposits. Material transported by the wind.

Eolian erosion. Mechanism of erosion or crater degradation caused by wind.

Eon. Largest subdivision of geologic time; the two main eons are the Precambrian (c. 4.6 billion years ago to 544 million years ago) and the Phanerozoic (c. 544 million years ago to the present).

Ephemeral stream. Watercourse that has water for only a DAY or so.

Epicontinental sea. Shallow SEAS that are located on the CONTINENTAL SHELF, such as the North Sea or Hudson Bay. Also called an EPEIRIC SEA.

Epifauna. Organisms that live on the seafloor.

Epilimnion. Warmer surface layer of water that occurs in a LAKE during summer stratification; during spring, warmer water rises from great depths, and it heats up through the summer SEASON.

Equal-area projection. MAP PROJECTION that maintains the correct area of surfaces on7 a MAP, although shape distortion occurs. The property of such a map is called equivalence.

Erg. Sandy DESERT, sometimes called a SEA of SAND. Erg deserts account for less than 30 percent of the world's deserts. "Erg" is an Arabic word.

Eruption, volcanic. Emergence of MAGMA (molten material) at the earth's surface as LAVA. There are various types of volcanic eruptions, depending on the chemistry of the magma and its viscosity. Scientists refer to effusive and explosive eruptions. Low-viscosity magma generally produces effusive eruptions, where the lava emerges gently, as in Hawaii and Iceland, although explosive events can occur at those SITES as well. Gently sloping SHIELD VOLCANOES are formed by effusive eruptions; FLOODS, such as the Columbia Plateau, can also result. Explosive eruptions are generally associated with SUBDUCTION. Much gas, including steam, is associated with magma formed from OCEANIC CRUST, and the compressed gas helps propel the explosion. COMPOSITE CONES, such as Mount Saint Helens, are created by explosive eruptions.

Escarpment. Steep slope, often almost vertical, formed by faulting. Sometimes called a FAULT SCARP.

Esker. Deposit of coarse gravels that has a sinuous, winding shape. An esker is formed by a STREAM of MELTWATER that flowed through a tunnel it formed under a CONTINENTAL GLACIER. Now that the continental glaciers have melted, eskers can be found exposed at the surface in many PLACES in North America.

Estuarine zone. Area near the COASTLINE that consists of estuaries and coastal saltwater WETLANDS.

Etesian winds. WINDS that blow from the north over the Mediterranean during July and August.

Ethnocentrism. Belief that one's own ETHNIC GROUP and its CULTURE are superior to any other group.

Ethnography. Study of different CULTURES and human societies.

Eustacy. Any change in global SEA LEVEL resulting from a change in the absolute volume of available sea water. Also known as eustatic sea-level change.

Eustatic movement. Changes in SEA LEVEL.

Exclave. Territory that is part of one COUNTRY but separated from the main part of that country by another country. Alaska is an exclave of the United States; Kaliningrad is an exclave of Russia. See also ENCLAVE.

Exfoliation. When GRANITE rocks cooled and solidified, removal of the overlyingrock that was present reduced the pressure on the granite mass, allowing it to expand and causing sheets or layers of rock to break off. An exfoliation DOME, such as Half Dome in Yosemite National Park, is the resultant LANDFORM.

Exotic stream. RIVER that has its source in an area of high RAINFALL and then flows through an arid REGION or DESERT. The Nile River is the most famous exotic STREAM. In the United States, the Colorado River is a good example of an exotic stream.

Expansion-contraction cycles. Processes of wetting-drying, heating-cooling, or freezing-thawing, which affect SOIL particles differently according to their size.

Extrusive rock. Fine-grained, or glassy, ROCK which was formed from a MAGMA that cooled on the surface of the earth.

Fall line. Edge of an area of uplifted land, marked by WATERFALLS where STREAMS flow over the edge.

Fata morgana. Large mirage. Originally, the name given to a multiple mirage phenomenon often observed over the Straits of Messina and supposed to be the work of the fairy ("fata") Morgana. Another famous fata morgana is located in Antarctica.

Fathometer. Instrument that uses sound waves or sonar to determine the depth of water or the depth of an object below the water.

Fault drag. Bending of ROCKS adjacent to a FAULT.

Fault line. Line of breakage on the earth's surface. FAULTS may be quite short, but many are extremely long, even hundreds of miles. The origin of the faulting may lie at a considerable depth below the surface. Movement along the fault line generates EARTHQUAKES.

Fault plane. Angle of a FAULT. When fault blocks move on either side of a fault or fracture, the movement can be vertical, steeply inclined, or sometimes horizontal. In a NORMAL FAULT, the fault plane is steep to almost vertical. In a REVERSE FAULT, one block rides over the other, forming an overhanging FAULT SCARP. The angle of inclination of the fault plane from the horizontal is called the dip. The inclination of a fault plane is generally constant throughout the length of the fault, but there can be local variations in slope. In a STRIKE-SLIP FAULT the movement is horizontal, so no fault scarp is produced, although the FAULT LINE may be seen on the surface.

Fault scarp. FAULTS are produced through breaking or fracture of the surface ROCKS of the earth's CRUST as a result of stresses arising from tectonic movement. A NORMAL FAULT, one in which the earth movement is predominantly vertical, produces a steep fault scarp. A STRIKE-SLIP FAULT does not produce a fault scarp.

Feldspar. Family name for a group of common MINERALS found in such ROCKS as GRANITE and composed of silicates of aluminum together with potassium, sodium, and calcium. Feldspars are the most abundant group of minerals within the earth's CRUST. There are many varieties of feldspar, distinguished by variations in chemistry and crystal structure. Although feldspars have some economic uses, their principal importance lies in their role as rock-forming minerals.

Felsic rocks. IGNEOUS ROCKS rich in potassium, sodium, aluminum, and SILICA, including GRANITES and related rocks.

Fertility rate. DEMOGRAPHIC MEASURE of the average number of children per adult female in any given POPULATION. Religious beliefs, education, and other cultural considerations influence fertility rates.

Fetch. Distance along a large water surface over which a WIND of almost uniform direction and speed blows.

Feudalism. Social and economic system that prevailed in Europe before the INDUSTRIAL REVOLUTION. The land was owned and controlled by a minority comprising noblemen or lords; all other people were peasants or serfs, who worked as agricultural laborers on the lords' land. The peasants were not free to leave, or to do anything without their lord's permission. Other REGIONS such as China and Japan also had a feudal system in the past.

Firn. Intermediate stage between SNOW and glacial ice. Firn has a granular TEXTURE, due to compaction. Also called NÉVÉ.

Fission, nuclear. Splitting of an atomic nucleus into two lighter nuclei, resulting in the release of neutrons and some of the binding ENERGY that held the nucleus together.

Fissure. Fracture or crack in ROCK along which there is a distinct separation.

Flash flood. Sudden rush of water down a STREAM CHANNEL, usually in the DESERT after a short but intense STORM. Other causes, such as a DAM failure, could lead to a flash flood.

Flood control. Attempts by humans to prevent flooding of STREAMS. Humans have consistently settled on FLOODPLAINS and DELTAS because of the fertile SOIL for AGRICULTURE, and attempts at flood control date back thousands of years. In strictly agricultural societies such as ancient Egypt, people built VILLAGES above the FLOOD levels, but transport and industry made riverside LOCATIONS desirabl and engineers devised technological means to try to prevent flood damage. Artificial LEVEES, RESERVOIRS, and DAMS of ever-increasing size were built on

RIVERS, as well as bypass CHANNELS leading to artificial floodplains. In many modern dam construction projects, the production of HYDRO-ELECTRIC POWER was more important than flood control. Despite modern TECHNOLOGY, floods cause the largest loss of human life of all natural disasters, especially in low-income countries such as Bangladesh.

Flood tide. Rising or incoming tide. Most parts of the world experience two flood TIDES in each 24-hour period.

Floodplain. Flat, low-lying land on either side of a STREAM, created by the DEPOSITION of ALLUVIUM from floods. Also called ALLUVIAL PLAIN.

Fluvial. Pertaining to running water; for example, fluvial processes are those in which running water is the dominant agent.

Fog deserts. Coastal DESERTS where FOG is an important source of moisture for plants, animals, and humans. The fog forms because of a cold OCEAN CURRENT close to the SHORE. The Namib Desert of southwestern Africa, the west COAST of California, and the Atacama Desert of Peru are coastal deserts.

Föhn wind. WIND warmed and dried by descent, usually on the LEE side of a MOUNTAIN. In North America, these winds are called the CHINOOK.

Fold mountains. ROCKS in the earth's CRUST can be bent by compression, producing folds. The Swiss Alps are an example of complex FOLDING, accompanied by faulting. Simple upward folds are ANTICLINES, downward folds are SYNCLINES; but subsequent EROSION can produce LANDSCAPES with synclinal MOUNTAINS.

Folding. Bending of ROCKS in the earth's CRUST, caused by compression. The rocks are deformed, sometimes pushed up to form mountain RANGES.

Foliation. TEXTURE or structure in which MINERAL grains are arranged in parallel planes.

Food web. Complex network of FOOD CHAINS. Food chains are interconnected, because many organisms feed on a variety of others, and in turn may be eaten by any of a number of predators.

Forced migration. MIGRATION that occurs when people are moved against their will. The Atlantic slave trade is an example of forced migration. People were shipped from Africa to countries in Europe, Asia, and the New World as forced immigrants. Within the United States, some NATIVE AMERICANS were forced by the federal government to migrate to new reservations.

Ford. Short shallow section of a RIVER, where a person can cross easily, usually by walking or riding a horse. To cross a STREAM in such a manner.

Formal region. Cultural REGION in which one trait, or group of traits, is uniform. LANGUAGE might be the basis of delineation of a formal cultural region. For example, the Francophone region of Canada constitutes a formal region based on one single trait. One might also identify a formal Mormon region centered on the STATE of Utah, combining RELIGION and LANDSCAPE as defining traits. Cultural geographers generally identify formal regions using a combination of traits.

Fossil fuel. Deposit rich in hydrocarbons, formed from organic materials compressed in ROCK layers—COAL, OIL, and NATURAL GAS.

Fossil record. Fossil record provides evidence that addresses fundamental questions about the origin and history of life on the earth: When life evolved; how new groups of organisms originated; how major groups of organisms are related. This record is neither complete nor without biases, but as scientists' understanding of the limits and potential of the fossil record grows, the interpretations drawn from it are strengthened.

Fossilization. Processes by which the remains of an organism become preserved in the ROCK record.

Fracture zones. Large, linear zones of the seafloor characterized by steep CLIFFS, irregular TOPOGRAPHY, and FAULTS; such zones commonly cross and displace oceanic RIDGES by faulting.

Free association. Relationship between sovereign NATIONS in which one nation—invariably the larger—has responsibility for the other nation's defense. The Cook Islands in the South Pacific have such a relationship with New Zealand.

Friction of distance. Distance is of prime importance in social, political, economic, and other relationships. Large distance has a negative effect on human activity. The time and cost of overcoming distance can be a deterrent to various activities. This has been called the friction of distance.

Frigid zone. Coldest of the three CLIMATE zones proposed by the ancient Greeks on the basis of their theories about the earth. There were two frigid zones, one around each POLE. The Greeks believed that human life was possible only in the TEMPERATE ZONE.

Fringing reef. Type of CORAL REEF formed at the SHORELINE, extending out from the land in shallow water. The top of the coral may be exposed at low TIDE.

Frontier Thesis. Thesis first advanced by the U.S. historian Frederick Jackson Turner, who declared that U.S. history and the U.S. character were shaped by the existence of empty, FRONTIER lands that led to exploration and westward expansion and DEVELOPMENT. The closing of the frontier occurred when transcontinental railroads linked the East and West Coasts and SETTLEMENTS spread across the United States. This thesis was used by later historians to explain the history of South Africa, Canada, and Australia. Critics of the Frontier Thesis point out that minorities and women were excluded from this view of history.

Frost wedging. Powerful form of PHYSICAL WEATHERING of ROCK, in which the expansion of water as it freezes in JOINTS or cracks shatters the rock into smaller pieces. Also known as frost shattering.

Fumarole. Crack in the earth's surface from which steam and other gases emerge. Fumaroles are found in volcanic areas and areas of GEOTHERMAL activity, such as Yellowstone National Park.

Fusion energy. Heat derived from the natural or human-induced union of atomic nuclei; in effect, the opposite of FISSION energy.

Gall's projection. MAP PROJECTION constructed by projecting the earth onto a cylinder that intersects the sphere at 45 DEGREES north and 45 degrees south LATITUDE. The resulting map has

less distortion of area than the more familiar CYLINDRICAL PROJECTION of Mercator.

Gangue. Apparently worthless ROCK or earth in which valuable gems or MINERALS are found.

Garigue. VEGETATION cover of small shrubs found in Mediterranean areas. Similar to the larger *maquis*.

Genus (plural, genera). Group of closely related species; for example, *Homo* is the genus of humans, and it includes the species *Homo sapiens* (modern humans) and *Homo erectus* (Peking Man, Java Man).

Geochronology. Study of the time SCALE of the earth; it attempts to develop methods that allow the scientist to reconstruct the past by dating events such as the formation of ROCKS.

Geodesy. Branch of applied mathematics that determines the exact positions of points on the earth's surface, the size and shape of the earth, and the variations of terrestrial GRAVITY and MAGNETISM.

Geoid. Figure of the earth considered as a MEAN SEA LEVEL surface extended continuously through the CONTINENTS.

Geologic terrane. Crustal block with a distinct group of ROCKS and structures resulting from a particular geologic history; assemblages of TERRANES form the CONTINENTS.

Geological column. Order of ROCK layers formed during the course of the earth's history.

Geomagnetic elements. MEASUREMENTS that describe the direction and intensity of the earth's MAGNETIC FIELD.

Geomagnetism. External MAGNETIC FIELD generated by forces within the earth; this force attracts materials having similar properties, inducing them to line up (point) along field lines of force.

Geostationary orbit. ORBIT in which a SATELLITE appears to hover over one spot on the PLANET's EQUATOR; this procedure requires that the orbit be high enough that its period matches the planet's rotational period, and have no inclination relative to the equator; for Earth, the ALTITUDE is 22,260 miles (35,903 km.).

Geostrophic. Force that causes directional change because of the earth's ROTATION.

Geotherm. Curve on a TEMPERATURE-depth graph that describes how temperature changes in the subsurface.

Geothermal power. Power having its source in the earth's internal heat.

Glacial erratic. ROCK that has been moved from its original position and transported by becoming incorporated in the ice of a GLACIER. Deposited in a new LOCATION, the rock is noteworthy because its geology is completely different from that of the surrounding rocks. Glacial erratics provide information about the direction of glacial movement and strength of the flow. They can be as small as PEBBLES, but the most interesting erratics are large boulders. Erratics become smoothed and rounded by the transport and EROSION.

Glaciation. This term is used in two senses: first, in reference to the cyclic widespread growth and advance of ICE SHEETS over the polar and high- to mid-LATITUDE REGIONS of the CONTINENTS; second, in reference to the effect of a GLACIER on the TERRAIN it transverses as it advances and recedes.

Global Positioning System (GPS). Group of SATELLITES that ORBIT Earth every twenty-four hours, sending out signals that can be used to locate PLACES on Earth and in near-Earth orbits.

Global warming. Trend of Earth CLIMATES to grow increasingly warm as a result of the GREENHOUSE EFFECT. One of the most dramatic effects of global warming is the melting of the POLAR ICE CAPS and a consequent rise the level of the world's OCEANS.

Gondwanaland. Hypothesized ancient CONTINENT in the SOUTHERN HEMISPHERE that geologists theorize broke into at least two large segments; one segment became India and pushed northward to collide with the Eurasian LANDMASS, while the other, Africa, moved westward.

Graben. Roughly symmetrical crustal depression formed by the lowering of a crustal block between two NORMAL FAULTS that slope toward each other.

Granules. Small grains or pellets.

Gravimeter. Device that measures the attraction of GRAVITY.

Gravitational differentiation. Separation of MINERALS, elements, or both as a result of the influence of a gravitational field wherein heavy phases sink or light phases rise through a melt.

Great circle. Largest circle that goes around a sphere. On the earth, all lines of LONGITUDE are parts of great circles; however, the EQUATOR is the only line of LATITUDE that is a great circle.

Green mud. SOILS that develop under conditions of excess water, or waterlogged soils, can display colors of gray to blue to green, largely because of chemical reactions involving iron. Fine CLAY soils and muds in areas such as BOGS or ESTUARIES can be called green mud. This soil-forming process is called gleization.

Greenhouse effect. Trapping of the SUN's rays within the earth's ATMOSPHERE, with a consequence rise in TEMPERATURES that leads to GLOBAL WARMING.

Greenhouse gas. Atmospheric gas capable of absorbing electromagnetic radiation in the infrared part of the spectrum.

Greenwich mean time. Also known as universal time, the solar mean time on the MERIDIAN running through Greenwich, England—which is used as the basis for calculating time throughout most of the world.

Grid. Pattern of horizontal and vertical lines forming squares of uniform size.

Groundwater movement. Flow of water through the subsurface, known as groundwater movement, obeys set principles that allow hydrologists to predict flow directions and rates.

Groundwater recharge. Water that infiltrates from the surface of the earth downward through SOIL and ROCK pores to the WATER TABLE, causing its level to rise.

Growth pole. LOCATION where high-growth economic activity is deliberately encouraged and promoted. Governments often establish growth poles by creating industrial parks, open cities, special economic zones, new TOWNS, and other incentives. The plan is that the new industries will further stimulate economic growth in a cumulative trend. Automobile plants are a traditional form of growth industry but have been overtaken by high-tech industries and BIOTECHNOLOGY. In France, the term "technopole" is used for a high-tech growth pole. A related concept is SPREAD EFFECTS.

Guyot. Drowned volcanic ISLAND with a flat top caused by WAVE EROSION or coral growth. A type of SEAMOUNT.

Gyre. Large semiclosed circulation patterns of OCEAN CURRENTS in each of the major OCEAN BASINS that move in opposite directions in the Northern and Southern hemispheres.

Haff. Term used for various WETLANDS or LAGOONS located around the southern end of the Baltic Sea, from Latvia to Germany. Offshore BARS of SAND and shingle separate the haffs from the open SEA. One of the largest is the Stettiner Haff, which covers the BORDER REGION between Germany and Poland and is separated from the Baltic by the low-lying ISLAND of Usedom. The Kurisches Haff (in English, the Courtland Lagoon) is located on the Lithuanian border.

Harmonic tremor. Type of EARTHQUAKE activity in which the ground undergoes continuous shaking in response to subsurface movement of MAGMA.

Headland. Elevated land projecting into a body of water.

Headwaters. Source of a RIVER. Also called headstream.

Heat sink. Term applied to Antarctica, whose cold CLIMATE causes warm AIR masses flowing over it to chill quickly and lose ALTITUDE, affecting the entire world's WEATHER.

Heterosphere. Major realm of the ATMOSPHERE in which the gases hydrogen and helium become predominant.

High-frequency seismic waves. EARTHQUAKE WAVES that shake the ROCK through which they travel most rapidly.

Histogram. Bar graph in which vertical bars represent frequency and the horizontal axis represents categories. A POPULATION PYRAMID, or age-sex pyramid, is a histogram, as is a CLIMOGRAPH.

Historical inertia. Term used by economic geographers when heavy industries, such as steelmaking and large manufacture, that require huge capital investments in land and plant continue in operation for long periods, even after they become out of date, uncompetitive, or obsolete.

Hoar frost. Similar to DEW, except that moisture is deposited as ice crystals, not liquid dew, on surfaces such as grass or plant leaves. When moist AIR cools to saturation level at TEMPERATURES below the freezing point, CONDENSATION occurs directly as ice. Technically, hoar frost is not the same as frozen dew, but it is difficult to distinguish between the two.

Hogback. Steeply sloping homoclinal RIDGE, with a slope of 45 DEGREES or more. The angle of the slope is the same as the dip of the ROCK STRATA. These LANDFORMS develop in REGIONS where the underlying rocks, usually SEDIMENTARY, have been folded into anticlinal ridges and synclinal VALLEYS. Differential EROSION causes softer rock layers to wear away more rapidly than the harder layers of rock that form the hogback ridge. A similar feature with a gentler slope is called a CUESTA.

Homosphere. Lower part of the earth's ATMOSPHERE. In this area, 60 miles (100 km.) thick, the component gases are uniformly mixed together, largely through WINDS and turbulent AIR CURRENTS. Above the homosphere is the REGION of the atmosphere called the HETEROSPHERE. There, the individual gases separate out into layers on the basis of their molecular weight. The lighter gases, hydrogen and helium, are at the top of the heterosphere.

Hook. A long, narrow deposit of SAND and SILT that grows outward into the OCEAN from the land is called a SPIT or sandspit. A hook forms when currents or WAVES cause the deposited material to curve back toward the land. Cape Cod is the most famous spit and hook in the United States.

Horse latitudes. Parts of the OCEANS from about 30 to 35 DEGREES north or south of the EQUATOR. In these latitudes, AIR movement is usually light WINDS, or even complete calm, because there are semipermanent high-pressure cells called ANTI-CYCLONES, which are marked by dry subsiding air and fine clear WEATHER. The atmospheric circulation of an anticyclone is divergent and CLOCKWISE in the NORTHERN HEMISPHERE, so to the north of the horse latitudes are the westerly winds and to the south are the northeast TRADE WINDS. In the SOUTHERN HEMISPHERE, the circulation is reversed, producing the easterly winds and the southeast trade winds. It is believed that the name originated because when ships bringing immigrants to the Americas were becalmed for any length of time, horses were thrown overboard because they required too much FRESH WATER. Also called the CALMS OF CANCER.

Horst. FAULT block or piece of land that stands above the surrounding land. A horst usually has been uplifted by tectonic forces, but also could have originated by downward movement or lowering of the adjacent lands. Movement occurs along the parallel faults on either side of a horst. If the land is downthrown instead of uplifted, a VALLEY known as a GRABEN is formed. "Horst" comes from the German word for horse, because the flat-topped feature resembles a vaulting horse used in gymnastics.

Hot spot. PLACE on the earth's surface where heat and MAGMA rise from deep in the interior, perhaps from the lower MANTLE. Erupting VOLCANOES may be present, as in the formation of the Hawaiian Islands. More commonly, the heat from the rising magma causes GROUNDWATER to form HOT SPRINGS, GEYSERS, and other thermal and HYDROTHERMAL features. Yellowstone National Park is located on a hot spot. Also known as a MANTLE PLUME.

Hot spring. SPRING where hot water emerges at the earth's surface. The usual cause is that the GROUNDWATER is heated by MAGMA. A GEYSER is a special type of hot spring at which the water heats under pressure and that periodically spouts hot water and steam. Old Faithful is the best known of many geysers in Yellowstone National Park. In some countries, GEOTHERMAL EN-ERGY from hot springs is used to generate electricity. Also called thermal spring.

Humus. Uppermost layer of a SOIL, containing decaying and decomposing organic matter such as leaves. This produces nutrients, leading to a fertile soil. Tropical soils are low in humus, because the rate of decay is so rapid. Soils of GRASSLANDS and DECIDUOUS FOREST develop thick layers of humus. In a SOIL PROFILE, the layer containing humus is the O Horizon.

Hydroelectric power. Electricity generated when falling water turns the blades of a turbine that converts the water's potential ENERGY to mechanical energy. Natural WATERFALLS can be used, but most hydroelectric power is generated by water from DAMS, because the flow of water from a dam can be controlled. Hydroelectric generation is a RENEWABLE, clean, cheap way to produce power, but dam construction inundates land, often displacing people, who lose their homes, VILLAGES, and farmland. Aquatic life is altered and disrupted also; for example, Pacific salmon cannot return upstream on the Columbia River to their spawning REGION. In a few coastal PLACES, TIDAL ENERGY is used to generate hydroelectricity; La Rance in France is the oldest successful tidal power plant.

Hydrography. Surveying of underwater features or those parts of the earth that are covered by water, especially OCEAN depths and OCEAN CURRENTS. Hydrographers make MAPS and CHARTS of the ocean floor and COASTLINES, which are used by mariners for NAVIGATION. For centuries, mariners used a leadline, a long rope with a lead weight at the bottom. The line was thrown overboard and the depth of water measured. The unit of MEASUREMENT was FATHOMS (6 feet/1.8 meters), which is one-thousandth of a NAUTICAL MILE. The invention of sonar (underwater echo sounding) has enabled mapping of large areas, and hydrographers currently use both television cameras and SATELLITE data.

Hydrologic cycle. Continuous circulation of the earth's HYDROSPHERE, or waters, through EVAPORATION, CONDENSATION, and PRECIPITATION.

Other parts of the hydrologic cycle include RUN-OFF, INFILTRATION, and TRANSPIRATION.

Hydrostatic pressure. Pressure imposed by the weight of an overlying column of water.

Hydrothermal vents. Areas on the OCEAN floor, typically along FAULT LINES or in the vicinity of undersea VOLCANOES, where water that has percolated into the ROCK reemerges much hotter than the surrounding water; such heated water carries various dissolved MINERALS, including metals and sulfides.

Hyetograph. Chart showing the DISTRIBUTION of RAINFALL over time. Typically, a hyetograph is constructed for a single STORM, showing the amount of total PRECIPITATION accumulating throughout the period. A hyetograph shows how rainfall intensity varies throughout the duration of a storm.

Hygrometer. Instrument for measuring the RELATIVE HUMIDITY of AIR, or the amount of water vapor in the ATMOSPHERE at any time.

Hypsometer. Instrument used for measuring ALTITUDE (height above SEA LEVEL), using boiling water that circulates around a THERMOMETER. Since ATMOSPHERIC PRESSURE falls with increased altitude, the boiling point of water is lower. The hypsometer relies on this difference in boiling point to calculate ELEVATION. A more common instrument for measuring altitude is the ALTIMETER.

Ice blink. Bright, usually yellowish-white glare or reflection on the underside of a CLOUD layer, produced by light reflected from an ice-covered surface such as pack ice. A similar phenomenon of reflection from a snow-covered surface is called snow blink.

Ice-cap climate. Earth's most severe CLIMATE, where the mean monthly TEMPERATURE is never above 32 DEGREES Fahrenheit (0 degrees Celsius). This climate is found in Greenland and Antarctica, which are high PLATEAUS, where KATABATIC WINDS blow strongly and frequently. At these high LATITUDES, INSOLATION (SOLAR ENERGY) is received for a short period in the summer months, but the high reflectivity of the ice and SNOW means that much is reflected back instead of being absorbed by the surface. No VEGETATION can grow, because the LANDSCAPE is permanently covered in ice and snow. Because AIR temperatures are so cold, PRECIPITATION is usually less than 5 inches (13 centimeters) annually. The POLES are REGIONS of stable, high-pressure air, where dry conditions prevail, but strong winds that blow the snow around are common. In the KÖPPEN CLIMATE CLASSIFICATION, the ice-cap climate is signified by the letters *EF*.

Ice sheet. Huge CONTINENTAL GLACIER. The only ice sheets remaining cover most of Antarctica and Greenland. At the peak of the last ICE AGE, around 18,000 years ago, ice covered as much as one-third of the earth's land surfaces. In the NORTHERN HEMISPHERE, there were two great ice sheets—the Laurentide ice sheet, covering North America, and the Scandinavian ice sheet, covering northwestern Europe and Scandinavia.

Ice shelf. Portion of an ICE SHEET extending into the OCEAN.

Ice storm. STORM characterized by a fall of freezing rain, with the formation of glaze on Earth objects.

Icefoot. Long, tapering extension of a GLACIER floating above the seawater where it enters the OCEAN. Eventually, it breaks away and forms an ICEBERG.

Igneous rock. ROCKS formed when molten material or MAGMA cools and crystallizes into solid rock. The type of rock varies with the composition of the magma and, more important, with the rate of cooling. Rocks that cool slowly, far beneath the earth's surface, are igneous INTRUSIVE ROCKS. These have large crystals and coarse grains. GRANITE is the most typical igneous intrusive rock. When cooling is more rapid, usually closer to or at the surface, finer-grained igneous EXTRUSIVE ROCKS such as rhyolite are formed. If the magma flows out to the surface as LAVA, it may cool quickly, forming a glassy rock called obsidian. If there is gas in the lava, rocks full of holes from bubbles of escaping gases form; PUMICE and BASALT are common igneous extrusive rocks.

Impact crater. Generally circular depression formed on the surface of a PLANET by the impact of a high-velocity projectile such as a METEORITE, ASTEROID, or COMET.

Impact volcanism. Process in which major impact events produce huge CRATERS along with MAGMA RESERVOIRS that subsequently produce volcanic activity. Such cratering is clearly visible on the MOON, Mars, Mercury, and probably Venus. It is assumed that Earth had similar craters, but EROSION has erased most of the evidence.

Import substitution. Economic process in which domestic producers manufacture or supply goods or services that were previously imported or purchased from overseas and foreign producers.

Index fossil. Remains of an ancient organism that are useful in establishing the age of ROCKS; index fossils are abundant and have a wide geographic DISTRIBUTION, a narrow stratigraphic RANGE, and a distinctive form.

Indian summer. Short period, usually not more than a week, of unusually warm WEATHER in late October or early November in the NORTHERN HEMISPHERE. Before the Indian summer, TEMPERATURES are cooler and there can be occurrences of FROST. Indian summer DAYS are marked by clear to hazy skies and calm to light WINDS, but nights are cool. The weather pattern is a high-pressure cell or ridge located for a few days over the East Coast of North America. The name originated in New England, referring to the practice of NATIVE AMERICANS gathering foods for winter storage over this brief spell. Similar weather in England is called an Old Wives' summer.

Infant mortality. DEMOGRAPHIC MEASURE calculated as the number of deaths in a year of infants, or children under one year of age, compared with the total number of live births in a COUNTRY for the same year. Low-income countries have high infant mortality rates, more than 100 infant deaths per thousand.

Infauna. Organisms that live in the seafloor.

Infiltration. Movement of water into and through the SOIL.

Initial advantage. In terms of economic DEVELOPMENT, not all LOCATIONS are suited for profitable investment. Some locations offer initial advantages, including an existing skilled labor pool, existing consumer markets, existing plants, and situational advantages. These advantages can also lead to clustering of a number of industries at a particular location and to further economic growth, which will provide the preconditions of initial advantage for further economic development.

Inlier. REGION of old ROCKS that is completely surrounded by younger rocks. These are often PLACES where ORES or MINERALS are found in commercial quantities.

Inner core. The innermost layer of the earth; the inner core is a solid ball with a radius of about 900 miles.

Inselberg. Exposed rocky HILL in a DESERT area, made of resistant ROCKS, rising steeply from the flat surrounding countryside. There are many inselbergs in Africa, but Uluru (Ayers Rock) in Australia is possibly the most famous inselberg. The word is German for "island mountain." A special type of inselberg is a bornhardt.

Insolation. ENERGY received by the earth from the SUN, which heats the earth's surface. The average insolation received at the top of the earth's ATMOSPHERE at an average distance from the Sun is called the SOLAR CONSTANT. Insolation is predominantly shortwave radiation, with wavelengths in the RANGE of 0.39 to 0.76 micrometers, which corresponds to the visible spectrum. Less than half of the incoming SOLAR ENERGY reaches the earth's surface-insolation is reflected back into space by CLOUDS; smaller amounts are reflected back by surfaces, absorbed, or scattered by the atmosphere. Insolation is not distributed evenly over the earth, because of Earth's curved surface. Where the rays are perpendicular, at the SUBSOLAR POINT, insolation is at the maximum. The word is a shortened form of incoming (or intercepted) SOLAR RADIATION.

Insular climate. Island climates are influenced by the fact that no PLACE is far from the SEA. Therefore, both the DIURNAL (daily) TEMPERATURE RANGE and the annual temperature range are small.

Insurgent state. STATE that arises when an uprising or guerrilla movement gains control of part of the territory of a COUNTRY, then establishes its own form of control or government. In effect, the insurgents create a state within a state. In Colombia, for example, the government and armed forces have been unable to control several REGIONS where insurgents have created their own domains. This is generally related to coca growing and the production of cocaine. Civilian farmers are unable to resist the drug-financed "armies."

Interfluve. Higher area between two STREAMS; the surface over which water flows into the stream. These surfaces are subject to RUNOFF and EROSION by RILL action and GULLYING. Over time, interfluves are lowered.

Interlocking spur. STREAM in a hilly or mountainous REGION that winds its way in a sinuous VALLEY between the different RIDGES, slowly eroding the ends of the spurs and straightening its course. The view of interlocking spurs looking upstream is a favorite of artists, as colors change with the receding distance of each interlocking spur.

Intermediate rock. IGNEOUS ROCK that is transitional between a basic and a silicic ROCK, having a SILICA content between 54 and 64 percent.

Internal migration. Movement of people within a COUNTRY, from one REGION to another. Internal MIGRATION in high-income economies is often urban-to-RURAL, such as the migration to the SUN BELT in the United States. In low-income economies, rural-to-URBAN migration is more common.

Intertillage. Mixed planting of different seeds and seedling crops within the same SWIDDEN or cleared patch of agricultural land. Potatoes, yams, corn, rice, and bananas might all be planted. The planting times are staggered throughout the year to increase the variety of crops or nutritional balance available to the subsistence farmer and his or her family.

Intrusive rock. IGNEOUS ROCK which was formed from a MAGMA that cooled below the surface of the earth; it is commonly coarse-grained.

Irredentism. Expansion of one COUNTRY into the territory of a nearby country, based on the residence of nationals in the neighboring country. Hitler used irredentist claims to invade Czechoslovakia, because small groups of German-speakers lived there in the Sudetenland. The term comes from Italian, referring to Italy's claims before World War I that all Italian-speaking territory should become part of Italy.

Isallobar. Imaginary line on a MAP or meteorological chart joining PLACES with an equal change in ATMOSPHERIC PRESSURE over a certain time, often three hours. Isallobars indicate a pressure tendency and are used in WEATHER FORECASTING.

Island arc. Chain of VOLCANOES next to an oceanic TRENCH in the OCEAN BASINS; an oceanic PLATE descends, or subducts, below another oceanic plate at ISLAND arcs.

Isobar. Imaginary line joining PLACES of equal ATMOSPHERIC PRESSURE. WEATHER MAPS show isobars encircling areas of high or low pressure. The spacing between isobars is related to the pressure gradient.

Isobath. Line on a MAP or CHART joining all PLACES where the water depth is the same; a kind of underwater CONTOUR LINE. This kind of map is a BATHYMETRIC CONTOUR.

Isoclinal folding. When the earth's CRUST is folded, the size and shape of the folds vary according to the force of compression and nature of the ROCKS. When the surface is compressed evenly so that the two sides of the fold are parallel, isoclinal folding results. When the sides or slopes of the fold are unequal or dissimilar in shape and angle, this can be an asymmetrical or overturned fold. See also ANTICLINE; SYNCLINE.

Isotherm. Line joining PLACES of equal TEMPERATURE. A world MAP with isotherms of average monthly temperature shows that over the OCEANS, temperature decreases uniformly from the EQUATOR to the POLES, and higher temperatures occur over the CONTINENTS in summer and lower temperatures in winter because of the unequal heating properties of land and water.

Isotropic surface. Hypothetical flat surface or PLAIN, with no variation in any physical attribute. An isotropic surface has uniform ELEVATION, SOIL type, CLIMATE, and VEGETATION. Economic geographic models study behavior on an isotropic surface before applying the results to the real world. For example, in an isotropic model, land value is highest at the CITY center and falls regularly with increasing distance from there. In the real world, land values are affected by elevation, water features, URBAN regulations, and other factors. The von Thuenen model of the Isolated State is based on a uniform plain or isotropic surface.

Isthmian links. Chains of ISLANDS between substantial LANDMASSES.

Isthmus. Narrow strip of land connecting two larger bodies of land. The Isthmus of Panama connects North and South America; the Isthmus of Suez connects Africa and Asia. Both of these have been cut by CANALS to shorten shipping routes.

Jet stream. WINDS that move from west to east in the upper ATMOSPHERE, 23,000 to 33,000 feet (7,000–10,000 meters) above the earth, at about 200 miles (300 km.) per hour. They are narrow bands, elliptical in cross section, traveling in irregular paths. Four jet streams of interest to earth scientists and meteorologists are the polar jet stream and the subtropical jet stream in the Northern and SOUTHERN HEMISPHERES. The polar jet stream is located at the TROPOPAUSE, the BOUNDARY between the TROPOSPHERE and the STRATOSPHERE, along the polar FRONT. There is a complex interaction between surface winds and jet streams. In winter the NORTHERN HEMISPHERE polar front can move as far south as Texas, bringing BLIZZARDS and extreme WEATHER conditions. In summer, the polar jet stream is located over Canada. The subtropical jet stream is located at the tropopause around 30 DEGREES north or south LATITUDE, but it also migrates north or south, depending on the SEASON.

At times, the polar and subtropical jet streams merge for a few DAYS. Aircraft take advantage of the jet stream, or avoid it, depending on the direction of their flight. Upper atmosphere winds are also known as GEOSTROPHIC winds.

Joint. Naturally occurring fine crack in a ROCK, formed by cooling or by other stresses. SEDIMENTARY ROCKS can split along bedding planes; other joints form at right angles to the STRATA, running vertically through the rocks. In IGNEOUS ROCKS such as GRANITE, the stresses of cooling and contraction cause three sets of joints, two vertical and one parallel to the surface, which leads to the formation of distinctive LANDFORMS such as TORS. BASALT often demonstrates columnar jointing, producing tall columns that are mostly hexagonal in section. The presence of joints in BEDROCK hastens WEATHERING, because water can penetrate into the joints. This is particularly obvious in LIMESTONE, where joints are rapidly enlarged by SOLUTION. FROST WEDGING is a type of PHYSICAL WEATHERING that can split large boulders through the expansion when water in a joint freezes to form ice. Compare with FAULTS, which occur through tectonic activity.

Jurassic. Second of the three PERIODS that make up

Kame. Small HILL of gravel or mixed-size deposits, SAND, and gravel. Kames are found in areas previously covered by CONTINENTAL GLACIERS or ICE SHEETS, near what was the outer edge of the ice. They may have formed by materials dropping out of the melting ice, or in a deltalike deposit by a STREAM of MELTWATER. These deposits of which kames are made are called drift. Small LAKES called KETTLES are often found nearby. A closely spaced group of kames is called a kame field.

Karst. LANDSCAPE of SINKHOLES, underground STREAMS and caverns, and associated features created by CHEMICAL WEATHERING, especially SOLUTION, in REGIONS where the BEDROCK is LIMESTONE. The name comes from a region in the southwest of what is now Slovenia, the Krs (Kras) Plateau, but the karst region extends south through the Dinaric Alps bordering the

Adriatic Sea, into Bosnia-Herzegovina and Montenegro. Where limestone is well jointed, RAINFALL penetrates the JOINTS and enters the GROUNDWATER, carrying the MINERALS, especially calcium, away in solution. Most of the famous CAVES and caverns of the world are found in karst areas. The Carlsbad Caverns in New Mexico are a good example. Kentucky, Tennessee, and Florida also have well-known areas of karst. In some tropical countries, a form called tower karst is found. Tall conical or steep-sided HILLS of limestone rise above the flat surrounding landscape. Around 15 percent of the earth's land surface is karst TOPOGRAPHY.

Katabatic wind. GRAVITY DRAINAGE WINDS similar to MOUNTAIN BREEZES but stronger in force and over a larger area than a single VALLEY. Cold AIR collects over an elevated REGION, and the dense cold air flows strongly downslope. The ICE-SHEETS of Antarctica and Greenland produce fierce katabatic winds, but they can occur in smaller regions. The BORA is a strong, cold, squally downslope wind on the Dalmatian COAST of Yugoslavia in winter.

Kettle. Small depression, often a small LAKE, produced as a result of continental GLACIATION. It is formed by an isolated block of ice remaining in the ground MORAINE after a GLACIER has retreated. Deposited material accumulates around the ice, and when it finally melts, a steep hole remains, which often fills with water. Walden Pond, made famous by writer Henry David Thoreau, is a glacial kettle.

Khamsin. Hot, dry, DUST-laden WIND that blows in the eastern Sahara, in Egypt, and in Saudi Arabia, bringing high TEMPERATURES for three or four DAYS. Winds can reach GALE force in intensity. The word Khamsin is Arabic for "fifty" and refers to the period between March and June when the khamsin can occur.

Knickpoint. Abrupt change in gradient of the bed of a RIVER or STREAM. It is marked by a WATER-FALL, which over time is eroded by FLUVIAL action, restoring the smooth profile of the riverbed. The knickpoint acts as a TEMPORARY BASE LEVEL for the upper part of the stream.

Knickpoints can occur where a hard layer of ROCK is slower to erode than the rocks downstream, for example at Niagara Falls. Other knickpoints and waterfalls can develop as a result of tectonic forces. UPLIFT leads to new ERO-SION by a stream, creating a knickpoint that gradually moves upstream. The bed of a tributary GLACIER is often considerably higher than the VALLEY of the main glacier, so that after the glaciers have melted, a waterfall emerges over this knickpoint from the smaller hanging valley to join the main stream. Yosemite National Park has several such waterfalls.

Köppen climate classification. Commonly used scheme of CLIMATE classification that uses statistics of average monthly TEMPERATURE, average monthly PRECIPITATION, and total annual precipitation. The system was devised by Wladimir Köppen early in the twentieth century.

La Niña. WEATHER phenomenon that is the opposite part of EL NIÑO. When the SURFACE WATER in the eastern Pacific Ocean is cooler than average, the southeast TRADE WINDS blow strongly, bringing heavy rains to countries of the western Pacific. Scientists refer to the whole RANGE of TEMPERATURE, pressure, WIND, and SEA LEVEL changes as the SOUTHERN OSCILLATION (ENSO). The term "El Niño" gained wide currency in the U.S. media after a strong ENSO warm event in 1997–1998. A weak ENSO cold event, or La Niña, followed it in 1998. Means "the little girl" in Spanish. Alternative terms are "El Viejo" and "anti-El Niño."

Laccolith. LANDFORM of INTRUSIVE volcanism formed when viscous MAGMA is forced between overlying sedimentary STRATA, causing the surface to bulge upward in a domelike shape.

Lahar. Type of mass movement in which a MUD-FLOW occurs because of a volcanic explosion or ERUPTION. The usual cause is that the heat from the LAVA or other pyroclastic material melts ice and SNOW at the VOLCANO'S SUMMIT, causing a hot mudflow that can move downslope with great speed. The eruption of Mount Saint Helens in 1985 was accompanied by a lahar.

Lake basin. Enclosed depression on the surface of the land in which SURFACE WATERS collect; BASINS are created primarily by glacial activity and tectonic movement.

Lakebed. Floor of a LAKE.

Land bridge. Piece of land connecting two CONTINENTS, which permits the MIGRATION of humans, animals, or plants from one area to another. Many former land bridges are now under water, because of the rise in SEA LEVEL after the last ICE AGE. The Bering Strait connecting Asia and North America was an important land bridge for the latter continent.

Land hemisphere. Because the DISTRIBUTION of land and water surfaces on Earth is quite asymmetrical on either side of the EQUATOR, the NORTHERN HEMISPHERE might well be called the land hemisphere. For many centuries, Europeans refused to believe that there was not an equal area of land in the SOUTHERN HEMISPHERE. Explorers such as James Cook were dispatched to seek such a "Great South Land."

Landmass. Large area of land—an ISLAND or a CONTINENT.

Landsat. Space-exploration project begun in 1972 to MAP the earth continuously with SATELLITE imaging. The satellites have collected data about the earth: its AGRICULTURE, FORESTS, flat lands, MINERALS, waters, and ENVIRONMENT. These were the first satellites to aid in Earth sciences, helping to produce the best maps available and assisting farmers around the world to improve their crop yields.

Language family. Group of related LANGUAGES believed to have originated from a common prehistoric language. English belongs in the Indo-European language family, which includes the languages spoken by half of the world's peoples.

Lapilli. Small ROCK fragments that are ejected during volcanic ERUPTIONS. A lapillus ranges from about the size of a pea to not larger than a walnut. Some lapilli form by accretion of VOLCANIC ASH around moisture droplets, in a manner similar to hailstone formation. Lapilli sometimes form into a textured rock called lapillistone.

Laterite. Bright red CLAY SOIL, rich in iron oxide, that forms in tropical CLIMATES, where both TEMPERATURE and PRECIPITATION are high year-round, as ROCKS weather. It can be used in brick making and is a source of iron. When the soil is rich in aluminum, it is called BAUXITE. When laterite or bauxite forms a hard layer at the surface, it is called duricrust. Australia and sub-Saharan Africa have large areas of duricrust, some of which is thought to have formed under previous conditions during the TRIASSIC period.

Laurasia. Hypothetical SUPERCONTINENT made up of approximately the present CONTINENTS of the NORTHERN HEMISPHERE.

Lava tube. Cavern structure formed by the draining out of liquid LAVA in a pahoehoe flow.

Layered plains. Smooth, flat REGIONS believed to be composed of materials other than sulfur compounds.

Leaching. Removal of nutrients from the upper horizon or layer of a SOIL, especially in the humid TROPICS, because of heavy RAINFALL. The remaining soil is often bright red in color because iron is left behind. Despite their bright color, tropical soils are infertile.

Leeward. Rear or protected side of a MOUNTAIN or RANGE is the leeward side. Compare to WINDWARD.

Legend. Explanation of the different colors and symbols used on a MAP. For example, a map of the world might use different colors for high-income, middle-income, and low-income economies. A historical map might use different colors for countries that were once colonies of Britain, France, or Spain.

Light year. Distance traveled by light in one year; widely used for measuring stellar distances, it is equal to roughly 6 trillion miles (9.5 million km.).

Lignite. Low-grade COAL, often called brown coal. It is mined and used extensively in eastern Germany, Slovakia, and the Moscow Basin.

Liquefaction. Loss in cohesiveness of water-saturated SOIL as a result of ground shaking caused by an EARTHQUAKE.

Lithification. Process whereby loose material is transformed into solid ROCK by compaction or cementation.

Lithology. Description of ROCKS, such as rock type, MINERAL makeup, and fluid in rock pores.

Lithosphere. Solid outermost layer of the earth. It varies in thickness from a few miles to more than 120 miles (200 km.). It is broken into pieces known as TECTONIC PLATES, some of which are extremely large, while others are quite small. The upper layer of the lithosphere is the CRUST, which may be CONTINENTAL CRUST or OCEANIC CRUST. Below the crust is a layer called the ASTHENOSPHERE, which is weaker and plastic, enabling the motion of tectonic plates.

Littoral. Adjacent to or related to a SEA.

Llanos. Grassy REGION in the Orinoco Basin of Venezuela and part of Colombia. SAVANNA VEGETATION gradually gives way to scrub at the outer edges of the *llanos*. The area is relatively undeveloped.

Loam. SOIL TEXTURE classification, indicating a soil that is approximately equal parts of SAND, SILT, and CLAY. Farmers generally consider a sandy loam to be the best soil texture because of its water-retaining qualities and the ease with which it can be cultivated.

Local sea-level change. Change in SEA LEVEL only in one area of the world, usually by land rising or sinking in that specific area.

Lode deposit. Primary deposit, generally a VEIN, formed by the filling of a FISSURE with MINERALS precipitated from a HYDROTHERMAL solution.

Loess. EOLIAN, or wind-blown, deposit of fine, silt-sized, light-colored material. Loess covers about 10 percent of the earth's land surface. The loess PLATEAU of China is good agricultural land, although susceptible to EROSION. Loess has the property of being able to form vertical CLIFFS or BLUFFS, and many people have built dwellings in the steep cliffs above the Huang He (Yellow) River. In the United States, loess deposits are found in the VALLEYS of the Platte, Missouri, Mississippi, and Ohio Rivers, and on the Columbia Plateau. A German word, meaning loose or unconsolidated, which comes from loess deposits along the Rhine River.

Longitudinal bar. Midchannel accumulation of SAND and gravel with its long end oriented roughly parallel to the RIVER flow.

Longshore current. Current in the OCEAN close to the SHORE, in the surf zone, produced by WAVES approaching the COAST at an angle. Also called a LITTORAL current. The longshore current combined with wave action can move large amounts of SAND and other BEACH materials down the coast, a process called LONGSHORE DRIFT.

Longshore drift. The movement of SEDIMENT parallel to the BEACH by a LONGSHORE CURRENT.

Maar. Explosion vent at the earth's surface where a volcanic cone has not formed. A small ring of pyroclastic materials surrounds the maar. Often a LAKE occupies the small CRATER of a maar. A larger form is called a TUFF RING.

Macroburst. Updrafts and downdrafts within a CUMULONIMBUS CLOUD or THUNDERSTORM can cause severe TURBULENCE. A DOWNBURST within a thunderstorm when windspeeds are greater than 130 miles (210 km.) per hour and over areas of 2.5 square miles (5 sq. km.) or more is called a macroburst. See also MICROBURST.

Magnetic poles. Locations on the earth's surface where the earth's MAGNETIC FIELD is perpendicular to the surface. The magnetic poles do not correspond exactly to the geographic North Pole and South Pole, or earth's AXIS; the difference is called magnetic variation or DECLINATION.

Magnetic reversal. Change in the earth's MAGNETIC FIELD from the North Pole to the South MAGNETIC POLE.

Magnetic storm. Rapid changes in the earth's MAGNETIC FIELD as a result of the bombardment of the earth by electrically charged particles from the SUN.

Magnetosphere. REGION surrounding a PLANET where the planet's own MAGNETIC FIELD predominates over magnetic influences from the SUN or other planets.

Mantle convection. Thermally driven flow in the earth's MANTLE thought to be the driving force of PLATE TECTONICS.

Mantle plume. Rising jet of hot MANTLE material that produces tremendous volumes of basaltic LAVA. See also HOT SPOT.

Map projection. Mathematical formula used to transform the curved surface of the earth onto a flat plane or sheet of paper. Projections are divided into three classes: CYLINDRICAL, CONICAL, and AZIMUTHAL.

Marchland. FRONTIER area where boundaries are poorly defined or absent. The marches themselves were a type of BOUNDARY REGION. Marchlands have changed hands frequently throughout history. The name is related to the fact that armies marched across them.

Mass balance. Summation of the net gain and loss of ice and SNOW mass on a GLACIER in a year.

Mass extinction. Die-off of a large percentage of species in a short time.

Mass wasting. Downslope movement of Earth materials under the direct influence of GRAVITY.

Massif. French term used in geology to describe very large, usually IGNEOUS INTRUSIVE bodies.

Meandering river. RIVER confined essentially to a single CHANNEL that transports much of its SEDIMENT load as fine-grained material in SUSPENSION.

Mechanical weathering. Another name for PHYSICAL WEATHERING, or the breaking down of ROCK into smaller pieces.

Mechanization. Replacement of human labor with machines. Mechanization occurred in AGRICULTURE as tractors, reapers, picking machinery, and similar technological inventions took the place of human farm labor. Mechanization in industry was part of the INDUSTRIAL REVOLUTION, as spinning and weaving machines were introduced into the textile industry.

Medical geography. Branch of geography specializing in the study of health and disease, with a particular emphasis on the areal spread or DIFFUSION of disease. The spatial perspective of geography can lead to new medical insights. Geographers working with medical researchers in Africa have made great contributions to understanding the role of disease on that CONTINENT. John Snow's studies of the origin and spread of cholera in London in 1854 mark the beginnings of medical geography.

Megalopolis. Conurbation formed when large cities coalesce physically into one huge built-up area. Originally coined by the French geographer Jean Gottman in the early 1960s for the northeastern part of the United States, from Boston to Washington, D.C.

Mesa. Flat-topped HILL with steep sides. EROSION removes the surrounding materials, while the mesa is protected by a cap of harder, more resistant ROCK. Usually found in arid REGIONS. A larger LANDFORM of this type is a PLATEAU; a smaller feature is a BUTTE. The Colorado Plateau and Grand Canyon in particular are rich in these landforms. From the Spanish word for table.

Mesosphere. Atmospheric layer above the STRATOSPHERE where TEMPERATURE drops rapidly.

Mestizo. Person of mixed European and Amerindian ancestry, especially in countries of LATIN AMERICA.

Metamorphic rock. Any ROCK whose mineralogy, MINERAL chemistry, or TEXTURE has been altered by heat, pressure, or changes in composition; metamorphic rocks may have IGNEOUS, SEDIMENTARY, or other, older metamorphic rocks as their precursors.

Metamorphic zone. Areas of ROCK affected by the same limited RANGE of TEMPERATURE and pressure conditions, commonly identified by the presence of a key individual MINERAL or group of minerals.

Meteor. METEOROID that enters the ATMOSPHERE of a PLANET and is destroyed through frictional heating as it comes in contact with the various gases present in the atmosphere.

Meteorite. Fragment of an ASTEROID that survives passage through the ATMOSPHERE and strikes the surface of the earth.

Meteoroid. Small planetary body that enters Earth's ATMOSPHERE because its path intersects the earth's ORBIT. Friction caused by the earth's atmosphere on the meteoroid creates a glowing METEOR, or "shooting star." This is a common phenomenon, and most meteors burn away completely. Those that are large enough to reach the ground are called METEORITES.

Microburst. Brief but intense downward WIND, lasting not more than fifteen minutes over an area of 0.6 to 0.9 square mile (1.5–8 sq. km.). Usually associated with THUNDERSTORMS, but are quite unpredictable. The sudden change in wind direction associated with a microburst can create wind shear that causes airplanes to crash, especially if it occurs during takeoff or landing. See also MACROBURST.

Microclimate. CLIMATE of a small area, at or within a few yards of the earth's surface. In this REGION, variations of TEMPERATURE, PRECIPITATION, and moisture can have a pronounced effect on the bioclimate, influencing the growth or well-being of plants and animals, including humans. DEW or FROST, RAIN SHADOW effects, wind-tunneling between tall buildings, and similar phenomena are studied by microclimatologists. Horticulturists know the variations in aspect that affect INSOLATION and temperature, so that certain plants grow best on south-facing walls, for example. The growing of grapes for wine production is a major industry where microclimatology is essential. The study of microclimatology was pioneered by the German meteorologist Rudolf Geiger.

Microcontinent. Independent LITHOSPHERIC PLATE that is smaller than a CONTINENT but possesses continental-type CRUST. Examples include Cuba and Japan.

Microstates. Tiny countries. In 2000, seventeen independent countries each had an area of less than 200 square miles (520 sq. km.). The smallest microstate is Vatican City, with an area of 0.2 square miles (0.5 sq. km.). Most of the world's microstates are island NATIONS, including Nauru, Tuvalu, Marshall Islands, Saint Kitts and Nevis, Seychelles, Maldives, Malta, Grenada, Saint Vincent and the Grenadines, Barbados, Antigua and Barbuda, and Palau.

Mineral species. Mineralogic division in which all the varieties in any one species have the same basic physical and chemical properties.

Monadnock. Isolated HILL far from a STREAM, composed of resistant BEDROCK. Monadnocks are found in humid temperate REGIONS. A similar LANDFORM in an arid region is an INSELBERG.

Monogenetic. Pertaining to a volcanic ERUPTION in which a single vent is used only once.

Moraine. Materials transported by a GLACIER, and often later deposited as a RIDGE of unsorted ROCKS and smaller material. Lateral moraine is found at the side of the glacier; medial moraine occurs when two glaciers join. Other types of moraine include ABLATION moraine, ground moraine, and push, RECESSIONAL, and TERMINAL MORAINE.

Mountain belts. Products of PLATE TECTONICS, produced by the CONVERGENCE of crustal PLATES. Topographic MOUNTAINS are only the surficial expression of processes that profoundly deform and modify the CRUST. Long after the mountains themselves have been worn away, their former existence is recognizable from the structures that mountain building forms within the ROCKS of the crust.

Nappe. Huge sheet of ROCK that was the upper part of an overthrust fold, and which has broken and traveled far from its original position due to the tremendous forces. The Swiss Alps have nappes in many LOCATIONS.

Narrows. STRAIT joining two bodies of water.

Nation-state. Political entity comprising a COUNTRY whose people are a national group occupying the area. The concept originated in eighteenth century France; in practice, such cultural homogeneity is rare today, even in France.

Natural increase, rate of. DEMOGRAPHIC MEASURE of POPULATION growth: the difference between births and deaths per year, expressed as a percentage of the POPULATION. The rate of natural increase for the United States in 2000 was 0.6 percent. In countries where the population is decreasing, the DEATH RATE is greater than the BIRTH RATE.

Natural selection. Main process of biological evolution; the production of the largest number of offspring by individuals with traits that are best adapted to their ENVIRONMENTS.

Nautical mile. Standard MEASUREMENT at SEA, equalling 6,076.12 feet (1.85 km.). The mile used for land measurements is called a statute mile and measures 5,280 feet (1.6 km.).

Neap tide. TIDE with the minimum RANGE, or when the level of the high tide is at its lowest.

Near-polar orbit. Earth ORBIT that lies in a plane that passes close to both the north and south POLES.

Nekton. PELAGIC organisms that can swim freely, without having to rely on OCEAN CURRENTS or WINDS. Nekton includes shrimp; crabs; oysters; MARINE reptiles such as turtles, crocodiles, and snakes; and even sharks; porpoises; and whales.

Net migration. Net balance of a COUNTRY or REGION's IMMIGRATION and EMIGRATION.

Nomadism. Lifestyle in which pastoral people move with grazing animals along a defined route, ensuring adequate pasturage and water for their flocks or herds. This lifestyle has decreased greatly as countries discourage INTERNATIONAL MIGRATION. A more restricted form of nomadism is TRANSHUMANCE.

North geographic pole. Northernmost REGION of the earth, located at the northern point of the PLANET's AXIS of ROTATION.

North magnetic pole. Small, nonstationary area in the Arctic Circle toward which a COMPASS needle points from any LOCATION on the earth.

Notch. Erosional feature found at the base of a SEA CLIFF as a result of undercutting by WAVE EROSION, bioabrasion from MARINE organisms, and dissolution of ROCK by GROUNDWATER seepage. Also known as a nip.

Nuclear energy. ENERGY produced from a naturally occurring isotope of uranium. In the process of nuclear FISSION, the unstable uranium isotope absorbs a neutron and splits to form tin and molybdenum. This releases more neurons, so a chain reaction proceeds, releasing vast amounts of heat energy. Nuclear energy was seen in the 1950s as the energy of the future, but safety fears and the problem of disposal of radioactive nuclear waste have led to public condemnation of nuclear power plants.

Nuée ardente. Hot cloud of ROCK fragments, ASH, and gases that suddenly and explosively erupt from some VOLCANOES and flow rapidly down their slopes.

Nunatak. Isolated MOUNTAIN PEAK or RIDGE that projects through a continental ICE SHEET. Found in Greenland and Antarctica.

Obduction. Tectonic collisional process, opposite in effect to SUBDUCTION, in which heavier OCEANIC CRUST is thrust up over lighter CONTINENTAL CRUST.

Oblate sphere. Flattened shape of the earth that is the result of ROTATION.

Occultation. ECLIPSE of any astronomical object other than the SUN or the MOON caused by the Moon or any PLANET, SATELLITE, or ASTEROID.

Ocean basins. Large worldwide depressions that form the ultimate RESERVOIR for the earth's water supply.

Ocean circulation. Worldwide movement of water in the SEA.

Ocean current. Predictable circulation of water in the OCEAN, caused by a combination of WIND friction, Earth's ROTATION, and differences in TEMPERATURE and density of the waters. The five great oceanic circulations, known as GYRES, are in the North Pacific, North Atlantic, South Pacific, South Atlantic, and Indian Oceans. Because of the CORIOLIS EFFECT, the direction of circulation is CLOCKWISE in the NORTHERN HEMISPHERE and COUNTERCLOCKWISE in the SOUTHERN HEMISPHERE, except in the Indian Ocean, where the direction changes annually with the pattern of winds associated with the Asian MONSOON. Currents flowing toward the EQUATOR are cold currents; those flowing away from the equator are warm currents. An important current is the warm Gulf Stream, which flows north from the Gulf of Mexico along the East Coast of the United States; it crosses the North Atlantic, where it is called the North Atlantic Drift, and brings warmer conditions to the western parts of Europe. The West Coast of the United States is affected by the cool, south-flowing California Current. The cool Humboldt, or Peru, Current, which flows north along the South American coast, is an important indicator of whether there will be an EL NIÑO event. Deep currents, below 300 feet (100 meters), are extremely complicated and difficult to study.

Oceanic crust. Portion of the earth's CRUST under its OCEAN BASINS.

Oceanic island. ISLANDS arising from seafloor volcanic ERUPTIONS, rather than from continental shelves. The Hawaiian Islands are the best-known examples of oceanic islands.

Off-planet. Pertaining to REGIONS off the earth in orbital or planetary space.

Ore deposit. Natural accumulation of MINERAL matter from which the owner expects to extract a metal at a profit.

Orogeny. MOUNTAIN-building episode, or event, that extends over a period usually measured in tens of millions of years; also termed a revolution.

Orographic precipitation. Phenomenon caused when an AIR mass meets a topographic barrier, such as a mountain RANGE, and is forced to rise; the air cools to saturation, and orographic precipitation falls on the WINDWARD side as rain or snow. The lee side is a RAIN SHADOW. This effect is noticeable on the West Coast of the United States, which has RAIN FOREST on the windward side of the MOUNTAINS and DESERTS on the lee.

Orography. Study of MOUNTAINS that incorporates assessment of how they influence and are affected by WEATHER and other variables.

Oscillatory flow. Flow of fluid with a regular back-and-forth pattern of motion.

Overland flow. Flow of water over the land surface caused by direct PRECIPITATION.

Oxbow lake. LAKE created when floodwaters make a new, shorter CHANNEL and abandon the loop of a MEANDER. Over time, water in the oxbow lake evaporates, leaving a dry, curving, low-lying area known as a meander scar. Oxbow lakes are common on FLOODPLAINS. Another name for this feature is a cut-off.

Ozone hole. Decrease in the abundance of ANTARCTIC OZONE as sunlight returns to the POLE in early springtime

Ozone layer. Narrow band of the STRATOSPHERE situated near 18 miles (30 km.) above the earth's surface, where molecules of OZONE are concentrated. The average concentration is only one in 4 million, but this thin layer protects the earth by absorbing much of the ultraviolet light from the SUN and reradiating it as longer-wavelength radiation. Scientists were disturbed to discover that the ozonosphere was being destroyed by photochemical reaction with CHLOROFLUOROCARBONS (CFCs). The OZONE HOLES over the South and North Poles negatively affect several animal species, as well as humans; skin cancer risk is increasing rapidly as a consequence of depletion of the ozone layer. Stratospheric ozone should not be confused with ozone at lower levels, which is a result of PHOTOCHEMICAL SMOG. Also called the ozonosphere.

P wave. Fastest elastic wave generated by an EARTHQUAKE or artificial ENERGY source; basically an acoustic or shock wave that compresses and stretches solid material in its path.

Pangaea. Name used by Alfred Wegener for the SUPERCONTINENT that broke apart to create the present CONTINENTS.

Parasitic cone. Small volcanic cone that appears on the flank of a larger VOLCANO, or perhaps inside a CALDERA.

Particulate matter. Mixture of small particles that adversely affect human health. The particles may come from smoke and DUST and are in their highest concentrations in large URBAN AREAS, where they contribute to the "DUST DOME." Increased occurrences of illnesses such as asthma and bronchitis, especially in children, are related to high concentrations of particulate matter.

Pastoralism. Type of AGRICULTURE involving the raising of grazing animals, such as cattle, goats, and sheep. Pastoral nomads migrate with their domesticated animals in order to ensure sufficient grass and water for the animals.

Paternoster lakes. Small circular LAKES joined by a STREAM. These lakes are the result of glacial EROSION. The name comes from the resemblance to rosary beads and the accompanying prayer (the Our Father).

Pedestal crater. A CRATER that has assumed the shape of a pedestal as a result of unique shaping processes caused by WIND.

Pedology. Scientific study of SOILS.

Pelagic. Relating to life-forms that live on or in open SEAS, rather than waters close to land.

Peneplain. In the geomorphic CYCLE, or cycle of LANDFORM development, described by W. M. Davis, the final stage of EROSION led to the creation of an extensive land surface with low RELIEF. Davis named this a peneplain, meaning "almost a plain." It is now known that tectonic forces are so frequent that there would be insufficient time for such a cycle to complete all stages required to complete this landform.

Percolation. Downward movement of part of the water that falls on the surface of the earth, through the upper layers of PERMEABLE SOIL and ROCKS under the influence of GRAVITY. Eventually, it accumulates in the zone of SATURATION as GROUNDWATER.

Perforated state. STATE whose territory completely surrounds another state. The classic example of a perforated state is South Africa, within which lies the COUNTRY of Lesotho. Technically, Italy is perforated by the MICROSTATES of San Marino and Vatican City.

Perihelion. Point in Earth's REVOLUTION when it is closest to the SUN (usually on January 3). At perihelion, the distance between the earth and the Sun is 91,500,000 miles (147,255,000 km.). The opposite of APHELION.

Periodicity. The recurrence of related phenomena at regular intervals.

Permafrost. Permanently frozen SUBSOIL. The condition occurs in perennially cold areas such as the ARCTIC. No trees can grow because their roots cannot penetrate the permafrost. The upper portion of the frozen SOIL can thaw briefly in the summer, allowing many smaller plants to thrive in the long daylight. Permafrost occurs in about 25 percent of the earth's land surface, and the condition even hampers construction in REGIONS such as Siberia and ARCTIC Canada.

Perturb. To change the path of an orbiting body by a gravitational force.

Petrochemical. Chemical substance obtained from NATURAL GAS or PETROLEUM.

Petrography. Description and systematic classification of ROCKS.

Photochemical smog. Mixture of gases produced by the interaction of sunlight on the gases emanating from automobile exhausts. The gases include OZONE, nitrogen dioxide, carbon monoxide, and peroxyacetyl nitrates. Many large cities suffer from poor AIR quality because of photochemical smog. Severe health problems arise from continued exposure to photochemical smog.

Photometry. Technique of measuring the brightness of astronomical objects, usually with a photoelectric cell.

Phylogeny. Study of the evolutionary relationships among organisms.

Phylum. Major grouping of organisms, distinguished on the basis of basic body plan, grade of anatomical complexity, and pattern of growth or development.

Physiography. The PHYSICAL GEOGRAPHY of a PLACE—the LANDFORMS, water features, CLIMATE, SOILS, and VEGETATION.

Piedmont glacier. GLACIER formed when several ALPINE GLACIERS join together into a spreading glacier at the base of a MOUNTAIN or RANGE. The Malaspina glacier in Alaska is a good example of a piedmont glacier.

Place. In geographic terms, space that is endowed with physical and human meaning. Geographers study the relationship between people, places, and ENVIRONMENTS. The five themes that geographers use to examine the world are LOCATION, place, human/environment interaction, movement, and REGIONS.

Placer. Accumulation of valuable MINERALS formed when grains of the minerals are physically deposited along with other, nonvaluable mineral grains.

Planetary wind system. Global atmospheric circulation pattern, as in the BELT of prevailing westerly WINDS.

Plantation. Form of AGRICULTURE in which a large area of agricultural land is devoted to the production of a single cash crop, for export. Many plantation crops are tropical, such as bananas, sugarcane, and rubber. Coffee and tea plantations require cooler CLIMATES. Formerly, slave labor was used on most plantations, and the owners were Europeans.

Plate boundary. REGION in which the earth's crustal PLATES meet, as a converging (SUBDUCTION ZONE), diverging (MID-OCEAN RIDGE), TRANSFORM FAULT, or collisional interaction.

Plate tectonics. Theory proposed by German scientist Alfred Wegener in 1910. Based on extensive study of ancient geology, STRATIGRAPHY, and CLIMATE, Wegener concluded that the CONTINENTS were formerly one single enormous LANDMASS, which he named PANGAEA. Over the past 250 million years, Pangaea broke apart, first into LAURASIA and GONDWANALAND, and subsequently into the present continents. Earth scientists now believe that the earth's CRUST is composed of a series of thin, rigid PLATES that are in motion, sometimes diverging, sometimes colliding.

Plinian eruption. Rapid ejection of large volumes of VOLCANIC ASH that is often accompanied by the collapse of the upper part of the VOLCANO. Named either for Pliny the Elder, a Roman naturalist who died while observing the ERUPTION of Mount Vesuvius in 79 CE, or for Pliny the Younger, his nephew, who chronicled the eruption.

Plucking. Term used to describe the way glacial ice can erode large pieces of ROCK as it makes its way downslope. The ice penetrates JOINTS, other openings on the floor, or perhaps the side wall, and freezes around the block of stone, tearing it away and carrying it along, as part of the glacial MORAINE. The rocks contribute greatly to glacial ABRASION, causing deep grooves or STRIATIONS in some places. The jagged torn surface left behind is subject to further plucking. ALPINE GLA-CIERS can erode steep VALLEYS called glacial TROUGHS.

Plutonic. IGNEOUS ROCKS made of MINERAL grains visible to the naked eye. These igneous rocks have cooled relatively slowly. GRANITE is a good example of a plutonic rock.

Pluvial period. Episode of time during which rains were abundant, especially during the last ICE AGE, from a few million to about 10,000 years ago.

Polar stratospheric clouds. CLOUDS of ice crystals formed at extremely low TEMPERATURES in the polar STRATOSPHERE.

Polder. Lands reclaimed from the SEA by constructing DIKES to hold back the sea and then pumping out the water retained between the dikes and the land. Before AGRICULTURE is possible, the SOIL must be specially treated to remove the SALT. Some polders are used for recreational land; cities also have been built on polders. The largest polders are in the Netherlands, where the northern part, known as the Low Netherlands, covers almost half of the total area of this COUNTRY.

Polygenetic. Pertaining to volcanism from several physically distinct vents or repeated ERUPTIONS from a single vent punctuated by long periods of quiescence.

Polygonal ground. Distinctive geological formation caused by the repetitive freezing and thawing of PERMAFROST.

Possibilism. Concept that arose among French geographers who rejected the concept of ENVIRONMENTAL DETERMINISM, instead asserting that the relationship between human beings and the ENVIRONMENT is interactive.

Potable water. FRESH WATER that is being used for domestic consumption.

Potholes. Circular depressions formed in the bed of a RIVER when the STREAM flows over BEDROCK. The scouring of PEBBLES as a result of water TURBULENCE wears away the sides of the depression, deepening it vertically and producing a smooth, rounded pothole. (In modern parlance, the term is also applied to holes in public roads.)

Primary minerals. MINERALS formed when MAGMA crystallizes.

Primary wave. Compressional type of EARTHQUAKE wave, which can travel in any medium and is the fastest wave.

Primate city. CITY that is at least twice as large as the next-largest city in that COUNTRY. The "law of the primate city" was developed by U.S. geographer Mark Jefferson, to analyze the phenomenon of countries where one huge city dominates the political, economic, and cultural life of that country. The size and dominance of a primate city is a PULL FACTOR and ensures its continuing dominance.

Principal parallels. The most important lines of LATITUDE. PARALLELS are imaginary lines, parallel to the EQUATOR. The principal parallels are the equator at zero DEGREES, the tropic of CANCER at 23.5 degrees North, the tropic of CAPRICORN at 23.5 degrees south, the Arctic Circle at 66.5 degrees north, and the Antarctic Circle at 66.5 degrees south.

Protectorate. COUNTRY that is a political DEPENDENCY of another NATION; similar to a COLONY, but usually having a less restrictive relationship with its overseeing power.

Proterozoic eon. Interval between 2.5 billion and 544 million years ago. During this PERIOD in the GEOLOGIC RECORD, processes presently active on Earth first appeared, notably the first clear evidence for PLATE TECTONICS. ROCKS of the Proterozoic eon also document changes in conditions on Earth, particularly an apparent increase in atmospheric oxygen.

Pull factors. Forces that attract immigrants to a new COUNTRY or LOCATION as permanent settlers. They include economic opportunities, educational facilities, land ownership, gold rushes, CLIMATE conditions, democracy, and similar factors of attraction.

Push factors. Forces that encourage people to migrate permanently from their HOMELANDS to settle in a new destination. They include war, persecution for religious or political reasons, hunger, and similar negative factors.

Pyroclasts. Materials that are ejected from a VOLCANO into the AIR. Pyroclastic materials return to Earth at greater or lesser distances, depending on their size and the height to which they are thrown by the explosion of the volcano. The largest pyroclasts are volcanic bombs. Smaller pieces are volcanic blocks and scoria. These generally fall back onto the volcano and roll down the sides. Even smaller pyroclasts are LAPILLI, cinders, and VOLCANIC ASH. The finest pyroclastic materials may be carried by WINDS for great distances, even completely around the earth, as was the case with DUST from the Krakatoa explosion in 1883 and the early 1990s explosions of Mount Pinatubo in the Philippines.

Qanat. Method used in arid REGIONS to bring GROUNDWATER from mountainous regions to lower and flatter agricultural land. A qanat is a long tunnel or series of tunnels, perhaps more than a mile long. The word *qanat* is Arabic, but the first qanats are thought to have been constructed in Farsi-speaking Persia more than 2,000 years ago. Qanats are still used there, as well as in Afghanistan and Morocco.

Quaternary sector. Economic activity that involves the collection and processing of information. The rapid spread of computers and the Internet caused a major increase in the importance of employment in the quaternary sector.

Radar imaging. Technique of transmitting radar toward an object and then receiving the reflected radiation so that time-of-flight MEASUREMENTS provide information about surface TOPOGRAPHY of the object under study.

Radial drainage. The pattern of STREAM courses often reveals the underlying geology or structure of a REGION. In a radial drainage pattern, streams radiate outward from a center, like spokes on a wheel, because they flow down the slopes of a VOLCANO.

Radioactive minerals. MINERALS combining uranium, thorium, and radium with other elements. Useful for nuclear TECHNOLOGY, these minerals furnish the basic isotopes necessary not only for nuclear reactors but also for advanced medical treatments, metallurgical analysis, and chemicophysical research.

Rain gauge. Instrument for measuring RAINFALL, usually consisting of a cylindrical container open to the sky.

Rain shadow. Area of low PRECIPITATION located on the LEEWARD side of a topographic barrier such as a mountain RANGE. Moisture-laden WINDS are forced to rise, so they cool ADIABATICALLY, leading to CONDENSATION and precipitation on the WINDWARD side of the barrier. When the AIR descends on the other side of the MOUNTAIN, it is dry and relatively warm. The area to the east of the Rocky Mountains is in a rain shadow.

Range, mountain. Linear series of MOUNTAINS close together, formed in an OROGENY, or mountain-building episode. Tall mountain ranges such as the Rocky Mountains are geologically much younger than older mountain ranges such as the Appalachians.

Rapids. Stretches of RIVERS where the water flow is swift and turbulent because of a steep and rocky CHANNEL. The turbulent conditions are called WHITE WATER. If the change in ELEVATION is greater, as for small WATERFALLS, they are called CATARACTS.

Recessional moraine. Type of TERMINAL MORAINE that marks a position of shrinkage or wasting or a GLACIER. Continued forward flow of ice is maintained so that the debris that forms the moraine continues to accumulate. Recessional moraines occur behind the terminal moraine.

Recumbent fold. Overturned fold in which the upper part of the fold is almost horizontal, lying on top of the nearest adjacent surface.

Reef (geology). VEIN of ORE, for example, a reef of gold.

Reef (marine). Underwater ridge made up of sand, rocks, or coral that rises near to the water's surface.

Refraction of waves. Bending of waves, which can occur in all kinds of waves. When OCEAN WAVES approach a COAST, they start to break as they approach the SHORE because the depth decreases.

The wave speed is retarded and the WAVE CREST seems to bend as the wavelength decreases. If waves are approaching a coast at an oblique angle, the crest line bends near the shore until it is almost parallel. If waves are approaching a BAY, the crests are refracted to fit the curve of the bay.

Regression. Retreat of the SEA from the land; it allows land EROSION to occur on material formerly below the sea surface.

Relative humidity. Measure of the HUMIDITY, or amount of moisture, in the ATMOSPHERE at any time and place compared with the total amount of moisture that same AIR could theoretically hold at that TEMPERATURE. Relative humidity is a ratio that is expressed as a percentage. When the air is saturated, the relative humidity reaches 100 percent and rain occurs. When there is little moisture in the air, the relative humidity is low, perhaps 20 percent. Relative humidity varies inversely with temperature, because warm air can hold more moisture than cooler air. Therefore, when temperatures fall overnight, the air often becomes saturated and DEW appears on grass and other surfaces. The human COMFORT INDEX is related to the relative humidity. Hot temperatures are more bearable when relative humidity is low. Media announcers frequently use the term "humidity" when they mean relative humidity.

Replacement rate. The rate at which females must reproduce to maintain the size of the POPULATION. It corresponds to a FERTILITY RATE of 2.1.

Reservoir rock. Geologic ROCK layer in which OIL and gas often accumulate; often SANDSTONE or LIMESTONE.

Retrograde orbit. ORBIT of a SATELLITE around a PLANET that is in the opposite sense (direction) in which the planet rotates.

Retrograde rotation. ROTATION of a PLANET in a direction opposite to that of its REVOLUTION.

Reverse fault. Feature produced by compression of the earth's CRUST, leading to crustal shortening. The UPTHROWN BLOCK overhangs the downthrown block, producing a FAULT SCARP where the overhang is prone to LANDSLIDES. When the movement is mostly horizontal, along a low angle FAULT, an overthrust fault is formed. This is commonly associated with extreme FOLDING.

Reverse polarity. Orientation of the earth's MAGNETIC FIELD so that a COMPASS needle points to the SOUTHERN HEMISPHERE.

Ria coast. Ria is a long narrow ESTUARY or RIVER MOUTH. COASTS where there are many rias show the effects of SUBMERGENCE of the land, with the SEA now occupying former RIVER VALLEYS. Generally, there are MOUNTAINS running at an angle to the coast, with river valleys between each RANGE, so that the ria coast is a succession of estuaries and promontories. The submergence can result from a rising SEA LEVEL, which is common since the melting of the PLEISTOCENE GLACIERS, or it can be the result of SUBSIDENCE of the land. There is often a great TIDAL RANGE in rias, and in some, a tidal BORE occurs with each TIDE. The eastern coast of the United States, from New York to South Carolina, is a ria coast. The southwest coast of Ireland is another. The name comes from Spain, where rias occur in the south.

Richter scale. SCALE used to measure the magnitude of EARTHQUAKES; named after U.S. physicist Charles Richter, who, together with Beno Gutenberg, developed the scale in 1935. The scale is a quantitative measure that replaced the older MERCALLI SCALE, which was a descriptive scale. Numbers range from zero to nine, although there is no upper limit. Each whole number increase represents an order of magnitude, or an increase by a factor of ten. The actual MEASUREMENT was logarithm to base 10 of the maximum SEISMIC WAVE amplitude (in thousandths of a millimeter) recorded on a standard SEISMOGRAPH at a distance of 60 miles (100 km.) from the earthquake EPICENTER.

Rift valley. Long, low REGION of the earth's surface; a VALLEY or TROUGH with FAULTS on either side. Unlike valleys produced by EROSION, rift valleys are produced by tectonic forces that have caused the faults or fractures to develop in the ROCKS of Earth's CRUST. TENSION can lead to the block of land between two faults dropping in ELEVATION compared to the surrounding blocks, thus forming the rift valley. A small LANDFORM produced in this way is called a GRABEN. A rift valley is a much larger feature. In Africa, the Great Rift Valley is partially occupied by Lake Malawi and Lake Tanganyika, as well as by the Red Sea.

Ring dike. Volcanic LANDFORM created when MAGMA is intruded into a series of concentric FAULTS. Later EROSION of the surrounding material may reveal the ring dike as a vertical feature of thick BASALT rising above the surroundings.

Ring of Fire. Zone of volcanic activity and associated EARTHQUAKES that marks the edges of various TECTONIC PLATES around the Pacific Ocean, especially those where SUBDUCTION is occurring.

Riparian. Term meaning related to the BANKS of a STREAM or RIVER. Riparian VEGETATION is generally trees, because of the availability of moisture. RIPARIAN RIGHTS allow owners of land adjacent to a river to use water from the river.

River terraces. LANDFORMS created when a RIVER first produces a FLOODPLAIN, by DEPOSITION of ALLUVIUM over a wide area, and then begins downcutting into that alluvium toward a lower BASE LEVEL. The renewed EROSION is generally because of a fall in SEA LEVEL, but can result from tectonic UPLIFT or a change in CLIMATE pattern due to increased PRECIPITATION. On either side of the river, there is a step up from the new VALLEY to the former alluvium-covered floodplain surface, which is now one of a pair of river terraces. This process may occur more than once, creating as many as three sets of terraces. These are called depositional terraces, because the terrace is cut into river deposits. Erosional terraces, in contrast, are formed by lateral migration of a river, from one part of the valley to another, as the river creates a floodplain. These terraces are cut into BEDROCK, with only a thin layer of alluvium from the point BAR deposits, and they do not occur in matching pairs.

River valleys. VALLEYS in which STREAMS flow are produced by those streams through long-term EROSION and DEPOSITION. The LANDFORMS produced by FLUVIAL action are quite diverse, ranging from spectacular CANYONS to wide, gently sloping valleys. The patterns formed by stream networks are complex and generally reflect the BEDROCK geology and TERRAIN characteristics.

Rock avalanche. Extreme case of a rockfall. It occurs when a large mass of ROCK moves rapidly down a steeply sloping surface, taking everything that lies in its path. It can be started by an EARTHQUAKE, rock-blasting operations, or vibrations from thunder or artillery fire.

Rock cycle. Cycle by which ROCKS are formed and reformed, changing from one type to another over long PERIODS of geologic time. IGNEOUS ROCKS are formed by cooling from molten MAGMA. Once exposed at the surface, they are subject to WEATHERING and EROSION. The products of erosion are compacted and cemented to form SEDIMENTARY ROCKS. The heat and pressure accompanying a volcanic intrusion causes adjacent rocks to be altered into METAMORPHIC ROCKS.

Rock slide. Event that occurs when water lubricates an unconsolidated mass of weathered ROCK on a steep slope, causing rapid downslope movement. In a RIVER VALLEY where there are steep SCREE slopes being constantly carried away by a swiftly flowing STREAM, the undercutting at the base can lead to constant rockslides of the surface layer of rock. A large rockslide is a ROCK AVALANCHE.

S waves. Type of SEISMIC disturbance of the earth when an EARTHQUAKE occurs. In an S wave, particles move about at right angles to the direction in which the wave is traveling. S waves cannot pass through the earth's CORE, which is why scientists believe the INNER CORE is liquid. Also called transverse wave, shear wave, or secondary wave.

Sahel. Southern edge of the Sahara Desert; a great stretch of semiarid land extending from the Atlantic Ocean in Senegal and Mauritania through Mali, Burkina Faso, Nigeria, Niger, Chad, and Sudan. Northern Ethiopia, Eritrea, Djibouti, and Somalia usually are included also. This transition zone between the hot DESERT and the tropical SAVANNA has low summer RAINFALL of less than 8 inches (200 millimeters) and a natural

VEGETATION of low grasses with some small shrubs. The REGION traditionally has been used for PASTORALISM, raising goats, camels, and occasionally sheep. Since a prolonged DROUGHT in the 1970s, DESERTIFICATION, SOIL EROSION, and FAMINE have plagued the Sahel. The narrow band between the northern Sahara and the Mediterranean North African COAST is also called Sahel. "Sahel" is the Arabic word for edge.

Saline lake. LAKE with elevated levels of dissolved solids, primarily resulting from evaporative concentration of SALTS; saline lakes lack an outlet to the SEA. Well-known examples include Utah's Great Salt Lake, California's Mono Lake and Salton Sea, and the Dead Sea in the Middle East.

Salinization. Accumulation of SALT in SOIL. When IRRIGATION is used to grow crops in semiarid to arid REGIONS, salinization is frequently a problem. Because EVAPORATION is high, water is drawn upward through the soil, depositing dissolved salts at or near the surface. Over years, salinization can build up until the soil is no longer suitable for AGRICULTURE. The solution is to maintain a plentiful flow of water while ensuring that the water flows through the soil and is drained away.

Salt domes. Formations created when deeply buried salt layers are forced upwards. SALT under pressure is a plastic material, one that can flow or move slowly upward, because it is lighter than surrounding SEDIMENTARY ROCKS. The salt forms into a plug more than a half mile (1 km.) wide and as much as 5 miles (8 km.) deep, which passes through overlying sedimentary rock layers, pushing them up into a dome shape as it passes. Some salt domes emerge at the earth's surface; others are close to the surface and are easy to mine for ROCK SALT. OIL and NATURAL GAS often accumulate against the walls of a salt dome. Salt domes are numerous around the COAST of the Gulf of Mexico, in the North Sea REGION, and in Iran and Iraq, all of which are major oil-producing regions.

Sand dunes. Accumulations of SAND in the shape of mounds or RIDGES. They occur on some COASTS and in arid REGIONS. Coastal dunes are formed when the prevailing WINDS blow strongly on-shore, piling up sand into dunes, which may become stabilized when grasses grow on them. DESERT sand dunes are a product of DEFLATION, or wind EROSION removing fine materials to leave a DESERT PAVEMENT in one region and sand deposits in another. Sand dunes are classified by their shape into barchans, or crescent-shaped dunes; seifs or LONGITUDINAL DUNES; TRANSVERSE DUNES; star dunes; and sand drifts or sand sheets.

Sapping. Natural process of EROSION at the bases of HILL slopes or CLIFFS whereby support is removed by undercutting, thereby allowing overlying layers to collapse; SPRING SAPPING is the facilitation of this process by concentrated GROUNDWATER flow, generally at the heads of VALLEYS.

Saturation, zone of. Underground REGION below the zone of AERATION, where all pore space is filled with water. This water is called GROUNDWATER; the upper surface of the zone of saturation is the WATER TABLE.

Scale. Relationship between a distance on a MAP or diagram and the same distance on the earth. Scale can be represented in three ways. A linear, or graphic, scale uses a straight line, marked off in equally spaced intervals, to show how much of the map represents a mile or a kilometer. A representative fraction (RF) gives this scale as a ratio. A verbal scale uses words to explain the relationship between map size and actual size. For example, the RF 1:63,360 is the same as saying "one inch to the mile."

Scarp. Short version of the word "ESCARPMENT," a short steep slope, as at the edge of a PLATEAU. EARTHQUAKES lead to the formation of FAULT SCARPS.

Schist. METAMORPHIC ROCK that can be split easily into layers. Schist is commonly produced from the action of heat and pressure on SHALE or SLATE. The rock looks flaky in appearance. Mica-schists are shiny because of the development of visible mica. Other schists include talc-schist, which contains a large amount of talc, and hornblende-schist, which develops from basaltic rocks.

Scree. Broken, loose ROCK material at the base of a slope or CLIFF. It is often the result of FROST WEDGING of BEDROCK cliffs, causing rockfall. Another name for scree is TALUS.

Sedimentary rocks. ROCKS formed from SEDIMENTS that are compressed and cemented together in a process called LITHIFICATION. Sedimentary rocks cover two-thirds of the earth's land surface but are only a small proportion of the earth's CRUST. SANDSTONE is a common sedimentary rock. Sedimentary rocks form STRATA, or layers, and sometimes contain FOSSILS.

Seif dunes. Long, narrow RIDGES of SAND, built up by WINDS blowing at different times of year from two different directions. Seif dunes occur in parallel lines of sand over large areas, running for hundreds of miles in the Sahara, Iran, and central Australia. Another name for seif dunes is LONGITUDINAL DUNES. The Arabic word means sword.

Seismic activity. Movements within the earth's CRUST that often cause various other geological phenomena to occur; the activity is measured by SEISMOGRAPHS.

Seismology. The scientific study of EARTHQUAKES. It is a branch of GEOPHYSICS. The study of SEISMIC WAVES has provided a great deal of knowledge about the composition of the earth's interior.

Shadow zone. When an EARTHQUAKE occurs at one LOCATION, its waves travel through the earth and are detected by SEISMOGRAPHS around the world. Every earthquake has a shadow zone, a band where neither P nor S WAVES from the earthquake will be detected. This shadow zone leads scientists to draw conclusions about the size, density, and composition of the earth's CORE.

Shale oil. SEDIMENTARY ROCK containing sufficient amounts of hydrocarbons that can be extracted by slow distillation to yield OIL.

Shallow-focus earthquakes. EARTHQUAKES having a focus less than 35 miles (60 km.) below the surface.

Shantytown. URBAN SQUATTER SETTLEMENT, usually housing poor newcomers.

Shield. Large part of the earth's CONTINENTAL CRUST, comprising very old ROCKS that have been eroded to REGIONS of low RELIEF. Each CONTINENT has a shield area. In North America, the Canadian Shield extends from north of the Great Lakes to the Arctic Ocean. Sometimes known as a CONTINENTAL SHIELD.

Shield volcano. VOLCANO created when the LAVA is quite viscous or fluid and highly basaltic. Such lava spreads out in a thin sheet of great radius but comparatively low height. As flows continue to build up the volcano, a low DOME shape is created. The greatest shield volcanoes on Earth are the ISLANDS of Hawaii, which rise to a height of almost 30,000 feet (10,000 meters) above SEA LEVEL.

Shock city. CITY that typifies disturbing changes in social and cultural conditions or in economic conditions. In the nineteenth century, the shock city of the United States was Chicago.

Sierra. Spanish word for a mountain RANGE with a serrated crest. In California, the Sierra Nevada is an important range, containing Mount Whitney, the highest PEAK in the continental United States.

Sill. Feature formed by INTRUSIVE volcanic activity. When LAVA is forced between two layers of ROCK, it can form a narrow horizontal layer of BASALT, parallel with the adjacent beds. Although it resembles a windowsill in its flatness, a sill may be hundreds of miles long and can range in thickness from a few centimeters to considerable thickness.

Siltation. Build-up of SILT and SAND in creeks and waterways as a result of SOIL EROSION, clogging water courses and creating DELTAS at RIVER MOUTHS. Siltation often results from DEFORESTATION or removal of tree cover. Such ENVIRONMENTAL DEGRADATION causes loss of agricultural productivity, worsening of water supply, and other problems.

Sima. Abbreviation for SILICA and *ma*gnesium. These are the two principal constituents of heavy ROCKS such as BASALT, which forms much of the OCEAN floor. Lighter, more abundant rock is SIAL.

Sinkhole. Circular depression in the ground surface, caused by WEATHERING of LIMESTONE, mainly through the effects of SOLUTION on JOINTS in the ROCK. If a STREAM flows above ground and then disappears down a sinkhole, the feature is called a swallow hole. In everyday language, many events that cause the surface to collapse are called sinkholes, even though they are rarely in limestone and rarely caused by weathering.

Sinking stream. STREAM or RIVER that loses part or all of its water to pathways dissolved underground in the BEDROCK.

Situation. Relationship between a PLACE, such as a TOWN or CITY, and its RELATIVE LOCATION within a REGION. A situation on the COAST is desirable in terms of overseas TRADE.

Slip-face. LEEWARD side of a SAND DUNE. As the WIND piles up sand on the WINDWARD side, it then slips down the rear or slip-face. The angle of the slip-face is gentler than the angle of the windward slope.

Slump. Type of LANDSLIDE in which the material moves downslope with a rotational motion, along a curved slip surface.

Snout. Terminal end of a GLACIER.

Snow line. The height or ELEVATION at which snow remains throughout the year, without melting away. Near the EQUATOR, the snow line is more than 15,000 feet (almost 5,000 meters); at higher LATITUDES, the snow line is correspondingly lower, reaching SEA LEVEL at the POLES. The actual snow line varies with the time of year, retreating in summer and coming lower in winter.

Soil horizon. SOIL consists of a series of layers called horizons. The uppermost layer, the O horizon, contains organic materials such as decayed leaves that have been changed into HUMUS. Beneath this is the A horizon, the TOPSOIL, where farmers plow and plant seeds. The B HORIZON often contains MINERALS that have been washed downwards from the A horizon, such as calcium, iron, and aluminum. The A and B horizons together comprise a solum, or true soil. The C horizon is weathered BEDROCK, which contains pieces of the original ROCK from which the soil formed. Another name for the C horizon is REGOLITH. Beneath this is the R horizon, or bedrock.

Soil moisture. Water contained in the unsaturated zone above the WATER TABLE.

Soil profile. Vertical section of a SOIL, extending through its horizon into the unweathered parent material.

Soil stabilization. Engineering measures designed to minimize the opportunity and/or ability of EXPANSIVE SOILS to shrink and swell.

Solar energy. One of the forms of ALTERNATIVE or RENEWABLE ENERGY. In the late 1990s, the world's largest solar power generating plant was located at Kramer Junction, California. There, solar energy heats huge OIL-filled containers with a parabolic shape, which produces steam to drive generating turbines. An alternative is the production of energy through photovoltaic cells, a TECHNOLOGY that was first developed for space exploration. Many individual homes, especially in isolated areas, use this technology.

Solar system. SUN and all the bodies that ORBIT it, including the PLANETS and their SATELLITES, plus numerous COMETS, ASTEROIDS, and METEOROIDS.

Solar wind. Gases from the SUN's ATMOSPHERE, expanding at high speeds as streams of charged particles.

Solifluction. Word meaning flowing SOIL. In some REGIONS of PERMAFROST, where the ground is permanently frozen, the uppermost layer thaws during the summer, creating a saturated layer of soil and REGOLITH above the hard layer of frozen ground. On slopes, the material can flow slowly downhill, creating a wavy appearance along the hillslope.

Solution. Form of CHEMICAL WEATHERING in which MINERALS in a ROCK are dissolved in water. Most substances are soluble, but the combination of water with CARBON DIOXIDE from the ATMOSPHERE means that RAINFALL is slightly acidic, so

that the chemical reaction is often a combination of solution and carbonation.

Sound. Long expanse of the SEA, close to the COAST, such as a large ESTUARY. It can also be the expanse of sea between the mainland and an ISLAND.

Source rock. ROCK unit or bed that contains sufficient organic carbon and has the proper thermal history to generate OIL or gas.

Spatial diffusion. Notion that things spread through space and over time. An understanding of geographic change depends on this concept. Spatial diffusion can occur in various ways. Geographers distinguish between expansion diffusion, relocation diffusion, and hierarchical diffusion.

Spheroidal weathering. Form of ROCK WEATHERING in which layers of rock break off parallel to the surface, producing a rounded shape. It results from a combination of physical and CHEMICAL WEATHERING. Spheroidal weathering is especially common in GRANITE, leading to the creation of TORS and similar rounded features. Onion-skin weathering is a term sometimes used, especially when this is seen on small rocks.

Spring tide. TIDE of maximum RANGE, occurring when lunar and solar tides reinforce each other, a few DAYS after the full and new MOONS.

Squall line. Line of vigorous THUNDERSTORMS created by a cold downdraft that spreads out ahead of a fast-moving COLD FRONT.

Stacks. Pieces of ROCK surrounded by SEA water, which were once part of the mainland. WAVE EROSION has caused them to be isolated. Also called sea stacks.

Stalactite. Long, tapering piece of calcium carbonate hanging from the roof of a LIMESTONE CAVE or cavern. Stalactites are formed as water containing the MINERAL in solution drips downward. The water evaporates, depositing the dissolved minerals.

Stalagmite. Column of calcium carbonate growing upward from the floor of a LIMESTONE CAVE or cavern.

Steppe. Huge REGION of GRASSLANDS in the midlatitudes of Eurasia, extending from central

Europe to northeast China. The region is not uniform in ELEVATION; most of it is rolling PLAINS, but some mountain RANGES also occur. These have not been a barrier to the migratory lifestyle of the herders who have occupied the steppe for many centuries. The Asian steppe is colder than the European steppe, because of greater elevation and greater continentality. The best-known rulers from the steppe were the Mongols, whose empire flourished in the thirteenth and fourteenth centuries. Geographers speak of a steppe CLIMATE, a semiarid climate where the EVAPORATION rate is double that of PRECIPITATION. South of the steppe are great DESERTS; to the north are midlatitude mixed FORESTS. In terms of climate and VEGETATION, the steppe is like the short-grass PRAIRIE vegetation west of the Mississippi River. Also called steppes.

Storm surge. General rise above normal water level, resulting from a HURRICANE or other severe coastal STORM.

Strait. Relatively narrow body of water, part of an OCEAN or SEA, separating two pieces of land. The world's busiest SEAWAY is the Johore Strait between the Malay Peninsula and the island of Sumatra.

Strata. Layers of SEDIMENT deposited at different times, and therefore of different composition and TEXTURE. When the sediments are laid down, strata are horizontal, but subsequent tectonic processes can lead to tilting, FOLDING, or faulting. Not all SEDIMENTARY ROCKS are stratified. Singular form of the word is stratum.

Stratified drift. Material deposited by glacial MELTWATERS; the water separates the material according to size, creating layers.

Stratigraphy. Study of sedimentary STRATA, which includes the concept of time, possible correlation of the ROCK units, and characteristics of the rocks themselves.

Stratovolcano. Type of VOLCANO in which the ERUPTIONS are of different types and produce different LAVAS. Sometimes an eruption ejects cinder and ASH; at other times, viscous lava flows down the sides. The materials flow, settle, and fall to produce a beautiful symmetrical LANDFORM with a broad circular base and concave slopes tapering upward to a small circular CRATER. Mount Rainier, Mount Saint Helens, and Mount Fuji are stratovolcanoes. Also known as a COMPOSITE CONE.

Streambed. Channel through which a STREAM flows. Dry streambeds are variously known as ARROYOS, DONGAS, WASHES, and WADIS.

Strike. Term used when earth scientists study tilted or inclined beds of SEDIMENTARY ROCK. The strike of the inclined bed is the direction of a horizontal line along a bedding plane. The strike is at right angles to the dip of the rocks.

Strike-slip fault. In a strike-slip fault, the surface on either side of the fault moves in a horizontal plane. There is no vertical displacement to form a FAULT SCARP, as there is with other types of faults. The San Andreas Fault is a strike-slip fault. Also called a transcurrent fault.

Subduction zone. CONVERGENT PLATE BOUNDARY where an oceanic PLATE is being thrust below another plate.

Sublimation. Process by which water changes directly from solid (ice) to vapor, or vapor to solid, without passing through a liquid stage.

Subsolar point. Point on the earth's surface where the SUN is directly overhead, making the Sun's rays perpendicular to the surface. The subsolar point receives maximum INSOLATION, compared with other PLACES, where the Sun's rays are oblique.

Sunspots. REGIONS of intense magnetic disturbances that appear as dark spots on the solar surface; they occur approximately every eleven years.

Supercontinent. Vast LANDMASS of the remote geologic past formed by the collision and amalgamation of crustal PLATES. Hypothesized supercontinents include PANGAEA, GONDWANALAND, and LAURASIA.

Supersaturation. State in which the AIR'S RELATIVE HUMIDITY exceeds 100 percent, the condition necessary for vapor to begin transformation to a liquid state.

Supratidal. Referring to the SHORE area marginal to shallow OCEANS that are just above high-tide level.

Swamp. WETLAND where trees grow in wet to water-logged conditions. Swamps are common close to the RIVER on FLOODPLAINS, as well as in some coastal areas.

Swidden. Area of land that has been cleared for SUBSISTENCE AGRICULTURE by a farmer using the technique of slash-and-burn. A variety of crops is planted, partly to reduce the risk of crop failure. Yields are low from a swidden because SOIL fertility is low and only human labor is used for CLEARING, planting, and harvesting. See also INTERTILLAGE.

Symbolic landscapes. LANDSCAPES centered on buildings or structures that are so visually emblematic that they represent an entire CITY.

Syncline. Downfold or TROUGH shape that is formed through compression of ROCKS. An upfold is an ANTICLINE.

Tableland. Large area of land with a mostly flat surface, surrounded by steeply sloping sides, or ESCARPMENTS. A small PLATEAU.

Taiga. Russian name for the vast BOREAL FORESTS that cover Siberia. The marshy ground supports a tree VEGETATION in which the trees are CONIFEROUS, comprising mostly pine, fir, and larch.

Talus. Broken and jagged pieces of ROCK, produced by WEATHERING of steep slopes, that fall to the base of the slope and accumulate as a talus cone. In high MOUNTAINS, a ROCK GLACIER may form in the talus. See also SCREE.

Tarn. Small circular LAKE, formed in a CIRQUE, which was previously occupied by a GLACIER.

Tectonism. The formation of MOUNTAINS because of the deformation of the CRUST of the earth on a large scale.

Temporary base level. STREAMS or RIVERS erode their beds down toward a BASE LEVEL—in most cases, SEA LEVEL. A section of hard ROCK may slow EROSION and act as a temporary, or local, base level. Erosion slows upstream of the temporary base level. A DAM is an artificially constructed temporary base level.

Tension. Type of stress that produces a stretching and thinning or pulling apart of the earth's CRUST. If the surface breaks, a NORMAL FAULT is created, with one side of the surface higher than the other.

Tephra. General term for volcanic materials that are ejected from a vent during an ERUPTION and transported through the AIR, including ASH (volcanic), BLOCKS (volcanic), cinders, LAPILLI, scoria, and PUMICE.

Terminal moraine. RIDGE of unsorted debris deposited by a GLACIER. When a glacier erodes it moves downslope, carrying ROCK debris and creating a ground MORAINE of material of various sizes, ranging from big angular blocks or boulders down to fine CLAY. At the terminus of the glacier, where the ice is melting, the ground moraine is deposited, building the ridge of unsorted debris called a terminal moraine.

Terrain. Physical features of a REGION, as in a description of rugged terrain. It should not be confused with TERRANE.

Terrane. Piece of CONTINENTAL CRUST that has broken off from one PLATE and subsequently been joined to a different plate. The terrane has quite different composition and structure from the adjacent continental materials. Alaska is composed mostly of terranes that have accreted, or joined, the North American plate.

Terrestrial planet. Any of the solid, rocky-surfaced bodies of the inner SOLAR SYSTEM, including the PLANETS Mercury, Venus, Earth, and Mars and Earth's SATELLITE, the MOON.

Terrigenous. Originating from the WEATHERING and EROSION of MOUNTAINS and other land formations.

Texture. One of the properties of SOILS. The three textures are SAND, SILT, and CLAY. Texture is measured by shaking the dried soil through a series of sieves with mesh of reducing diameters. A mixture of sand, silt, and clay gives a LOAM soil.

Thermal equator. Imaginary line connecting all PLACES on Earth with the highest mean daily TEMPERATURE. The thermal equator moves south of the EQUATOR in the SOUTHERN HEMISPHERE summer, especially over the CONTINENTS of South America, Africa, and Australia. In the northern summer, the thermal equator moves far into Asia, northern Africa, and North America.

Thermal pollution. Disruption of the ECOSYSTEM caused when hot water is discharged, usually as a thermal PLUME, into a relatively cooler body of water. The TEMPERATURE change affects the aquatic ecosystem, even if the water is chemically pure. Nuclear power-generating plants use large volumes of water in the process and are important sources of thermal pollution.

Thermocline. Depth interval at which the TEMPERATURE of OCEAN water changes abruptly, separating warm SURFACE WATER from cold, deep water.

Thermodynamics. Area of science that deals with the transformation of ENERGY and the laws that govern these changes; equilibrium thermodynamics is especially concerned with the reversible conversion of heat into other forms of energy.

Thermopause. Outer limit of the earth's ATMOSPHERE.

Thermosphere. Atmospheric zone beyond the MESOSPHERE in which TEMPERATURE rises rapidly with increasing distance from the earth's surface.

Thrust belt. Linear BELT of ROCKS that have been deformed by THRUST FAULTS.

Thrust fault. FAULT formed when extreme compression of the earth's CRUST pushes the surface into folds so closely spaced that they overturn and the ROCK then fractures along a fault.

Tidal force. Gravitational force whose strength and direction vary over a body and thus act to deform the body.

Tidal range. Difference in height between high TIDE and low tide at a given point.

Tidal wave. Incorrect name for a TSUNAMI.

Till. Mass of unsorted and unstratified SEDIMENTS deposited by a GLACIER. Boulders and smaller rounded ROCKS are mixed with CLAY-sized materials.

Timberline. Another term for tree line, the BOUNDARY of tree growth on MOUNTAIN slopes. Above the timberline, TEMPERATURES are too cold for tree growth.

Tombolo. Strip of SAND or other SEDIMENT that connects an ISLAND or SEA stack to the mainland. Mont-Saint-Michel is linked to the French mainland by a tombolo.

Topography. Description of the natural LANDSCAPE, including LANDFORMS, RIVERS and other waters, and VEGETATION cover.

Topological space. Space defined in terms of the connectivity between LOCATIONS in that space. The nature and frequency of the connections are measured, while distance between locations is not considered an important factor. An example of topological space is a transport network diagram, such as a bus route or a MAP of an underground rail system. Networks are most concerned with flows, and therefore with connectivity.

Toponyms. PLACE names. Sometimes, names of features and SETTLEMENTS reveal a good deal about the history of a REGION. For example, the many names starting with "San" or "Santa" in the Southwest of the United States recall the fact that Spain once controlled that area. The scientific study of place names is toponymics.

Tor. Rocky outcrop of blocks of ROCK, or corestones, exposed and rounded by WEATHERING. Tors frequently form in GRANITE, where three series of JOINTS often developed as the rock originally cooled when it was formed.

Transform faults. FAULTS that occur along DIVERGENT PLATE boundaries, or SEAFLOOR SPREADING zones. The faults run perpendicular to the spreading center, sometimes for hundreds of miles, some for more than five hundred miles. The motion along a transform fault is lateral or STRIKE-SLIP.

Transgression. Flooding of a large land area by the SEA, either by a regional downwarping of continental surface or by a global rise in SEA LEVEL.

Transmigration. Policy of the government of Indonesia to encourage people to move from the densely overcrowded ISLAND of Java to the sparsely populated other islands.

Transverse bar. Flat-topped body of SAND or gravel oriented transverse to the RIVER flow.

Trophic level. Different types of food relations that are found within an ECOSYSTEM. Organisms that derive food and ENERGY through PHOTOSYNTHESIS are called autotrophs (self-feeders) or producers. Organisms that rely on producers as their source of energy are called heterotrophs (feeders on others) or consumers. A third trophic level is represented by the organisms known as decomposers, which recycle organic waste.

Tropical cyclone. STORM that forms over tropical OCEANS and is characterized by extreme amounts of rain, a central area of calm AIR, and spinning WINDS that attain speeds of up to 180 miles (300 km.) per hour.

Tropical depression. STORM with WIND speeds up to 38 miles (64 km.) per hour.

Tropopause. BOUNDARY layer between the TROPOSPHERE and the STRATOSPHERE.

Troposphere. Lowest and densest of Earth's atmospheric layers, marked by considerable TURBULENCE and a decrease in TEMPERATURE with increasing ALTITUDE.

Tsunami. SEISMIC SEA WAVE caused by a disturbance of the OCEAN floor, usually an EARTHQUAKE, although undersea LANDSLIDES or volcanic ERUPTIONS can also trigger tsunami.

Tufa. LIMESTONE or calcium carbonate deposit formed by PRECIPITATION from an alkaline LAKE. Mono Lake is famous for the dramatic tufa towers exposed by the lowering of the level of lake water. Also known as TRAVERTINE.

Tumescence. Local swelling of the ground that commonly occurs when MAGMA rises toward the surface.

Tunnel vent. Central tube in a volcanic structure through which material from the earth's interior travels.

U-shaped valley. Steep-sided VALLEY carved out by a GLACIER. Also called a glacial TROUGH.

Ubac slope. Shady side of a MOUNTAIN, where local or microclimatic conditions permit lower TIMBERLINES and lower SNOW LINES than occur on a sunny side.

Ultimate base level. Level to which a STREAM can erode its bed. For most RIVERS, this is SEA LEVEL. For streams that flow into a LAKE, the ultimate base level is the level of the lakebed.

Unconfined aquifer. AQUIFER whose upper BOUNDARY is the WATER TABLE; it is also called a water table aquifer.

Underfit stream. STREAM that appears to be too small to have eroded the VALLEY in which it flows. A RIVER flowing in a glaciated valley is a good example of underfit.

Uniformitarianism. Theory introduced in the early nineteenth century to explain geologic processes. It used to be believed that the earth was only a few thousand years old, so the creation of LANDFORMS would have been rapid, even catastrophic. This theory, called CATASTROPHISM, explained most landforms as the result of the Great Flood of the Bible, when Noah, his family, and animals survived the deluge. Uniformitarian- ism, in contrast, stated that the processes in operation today are slow, so the earth must be immensely older than a mere few thousand years.

Universal time (UT). See GREENWICH MEAN TIME.

Universal Transverse Mercator. Projection in which the earth is divided into sixty zones, each six DEGREES of LONGITUDE wide. In a traditional Mercator projection, the earth is seen as a sphere with a cylinder wrapped around the EQUATOR. UTM can be visualized as a series of six-degree side strips running transverse, or north-south.

Unstable air. Condition that occurs when the AIR above rising air is unusually cool so that the rising air is warmer and accelerates upward.

Upthrown block. When EARTHQUAKE motion produces a FAULT, the block of land on one side is displaced vertically relative to the other. The higher is the upthrown block; the lower is the downthrown block.

Upwelling. OCEAN phenomenon in which warm SURFACE WATERS are pushed away from the

COAST and are replaced by cold waters that carry more nutrients up from depth.

Urban heat island. Cities experience a different MICROCLIMATE from surrounding REGIONS. The CITY TEMPERATURE is typically higher by a few DEGREES, both DAY and night, because of factors such as surfaces with higher heat absorption, decreased WIND strength, human heat-producing activities such as power generation, and the layer of AIR POLLUTION (DUST DOME).

Vadose zone. The part of the SOIL also known as the zone of AERATION, located above the WATER TABLE, where space between particles contains AIR.

Valley train. Fan-shaped deposit of glacial MORAINE that has been moved down-valley and redeposited by MELTWATER from the GLACIER.

Van Allen radiation belts. Bands of highly energetic, charged particles trapped in Earth's MAGNETIC FIELD. The particles that make up the inner BELT are energetic protons, while the outer belt consists mainly of electrons and is subject to DAY-night variations.

Varnish, desert. Shiny black coating often found over the surface of ROCKS in arid REGIONS. This is a form of OXIDATION or CHEMICAL WEATHERING, in which a coating of manganese oxides has formed over the exposed surface of the rock.

Varve. Pair of contrasting layers of SEDIMENT deposited over one year's time; the summer layer is light, and the winter layer is dark.

Ventifacts. PEBBLES on which one or more sides have been smoothed and faceted by ABRASION as the WIND has blown SAND particles.

Volcanic island arc. Curving or linear group of volcanic ISLANDS associated with a SUBDUCTION ZONE.

Volcanic rock. Type of IGNEOUS ROCK that is erupted at the surface of the earth; volcanic rocks are usually composed of larger crystals inside a fine-grained matrix of very small crystals and glass.

Volcanic tremor. Continuous vibration of long duration, detected only at active VOLCANOES.

Volcanology. Scientific study of VOLCANOES.

Voluntary migration. Movement of people who decide freely to move their place of permanent residence. It results from PULL FACTORS at the chosen destination, together with PUSH FACTORS in the home situation.

Warm temperate glacier. GLACIER that is at the melting TEMPERATURE throughout.

Water power. Generally means the generation of electricity using the ENERGY of falling water. Usually a DAM is constructed on a RIVER to provide the necessary height difference. The potential energy of the falling water is converted by a water turbine into mechanical energy. This is used to power a generator, which produces electricity. Also called HYDROELECTRIC POWER. Another form of water power is tidal power, which uses the force of the incoming and outgoing TIDE as its source of energy.

Water table. The depth below the surface where the zone of AERATION meets the zone of SATURATION. Above the water table, there may be some SOIL MOISTURE, but most of the pore space is filled with air. Below the water table, pore space of the ROCKS is occupied by water that has percolated down through the overlying earth material. This water is called GROUNDWATER. In practice, the water table is rarely as flat as a table, but curved, being far below the surface in some PLACES and even intersecting the surface in others. When GROUNDWATER emerges at the surface, because it intersects the water table, this is called a SPRING. The depth of the water table varies from SEASON to season, and with pumping of water from an AQUIFER.

Watershed. The whole surface area of land from which RAINFALL flows downslope into a STREAM. The watershed comprises the STREAMBED or CHANNEL, together with the VALLEY sides, extending up to the crest or INTERFLUVE, which separates that watershed from its neighbor. Each watershed is separated from the next by the drainage DIVIDE. Also called a DRAINAGE BASIN.

Waterspout. TORNADO that forms over water, or a tornado formed over land which then moves over water. The typical FUNNEL CLOUD, which reaches down from a CUMULONIMBUS CLOUD, is a narrow rotating STORM, with WIND speeds reaching hundreds of miles per hour.

Wave crest. Top of a WAVE.

Wave-cut platform. As SEA CLIFFS are eroded and worn back by WAVE attack, a wave-cut platform is created at the base of the cliffs. ABRASION by ROCK debris from the cliffs scours the platform further, as waves wash to and fro and TIDES ebb and flow. The upper part of the wave-cut platform is exposed at high tide. These areas contain rockpools, which are rich in interesting MARINE life-forms. Offshore beyond the platform, a wave-built TERRACE is formed by DEPOSITION.

Wave height. Vertical distance between one WAVE CREST and the adjacent WAVE TROUGH.

Wave length. Distance between two successive WAVE CRESTS or two successive WAVE TROUGHS.

Wave trough. The low part of a WAVE, between two WAVE CRESTS.

Weather analogue. Approach to WEATHER FORECASTING that uses the WEATHER behavior of the past to predict what a current weather pattern will do in the future.

Weather forecasting. Attempt to predict WEATHER patterns by analysis of current and past data.

Wilson cycle. Creation and destruction of an OCEAN BASIN through the process of SEAFLOOR SPREADING and SUBDUCTION of existing ocean basins.

Wind gap. Abandoned WATER GAP. The Appalachian Mountains contain both wind gaps and water gaps.

Windbreak. Barrier constructed at right angles to the prevailing WIND direction to prevent damage to crops or to shelter buildings. Generally, a row of trees or shrubs is planted to form a windbreak. The feature is also called a shelter belt.

Windchill. MEASUREMENT of apparent TEMPERATURE that quantifies the effects of ambient WIND and temperature on the rate of cooling of the human body.

World Aeronautical Chart. International project undertaken to map the entire world, begun during World War II.

World city. CITY in which an extremely large part of the world's economic, political, and cultural activity occurs. In the year 2018, the top ten world cities were London, New York City, Tokyo, Paris, Singapore, Amsterdam, Seoul, Berlin, Hong Kong, and Sydney.

Xenolith. Smaller piece of ROCK that has become embedded in an IGNEOUS ROCK during its formation. It is a piece of older rock that was incorporated into the fluid MAGMA.

Xeric. Description of SOILS in REGIONS with a MEDITERRANEAN CLIMATE, with moist cool winters and long, warm, dry summers. Since summer is the time when most plants grow, the lack of SOIL MOISTURE is a limiting factor on plant growth in a xeric ENVIRONMENT.

Xerophytic plants. Plants adapted to arid conditions with low PRECIPITATION. Adaptations include storage of moisture in tissue, as with cactus plants; long taproots reaching down to the WATER TABLE, as with DESERT shrubs; or tiny leaves that restrict TRANSPIRATION.

Yardangs. Small LANDFORMS produced by WIND EROSION. They are a series of sharp RIDGES, aligned in the direction of the wind.

Yazoo stream. TRIBUTARY that flows parallel to the main STREAM across the FLOODPLAIN for a considerable distance before joining that stream. This occurs because the main stream has built up NATURAL LEVEES through flooding, and because RELIEF is low on the floodplain. The yazoo stream flows in a low-lying wet area called backswamps. Named after the Yazoo River, a tributary of the Mississippi.

Zero population growth. Phenomenon that occurs when the number of deaths plus EMIGRATION is matched by the number of births plus IMMIGRATION. Some European countries have reached zero population growth.

BIBLIOGRAPHY

THE NATURE OF GEOGRAPHY

Adams, Simon, Anita Ganeri, and Ann Kay. *Geography of the World*. London: DK, 2010. Print.

Harley, J. B., and David Woodward, eds. *The History of Cartography: Cartography in the Traditional Islamic and South Asian Societies*. Vol. 2, book 1. Chicago: University of Chicago Press, 1992. Offers a critical look at maps, mapping, and mapmakers in the Islamic world and South Asia.

_____, eds. *The History of Cartography: Cartography in the Traditional East and Southeast Asian Societies*. Vol. 2, book 2. Chicago: University of Chicago Press, 1994. Similar in thrust and breadth to volume 2, book 1.

Marshall, Tim, and John Scarlett. *Prisoners of Geography: Ten Maps That Tell You Everything You Need to Know About Global Politics*. London : Elliott and Thompson Limited, 2016. Print

Nijman, Jan. *Geography: Realms, Regions, and Concepts*. Hoboken, NJ : Wiley, 2020. Print.

Snow, Peter, Simon Mumford, and Peter Frances. *History of the World Map by Map*. New York: DK Smithsonian, 2018.

Woodward, David, et al., eds. *The History of Cartography: Cartography in the Traditional African, American, Arctic, Australian, and Pacific Societies*. Vol. 2, book 3. Chicago: University of Chicago Press, 1998. Investigates the roles that maps have played in the wayfinding, politics, and religions of diverse societies such as those in the Andes, the Trobriand Islanders of Papua-New Guinea, the Luba of central Africa, and the Mixtecs of Central America.

PHYSICAL GEOGRAPHY

Christopherson, Robert W, and Ginger H. Birkeland. *Elemental Geosystems*. Hoboken, NJ : Wiley, 2016. Print.

Lutgens, Frederick K., and Edward J. Tarbuck. *Foundations of Earth Science*. Upper Saddle River, N.J.: Prentice-Hall, 2017. Undergraduate text for an introductory course in earth science, consisting of seven units covering basic principles in geology, oceanography, meteorology, and astronomy, for those with little background in science.

McKnight, Tom. Physical Geography: A Landscape Appreciation. 12th ed. New York: Prentice Hall, 2017. Now-classic college textbook that has become popular because of its illustrations, clarity, and wit. Comes with a CD-ROM that takes readers on virtual-reality field trips.

Robinson, Andrew. *Earth Shock: Climate Complexity and the Force of Nature*. New York: W. W. Norton, 1993. Describes, illustrates, and analyzes the forces of nature responsible for earthquakes, volcanoes, hurricanes, floods, glaciers, deserts, and drought. Also recounts

how humans have perceived their relationship with these phenomena throughout history.

Weigel, Marlene. *UxL Encyclopedia of Biomes*. Farmington Hills, Mich.: Gale Group, 1999. This three-volume set should meet the needs of seventh grade classes for research. Covers all biomes such as the forest, grasslands, and desert. Each biome includes sections on development of that particular biome, type, and climate, geography, and plant and animal life.

Woodward, Susan L. *Biomes of Earth*. Westport, CT: Greenwood Press, 2003. Print.

HUMAN GEOGRAPHY

Blum, Richard C, and Thomas C. Hayes. *An Accident of Geography: Compassion, Innovation, and the Fight against Poverty*. Austin, TX: Greenleaf Book Group, 2016. Print.

Dartnell, Lewis. *Origins: How the Earth Made Us*. New York: Hachette Book Group, 2019. Print.

Glantz, Michael H. *Currents of Change: El Niño's Impact on Climate and Society*. New York: Cambridge University Press, 1996. Aids readers in understanding the complexities of the earth's weather pattern, how it relates to El Niño, and the impact upon people around the globe.

Morland, Paul. *The Human Tide: How Population Shaped the Modern World*. New York: PublicAffairs, 2019. Print.

Novaresio, Paolo. *The Explorers: From the Ancient World to the Present*. New York: Stewart, Tabori and Chang, 1996. Describes amazing journeys and exhilarating discoveries from the earliest days of seafaring to the first landing on the moon and beyond.

Rosin, Christopher J, Paul Stock, and Hugh Campbell. *Food Systems Failure: The Global Food Crisis and the Future of Agriculture*. New York: Routledge, 2014. Print.

ECONOMIC GEOGRAPHY

Diamond, Jared M. *Guns, Germs, and Steel: The Fates of Human Societies*. New York: Norton, 2011. Print.

Esping-Andersen, Gosta. *Social Foundations of Postindustrial Economies*. New York: Cambridge University Press, 1999. Examines such topics as social risks and welfare states, the structural bases of postindustrial employment, and recasting welfare regimes for a postindustrial era.

Michaelides, Efstathios E. S.*Alternative Energy Sources*. Berlin: Springer Berlin, 2014. Print. This book offers a clear view of the role each form of alternative energy may play in supplying energy needs in the near future. It details the most common renewable energy sources as well as examines nuclear energy by fission and fusion energy.

Robertson, Noel, and Kenneth Blaxter. *From Dearth to Plenty: The Modern Revolution in Food Production*. New York: Cambridge University Press, 1995. Tells a story of scientific discovery and its exploitation for techno-logical advance in agriculture. It encapsulates the history of an important period, 1936-86, when government policy sought to aid the competitiveness of the agricultural industry through fiscal measures and by encouraging scientific and technical innovation.

REGIONAL GEOGRAPHY

Biger, Gideon, ed. *The Encyclopedia of International Boundaries*. New York: Facts on File, 1995. Entries for approximately 200 countries are arranged alphabetically, each beginning with introductory information describing demographics, political structure, and political and cultural history. The boundaries of each state are then described with details of the geographical setting, historical background, and present political situation, including unresolved claims and disputes.

Leinen, Jo, Andreas Bummel, and Ray Cunningham. *A World Parliament: Governance and Democracy in the 21st Century*. Berlin Democracy Without Borders, 2018. Print.

Pitts, Jennifer. *Boundaries of the International: Law and Empire*. Cambridge, Mass: Harvard University Press, 2018. Print.

SOUTH AMERICA AND CENTRAL AMERICA

PHYSICAL GEOGRAPHY

Blouet, Brian W, and Olwyn M. Blouet. *Latin America and the Caribbean: a Systematic and Regional Survey*, 7th Ed. John Wiley & Sons, 2015. Print.

Georges, D. V. *South America*. Danbury, Conn.: Children's Press, 1986. Discusses characteristics of various sections of South America such as the Andes, the Amazon rain forest, and the pampas.

Matthews, Down, and Kevin Schaefer. *Beneath the Canopy: Wildlife of the Latin American Rain Forest*. San Francisco: Chronicle Books, 2007. Kevin Schaefer's photographs offer a rare, up-close look at the beautiful and elusive creatures that make their home in this natural paradise—from its leafy shadows to the forest canopy. Captions and text by nature writer Matthews give further insight into the lives of these amazing animals.

HUMAN GEOGRAPHY

Early, Edwin, et al., eds. *The History Atlas of South America*. Foster City, Calif.: IDG Books Worldwide, 1998. Describes South America's history, which is a rich tapestry of complex ancient civilizations, colonial clashes, and modern growth, economic challenges, and cultural vibrancy.

Kelly, Philip. *Checkerboards and Shatterbelts: The Geopolitics of South America*. Austin: University of Texas Press, 1997. Uses the geographical concepts of "checkerboards" and "shatterbelts" to characterize much of South America's geopolitics and to explain why the continent has never been unified or dominated by a single nation.

Levine, Robert M., and John J. Crocitti, eds. *The Brazil Reader: History, Culture, Politics*. Durham, N.C.: Duke University Press, 2019. Selections range from early colonization to the present day, with sections on imperial and republican Brazil, the days of slavery, the Vargas years, and the more recent return to democracy.

Levinson, David, ed. *The Encyclopedia of World Culture, Vol. 7: South America*. Indianapolis, Ind.: Macmillan, 1994. Addresses the diverse cultures of South America south of Panama, with an emphasis on the American Indian cultures, although the African-American culture and the European and Asian immigrant cultures are also covered. Linguistics, historical and cultural relations, economy, kinship, marriage, sociopolitical organizations, and religious beliefs are among the topics discussed for each culture.

Webster, Donovan. "Orinoco River" *National Geographic* 193, no. 4 (April, 1998): 2-31. Examination of the Orinoco River Basin in the Amazonian portion of Venezuela. The article focuses on the fauna, flora, and the Yanomani, Yekwana, and Piaroa tribes and "tropical cowboys."

ECONOMIC GEOGRAPHY

Biondi-Morra, Brizio. *Hungry Dreams: The Failure of Food Policy in Revolutionary Nicaragua*. Ithaca, N.Y.: Cornell University Press, 1993. Examines how food policy was formulated in Nicaragua and the effects on foreign exchange, food prices, and the relationship to wages and credit.

Cupples, Julie, Marcela Palomino-Schalscha, and Manuel Prieto. *The Routledge Handbook of Latin American Development*. New York: Routledge, 2019. Print.

Folch, Christine. *Hydropolitics: The Itaipu Dam, Sovereignty, and the Engineering of Modern South America*. Princeton, NJ: Princeton University Press, 2019. Print.

Wilken, Gene C. *Good Farmers: Traditional Agriculture Resource Management in Mexico and Central America*. Berkeley, Calif.: University of California Press, 1987. Focusing on the farming practices of Mexico and Central America, this book examines in detail the effectiveness of sophisticated traditional methods of soil, water, climate, slope, and space management that rely primarily on human and animal power.

REGIONAL GEOGRAPHY

Frank, Zephyr L, Frederico Freitas, and Jacob Blanc. *Big Water: The Making of the Borderlands between Brazil, Argentina, and Paraguay*. Tucson: The University of Arizona Press, 2018. Internet resource.

Gheerbrandt, Alain. *The Amazon: Past, Present and Future*. New York: Harry N. Abrams, 1992. Presents the past, present, and uncertain future of the Amazon rain forest and its inhabitants. It includes spectacular illustrations and a section of historical documents.

Kent, Robert B. *Latin America: Regions and People*. New York: The Guilford Press, 2016. Print.

McClain, Michael E, Jeffrey E. Richey, and Reynaldo L. Victoria. *The Biogeochemistry of the Amazon Basin*. Oxford: Oxford University Press, 2001. Print.

Pulsipher, Lydia Mihelic. *World Regional Geography*. W H Freeman, 2019. Chacterizing global issues through the daily lives of individuals helps readers to better grasp circumstances affecting regions around the world such as environment; gender and population; urbanization; globalization and development; and power and politics.

INDEX

and plant diversity, 109, 367
Degradation, 476
Degree (geography), 476
Degree (temperature), 476
Delhi, 267
Demilitarized Zone, 28, 267
Demography, 476
Dendritic drainage, 476
Dengue fever, 73 - 74
Denmark
    bridges, 394
Denmark Strait, 433
Denudation, 476
Deposition, 477
Deranged drainage, 477
Derivative maps, 477
Desalination plants, 230
Desalinization, 477
Desert climate, 477
Desert pavement, 477
Desertification, 139, 141, 199, 477
    Asia, 53, 90
    and siltation, 505
Deserts, 347 - 348, 462
    Asia, 39, 50, 77, 90, 94, 125, 246
    Middle East, 55, 83, 91, 157, 245
    biomes, 369
Detrital rock, 477
Devils Postpile, 468
Devolution, 477
Devonian period, 334
Dew point, 360, 363, 477
Dhaka, 267
Diagenesis, 477
Dias, Bartolomeu, 174, 238, 378
Diaspora, 477
Diastrophism, 477
Diatom ooze, 477
Dike (geology), 477
Dike (water), 478
Dinaric Alps, 490
Disease
    and air travel, 386
    and climate, 72 - 74
    and flooding, 73
Distance-decay function, 478
Distributary, 478
Diurnal range, 478
Divergent boundary, 478
Divergent margin, 478
Dogukaradeniz Mountains, 83
Doline, 478
Dolomite, 478
Dolphins, 109
    freshwater, 17
Donbass, War in, 424
Dover, Strait of, 433
Downwelling, 478
Drainage basin, 478
Drainage basins, 342
    Asia, 306
Drake Passage, 433
Dravidians, 40
Drift ice, 478

Drumlin, 478
Dunes, 347, 466, 505
    sand, 503
    seif, 504
    types, 347
Dust devil, 478
Dust dome, 478
Débâcle, 475

E

Earth, 317 - 318
    age of, 333
    atmosphere, 351 - 354
    biomes, 367 - 370
    circumference, 3, 7
    external processes, 339 - 341
    formation of, 321
    internal geological processes, 337 - 339
    internal structure, 321 - 323
    layers, 321
    orbit around Sun, 317
    plate tectonics, 323 - 328
    population of, 142 - 144
Earth pillar, 478
Earth radiation, 478
Earth tide, 478
Earthflow, 478
Earthquakes, 323 - 324, 326 - 328, 504
    China, 16 - 17
    and Richter scale, 502
    aftershocks, 465
    deep-focus, 476
    shallow-focus, 504
    tremors, 485
    waves, 497, 500, 503
Earth's axis, 318
    and seasons, 320
Earth's core, 321 - 322, 337
    boundary of, 323
Earth's heat budget, 478
Earth's mantle, 321 - 327, 337
    boundary of, 323
    plumes, 328
East Asia, 267
East China Sea, 15 - 16, 267, 433
East Korea Warm Current, 267
East Pacific Rise, 338 - 339, 433
East Pakistan, 262
East Siberian Sea, 433
East Timor
    independence of, 186
Eastern Ghats, 39, 81, 267
Ebola virus, 386
Eclipses, 318, 496
    lunar, 479
    solar, 479
Ecliptic
    plane of, 479
Eco-fallow farming, 203
Ecology, 139
Effective temperature, 479
Egypt

early mapmakers, 7
    urbanization, 144
Ejecta, 479
Ekman layer, 479
El Niño, 207, 433, 471
    Southeast Asia, 115
    and disease, 74
El Niño-Southern Oscillation, 433, 491
Elbrus, Mount, 268
Elburz Mountains, 55, 59, 83, 268
Elephants
    Africa, 418
    Asia, 40, 134, 136, 380
    Philippines, 132
Elevation
    and climate, 355
Eluviation, 479
Emperor Seamount Chain, 433
Enclave, 479
Endangered species
    Asia, 128 - 132, 134
Endemic species, 479
Enderby Land, 434
Endogenic sediment, 479
Energy
    alternative sources, 391 - 393
    and pollution, 390
    sources, 387 - 391
    tidal power, 392
    and warfare, 389
    wind power, 392
Engineering projects, 393 - 396
    Asia, 227 - 236
    China, 19
    environmental problems, 394
English Channel, 434
    and Chunnel, 395
Environmental degradation, 479
Environmental determinism, 5 - 6, 479
Eolian, 479
Eolian deposits, 479
Eolian erosion, 479
Eon, 479
Epicontinental sea, 480
Epifauna, 480
Epilimnion, 480
Equal-area projection, 480
Equator
    and Asia, 31, 93
    and climate, 65, 85, 356, 369, 429
    and seasons, 319 - 320, 355
Equatorial Current, 434
Equinoxes, 318, 320
Eratosthenes, 3, 7
Erg, 480
Erie Canal, 394
Eritrea, 424
Erosion, 339, 342 - 343, 348, 372
    and agriculture, 201 - 203, 387
    and deforestation, 390
    eolian, 347
    and glaciation, 345
    and mountains, 337 - 338